Chemical Theory beyond the Born-Oppenheimer Paradigm

Nonadiabatic Electronic and Nuclear Dynamics in Chemical Reactions

Chemical Theory beyond the Born-Oppenheimer Paradigm

Nonadiabatic Electronic and Nuclear Dynamics in Chemical Reactions

Kazuo Takatsuka • Takehiro Yonehara
Kota Hanasaki • Yasuki Arasaki

The University of Tokyo, Japan

World Scientific

NEW JERSEY • LONDON • SINGAPORE • BEIJING • SHANGHAI • HONG KONG • TAIPEI • CHENNAI

Published by

World Scientific Publishing Co. Pte. Ltd.

5 Toh Tuck Link, Singapore 596224

USA office: 27 Warren Street, Suite 401-402, Hackensack, NJ 07601

UK office: 57 Shelton Street, Covent Garden, London WC2H 9HE

Library of Congress Cataloging-in-Publication Data

Chemical theory beyond the Born-Oppenheimer paradigm : nonadiabatic electronic and nuclear dynamics in chemical reactions / Kazuo Takatsuka, Takehiro Yonehara, Kota Hanasaki, Yasuki Arasaki, The University of Tokyo, Japan.

pages cm

Includes bibliographical references and index.

ISBN 978-9814619646 (hardcover : alk. paper)

1. Born-Oppenheimer approximation. 2. Chemical reactions. 3. Charge exchange. I. Takatsuka, Kazuo. II. Yonehara, Takehiro. III. Hanasaki, Kota. IV. Arasaki, Yasuki.

QC174.17.A66C44 2015

541'.3901--dc23

2014037595

British Library Cataloguing-in-Publication Data

A catalogue record for this book is available from the British Library.

Contents

Preface

Since the publication of the seminal paper by Born and Oppenheimer in 1927, the theoretical foundation of molecular science in the 20th century, and even to date, had been dominated by the so-called Born–Oppenheimer approximation, which dynamically separates electronic and nuclear motions under an assumption that electrons can follow the nuclear dynamics almost instantaneously. This idea led to the fixed nuclei approximation and the notion of electronic stationary-states that adjust themselves to any nuclear configurations in space.

However, nonadiabatic interactions and associated quantum transitions arising from the breakdown of the Born–Oppenheimer separation are critically important in chemical reaction dynamics. This is because almost all the interesting chemical and even biological processes involve nonadiabatic transition events in them. In particular, dynamics of molecules optically pumped to electronically excited states shows very characteristic quantum phenomena, which are vital to understanding the nature of chemistry and chemical reactions. Many theoretical methods for treating such nonadiabatic transitions have been proposed, most of which are about quantum transition of nuclear dynamics between nonadiabatically coupled electronic-energy surfaces (the so-called potential energy surfaces). With no doubt their contributions to the progress of elementary dynamical processes of chemical reactions have been and continue to be critical for developments in chemical science.

There have been already published so many excellent books on the standard theories of nonadiabatic transitions, and we are not trying to add a new volume into the list of them.

Instead, the present book has been written based on our thought about chemical dynamics, which is briefly summarized as follows. (i) Chemical properties and dynamical processes of molecules are primarily determined by the electronic states in them. Hence the primary aim of chemical reaction theories is to track the qualitative change of electronic states in courses of chemical reactions. This is why quantum chemistry (molecular electronic structure theory) is so important. (ii) However, it should not be disregarded that electronic states themselves couple with nuclear motions not only dynamically (through potential energy) but also kinematically (through the quantum momentum operators). Besides, electrons move with finite speed in molecules. This is an obvious fact, not to mention the theory of relativity, and even energetics is under control by the energy-time uncertainty principle through such an elapsed time taken by electronic-state change. To place a chemical theory in a correct context, therefore, the study on dynamics of electrons is indeed critical. (iii) Thus, chemistry can be better understood in terms of the direct notion of nonadiabatic dynamical electrons, that is, a theory of time-dependent electron wavepacket dynamics that nonadiabatically couples with nuclear motions.

A little more concrete reasoning behind the necessity of nonadiabatic dynamical electron theory is as follows. In the current status of the experimental progress, we are often faced with complicated and difficult situations that have not been studied before. For instance, due to advances in laser technology, an intense electromagnetic vector potential has become available, which now makes it possible to modify the molecular electronic states easily, which in turn induces novel nonadiabatic couplings. Also, chemistry of molecules in highly fluctuating electronic states due to high quasi-degeneracy, being coupled with nuclear motions, will become one of the most exciting fields in the next era of chemistry, which can be termed as non-Born–Oppenheimer chemistry or beyond-Born–Oppenheimer chemistry.

Thus this book describes the recent theories of chemical dynamics beyond the Born–Oppenheimer framework from a fundamental perspective of quantum wavepacket dynamics. To formulate these issues on a clear theoretical basis and to develop the novel theories beyond the Born–Oppenheimer approximation, however, we should first learn a basic classical and quantum nuclear dynamics on an adiabatic (the Born–Oppenheimer) potential energy surface. So we learn much from the classic theories of nonadiabatic transition such as the Landau-Zener theory and its variants.

Subsequently our characteristic way of discussion on nonadiabatic transition enters, paying a particular attention to the direct (experimental) observation of the nuclear wavepacket bifurcation in the passage of nonadiabatic regions. We show that the instant of nonadiabatic transition can be indeed observed experimentally and the wavepacket bifurcation caused by quantum entanglement between electronic and nuclear motions. We claim that "the wavepacket bifurcation" represents the very essential feature of nonadiabatic transitions.

Then the recent notion of nonadiabaticity in electron dynamics is introduced. To be consistent with the wavepacket bifurcation, we introduce the method of electron wavepacket dynamics that undergoes bifurcation while being carried along the so-called non-Born–Oppenheimer paths, which also branch due to nonadiabatic interactions. We will further proceed to the discussion about the interaction of molecular nonadiabatic states with intense laser fields. In this way, we penetrate on one hand into unknown domains of molecular properties such as (1) electron-nuclear quantum entanglement due to nonadiabatic transitions and its experimental observation, (2) coherence and decoherence of electron and nuclear wavepackets, which qualitatively dominate the quantum mechanical probabilities of quantum transition dynamics, (3) characteristic phenomena arising from the time-dependent fluctuation of molecular electronic states, (4) the physics of interference between the nonadiabatic dynamics and external fields, and so on.

On the other hand, the nonadiabatic electron wavepacket dynamics is extremely vital not only in chemical dynamics in laser fields but also for the theory of chemical reactivity in the absence of laser field. In particular, for analysis of chemical reactions in electronically excited states the nonadiabatic electron dynamics is indispensable. Moreover, there exist rather ubiquitously systems that have manifolds composed of highly quasi-degenerate electronic state, for which it is almost meaningless to single out each potential energy surface as usually done in the Born–Oppenheimer approximation. This is what we call chemistry without potential energy surfaces. Even in these extreme cases, the nonadiabatic electron dynamics evolves in time along non-Born–Oppenheimer branching paths.

Finally we note that one of the vital factors that are necessary to actually perform deeper research beyond the textbook knowledge about formal theories is practical capability to cope with materials in hand. Theoretical understanding alone is quite often not sufficient to knock the doors of and penetrate into the secrets of nature. We therefore describe in this book technical issues to some extent, which we think are useful in actual applications.

The authors believe that only through such quests for the new and fundamental approaches along with technical expertise such as the art of computing, one of the doors to the next stage of chemistry will open.

Acknowledgments

This book has been made up with many scientific materials and computational results that have been studied and carried out by one of the authors (KT) and his colleages and former graduate students. He deeply appreciates their sincere and valiant attitudes towards a new field of chemical dynamics. Consequently some of the materials chosen for this book are reorganized and instructive excerpts from those original works.

The authors are grateful to Professors Hiroshi Ushiyama and Satoshi Takahashi not only for their scientific contributions but to their efforts in management of the research enviroments.

Very special thanks are due to Professor Vincent McKoy at California Institute of Technology. Most of the theoretical results from the studies of time-resolved photoelectron spectroscopy in Section 5 had been attained only by very close collaboration between the McKoy and Takatsuka groups.

We are deeply grateful to Professor Jörn Manz for his long standing collaboration in scientific activities, in particular in the studies of electron dynamics and electron flux in chemical reactions.

Last but not least, writing this was supported in part by a Grant-in-Aid for Scientific Research from the Ministry of Education and Science in Japan (MEXT). We thank the MEXT for the long-standing supports of our scientific activities.

Chapter 1

The Aim of This Book: Where Are We?

1.1 Potential energy surfaces and nonadiabatic transitions

1.1.1 *Electronic state theory*

In the stationary molecular electronic state theory, alias quantum chemistry, a great success has been achieved historically, and still active studies are going on mainly towards very accurate calculations of molecular properties and/or for the development of methodology to treat large molecules. The theoretical framework that basically supports quantum chemistry is the so-called Born–Oppenheimer approximation, in which time variable is eliminated by freezing the nuclear motions (molecular geometry). Quantum chemistry can now provide an extremely useful method to study rather mild chemical reactions by identifying the molecular structures not only of stable configurations but also those at transitions states, which are usually invisible experimentally. Various static molecular properties are also estimated accurately, in particular, energetics reflecting the global feature of potential energy surfaces is extremely useful to understand and even predict chemical reactions under study.

The electronic energy thus computed at each molecular shape serves as a potential function working on nuclei, called (adiabatic) potential energy surface (PES), which drives nuclear wavepackets on it, and only in this stage time-variable is retrieved, to the time scale of nuclear dynamics mostly of the order of femtosecond. This is the standard theoretical framework for the study of the dynamics of molecules [59]. Very well structured and fast computer codes for quantum chemistry are now available, which can serve even as an alternative for experimental apparatus.

1.1.2 *Nonadiabatic transitions—A brief overview*

It is well-known however that in case two or more adiabatic potential energy surfaces come close to each other in energy, the Born–Oppenheimer approximation breaks down and nonadiabatic transition takes place as an important quantum effect (see [28, 84, 85, 117, 198, 291] for extensive reviews). The theory of nonadiabatic transition was first proposed in 1932 by Landau [229], Zener [505], Stueckelberg [395], and London [246], to study phenomena including electron transfer between two atoms. Since then, nonadiabatic dynamics has been found in many other phenomena mostly in chemical reactions. Among these the dynamics across conical intersections is one of the most important subjects in current chemistry. These classic theories, however, suffer from severe limitations and/or drawbacks from their theoretical structures. For instance, the Landau–Zener formula assumes linearly crossed diabatic (one-dimensional) energy curves with a constant nonadiabatic transition coupling, thereby allowing a transition only at the crossing point. However, these assumptions are far from the reality in many systems. To overcome these difficulties, many theories have been proposed in the literature. For example, the Zhu–Nakamura theory [513] is regarded as an ultimate theory within the Landau–Zener type dynamics.

Another basic theory of nonadiabatic transitions is the semiclassical Ehrenfest theory (SET). Although it can cope with multidimensional nonadiabatic electronic-state mixing, it inevitably produces a nuclear path that runs on an averaged potential energy surface after having passed across the nonadiabatic region, which is totally unphysical. Unfortunately, since SET seems intuitively correct, a naive and conventional derivation of this theory obscures how this critical difficulty arises.

Surface Hopping Model (SHM) first proposed by Tully and Preston [444] is a practical method to cope with nonadiabatic transition. It is actually not a theory but an intuitive prescription to take account of "quantum coherent jump" by replacing with a classical hop from one potential energy surface to another with a transition probability that is borrowed from other theories of semiclassical (or full quantum mechanical) nonadiabatic transitions state theory such as Zhu–Nakamura method. The fewest switch surface hopping method [445] and the theory of natural decay of mixing [197, 452, 509, 515] are among the most advanced methodologies so far proposed to practically resolve the critical difficulty of SET and the primitive version of SHM.

Recognizing that these classic standard theories have made and will continue to make great contributions to the progress of chemical dynamics, we on the other hand faithfully admit that there are many situations that are not even considered by such state-of-the-art theories. For instance, recent advances in laser technology enables drastic modification in the molecular electronic states, which in turn can induce novel nonadiabatic coupling in addition to the native nonadiabatic transition. Besides, the ultrashort laser pulses currently available (shorter than 100 attoseconds (1 as $= 10^{-18}$ s)) is faster the typical motions of valence electrons in molecules. Such nonadiabatic chemistry in intense laser fields and ultrafast electron dynamics make the current situation in science totally different from that of traditional theories of nonadiabatic transition. Another new perspective of excited state chemistry is in the study of the properties and chemical dynamics of molecules having densely quasi-degenerate electronic states. In those systems the continuous nonadiabatic couplings induce a large fluctuation of electronic states, which should be described as electron wavepacket dynamics beyond the Born–Oppenheimer framework of chemistry. It seems therefore obvious that we should reconsider the theory of nonadiabatic transitions from a deeper point of view rather than try to technically augment the classic theories that rest on sometimes naive assumptions.

1.2 Necessity of nonadiabatic dynamical electron theory

Let us look into a little more detailed aspects of the current and future perspective for chemical dynamics. As noted above, the foundations of theoretical chemistry were already established in the 1920's (both the papers of Born–Oppenheimer and Heitler London were published in 1927) and 1930's (Landau and Zener published in 1932, and the transition state theory of Eyring and Evans–Polanyi was almost simultaneously launched in 1935), and even today the basic framework remains essentially the same. However, there are many reasons we need to promote the electronic-state theory into the realm of dynamical electron theory by taking explicit account of time t in it. Below are listed some of the current attempts to achieve this goal.

1.2.1 *Progress in laser chemistry*

1.2.1.1 *Tracking attosecond electron dynamics*

The most powerful driving force to demand for the advancement of the dynamical theory of electronic states comes from the progress being made in laser technology of which there are two aspects. The first is in the rapid advances of ultra-short pulse lasers, whose width is shorter than 100 attoseconds. This is comparable with or shorter than the time scale of the valence electrons in a molecule [30, 72, 97, 179, 212, 222, 254, 282, 300, 323, 345]. For many years, nothing that could be experimentally generated was faster than the time scale of electrons. But that has changed and now the exploration of electron wavepacket dynamics in chemical reactions of polyatomic molecules becomes more and more realistic using such ultrafast laser pulse technology. To date most of the full quantal numerical studies on electronic–nuclear entire dynamics are made for the hydrogen molecule or its cation. Those relevant works include Ref. [33, 79, 80, 88, 89, 169, 170, 194, 204, 205, 216, 258, 264, 300, 313, 314, 344, 390, 440].

However, to achieve further progress, more general ab initio methods need to be developed for the treatment of multi-electronic and polyatomic molecules.

1.2.1.2 *Modulating electronic states and control of chemical reactions*

Intense lasers have brought about a revolutionary change in modern molecular spectroscopy. In the past, in the field of photochemistry only a weak and/or almost resonant perturbation was applied to a molecule to observe its response. This meant that only the static states were characterized. On the other hand, a laser field which is more intense than 10^{16} W cm^{-2} can apply as strong forces to nuclei and electrons as their original interactions do. The result therefore is that it can readily modulate the electronic states directly by creating a new wavepacket state. It can also induce nonadiabatic interactions in addition to the original ones. For instance, an efficient way of inducing an electronic excitation can be possible through the use of vibrational excitation using IR lasers [425]. One of the ultimate aims in this context would be to create new electronic states, through which the control of chemical reactions could be achieved.

1.2.1.3 *Dynamics of internal and external electrons*

Similarly, intense lasers make it possible to study the early stages in the multiphoton ionization of molecules through multiply excited states. Above-threshold ionization gives rise to a quasi-free electronic wavepacket state, and the recombination (collision) of such an electronic wavepacket with the remaining cation species results in the high harmonic generation [96, 114, 193, 237, 303, 317, 375]. These issues give rise an extremely interesting challenge on how to describe the electronic wavepacket states.

1.2.1.4 *Secondary effects of an induced electromagnetic field by a molecular electron current in external laser fields*

The electron current within a molecule driven by an intense laser field should generate induced electromagnetic field. These in turn will bring about the nontrivial secondary effects [39, 213, 298, 299, 334, 488]. The more intense external field is, the more prominent these secondary effects should be. However, they can be studied only by solving the coupled equations of quantum electronic wavepacket and the Maxwell equations in a self-consistent manner [249]. To successfully achieve this project, a very good theory for nonadiabatic electron wavepacket theory must be constructed first.

1.2.2 *Chemistry without potential energy surfaces: Highly quasi-degenerate electronic states*

The dynamics in densely degenerate electronic states such as those commonly found in highly excited states and metal-like states [474] are expected to undergo large fluctuation among the many electronic states involved. In addition, the energy-time uncertainty relation precludes dynamics that are confined to a single potential energy surface, and hence the notion of a separated potential energy surface may lose any sense. This is what is called "chemistry without potential energy surfaces". Since it is a meaningless effort to single out an individual global adiabatic potential energy surface in such a situation, only the so-called "on-the-fly" method should be able to track such complicated nonadiabatic dynamics. (See Chap. 7 for the definition of on-the-fly method.)

1.2.2.1 *Nonadiabatic dynamical electron theory for chemical reactions*

It turns out through our extensive studies that the electron wavepacket description of chemical reactions offers an interesting and novel way of comprehending chemical phenomena, in which the shift of nuclear configuration induces a qualitative change in the electronic states through nonadiabatic interactions. This situation is typically observed in chemical reactions that are non-concerted reactions or Woodward–Hoffmann symmetry forbidden reactions, proton transfer dynamics associated with keto-enol tautomerization, fast relaxation dynamics of excited states through conical intersections, and so on. Electron flow dynamics and their resultant forces working on the nuclei will provide new and deeper insights into the chemical reactivity theory.

In addition, the influence of truly multidimensional nonadiabatic effects in the study of chemical reactions is quite important. So far most of the nonadiabatic theories have been limited to one-dimensional systems. They are applied to multidimensional systems by locally slicing the potential energy surface in a direction parallel to that of a classical trajectory under study. However, essentially multidimensional effects of nonadiabatic dynamics certainly exist. Conical intersection and the so-called geometric phase (for instance, the Herzberg–Longuet-Higgins (HLH) phase [247], the Berry phase [48], as well as the phase in the molecular Aharonov–Bohm effect [267, 268] among others) are illuminating examples. Moreover, it is anticipated that other essentially multidimensional effects will be discovered in future. In this book, in Chap. 7, we will present a case study of electron wavepacket dynamics around a conical intersection, thereby showing some of its new features. Also, attention is drawn to a theoretical parallelism between the electromagnetic vector potential and the nonadiabatic interaction, which should give a clue to finding multidimensional effects of nonadiabatic interaction.

1.2.3 *General theory of mixed quantum and classical dynamics*

Let us suppose it is possible to treat the nuclear subsystem in a molecule "classically" and electronic subsystem quantum mechanically. This type of theoretical framework is called a mixed quantum–classical representation. Such a mixed representation can find many applications in science. For instance, a fast mode such as the proton dynamics in a protein should be considered as a quantum subsystem, while the rest of the skeletal structure can be treated as a "classical" subsystem [3, 484, 485]. It is quite important in this context to establish the correct equations of motion for each of the subsystems and to ask what are their rigorous solutions and how the quantum effects penetrate into the classical subsystems. By studying the quantum-electron and classical-nucleus nonadiabatic dynamics as deeply as possible, we will see how such rigorous solutions, if any, look like qualitatively and quantitatively. This is one of the main aims of this book.

1.3 Structure of this book

This book has been written with a hope that the theory of nonadiabatic transition and nonadiabatic electron wavepacket dynamics can be rebuilt from a fundamental perspective. There have been published many excellent review articles on the theory of nonadiabatic transition, among which Refs. [28, 84, 85, 117, 178, 198, 291, 295] are particularly recommended. However, we begin the presentation with some very basic and classic theories which are among the intellectual assets of human beings, and then we proceed beyond them to be equipped to explore the realm of non-Born–Oppenheimer. With these new fundamental theories along with further innovation to be brought in by readers, the unknown molecular properties and dynamical processes will be investigated, which opens the gate of the next generation of chemical science. The structure of this book is as follows:

In Chap. 2, we formulate our basic framework of the chemical theory based on the Born–Oppenheimer approximation. We briefly discuss how valid or how accurate the Born–Oppenheimer approximation for bound states is. Also a theory of electron scattering by polyatomic molecules within the Born–Oppenheimer framework (or the so-called fixed nuclei approximation) is presented. This is one of the typical theories of electron dynamics, along with the theory of molecular photoionization.

Chapter 3 treats nuclear motions on the adiabatic potential energy surfaces (PES). One of the most powerful and simplest means to study chemical dynamics is the so-called ab initio molecular dynamics (or the first principle dynamics), in which nuclear motion is described in terms of the Newtonian dynamics on an ab initio PES. Next, we review some of the representative time-dependent quantum theory for nuclear wavepackets such as the multiconfigurational time-dependent Hartree approach. Then, we show how such nuclear wavepacket dynamics of femtosecond time scale can be directly observed with pump-probe photoelectron spectroscopy.

After these studies we proceed into the world of nonadiabatic theories. We first review briefly in Chap. 4 the very basic and classic theories of nonadiabatic transitions and the ideas behind them. As stressed above, the theories shown in this chapter were developed in the early stage of theoretical chemistry and do not necessarily care about the recent experimental progress in nonadiabatic processes associated with nonadiabatic electron dynamics in laser fields. But studying these classic theories and the ideas behind is very instructive.

Then in Chap. 5, we show that nonadiabatic transition and the associated wavepacket bifurcation are never a product of mathematical imagination but can be indeed observed experimentally. The direct observation of the instant of nonadiabatic transition (namely, the passage of a wavepacket across a nonadiabatic region) has been achieved only recently. This chapter also discuss a possible control of nonadiabatic transitions on the basis of the notion of wavepacket bifurcation.

Next we develop the theory of nonadiabatic electron wavepacket dynamics in Chap. 6. We design the theory so as to make full use of the great technological progress in quantum chemistry (electronic structure theory). Furthermore, bifurcation of both electron and nuclear wavepackets, or quantum mechanical entanglement between electrons and nuclei as stressed in Chap. 5 as the essential feature of nonadiabatic transition, should be appropriately reflected in the theory. In doing so, we propose path-branching representation based on the matrix force, which is a generalization of the force in Newtonian mechanics.

Chapter 7 continues the presentation of nonadiabatic electron wavepacket dynamics as applied in various chemical reactions, mainly in electronically excited states. Quantization the branching paths (non-Born–Oppenheimer paths) will be also discussed. Likewise, in Chap. 8, the electron wavepacket dynamics is considered for molecules placed in laser fields. In addition to the ordinary nonadiabatic transitions due to the Born–Oppenheimer approximation, novel nonadiabatic transitions due to optical interactions appear to need special cares. This chapter is to be continued to future studies of laser design of electronic states and concomitant control of chemical reactions.

Chapter 2

Basic Framework of Theoretical Chemistry

2.1 Born–Huang expansion

Immediately after Heisenberg and Schrödinger established quantum mechanics, in 1927, Born and Oppenheimer developed a perturbation theory to comprehend how a molecule can exist in a stable state [59]. The notion of separation of the nuclear and electronic motions essentially emerged from this study along with the so-called Born–Oppenheimer fixed nuclear approximation (more simply referred to as the Born–Oppenheimer (BO) approximation). The BO approximation is so successful, actually even more successful than originally anticipated, it is quite often regarded as a robust paradigm of molecular science. Incidentally, in the same year, Heitler and London [175] proposed the quantum theory of valence, which was followed by Pauling with his powerful concepts of valence bond, resonance, elec-tonegativity and so on. In the theory of rate process in chemical reactions, Eyring [127] and Evans and Polanyi [126] established the transition-state theory in 1935, and the early theories of nonadiabatic transition of Landau [229], Zener [505], Stueckelberg [395] date back to the year 1932. It is rather astonishing, therefore, that the theoretical foundations of present chemistry were established in such early days of quantum theory.

In order to discuss as accurately as possible these complex theories, we will now briefly review the traditional and canonical way of describing a molecular system. Within the nonrelativistic scheme, the quantum-mechanical molecular Hamiltonian is generally written as

$$\begin{aligned} H(\mathbf{r}, \mathbf{R}) &= T_N + H^{el}(\mathbf{r}; \mathbf{R}) \\ &= \frac{1}{2} \sum_k \frac{\hat{P}_k^2}{M_k} + H^{(el)}(\mathbf{r}; \mathbf{R}), \end{aligned} \tag{2.1}$$

where many-body electronic Hamiltonian is defined as

$$H^{(el)}(\mathbf{r};\mathbf{R}) = T_e + V_c(\mathbf{r};\mathbf{R}) = \frac{1}{2m}\sum_j \hat{p}_j^2 + V_c(\mathbf{r};\mathbf{R}). \tag{2.2}$$

Here and in what follows, $\mathbf{r}$ and $\mathbf{R}$ denote the electronic and nuclear coordinates, respectively, and $\hat{p}_j$ and $\hat{P}_k$ are the operators of their conjugate momenta of the jth and kth component of $\mathbf{r}$ (denoted as r_j) and $\mathbf{R}$ (R_k), respectively. $V_c(\mathbf{r};\mathbf{R})$ represents collectively the Coulombic interactions among electrons and nuclei, which is

$$V_c(\mathbf{r};\mathbf{R}) = \sum_{a<b} \frac{e^2}{|\mathbf{r}_a - \mathbf{r}_b|} - \sum_a \sum_A \frac{Z_A e^2}{|\mathbf{r}_a - \mathbf{R}_A|} + \sum_{A<B} \frac{Z_A Z_B e^2}{|\mathbf{R}_A - \mathbf{R}_B|}, \tag{2.3}$$

where $\mathbf{r}_a$ and $\mathbf{R}_A$ are the ath electron and Ath nuclei, respectively. Because nuclei move far more slowly than electrons due to their heavy masses, it is standard practice to expand the total wavefunction $\Psi(\mathbf{r},\mathbf{R},t)$ in stationary electronic basis functions $\{\Phi_I(\mathbf{r};\mathbf{R})\}$, which are determined at each nuclear position $\mathbf{R}$, in such a way that

$$\Psi(\mathbf{r},\mathbf{R},t) = \sum_I \chi_I(\mathbf{R},t)\Phi_I(\mathbf{r};\mathbf{R}). \tag{2.4}$$

This is often referred to as the Born–Huang expansion [58]. We assume that the electronic basis set $\{|\Phi_I\rangle\}$, either adiabatic or diabatic, are orthonormal at each nuclear configuration as

$$\langle \Phi_I(\mathbf{R})|\Phi_J(\mathbf{R})\rangle = \delta_{IJ}. \tag{2.5}$$

In what follows, the bra-ket inner product represents integration over the electronic coordinates only.

Inserting this form of the total wavefunction into the time-dependent Schrödinger equation and using the above orthonormal property, one obtains the standard coupled equations of motion for nuclear wavefunctions

$$i\hbar\frac{\partial}{\partial t}\chi_I = \frac{1}{2}\sum_k \frac{\hat{P}_k^2}{M_k}\chi_I - i\hbar\sum_k\sum_J \frac{X_{IJ}^k \hat{P}_k}{M_k}\chi_J - \frac{\hbar^2}{2}\sum_k\sum_J \frac{Y_{IJ}^k}{M_k}\chi_J + \sum_J H_{IJ}^{el}\chi_J, \tag{2.6}$$

where

$$X_{IJ}^k = \left\langle \Phi_I \Big| \frac{\partial \Phi_J}{\partial R_k} \right\rangle, \quad Y_{IJ}^k = \left\langle \Phi_I \Big| \frac{\partial^2 \Phi_J}{\partial {R_k}^2} \right\rangle, \tag{2.7}$$

and

$$H_{IJ}^{el}(\mathbf{R}) = \langle \Phi_I | H^{el} | \Phi_J \rangle . \tag{2.8}$$

These quantities are the main players in the theory of nonadiabaticity. In addition to these, H^{el} can contain spin-orbit couplings. We do not, however, discuss the relativistic effects any further, since it is beyond the scope of this book. Interested readers are referred to Ref. [103] This multi-state coupled Schrödinger equation can be transformed into a compact matrix form as shown below

$$i\hbar\frac{\partial}{\partial t}\chi_I = \sum_J \left[\sum_k \frac{1}{2M_k} \left[\left(\mathbf{I}\hat{P}_k - i\hbar\mathbf{X}^k \right)^2 \right]_{IJ} + H_{IJ}^{el} \right] \chi_J, \tag{2.9}$$

where $[\mathbf{I}]_{IJ} = \delta_{IJ}$ and $[\mathbf{X}^k]_{IJ} = X_{IJ}^k$ [28, 198]. The above reduction to Eq. (2.6) is mathematically complete, but these close coupling equations practically require much labor in solving even for small systems [91, 103, 139, 275–277, 359, 439, 441, 479].

2.2 Born–Oppenheimer approximation

On the other hand, nuclear dynamics on a given potential energy surface (PES) $V_I(\mathbf{R})$ within the BO separation scheme is reduced to solving the following equation

$$i\hbar\frac{\partial}{\partial t}\chi_I(\mathbf{R},t) = [T_N + V_I(\mathbf{R})]\,\chi_I(\mathbf{R},t). \tag{2.10}$$

The full quantum wavepacket dynamics, which is shown in Eq. (2.6), is based on the Born–Huang expansions and if it were perfectly achieved, appears to be mathematically exact. However, since the wavelength of matter waves for heavy particles such as nuclei is much shorter in general than that of an electron, it is never easy to solve even Eq. (2.10) unless certain constraints are additionally imposed. Classical alternatives are therefore frequently adopted with the understanding that quantum effects are totally neglected. It is commonly believed that semiclassical mechanics compromises two conflicting factors; computability and the taking into account of quantum effects. However, to the best of our knowledge, semiclassical mechanics remains a tough challenge. It should be also noted that these studies on nuclear dynamics, including the close coupling equations as shown in Eq. (2.6), are quite often performed without any deep concern about the change of electronic states. The electronic energies, together with the nonadiabatic coupling elements, are supposedly given independently by quantum chemistry. Indeed, this seems to be largely the case in the actual performance of chemical dynamics. Consequently, these two major fields have been separated from each other.

2.2.1 *Bound states and notion of potential energy surface*

2.2.1.1 *Potential energy hyper surfaces*

It is generally accepted that one of the most successful fields in theoretical chemistry is the so-called quantum chemistry. This is an electronic stationary-state structure theory based on the fixed nuclei (Born–Oppenheimer) approximation, which is the eigenvalue problem with respect to the electronic Hamiltonian $H^{el}(\mathbf{r};\mathbf{R})$ at each nuclear configuration, that is

$$H^{el}(\mathbf{r};\mathbf{R})\Phi_I(\mathbf{r};\mathbf{R}) = V_I(\mathbf{R})\Phi_I(\mathbf{r};\mathbf{R}). \tag{2.11}$$

Quantum chemistry serves as quite a reliable tool for the interpretation and prediction of complex chemical phenomena and can aid in designing novel molecules prior to experimental synthesis. The result is a great contribution to chemistry and thereby human society. Techniques have been developed there that are highly sophisticated and are aimed at the study of larger molecular systems with faster and more accurate computation so as to make it possible treat the more realistic molecules found in nature.

There are many excellent monographs for quantum chemistry such as Ref. [176] and one of the latest and extensive review articles is Ref. [517].

2.2.1.2 *Molecular orbitals and the Slater determinant*

The first approach to the exact electronic ground state $\Phi_0(\mathbf{r};\mathbf{R})$ is usually the Hartree–Fock method using a single Slater determinant that is composed of molecular orbitals, that is

$$\Phi_0(\mathbf{r};\mathbf{R}) \simeq \mathcal{A}\left[\phi_1(\mathbf{r}_1;\mathbf{R})\phi_2(\mathbf{r}_2;\mathbf{R})\cdots\phi_N(\mathbf{r}_N;\mathbf{R})\right], \tag{2.12}$$

where $\phi_i(\mathbf{r}_i;\mathbf{R})$ are molecular orbitals determined in the so-called self-consistent field to minimize the total energy, and $\mathcal{A}$ is the antisymmetrizer, which makes $\Phi_0(\mathbf{r};\mathbf{R})$ equivalent to the Slater determinant. Here and in what follows we omit the spin variables of electrons unless we need to specify them. It is a usual practice to expand the molecular orbitals in a given set of basis functions like the Cartesian Gaussian functions centered at the positions of nuclei. Many a variant to calculate the molecular orbitals and technique for faster convergence in the SCF process are available to date. However, we do not discuss the technical details at all in this book.

The density functional theory (DFT) is now known as a method to give far better energy yet with the same functional form as Eq. (2.12). This very attractive feature is brought about by using an "effective" one-body interaction Hamiltonian rather than the standard Coulomb and exchange repulsions naturally arising from Eq. (2.2). However, we basically refer to the standard molecular orbitals throughout this book.

2.2.1.3 *Method of configuration interactions*

The most straightforward way to go beyond the Hartree–Fock level approximation is the method of configuration interactions (CI). The total electronic wavefunctions, not only of the ground state but for any excited states in principle, can be represented as

$$\Phi_I(\mathbf{r};\mathbf{R}) \simeq \sum_{\{k\}} C_{I\{k\}}\mathcal{A}\left[\phi_{k1}(\mathbf{r}_1;\mathbf{R})\phi_{k2}(\mathbf{r}_2;\mathbf{R})\cdots\phi_{kN}(\mathbf{r}_N;\mathbf{R})\right], \tag{2.13}$$

where $\phi_{ki}(\mathbf{r}_i;\mathbf{R})$ is the ith molecular orbital picked up in the kth electronic configuration, and the coefficients $C_{I\{k\}}$ are to be determined variationally or in the scheme of perturbation theories. Usually, the choice of the electronic configuration is made by replacing one of more molecular orbitals in Eq. (2.12) with the virtual orbitals. Those virtual orbitals are automatically produced in the Hartree–Fock equation to determine the occupied molecular orbitals. Yet, there have been proposed methods to improve such naive virtual orbitals. Also, there are many theories to describe the accurate electronic wavefunctions without use of the primitive expansion of Eq. (2.13). In the present book, however, we describe the the electron wavepackets in terms of the configuration expansion.

2.2.1.4 *Electron wavepacket dynamics in Born–Oppenheimer framework*

A primitive electron wavepacket can be constructed as

$$\Phi(\mathbf{r},t;\mathbf{R}) = \sum_I c_I \Phi_I(\mathbf{r};\mathbf{R}) \exp\left(-\frac{i}{\hbar}V_I(\mathbf{R})t\right) \tag{2.14}$$

when the nuclear geometry $\mathbf{R}$ is frozen for a very short time. There are certainly cases in which this wavepacket description can be useful. For instance, if an electronic wavepacket is created by optically pumping with a very short pulse, the short-time dynamics after the pulse is switched off will be conveniently described in this way. However, this situation is far from what we are going to challenge in this book, since the nonadiabatic interactions are not taken into account.

2.2.2 *Stationary-state scattering theory for electrons by molecules*

2.2.2.1 *Electronically inelastic scattering for polyatomic molecules*

The fixed nuclear approximation is extremely useful, not only for the bound states, but also for the treatment of electronic scattering by molecules. For instance, stationary-state scattering theories within the fixed nuclear approximation have been extensively developed for molecular photoionization [18, 250, 336, 516] and electron scattering from polyatomic molecules [210, 418, 419]. These scattering phenomena are quite important in that they are widely found in nature as elementary processes and even in industrial applications using plasma processes. These scatterings may be referred to as stationary-state *electron dynamics* in fixed nuclei approximation.

As usual, the total electronic Hamiltonian H we consider can be written as

$$H^{(el)} = (H_N + T_{N+1}) + V, \tag{2.15}$$

where H_N is the target electronic Hamiltonian at a molecular geometry and $T_{N+1}$is the kinetic energy operator for the incident electron and V is the interaction potential between the incident electron and the target, i.e.,

$$V = \sum_{i=1}^{N} \frac{1}{r_{i,N+1}} - \sum_{I=1}^{Nuclei} \frac{1}{r_{I,N+1}} \tag{2.16}$$

In Eq. (2.16) the first and second terms are the electron repulsion and electron-nuclei attraction, respectively. We would like to solve the stationary state Schrödinger equation

$$H\psi^{(+)} = E\psi^{(+)}. \tag{2.17}$$

The major differences of this equation from Eq. (2.11) is that (1) E of Eq. (2.17) is not an eigenvalue but a parameter that specifies the experimental collision energy, and (2) $\psi^{(+)}$ has a nonzero component in the asymptotic region ($|\mathbf{r}| \to \infty$) with an appropriate boundary condition representing the collision status. Therefore the scattering wavefunction $\psi^{(+)}$ is not square-integrable and thereby does not belong to the Hilbert space.

To solve the problem, we first define a projection operator P which specifies the open target eigenstates of the collision. By open channels we mean the states that are accessible by the collision energy. P is an N-body operator and, unlike the projection operator in the Feshbach formalism [130], is unique. With Φ_m as a target eigenstate P is simply

$$P = \sum_{m}^{\text{open channels}} |\Phi_m\rangle \langle \Phi_m| \,. \tag{2.18}$$

With this operator we obtain a projected Lippmann–Schwinger equation for the scattering state

$$P\psi_m^{(+)} = S_m + G_P^{(+)} V \psi_m^{(+)}, \tag{2.19}$$

where $\psi_m^{(+)}$ has the asymptotic form of an incoming plane wave and a scattered component. S_m is hence a product of the target wavefunction Φ_m and an incident plane wave, i.e.,

$$S_m = \Phi_m (1, 2, \cdots, N) \exp\left(i\mathbf{k}_m \cdot \mathbf{r}_{N+1}\right) \tag{2.20}$$

The Green's function of Eq. (2.19) is uniquely defined in terms of the open channel components

$$G_P^{(+)} = \sum_{m}^{\text{open channels}} |\Phi_m\rangle \, g_m^{(+)} \left(\mathbf{r}_{N+1}, \mathbf{r}'_{N+1}\right) \langle \Phi_m| \,. \tag{2.21}$$

In this equation $g_m^{(+)} \left(\mathbf{r}_{N+1}, \mathbf{r}'_{N+1}\right)$ denotes the outgoing-wave Green's function for T_{N+1} at energy $E - E_m$, where E and E_m are the total energy and eigenvalue of Φ_m, respectively.

We next multiply the potential V on both sides of Eq. (2.19) such that

$$\left(VP - VG_P^{(+)}V\right) \psi_m^{(+)} = \psi_m^{(+)} S_m. \tag{2.22}$$

Unlike the original Schwinger principle the operator on the left-hand side of this equation is not symmetric. This is simply because we have already projected the closed-channel components out of Eq. (2.19). The closed-channel contributions can, however, be taken into account through a projected Schrodinger equation

$$\left[\hat{H} - a\left(P\hat{H} + \hat{H}P\right)\right] \psi_m^{(+)} = a\left(VP - PV\right) \psi_m^{(+)} \tag{2.23}$$

where, at this stage, the parameter a is arbitrary. We note below that combining Eqs. (2.22) and (2.23) can lead to an integro-differential equation with a symmetric operator, i.e.,

$$\left[\frac{1}{2}(PV+VP)-VG_P^{(+)}V+\frac{1}{2a}\left\{\hat{H}-a\left(P\hat{H}+\hat{H}P\right)\right\}\right]\psi_m^{(+)}=VS_m \tag{2.24}$$

Closer examination of the operators in this equation shows that for Fermion collisions the parameter a must be chosen to be $(N+1)/2$ to make the operator $\hat{H}-a\left(P\hat{H}+\hat{H}P\right)$ on the left-hand side Hermitian.

In a general inhomogeneous equation as in Eq. (2.24)

$$A^{(+)}x^{(+)}=b, \tag{2.25}$$

where $A^{(+)}$ is a scattering operator having outgoing-wave boundary conditions and $x^{(+)}$ is an unknown function, if the operator $A^{(+)}$ does not satisfy the condition

$$\left\langle x^{(-)}\middle| A^{(+)}x^{(+)}\right\rangle=\left\langle A^{(-)}x^{(-)}\middle| x^{(+)}\right\rangle, \tag{2.26}$$

the corresponding variational functional

$$[X]=\frac{\left\langle x^{(-)}\middle| b\right\rangle\left\langle b\middle| x^{(+)}\right\rangle}{\left\langle x^{(-)}\middle|A^{(+)}\middle| x^{(+)}\right\rangle} \tag{2.27}$$

is not stable and consequently no unique solutions are given in general. The choice of $a=(N+1)/2$ ensures that the above condition, Eq. (2.26), is satisfied.

Based on Eq. (2.24) we can write a multichannel variational functional for the scattering amplitude

$$[f_{mn}] = -\frac{1}{2\pi}\frac{\left\langle\psi_m^{(-)}\middle|V\middle|S_n\right\rangle\left\langle S_m\middle|V\middle|\psi_n^{(+)}\right\rangle}{\left\langle\psi_m^{(-)}\middle|\left[\frac{1}{2}(PV+VP)-VG_P^{(+)}V+\frac{1}{2a}\left\{\hat{H}-a\left(P\hat{H}+\hat{H}P\right)\right\}\right]\middle|\psi_n^{(+)}\right\rangle}, \tag{2.28}$$

which is an extension of the usual Schwinger principle for electron collision by a multi-electronic-state molecule. The variational expression for the scattering amplitude can then be written as

$$f_{mn}=-\frac{1}{2\pi}\sum_{i,j}\langle S_m|V|\Phi_i\rangle\,\Delta_{ij}^{-1}\,\langle\Phi_j|V|S_n\rangle, \tag{2.29}$$

where the Φ_is are basis functions defined by $(N+1)$-body Slater determinants. The Δ is the matrix representation of the operator of the denominator of Eq. (2.28) in this basis. This formalism is obviously based on the variation of the total wavefunction rather than that of a specific scattering orbital.

As an example of the use of this functional, we now use it to obtain the Schwinger variational principle for the static-exchange potential. The total wavefunction is taken as

$$\psi_0^{(+)} = \mathcal{A}\left[\Phi_{HF}\left(1, 2, \cdots, N\right) \mu^{(+)}(\mathbf{r}_{N+1})\right] \tag{2.30}$$

where Φ_{HF} is the Hartree–Fock wavefunction for the target state and $\mu^{(+)}(\mathbf{r}_{N+1})$ denotes the scattering orbital. $\mathcal{A}$ is the antisymmetrizing operator. The matrix elements in Eq. (2.28) arising from the Hamiltonian in the denominator vanish identically. The resultant equation is just the Schwinger variational principle for elastic scattering by the static-exchange potential (Hartree–Fock potential from the target molecule).

The above stationary-state scattering theory looks rather complicated but it is actually quite powerful to determine the scattering amplitude and cross sections for electron scattering for polyatomic molecules. Many such prominent examples are seen in the series of very active works by the Vincent McKoy group.

2.2.2.2 *Molecular photoionization*

Another important electron dynamics induced by photons is molecular photoionization. The theory we use will be rather precisely described later in Chap. 3.

2.3 Validity of the BO approximation

There is no doubt that the Born–Oppenheimer approximation captures the essential feature of the molecular properties of stable molecules. (So many papers have been published on the mathematical and numerical analyses of the Born–Oppenheimer theory, but it is beyond the scope of this book. See Refs. [181, 405, 409] for relevant recent literature.) The validity or the level of accuracy of the Born–Oppenheimer separation is roughly assessed as follows. Let m be the mass of an electron and let M_l be the mass of the l-th nucleus. The perturbation parameter κ is taken as

$$\kappa = \sqrt[4]{\frac{m}{M}}, \tag{2.31}$$

where M is a kind of average of the M_l's such that

$$M_l = M\frac{1}{\mu_l} = \frac{m}{\kappa^4 \mu_l}, \tag{2.32}$$

with μ_l being a dimensionless number of order $O(1)$. Born and Oppenheimer [59] studied the stationary Schrödinger equation

$$H(\mathbf{r}, \mathbf{R})\psi(\mathbf{r}, \mathbf{R}) = E\psi(\mathbf{r}, \mathbf{R}), \tag{2.33}$$

using a perturbation theory with the total Hamiltonian represented by

$$H = H_0 + \kappa^4 H_1, \tag{2.34}$$

where the zeroth order Hamiltonian is the electronic Hamiltonian and the perturbation interaction $\kappa^4 H_1$ is the nuclear kinetic energy operator

$$T_N = \kappa^4 H_1. \tag{2.35}$$

They expanded the total energy according to the order of the perturbation interaction such that

$$W = W^0 + \kappa W^{(1)} + \kappa^2 W^{(2)} + \kappa^3 W^{(3)} + \kappa^4 W^{(4)} \cdots . \tag{2.36}$$

They showed that W^0 is the electronic energy, $\kappa^2 W^{(2)}$ corresponds to the harmonic nuclear vibrational energy, and $\kappa^4 W^{(4)}$ gives the anharmonic vibrational energy with molecular rotational energy. They also explicitly proved that

$$W^{(1)} = W^{(3)} = 0. \tag{2.37}$$

The energy terms up to $\kappa^4 W^{(4)}$ constitute the so-called Born–Oppenheimer energy of a molecule. The higher terms should represent the various couplings between electronic, vibrational, and rotational modes.

The critical question is how accurate (or inaccurate) the Born–Oppenheimer energy of a molecule defined as above is. Recently Takahashi and Takatsuka have addressed this question and reported a semiclassical analysis of this matter [413]. They explicitly showed that the fifth order term $W^{(5)}$ is also exactly zero, and therefore the lowest-order correction to the Born–Oppenheimer energy must be of the order of

$$(m/M)^{6/4}. \tag{2.38}$$

This fundamental result explicitly highlights that the Born–Oppenheimer approximation is far better than what would be expected from a simple mass ratio m/M. For example, if $m/M \sim 10^{-6}$, the error term is $(m/M)^{1.5} \sim 10^{-9}$.

Note however that the present order estimate has been made on the basis of perturbation expansion as above, which is valid only around the stable geometry of a molecule. It is obvious that once a degeneracy takes place in the electronic state manifold, the Born–Oppenheimer expansion loses the sense as we will show in Chap. 5.

2.4 Generalization of the adiabatic electronic states

Equation (2.11) defines the standard adiabatic electronic states and the potential energy surfaces associated with them. Yet, there can exist other definitions depending on the choice of the "electronic" Hamiltonian.

In his early study of nonadiabatic transitions of a diatomic molecule due to the Coriolis coupling, H. Nakamura proposed a new electronic Hamiltonian [290], in which a part of the rotational interactions is explicitly taken into account as

$$h_{IJ}^{el} = H_{IJ}^{el} + H_{IJ}^{rot}, \tag{2.39}$$

where H^{rot} is the rotational Hamiltonian of the corresponding rotor (see Ref. [291] for details). Nakamura refers this representation to dynamical state representation.

Likewise, recalling Eq. (2.6), one may want to define an electronic Hamiltonian matrix

$$h_{IJ}^{el} = H_{IJ}^{el} - \frac{\hbar^2}{2}\sum_k \frac{Y_{IJ}^k}{M_k} \tag{2.40}$$

in place of H_{IJ}^{el} alone. A new set of "adiabatic" basis functions diagonalizing this Hamiltonian should obviously simplify the coupled equations Eq. (2.6), and the nonadiabatic coupling elements are reduced only to X_{IJ}^k in this particular representation.

Another interesting idea to generalize the adiabatic states arises from Eq. (2.9). By replacing the quantum mechanical momentum operators for nuclei ($\hat{P}_k$) with their classical counterparts (P_k), one obtains another electronic Hamiltonian

$$h_{IJ}^{el} = \sum_k \frac{1}{2M_k}\left[\left(\mathbf{I}P_k - i\hbar\mathbf{X}^k\right)^2\right]_{IJ} + H_{IJ}^{el}. \tag{2.41}$$

The adiabatic states that diagonalize this Hamiltonian seem to dramatically reduce the magnitude of the nonadiabatic couplings among them. In return, these adiabatic electronic states depend not only on the nuclear positions but the momenta [368].

Chapter 3

Nuclear Dynamics on Adiabatic Electronic Potential Energy Surfaces

On the very affirmative basis about the accuracy of the adiabatic potential energy surfaces as shown above, one can study the dynamics of nuclear motions on those potential surfaces to explore the dynamical aspects of chemical reactions like rate process and product distribution, which are not determined by the energetics within quantum chemistry. In this class of dynamics, everything begins from Eq. (2.10), which declares that an adiabatic potential energy surface arising from the electronic energy serves as a potential energy function for the nuclei involved.

The hardest obstacle in an attempt to solve Eq. (2.10) is originated from the general fact that the wavelengths of the matter waves for nuclei are generally very short, much shorter than those of electrons compared with the scale of a molecule. This is simply because the nuclear masses are much heavier than those of electrons and therefore the theoretical foundation of the Born–Oppenheimer approximation and the difficulty of nuclear dynamics share the same root. This difficulty in quantum mechanical treatise of nuclear dynamics is essential and remains among one of the most difficult challenges in theoretical chemistry.

3.1 Classical nuclear dynamics: Ab initio molecular dynamics

The fact that the nuclear wavelengths are much shorter compared to the molecular sizes suggests that the wave characters like diffraction and interference are not significant and instead the particle nature should dominate the relevant dynamics. This implies that the quantum dynamics may be well approximated by classical mechanics. That is, Eq. (2.10) may be approximately replaced with

$$M_i \ddot{\mathbf{R}}_i(t) = -\frac{\partial V_I(\mathbf{R})}{\partial \mathbf{R}_i}, \tag{3.1}$$

where $\mathbf{R}_i$ is the ith component of the nuclear position, which is now treated as a function of time, and M_i is the mass associated with this coordinate. It is the standard practice to solve Eq. (3.1) in the form of the Hamilton canonical equations of motion.

Then the difficult part in solving the classical dynamics is the calculation of the $\partial V_I(\mathbf{R})/\partial \mathbf{R}_i$, the energy gradient of PES. Fortunately, an astonishing progress in quantum chemistry has made it possible to calculate the energy gradient *analytically* at each nuclear position [330]. Therefore there is no need to apply the numerical methods such as the method of finite difference. Most of the current advanced methods of quantum chemistry are equipped with the algorithm of such analytical derivative of ab initio PES. The study of chemical dynamics based on the ab initio energy gradient is called ab initio molecular dynamics or the first principle dynamics.

Although the nuclear quantum effects are totally neglected, the ab initio molecular dynamics provides us with a very powerful methodology to analyze chemical dynamics. Here we will show an example of such applications.

3.1.1 *Coupling of electronic and nuclear motion in tautomerization dynamics*

Proton transfer is one of the most fundamental chemical reactions and a very critical element in biological processes and material sciences [45, 60, 242, 266]. In their studies on the fluorescence excitation and hole-burning spectra in S_1-S_0 region of 5-methyltropolone (5MTR, see Fig. 3.1), Nishi et al. have found that the excitation of the internal rotational levels of methyl group promotes proton tunneling [301, 302]. This remarkable finding suggests the existence of a very long-range interaction between proton tunneling and methyl internal rotation in the excited state, which is quite unusual in chemical dynamics.

Inspired by this discovery, Ushiyama and Takatsuka have carried out the full-dimensional ab initio molecular dynamics at the RHF level of approximation [56, 357] for the proton transfer dynamics on the *ground state* potential surface of 5MTR and other relevant molecules seeking for a long-range mechanical interaction in molecules. As a result they have found that the proton transfer does induce the methyl group rotation, not the other

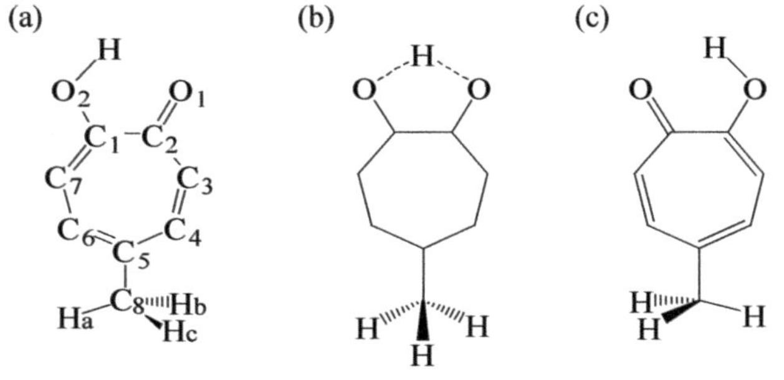

Fig. 3.1 (a) Stable structure of 5MTR. (b) The transition state. (c) The mirror image of (a). The torsional angle θ of the methyl group is defined as a dihedral angle between the planes of C_4-C_5-C_8 and C_5-C_8-H_a. $\theta = \pi$ and 0 for the structures (a) and (c), respectively. The irrelevant hydrogen atoms are ignored throughout. (Reprinted with permission from H. Ushiyama *et al.*, Angew. Chem. Int. Ed. 44, 1237 (2005)).

way around, through the electronic-state change called tautomerization: The very basic characteristic of the electronic structure is reflected in the bond lengths. See Fig. 3.1. The single bonds (C_2-C_3, C_4-C_5 and C_6-C_7) have about 1.43 Å, while the double bonds (C_3-C_4, C_5-C_6 and C_7-C_1) are shorter to about 1.35 Å. Bond alternation follows the proton transfer changing the sites of single and double bonds to each other. This is called tautomerization. In other words, the proton transfer has to surmount the potential barrier to perform such a tautomerization and such a large molecular deformation [449]. The bond C_1-C_2 is exceptional from this deformation, since it remains single and long during the proton transfer.

Figure 3.2 exhibits the change of the torsional angle of the methyl group after proton transfer, which has been averaged over sampled trajectories. Around 30-40 fs (marked with an arrow in Fig. 3.2) after the proton transfer the methyl begins to rotate, changing the angle with a speed of 60 degrees for about 100 fs. The torsional angle remains mostly constant in this time scale if proton transfer does not occur. Since the tautomerization takes 10-15 fs after the proton transfer, the methyl group begins to rotate after the tautomerization is well finished. Although these estimated values should depend on the quality of the adopted approximation, it is thus confirmed that the methyl rotation follows the proton transfer. It turns out therefore that a quantum-mechanical mechanism is involved in this molecule so that the in-plane reciprocating motion of the proton is unidirectionally transmitted to the rotational motion of the methyl group at a remote site.

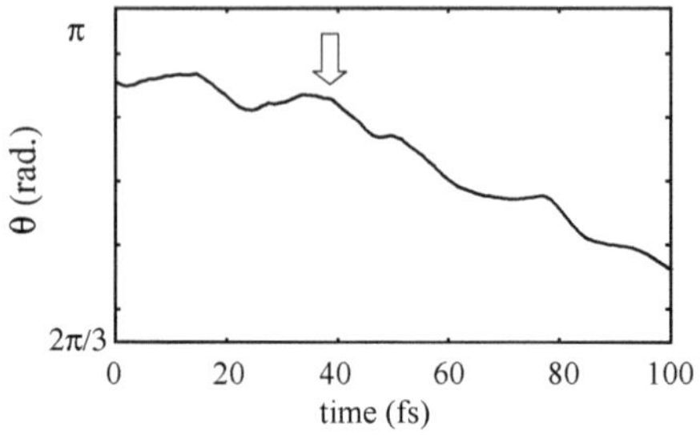

Fig. 3.2 Time dependence of the torsional angle θ of methyl group after proton transfers (see Fig. 3.1). The angles are statistically averaged over the different occasions of proton transfer. θ larger than π has been transformed to $2\pi - \theta$. (Reprinted with permission from H. Ushiyama *et al.*, Angew. Chem. Int. Ed. 44, 1237 (2005)).

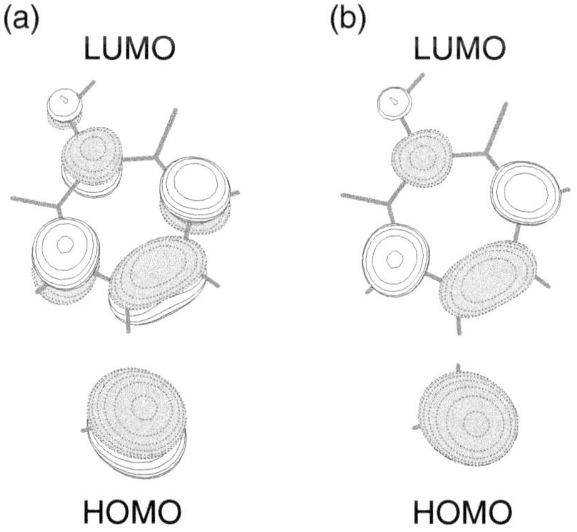

Fig. 3.3 HOMO (methyl group)–LUMO (tropolone) interaction representing a coupling between hyperconjugation and tautomerization. (a) a perspective view of 5MTR of Fig. 3.1(a) from the side of the methyl group at an angle. (b) projection of the panel (a) onto the molecular plane. (Reprinted with permission from H. Ushiyama *et al.*, Angew. Chem. Int. Ed. 44, 1237 (2005)).

This mechanism can be comprehended in terms of the coupling of hyperconjugation with the tautomerization. Hyperconjugation is a classic concept in organic chemistry established by Dewar many years ago [113, 394]: A minus linear combination of the 1s atomic orbitals on two hydrogen atoms of a methyl group can behave somewhat like a 2p orbital and thereby participates in a nearby π-conjugation at an appropriate orientation. It is therefore quite natural to consider that this hyperconjugation should couple with the π-system of the seven membered ring of tropolone.

Since the methyl group is well-known as an electron donating functional group, the primary interaction between the methyl and π-system should arise from methyl's HOMO (Highest Occupied Molecular Orbital) and the LUMO (Lowest Unoccupied Molecular Orbital) of tropolone. Practically, the methyl group is represented in terms of methane of the same size and same orientation. (As for another HOMO-LUMO interaction, the LUMO of methane at the molecular structure in Fig. 3.3 is symmetric with respect to the molecular plane, while the HOMO of tropolone, which is a π-orbital, is antisymmetric. This results in virtually zero interaction between these two orbitals.) Figure 3.3, viewing 5MTR in Fig. 3.1(a) from the side of the methyl group, shows such a HOMO-LUMO interaction. In this figure, one of the HOMO's of methane, representing the methyl group, is primarily composed of two 1s orbitals on two hydrogen atoms, which stand perpendicular to the molecular plane, just looks like a 2p orbital. (Two hydrogen atoms explicitly seen in the figure lie on the molecular plane and have nothing to do with hyperconjugation. One of them is responsible for C-C bond between tropolone and the methyl group.) As is clearly seen in the panel (b) of Fig. 3.3, the LUMO of tropolone has a large amplitude on the bond of C_4-C_5 bearing a right phase to have a good overlap with the HOMO of the methyl. In this way, the methyl group sits comfortably with the tropolone π-system. Once the proton transfer takes place being associated with tautomerization, the HOMO of the methyl becomes out of phase with the LUMO of tropolone, resulting in the higher electron repulsion. (The LUMO of MTR after the proton-transfer can be readily obtained just by making use of the mirror image of the HOMO in Fig. 3.3.) To avoid this repulsion, the methyl rotates by 60 degrees to establish a new stable HOMO-LUMO interaction in the site of C_5-C_6 as in Fig. 3.1(c).

Having a precise look at Fig. 3.3, one can predict a couple of phenomena based on the mechanism of coupling between tautomerization and hyperconjugation. For instance, the LUMO of tropolone has a wide distribution over C_4-C_5 (single bond), localized $2p_z$-like components on C_1, C_3, and C_6, and virtually no component on C_7. Therefore, neglecting all the possible secondary effects, we anticipate the following events: (1) A methyl group substituted at C_4 can stabilize this structure (4MTR) as well as 5MTR. (Two hydrogen atoms in the methyl group should be placed vertical to the molecular plane and in between C_4 and C_5.) (2) However, even if proton transfer takes place in 4MTR, the internal rotation of the methyl group is not induced so strongly as in 5MTR. This is because the component of LUMO at C_4 after proton-transfer should look like that at C_6 before proton-transfer. (3) It is quite likely that a methyl group placed at C_7 (7MTR) will not significantly couple with the π-conjugation of tropolone, resulting in freer rotation. Proton-transferred 7MTR will have a little stronger HOMO-LUMO interaction but should be not so as in 5MTR and 4MTR. In this way, one can choose a position for the methyl rotation to be enhanced most efficiently.

3.1.2 *Molecular machine?*

The present mechanical interaction through the coupling between tautomerization and hyperconjugation is very general and can be seen in other molecular systems. The reader should consult Ref. [450]. The present mechanism can thus realize a very long range dynamical interaction between two functional groups under such the similar condition. This kind of deterministic dynamics in a long distance is never made possible by the ordinary mechanism of energy transfer in a molecule, which is usually very stochastic.

This dynamics is regarded as a quantum-mechanical transmission of the reciprocating motion of the proton transfer to the rotation of methyl group, which can become one of the elements of the so-called molecular machine.

3.2 Nuclear quantum dynamics on an adiabatic potential surface

In contrast to the above shown ab initio molecular dynamics approach, quantum wavepacket methods can provide a rigorous description for any chemically important properties, thereby properly taking account of the quantum effects of nuclear dynamics. However, the direct applications of the full quantum theories are often prohibitively difficult for many dimensional systems. Also, the time evolution of quantum wavefunctions generally requires the knowledge of the global PES beforehand, in contrast to the trajectory-based methods such as ab initio molecular dynamics, which demands only local information along the paths used. The latter is often referred to as on-the-fly method.

There are two classes in applications of quantum nuclear dynamics: one is the stationary-state scattering theory to treat reactive scattering (chemical reactions), and the other is time-dependent wavepacket method. Here in this section, we briefly outline some of the standard methods developed for chemical purposes [29, 43, 91, 437, 480, 507].

3.2.1 *Time-propagation with split operator formalism*

The time-dependent vibrational Schrödinger equation,

$$i\hbar\frac{\partial}{\partial t}\chi(\boldsymbol{R},t) = \hat{H}\chi(\boldsymbol{R},t), \tag{3.2}$$

can be solved numerically to high accuracy when the number of degrees of freedom is small. Many of the methods make use of the fact that the Hamiltonian can be separated into two parts,

$$\hat{H} = \hat{T}_N + V(\boldsymbol{R}), \tag{3.3}$$

where the potential energy part $V(\boldsymbol{R})$ is diagonal in the chosen representation and therefore trivial to operate; the kinetic energy part $\hat{T}_N$ is the nondiagonal nontrivial part. For a kinetic energy operator of the form

$$\hat{T}_N = -\frac{\hbar^2}{2M}\nabla^2, \tag{3.4}$$

the fast Fourier transform may be employed as an efficient numerical method for computing the differential operator (also for a few degrees of freedom, computation costs allowing). This is because once one transforms a wavefunction in $\boldsymbol{R}$ to the associated momentum representation, the kinetic energy operator is also transformed to the form

$$T_N = -\frac{\hbar^2}{2M}\boldsymbol{P}\cdot\boldsymbol{P}, \tag{3.5}$$

where $\boldsymbol{P}$ is the classical momentum conjugate to $\boldsymbol{R}$. T_N is simply a scalar to be multiplied on the momentum-representation wavefunction. The quantity thus multiplied is to be Fourier back-transformed to the $\boldsymbol{R}$-representation [129, 219]. For one or two dimensions, the Hamiltonian represented at the grid-points for the Fourier transformation may be diagonalized to give eigenfunctions without resorting to the direct time-propagation [260].

To be more precise, this time propagation scheme consists of two ingredients: (i) application of the Trotter product formula to the time propagation operator, and (ii) an efficient use of the Fast Fourier transform (FFT). Recall that the increase of time to perform FFT scales to $N\log N$, where N is the number of grid points [219]. A time evolution operator for a short time interval Δt, which is composed of two noncommutable operators $\hat{A}$ and $\hat{B}$, can be approximately represented with the Trotter product (decomposition) formula as

$$e^{(\hat{A}+\hat{B})\Delta t} = e^{\hat{A}\Delta t/2}e^{\hat{B}\Delta t}e^{\hat{A}\Delta t/2} + O(\Delta t^3). \tag{3.6}$$

The third order error term includes the commutator of $\hat{A}$ and $\hat{B}$. By setting $\hat{A}$ and $\hat{B}$ to be the kinetic energy operator and potential energy operator respectively, this relation reads

$$e^{-\frac{i}{\hbar}(\hat{T}+\hat{V})\Delta t} = e^{-\frac{i}{\hbar}\hat{T}\Delta t/2}e^{-\frac{i}{\hbar}\hat{V}\Delta t}e^{-\frac{i}{\hbar}\hat{T}\Delta t/2} + O(\Delta t^3). \tag{3.7}$$

It is useful to employ the diagonal representation of the operator for an action on a vector. If $\hat{T}$ and $\hat{V}$ are the functions of momentum ($\boldsymbol{P}$) and position ($\boldsymbol{R}$) operators, respectively, and if $\mathcal{U}$ denotes a transformation operator from $\boldsymbol{P}$ to $\boldsymbol{R}$ representation and $\mathcal{U}^{-1}$ being its inverse, the expression (3.7) can be cast into

$$\mathcal{U}^{-1}e^{-\frac{i}{\hbar}(\hat{T}+\hat{V})\Delta t}\mathcal{U} = \mathcal{U}^{-1}e^{-\frac{i}{\hbar}\hat{T}\Delta t/2}\mathcal{U}e^{-\frac{i}{\hbar}\hat{V}\Delta t}\mathcal{U}^{-1}e^{-\frac{i}{\hbar}\hat{T}\Delta t/2}\mathcal{U} + O(\Delta t^3), \tag{3.8}$$

which gives us the split-operator short-time propagation method. It is usual practice for small dimensional systems to take $\mathcal{U}$ to be FFT from position to momentum space. Then $e^{-\frac{i}{\hbar}\hat{T}\Delta t/2}$ should be acted on the wavefunction in momentum representation, while $e^{-\frac{i}{\hbar}\hat{V}\Delta t}$ should be simply multiplied on the counterpart in the position-space representation. The order of $\hat{T}$ and $\hat{V}$ may be interchanged in the right hand side of Eq. (3.7) with an appropriate exchange of the places of $\mathcal{U}$ and $\mathcal{U}^{-1}$ in Eq. (3.8).

It is obvious that the shorter Δt gives the more accurate results, but with higher computational costs. Also, the split operator technique have been extended to the higher order product forms [410, 497], and the higher order schemes generally lead to the more accurate results for a given time interval Δt, again the more computational labor in return.

3.2.2 *An example of three dimensional dynamics*

As an example of the split-operator method, we here consider the time propagation of the vibrational wavefunction of a triatomic molecule, such as NO_2 (Sec. 5.4). The two bond lengths of the molecule are r_1 and r_2, and the bond angle is β. The total kinetic energy operator $\hat{T}_N$ is represented in a convenient form by taking the Jacobi coordinates (r, R, θ) as depicted in Fig. 3.4, in which the length R is the distance between the upper O atom and the centroid of the N and the lower O atoms, and r is the bond length of one NO moiety. It is

$$\hat{T}_N = \hat{T}_r + \hat{T}_R + \hat{T}_\theta, \tag{3.9}$$

$$\hat{T}_r = -\frac{\hbar^2}{M_{\mathrm{NO}}r^2}\frac{\partial}{\partial r}r^2\frac{\partial}{\partial r}, \qquad \hat{T}_R = -\frac{\hbar^2}{M_{\mathrm{NO}_2}R^2}\frac{\partial}{\partial R}R^2\frac{\partial}{\partial R}, \tag{3.10}$$

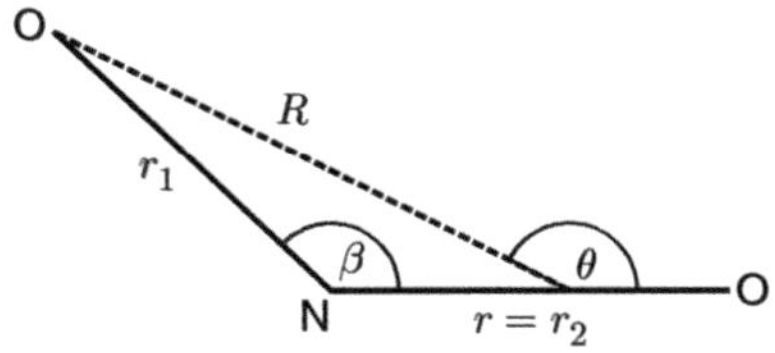

Fig. 3.4 Coordinate system for a triatomic molecule (NO_2).

and

$$\hat{T}_\theta = -\frac{\rho\hbar^2}{\sin\theta}\frac{\partial}{\partial\theta}\frac{1}{\sin\theta}\frac{\partial}{\partial\theta}. \tag{3.11}$$

The reduced masses are given by

$$M_{\mathrm{NO}} = \frac{m_{\mathrm{N}} m_{\mathrm{O}}}{m_{\mathrm{N}} + m_{\mathrm{O}}}, \qquad M_{\mathrm{NO_2}} = \frac{m_{\mathrm{O}}(m_{\mathrm{N}} + m_{\mathrm{O}})}{m_{\mathrm{N}} + 2m_{\mathrm{O}}}, \tag{3.12}$$

where m_{N} and m_{O} are atomic masses. Note that ρ appearing in $\hat{T}_\theta$,

$$\rho = \frac{1}{2M_{\mathrm{NO}} r^2} + \frac{1}{2M_{\mathrm{NO_2}} R^2}, \tag{3.13}$$

is dependent on both of r and R.

We apply the split-operator method Eq. (3.7) to separately handle the potential and kinetic energy terms of the Hamiltonian. We further apply the split-operator scheme to separate the kinetic energy term into the two exponentially noncommutative parts $\hat{T}_r + \hat{T}_R$ and $\hat{T}_\theta$ to obtain a numerical short-time propagation method.

We note that $\hat{T}_\theta$, Eq. (3.11) cannot be simply written in the form of Eq. (3.4). In such a case the general prescription is to introduce a suitable orthogonal basis set that diagonalizes the operator under study, or is to enable efficient evaluation of its operation (using the so-called discrete variable representation [240]). In the former approach, even a wavefunction spatially confined (such as would be the case when the angular coordinate represents molecular vibration) would be a coherent sum of a potentially large number of spatially global basis functions. Here we outline a spatially local method devised by Dateo and Metiu [106] based on the Fourier transform.

The time propagation step in the split-operator scheme using the operator $\hat{T}_\theta$ of Eq. (3.11) is approximated as

$$\chi(\theta, t + \Delta t) = e^{-i\hat{T}_\theta \Delta t/\hbar}\chi(\theta, t) \approx \frac{1 - \tau\hat{T}_\theta}{1 + \tau\hat{T}_\theta}\chi(\theta, t), \tag{3.14}$$

where Δt is a short time step, $\chi(\theta, t)$ is an intermediate wavefunction in the middle of being propagated with the split-operator method, and the variable $\tau = i\Delta t/(2\hbar)$ is introduced here for convenience of notation. This expression leads to

$$\chi(\theta, t+\Delta t) = \chi(\theta, t) - \tau \hat{T}_\theta \chi(\theta, t+\Delta t) - \tau \hat{T}_\theta \chi(\theta, t). \tag{3.15}$$

Introducing the (inverse) Fourier transformation

$$\chi(\theta) = \int \varphi(\kappa) e^{2\pi i \kappa \theta} d\kappa, \tag{3.16}$$

we see that

$$\sin\theta \chi(\theta) = \int \frac{1}{2i} \varphi(\kappa) e^{i(2\pi\kappa+1)\theta} d\kappa - \int \frac{1}{2i} \varphi(\kappa) e^{i(2\pi\kappa-1)\theta} d\kappa, \tag{3.17}$$

and

$$\begin{aligned} \frac{\partial}{\partial\theta} \sin\theta \frac{\partial}{\partial\theta} \chi(\theta) &= \int \pi i \kappa (2\pi\kappa+1) \varphi(\kappa) e^{i(2\pi\kappa+1)\theta} d\kappa \\ &\quad - \int \pi i \kappa (2\pi\kappa-1) \varphi(\kappa) e^{i(2\pi\kappa-1)\theta} d\kappa. \end{aligned} \tag{3.18}$$

Using Eqs. (3.17) and (3.18), Eq. (3.15) can be rewritten as

$$\begin{aligned} &[1 + 2\pi\kappa(2\pi\kappa - 1)\rho\tau]\varphi(\kappa - \frac{1}{2\pi}, t+\Delta t) \\ &= [1 + 2\pi\kappa(2\pi\kappa + 1)\rho\tau]\varphi(\kappa + \frac{1}{2\pi}, t+\Delta t) \\ &\quad + [1 - 2\pi\kappa(2\pi\kappa - 1)\rho\tau]\varphi(\kappa - \frac{1}{2\pi}, t) \\ &\quad - [1 - 2\pi\kappa(2\pi\kappa + 1)\rho\tau]\varphi(\kappa + \frac{1}{2\pi}, t). \end{aligned} \tag{3.19}$$

For a bound bandwidth limited wavefunction, φ must approach zero for large κ. Therefore a discretized version of Eq. (3.19) can be used iteratively to obtain $\varphi(t+\Delta t)$ for all κ from values at larger κ and $\varphi(t)$ of a previous time step.

3.2.3 *Polynomial expansion method*

We next describe another propagation method that is also frequently used. In this scheme, the better accuracy is attained by the higher expansion order of the propagator in terms of the energy scaled Hamiltonian in the spectral range between $[-1, 1]$ [480]. An advantage of this method is that it allows a relatively large time step Δt. This expansion is also applied to the Green operator to result in numerically exact accuracy, which is often useful in the evaluation of a spectral density operator.

As a useful polynomial expansion method we show the Chebyshev expansion scheme below. We consider the general case such as a dissociation process, in which a wavepacket may proceed out of the grid region set in advance. The time propagation operator of the wavepacket $|\chi\rangle$ is expanded in terms of the modified Chebyshev polynomial $\hat{Q}_n^\gamma$ [255],

$$e^{-\frac{i}{\hbar}\hat{H}\Delta t}|\chi\rangle = \sum_{n=0}^{M}(2-\delta_{n0})i^n e^{-\frac{i}{\hbar}\langle E\rangle\Delta t}J_n(\Delta E\Delta t/\hbar)\,|\xi_n^\gamma\rangle\,, \qquad (3.20)$$

where $J_n(\Delta E\Delta t/\hbar)$ is the first kind of Bessel function and $|\xi_n^\gamma\rangle$ is given by $|\xi_n^\gamma\rangle = \hat{Q}_n^\gamma(\hat{H}_{\text{norm}})\;|\chi\rangle$, with $\hat{H}_{\text{norm}} = (\hat{H} - \langle E\rangle)/\Delta E$ being the scaled Hamiltonian, and $\langle E\rangle = (H_{\text{max}} + H_{\text{min}})/2$ and $\Delta E = (H_{\text{max}} - H_{\text{min}})/2$, respectively. The energy region defined by the parameters H_{min} and H_{max} needs to include the spectral range of the Hamiltonian matrix.

To numerically absorb the wavepacket components at the boundaries, which are set beforehand to prevent the packet from flowing out to the asymptotic ranges, a damping function $e^{-\gamma}$ is introduced in the Chebyshev recursion scheme such that

$$\begin{aligned} |\xi_0^\gamma\rangle &= |\chi\rangle\,, \\ |\xi_1^\gamma\rangle &= e^{-\gamma}\hat{H}_{\text{norm}}\,|\chi\rangle\,, \\ |\xi_{n+1}^\gamma\rangle &= e^{-\gamma}\,(\,2\hat{H}_{\text{norm}}\,|\xi_n^\gamma\rangle\, - e^{-\gamma}\,|\xi_{n-1}^\gamma\rangle\,)) \qquad (n \geq 2), \end{aligned} \qquad (3.21)$$

where γ is taken to be a function of position coordinates.

The above time propagation scheme can be carried out "exactly" within a given numerical accuracy. An advantage of the Chebyshev expansion method is its rapid convergence. In particular, it shows one of the best performances among others when the time step Δt is taken to be long within a given tolerance.

3.2.4 *Multiconfigurational time-dependent Hartree (MCTDH) approach*

One strategy to treat nuclear quantum dynamics in systems of large dimensions is to use a compact basis to describe the total wavefunction Ψ. The optimization of the coefficients for those basis functions is usually performed with the Dirac–Frenkel time-dependent variational principle

$$\langle\delta\Psi|\hat{H} - i\hbar\partial/\partial t|\Psi\rangle = 0. \qquad (3.22)$$

As in the molecular orbital theory, the time-dependent Hartree (TDH) method provides a compact expression for the equations of motion of those coefficients by approximating Ψ in the Hartree product. In this formalism, the f-dimensional wavefunction is given by

$$\Phi(R_1, ..., R_f, t) = a(t) \prod_{k=1}^{f} \phi_k(R_k, t), \tag{3.23}$$

where R_k is the k-th nuclear coordinate. $\phi_k(R_k, t)$ are the associated orbitals (Hartree orbitals to describe nuclear motions) and $a(t)$ denotes a coefficient for the entire wavefunction, which are to be determined according to the above variational principle. The resultant equations of motion of these quantities under the constraint $i\hbar\langle\phi_k|\dot{\phi}_k\rangle \equiv g_k(t) = 0$ are

$$\begin{aligned} a(t) &= a(0) \exp\left(-i \int_0^t E(t')dt'/\hbar\right) \\ i\hbar\dot{\phi}_k &= [\mathcal{H}_k - E(t)]\phi_k = \rho_k \mathcal{H}_k \phi_k, \end{aligned} \tag{3.24}$$

where $E(t) = \langle\Psi|\hat{H}|\Psi\rangle/\langle\Psi|\Psi\rangle$. The effective Hamltonian $\mathcal{H}_k$ working on the orbital ϕ_k is defined as

$$\mathcal{H}_k \equiv \langle \prod_{i \neq k}^{f} \phi_i | \hat{H} | \prod_{j \neq k}^{f} \phi_j \rangle, \tag{3.25}$$

which arises naturally to derive the coupled equations, and

$$\rho_k \equiv \hat{1} - |\phi_k\rangle\langle\phi_k|. \tag{3.26}$$

The counterpart of Eq. (3.24) with the arbitrary constraint $i\hbar\langle\psi_k|\dot{\psi}_k\rangle = g_k(t)$ is written by

$$i\hbar\dot{a} = a\left(E - \sum_{k=1} g_k(t)\right) \tag{3.27}$$

and

$$i\hbar\dot{\psi}_k = \mathcal{H}_k\psi_k + (g_k - E)\psi_k. \tag{3.28}$$

As in the Hartree–Fock molecular orbital theory, which is based on the independent particle model, the above Hartree product method also lacks enough correlation among the orbitals, and thereby the resultant accuracy is limited. To overcome the drawback, one can take account of the interaction among possible configurations (or the Hartree products) as in the configuration interaction method and multiconfiguration SCF methods in electronic structure theory. The multiconfigulational time-dependent Hartree

(MCTDH) method [43, 272] is designed to realize such a spirit by employing the multiconfigurational form

$$|\Psi(q_1, .., q_f, t)\rangle = \sum_J A_J(t)|\Phi_J(t)\rangle = \sum_{j_1} .. \sum_{j_f} a_{j_1}..a_{j_f} \prod_{k=1}^{f} |\phi^k_{j_k}(t)\rangle, \quad (3.29)$$

where J denotes a composite index $J = (j_1, j_2..., j_f)$ and a configuration $\Phi_J(t)$ consists of the single particle functions $\phi^k_{j_k}(t)$ as $\Phi_J(t) = \prod_{k=1}^{f} \phi^k_{j_k}(t)$. We refer to $A_J = \prod_{k=1}^{f} a_{j_k}$ and $\phi^k_{j_k}$ as the coefficients of configurations and orbitals, respectively. In the equation of motion in the MCTDH approach, a single hole one particle function (SHF) plays an important role in determining the one particle orbitals. A SHF is defined as

$$|G^k_m(t)\rangle = \sum_{j_1} .. \sum_{j_{k-1}} \sum_{j_{k+1}} .. \sum_{j_f} A_{j_1, j_{k-1}, m, j_{k+1}, j_f}$$

$$\times |\phi^1_{j_1}(t)\rangle..|\phi^{k-1}_{j_{k-1}}(t)\rangle|\phi^{k+1}_{j_{k+1}}(t)\rangle..|\phi^f_{j_f}(t)\rangle, \quad (3.30)$$

which corresponds to a $(f-1)$ degree of freedom function created by removing the orbital in the kth degree of freedom from the total wavefunction. Note that this is assigned to the degree of freedom labeled k and the removed orbital label m in the kth degree of freedom. Use of the SHF is helpful in extracting the contribution of the kth single particle function in the total wavefunction. The total wavefunction is re-expressed for any k as

$$|\Psi(t)\rangle = \sum_n |\phi^k_n(t)\rangle|G^k_n(t)\rangle, \quad (3.31)$$

of which time derivative is given as

$$|\dot{\Psi}(t)\rangle = \sum_k \sum_n |\dot{\phi}^k_n(t)\rangle|G^k_n(t)\rangle + \sum_J \dot{A}_J(t)|\Phi_J(t)\rangle. \quad (3.32)$$

Notice that the time derivative of the total wavefunction is divided into two parts as above.

The variational principles with respect to the coefficients of configurations and orbitals as single particle functions, under the specific constraint of the time derivative overlap $\langle\phi^k_n|\dot{\phi}^k_m\rangle = 0$ (known as a numerically rather stable constraint),

$$\langle\delta\Psi|i\hbar\partial_t - \hat{H}|\Psi\rangle_A = 0, \quad (3.33)$$

$$\langle\delta\Psi|i\hbar\partial_t - \hat{H}|\Psi\rangle_\phi = 0, \quad (3.34)$$

yield the dynamics for A and orbitals ϕ such that

$$i\hbar\dot{A}_J(t) = \sum_L H_{JL}(t)A_L(t), \quad (3.35)$$

$$i\hbar\sum_m \rho^k_{nm}|\dot{\phi}^k_m(t)\rangle = (1 - \hat{P}_k)\sum_m \langle\hat{h}(t)\rangle^k_{nm}|\phi^k_m(t)\rangle, \quad (3.36)$$

where

$$H_{KL} \equiv \langle \Phi_K(t)|\hat{H}|\Phi_L(t)\rangle, \qquad \hat{P}_k = \sum_{j_k} |\phi^k_{j_k}(t)\rangle\langle\phi^k_{j_k}(t)|, \tag{3.37}$$

$$\rho^k_{nm}(t) \equiv \langle G^k_n|G^k_m\rangle, \qquad \langle\hat{h}(t)\rangle^k_{nm} \equiv \langle G^k_n|\hat{H}|G^k_m\rangle. \tag{3.38}$$

Further technical details along with numerical performances are found in Refs. [43, 272].

The MCTDH method allows for a treatment of quantum dynamics of (mainly) bound states in medium size molecular systems (with number of degrees of freedom up to $\sim$ 24). Many applications employing accurate potential energy surfaces are successfully achieved and compared with the relevant experimental spectral data, such as vibrational absorption, vibronic spectrum and so on. The method, however, is known not necessarily to work well for the evaluation of reaction rate in case of unimolecular dissociation with chaotic motions and hard recombination reaction systems. In these systems it is quite difficult to reproduce the dynamics accurately in terms of a small number of reference multireference orbitals.

Towards large molecular systems of dimensions up to about 1000 degrees of freedom, a multilayer MCTDH have been developed by Wang and Thoss [467]. This formalism utilizes a multilayer structure in primitive basis functions for the single particle functions and apply the time-dependent variational approach to optimize all the layer primitive basis functions. This procedure gives a compact basis set. The method seems to be useful to study the dynamics of a system embedded in an environment, which is important in nano device science.

3.2.5 *Eigenfunctions extracted from wavepacket dynamics: Energy screening*

As an example of the application of time propagating the quantum vibrational wavefunction and also as a practical means of obtaining selected vibrational eigenstates, we briefly take a look at the energy screening method [420].

Applying what we may call the energy screening operator

$$\Delta(E_{\min}, E_{\max}) = \int_{E_{\min}}^{E_{\max}} \delta(E - \hat{H}), \tag{3.39}$$

to a vibrational wavepacket $\chi(\boldsymbol{R}, t)$, we have

$$\begin{aligned}\chi_{\mathrm{ES}}(\boldsymbol{R}, t) &= \Delta(E_{\min}, E_{\max})\chi(\boldsymbol{R}, t) = \sum_{E_{\min} \le E_j \le E_{\max}} c_j \chi_j(\boldsymbol{R}) \\ &= \frac{1}{2\pi\hbar}\int_{E_{\min}}^{E_{\max}} dE \int_{-\infty}^{\infty} dt e^{i(E-\hat{H})t/\hbar}\chi(\boldsymbol{R}, t) \\ &= \frac{1}{2\pi}\int_{-\infty}^{\infty} dt \frac{1}{it}\left(e^{iE_{\max}t/\hbar} - e^{iE_{\min}t/\hbar}\right)e^{-i\hat{H}t/\hbar}\chi(\boldsymbol{R}, t),\end{aligned} \tag{3.40}$$

where $\chi_j(\boldsymbol{R})$ is a vibrational eigenstate with energy E_j. Thus, we can obtain a wavepacket $\chi_{\mathrm{ES}}(\boldsymbol{R}, t)$ composed of eigenstates in a given energy range by time integrating a time propagated arbitrary wavepacket. When there lies only one state in the given energy range, we obtain an eigenstate directly.

The factor $1/t$ accelerates convergence to an eigenstate (the uncertainty principle for time and energy range should be taken into account when making time divisions). We get the answer in an even shorter time if we make use of the fact that an eigenstate is real-valued. After a single run over a time division, we can take the spatial coordinate where the amplitude of the computed wavefunction $\chi_{\mathrm{ES}}(\boldsymbol{R}, t)$ has the largest absolute value, for example. Writing the phase as ϕ, we have $\chi'_{\mathrm{ES}}(\boldsymbol{R}, t) = e^{i(2\pi-\phi)}\chi_{\mathrm{ES}}(\boldsymbol{R}, t)$, which gives an almost real function if $\chi_{\mathrm{ES}}(\boldsymbol{R}, t)$ is nearly an eigenstate. The imaginary part of $\chi'_{\mathrm{ES}}(\boldsymbol{R}, t)$ indicates how near $\chi_{\mathrm{ES}}(\boldsymbol{R}, t)$ is to an eigenstate. If the imaginary part is small enough, we may simply take the imaginary part to be zero and use this as a better input wavepacket for the run over the next time integration.

The method of energy screening is known as a powerful means to extract eigenfunctions within a given energy range. Note that such eigenfunctions are obtained without imposing the orthogonal condition to the lower states.

3.3 Probing the dynamics with time-resolved photoelectron spectroscopy

We next study how the results of wavepacket dynamics can be tracked experimentally as a real-time history of chemical or physical events. Femtosecond time-resolved spectroscopy enables us to observe nuclear dynamics and to chart the path of chemical reactions in real time, and has been exploited in numerous applications ranging from fundamental studies of real-time motion in the photodissociation of diatomic molecules to studies of electron transfer in biological systems [104, 257, 506]. A pump pulse prepares an initial wavepacket which evolves in accordance with the time scales for vibrational ($\sim 10^{-13}$ s) and rotational ($\sim 10^{-10}$ s) motion. The evolution of the wavepacket is then probed by a time-delayed, probe pulse via excitation to a final state. Several techniques using different types of final states have been used to probe these wavepackets [82, 257, 309, 506]. These include the use of absorption, laser-induced fluorescence, multiphoton ionization, photoelectron spectroscopy, nonlinear degenerate four-wave mixing, time-resolved mass spectroscopy, and stimulated emission pumping.

Time-resolved ionization offers several advantages as a probe of these wavepackets [41, 42, 343, 360]. For example, the ground state of an ion is often better characterized than higher excited states of the neutral molecule, particularly for polyatomics. Ionization is also universal and hence there are no dark states. Furthermore, ionization provides both ions and photoelectrons and, while ion detection provides mass and kinetic-energy resolution in time-resolved studies [508], photoelectron spectra can provide complementary information on the evolution of the wavepacket [22, 63, 78, 132, 201, 270, 271, 362, 363, 377]. Its utility for real-time probing of molecular dynamics in the femtosecond regime has been nicely demonstrated in studies of wavepackets on excited states of Na_2 [22], on the B state of I_2 [132], and on the A state of NaI [201]. Femtosecond photoelectron–photoion coincidence imaging studies of photodissociation dynamics have been reported [107].

In this section we present the formulation and computation of femtosecond time-resolved photoelectron spectroscopy employing geometry- and energy-dependent photoionization amplitudes [15].

3.3.1 *Hamiltonian describing interaction with an external field*

A charged particle (of charge q and mass m positioned at $\boldsymbol{r}$) in a potential $V(\boldsymbol{r})$ interacting with a linearly polarized external field is described by the Hamiltonian,

$$\hat{H} = \frac{1}{2m}\left(\hat{\boldsymbol{p}} - \frac{q}{c}\boldsymbol{A}(\boldsymbol{r},t)\right)^2 + V(\boldsymbol{r}) - q\phi(\boldsymbol{r}), \tag{3.41}$$

where, in the Coulomb gauge, the vector potential $\boldsymbol{A}(\boldsymbol{r},t)$,

$$\boldsymbol{A}(\boldsymbol{r},t) = \boldsymbol{e}A_0\cos(\boldsymbol{k}\cdot\boldsymbol{r} - \omega t - \varphi), \tag{3.42}$$

and scalar potential $\phi(\boldsymbol{r})$ describe a freely propagating plane wave of frequency ω and wave vector $\boldsymbol{k}$ with polarization direction in the direction of the unit vector $\boldsymbol{e}$ (perpendicular to $\boldsymbol{k}$). φ is an arbitrary phase and c is the speed of light in vacuum.

Invoking the weak-field approximation (neglecting the term involving $\boldsymbol{A}\cdot\boldsymbol{A}$) and the long wavelength approximation ($e^{i\boldsymbol{k}\cdot\boldsymbol{r}} \approx 1$), we arrive at the familiar dipole approximation expression,

$$\hat{H} = -\frac{\hbar^2}{2m}\nabla^2 + V(\boldsymbol{r}) - \boldsymbol{\mu}(\boldsymbol{r})\cdot\boldsymbol{E}(t), \tag{3.43}$$

where the last term describes the particle's interaction with the field, with the dipole moment $\boldsymbol{\mu}(\boldsymbol{r})$,

$$\boldsymbol{\mu}(\boldsymbol{r}) = q\boldsymbol{r}, \tag{3.44}$$

the electric field $\boldsymbol{E}(t)$,

$$\boldsymbol{E}(t) = -\frac{1}{c}\frac{\partial}{\partial t}\boldsymbol{A}(t), \tag{3.45}$$

and the vector potential under the long-wavelength approximation now position independent as

$$\boldsymbol{A}(t) = \boldsymbol{e}A_0\cos(\omega t + \varphi). \tag{3.46}$$

3.3.2 *Coupled dynamics in electronic excitation*

The time evolution of a molecular system $\Psi(\boldsymbol{r}, \boldsymbol{R}, t)$ under (possibly several) external fields is given by the Schrödinger equation,

$$i\hbar\frac{\partial}{\partial t}\Psi(\boldsymbol{r}, \boldsymbol{R}, t) = \left[\hat{T}_N + \hat{H}^{\text{el}} + \sum_{p=1}^{N_p}\hat{V}_p(t)\right]\Psi(\boldsymbol{r}, \boldsymbol{R}, t), \tag{3.47}$$

where each $\hat{V}_p(t)$ represents the effect of external field p.

The total wavefunction of our system, $\Psi(\boldsymbol{r}, \boldsymbol{R}, t)$, may be decomposed as in the usual Born–Huang representation into several neutral electronic states $\Phi_n(\boldsymbol{r}; \boldsymbol{R})$,

$$\Psi(\boldsymbol{r}, \boldsymbol{R}, t) = \sum_n \chi_n(\boldsymbol{R}, t)\Phi_n(\boldsymbol{r}; \boldsymbol{R}). \tag{3.48}$$

Then, employing the dipole form of the external field interaction [Eq. (3.43)], the Schrödinger equation for the nuclear wavefunctions in the diabatic representation is given in the matrix form,

$$i\hbar\frac{\partial}{\partial t}\boldsymbol{\chi}(\boldsymbol{R}, t) = \left[\mathbf{1}\hat{T}_N + \mathbf{V}(\boldsymbol{R}) - \sum_p \boldsymbol{\mu}_p(\boldsymbol{R})\boldsymbol{E}_p(t)\right]\boldsymbol{\chi}(\boldsymbol{R}, t), \tag{3.49}$$

where $\boldsymbol{\chi}(\boldsymbol{R},t)$ is the column vector of all $\chi_n(\boldsymbol{R},t)$. $\mathbf{V}(\boldsymbol{R})$ is a diagonal matrix of the neutral potential energy functions $V_n(\boldsymbol{R})$. The dipole moment matrix in the direction of the external field polarization, $\boldsymbol{\mu}_p(\boldsymbol{R})$, includes matrix elements $\mu_{n_1,n_2}(\boldsymbol{R})$ between neutral states n_1 and n_2. $\mu_{n_1,n_2}(\boldsymbol{R})$ would be transition dipole moments for $n_1 \neq n_2$ and permanent dipole moments for $n_1 = n_2$ in the direction of polarization.

Within the split-operator scheme, for example, numerical short-time propagation would involve diagonalizing the interaction matrix (either together with the potential energy matrix or as a separate term) at each time step. We will see several other time-propagation methods in Sec. 5.2.1.

Time propagation with Eq. (3.49) will result in the time evolution of the population of each state,

$$P_n(t) = \int d\boldsymbol{R}\chi_n^*(\boldsymbol{R},t)\chi_n(\boldsymbol{R},t), \tag{3.50}$$

summarizing the effect of the external fields on the system. For example, if there are two states 1 and 2 where initially $P_1 = 1$ and $P_2 = 0$, a pump pulse interaction may have the effect of decreasing P_1 and increasing P_2.

3.3.3 *Nuclear wavepacket description of femtosecond time-resolved photoelectron spectroscopy*

To describe photoionization in a pump–probe context, the total wavefunction of our system, $\Psi(\boldsymbol{r},\boldsymbol{R},t)$, may be decomposed into several neutral electronic states $\Phi_n(\boldsymbol{r};\boldsymbol{R})$, whose dynamics we wish to probe, and several ionized states $\Phi_{c,\boldsymbol{k}}^{(-)}(\boldsymbol{r};\boldsymbol{R})$ we use for probing,

$$\Psi(\boldsymbol{r},\boldsymbol{R},t) = \sum_n \chi_n(\boldsymbol{R},t)\Phi_n(\boldsymbol{r};\boldsymbol{R}) + \sum_c \int d\boldsymbol{k}\chi_{c,\boldsymbol{k}}(\boldsymbol{R},t)\Phi_{c,\boldsymbol{k}}^{(-)}(\boldsymbol{r};\boldsymbol{R}), \tag{3.51}$$

where the first summation on the right hand side [the same as Eq. (3.48)] is the neutral system including the initial and pumped state. The additional, second summation is over any number of ionized electronic states labeled c. The ionized electronic state includes the photoelectron with incoming boundary condition, denoted with the $(-)$ superscript, and the photoelectron's continuous wave vector is denoted $\boldsymbol{k}$. Note that we must integrate over all $\boldsymbol{k}$ to describe the total ionized state.

The electronic wavefunctions represent the eigenstates of the electronic Hamiltonian $\hat{H}^{\mathrm{el}}$ at each nuclear position and are orthonormalized, e.g.,

$$\langle\Phi_{c_1,\boldsymbol{k}_1}^{(-)}|\Phi_{c_2,\boldsymbol{k}_2}^{(-)}\rangle = \delta_{c_1,c_2}\delta(\boldsymbol{k}_1 - \boldsymbol{k}_2). \tag{3.52}$$

The electronic wavefunctions are time-independent but depend parametrically on the nuclear coordinates, while the time dependence is explicit in the nuclear wavefunctions, $\chi_n(\boldsymbol{R}, t)$ and $\chi_{c,\boldsymbol{k}}(\boldsymbol{R}, t)$, associated with the neutral and ionized electronic states, respectively.

In the following, we explicitly write out the coupled Schrödinger equation for the case of one neutral and one ionized state; extension to any number of states is formally straightforward. Resorting to the dipole approximation, we aim to write the time-dependent Schrödinger equation for the nuclear wavefunctions in the form,

$$i\hbar\frac{\partial}{\partial t}\begin{pmatrix}\chi_n(\boldsymbol{R},t)\\ \chi_{c,\boldsymbol{k}}(\boldsymbol{R},t)\end{pmatrix} = \left[\mathbf{1}\hat{T}_N + \mathbf{V}(\boldsymbol{R}) - \mu_{nc,\boldsymbol{k}}(\boldsymbol{R})E_p(t)\right]\begin{pmatrix}\chi_n(\boldsymbol{R},t)\\ \chi_{c,\boldsymbol{k}}(\boldsymbol{R},t)\end{pmatrix}, \quad (3.53)$$

where the interaction with an external field $E_p(t)$ through the dipole moment $\mu_{nc,\boldsymbol{k}}(\boldsymbol{R})$ causing photoionization is added to the nuclear Hamiltonian in the form of a potential term. Though similar in form to Eq. (3.49), what Eq. (3.53) implies is much more involved, as we shall see in the following sections.

3.3.4 *Geometry-dependent photoionization matrix elements*

To explicitly obtain the spatial distribution of the photoelectrons for the final electronic state, we assume a frozen-core Hartree–Fock description where the final ionized electronic state is represented as an antisymmetrized product of a Hartree–Fock ion core wavefunction, $\Phi_{c,+}$, and a photoelectron orbital, $\phi^{(-)}_{c,\boldsymbol{k}}$,

$$\Phi^{(-)}_{c,\boldsymbol{k}} = \mathcal{A}(\Phi_{c,+} \cdot \phi^{(-)}_{c,\boldsymbol{k}}). \quad (3.54)$$

The photoelectron orbital is expanded in a single-center expansion using the spherical harmonics, $Y_{lm}(\theta, \phi)$, as

$$\phi^{(-)}_{c,\boldsymbol{k}}(\boldsymbol{r}'; \boldsymbol{R}) = \sum_{l\lambda} i^l e^{-i\eta_l} Y^*_{l\lambda}(\theta_k, \phi_k)\phi^{(-)}_{c,kl\lambda}(\boldsymbol{r}'; \boldsymbol{R}), \quad (3.55)$$

where $\boldsymbol{r}' = (r, \hat{r})$ is the photoelectron coordinates in a molecule-fixed center-of-mass spherical coordinate system, k and (θ_k, ϕ_k) are the magnitude and molecular frame angles of the wave vector, respectively, and η_l is the Coulomb phase shift. The nuclear wavefunction associated with the ionized electronic state is correspondingly expanded in spherical harmonics,

$$\chi_{c,\boldsymbol{k}}(\boldsymbol{R}, t) = \sum_{l\lambda} \chi_{c,kl\lambda}(\boldsymbol{R}, t) Y_{l\lambda}(\theta_k, \phi_k), \quad (3.56)$$

so that each photoelectron orbital partial-wave component, $\phi^{(-)}_{c,kl\lambda}(\boldsymbol{r}';\boldsymbol{R})$, has an associated nuclear wave component, $\chi_{c,kl\lambda}(\boldsymbol{R},t)$. The Schrödinger equation for the nuclear wavefunctions now take the coupled form,

$$i\hbar\frac{\partial}{\partial t}\chi_n(\boldsymbol{R},t) = \left[\hat{T}_N + V_n(\boldsymbol{R})\right]\chi_n(\boldsymbol{R},t) + E_p(t)\int k^2 dk \sum_{l\lambda} C^*_{nc,kl\lambda}(\boldsymbol{R},\hat{\Omega})\chi_{c,kl\lambda}(\boldsymbol{R},t), \quad (3.57)$$

and

$$i\hbar\frac{\partial}{\partial t}\chi_{c,kl\lambda}(\boldsymbol{R},t) = \left[\hat{T}_N + V_c(\boldsymbol{R}) + \varepsilon_k\right]\chi_{c,kl\lambda}(\boldsymbol{R},t) + E_p(t)C_{nc,kl\lambda}(\boldsymbol{R},\hat{\Omega})\chi_n(\boldsymbol{R},t), \quad (3.58)$$

where $\varepsilon_k = \hbar^2k^2/(2m_e)$ is the kinetic energy of the photoelectron with electron mass m_e. What we shall call the *photoionization matrix element*, $C_{nc,kl\lambda}(\boldsymbol{R},\hat{\Omega})$, is given by,

$$C_{nc,kl\lambda}(\boldsymbol{R},\hat{\Omega}) = \sqrt{\frac{4\pi}{3}}\sum_{\mu} I_{nc,kl\lambda\mu}(\boldsymbol{R})\mathcal{D}^1_{\mu 0}(\hat{\Omega}), \quad (3.59)$$

with the dipole operator given for linear polarization as

$$\hat{\mu} = \sqrt{\frac{4\pi}{3}}r\sum_{\mu}\mathcal{D}^1_{\mu 0}(\hat{\Omega})Y_{1\mu}(\hat{r}), \quad (3.60)$$

where $\mathcal{D}^l_{mm'}(\hat{\Omega})$ is the Wigner rotation matrix, $(r,\hat{r})$ are the magnitude and angle coordinates, respectively, of the photoelectron, and the angular coordinates $\hat{\Omega}$ orient the probe pulse with respect to the molecular frame.

The photoionization matrix elements, $C_{nc,kl\lambda}(\boldsymbol{R},\hat{\Omega})$, in Eq. (3.59) is explicitly dependent on the nuclear geometry and photoelectron momentum, and is the essential ingredient in describing the molecular photoionization dynamics. It is built from geometry-dependent partial-wave dipole matrix elements between the neutral and ionized states, $I_{nc,kl\lambda\mu}(\boldsymbol{R})$,

$$I_{nc,kl\lambda\mu}(\boldsymbol{R}) = (-i)^l e^{i\eta_l}\sum_{l_0\lambda_0}\langle\phi^{(-)}_{c,kl\lambda}|rY_{1\mu}(\hat{r})|\phi_{n,l_0\lambda_0}(r)Y_{l_0\lambda_0}(\hat{r})\rangle, \quad (3.61)$$

where $\phi_{n,l_0\lambda_0}(r;\boldsymbol{R})$ is a component of the single-center expansion of the orbital ϕ_n out of which ionization occurred. The angular momentum coupling inherent in molecular photoelectrons can be seen in a single-center expansion of $\phi^{(-)}_{c,kl\lambda}(\boldsymbol{r}';R)$ for a linear molecule,

$$\phi^{(-)}_{c,kl\lambda}(\boldsymbol{r}';R) = \sum_{l'} g^{(-)}_{c,ll'\lambda}(k,r,R)Y_{l'\lambda}(\hat{r}). \quad (3.62)$$

Use of photoelectron orbitals which correctly incorporate such angular momentum coupling is essential for a quantitative description of molecular photoionization.

3.3.5 *Calculation of the photoelectron orbitals*

To obtain the final state wavefunctions needed in evaluating the photoelectron matrix elements, we assume a frozen-core Hartree–Fock model in which the core orbitals are taken to be those of the ion for both ionized and neutral electronic states. The photoelectron orbital, $\phi^{(-)}_{c,kl\lambda}(\boldsymbol{r}';\boldsymbol{R})$, is obtained as a solution of a one-electron Schrödinger equation containing the Hartree–Fock potential of the molecular ion, $V_c(\boldsymbol{r},\boldsymbol{R})$,

$$\left[-\frac{\hbar^2}{2m_e}\nabla^2 + V_c(\boldsymbol{r},\boldsymbol{R}) - \varepsilon_k\right]\phi^{(-)}_{c,kl\lambda}(\boldsymbol{r}';\boldsymbol{R}) = 0. \tag{3.63}$$

We use an iterative procedure, based on the Schwinger variational principle, to solve the Lippmann–Schwinger equation associated with Eq. (3.63) [250]. The procedure begins by approximating the static-exchange potential of the relaxed ion core by a separable form

$$U_{\mathrm{SE}} \approx U_{\mathrm{S}}(\boldsymbol{r},\boldsymbol{r}') = \sum_{ij}\langle\boldsymbol{r}|U|\alpha_i\rangle(U^{-1})_{ij}\langle\alpha_j|U|\boldsymbol{r}'\rangle, \tag{3.64}$$

where the matrix U^{-1} is the inverse of the matrix with the elements $(U)_{ij} = \langle\alpha_i|U|\alpha_j\rangle$, the α's are discrete basis functions such as Cartesian or spherical Gaussian functions, and U is twice the static-exchange potential in Eq. (3.63) with the long-range Coulomb potential removed. The Lippmann–Schwinger equation with this separable potential $U_{\mathrm{S}}(\boldsymbol{r},\boldsymbol{r}')$ can be readily solved and provides approximate initial solutions to the photoelectron orbitals $\phi^{(-)}_{c,kl\lambda}(\boldsymbol{r}';\boldsymbol{R})$. These solutions can be iteratively improved to yield converged solutions to the Lippmann–Schwinger equation containing the exact static-exchange potential U_{SE}. Several iterations usually provide converged solutions of Eq. (3.63).

3.3.6 *Numerical time-propagation describing photoionization*

The integration over k in Eq. (3.57) may be approximated with a numerical quadrature over sampled k_j up to a maximum photoelectron momentum $k_{N_k} = k_{\max}$ with quadrature weights w_j,

$$\int k^2 dk \approx \sum_{j=1}^{N_k} w_j k_j^2, \tag{3.65}$$

and the vibrational partial-wave components may be redefined as

$$\tilde{\chi}_{c,k_jl\lambda}(\boldsymbol{R},t) = \sqrt{w_j}k_j\chi_{c,k_jl\lambda}(\boldsymbol{R},t), \tag{3.66}$$

to keep the Hamiltonian Hermite, resulting in a discretized form of the coupled equations of motion [Eqs. (3.57) and (3.58)],

$$i\hbar\frac{\partial}{\partial t}\chi_n(\boldsymbol{R},t) = \left[\hat{T}_N + V_n(\boldsymbol{R})\right]\chi_n(\boldsymbol{R},t) + E_p(t)\sum_{jl\lambda}\sqrt{w_j}k_j C^*_{nc,k_jl\lambda}(\boldsymbol{R},\hat{\Omega})\tilde{\chi}_{c,k_jl\lambda}(\boldsymbol{R},t), \quad (3.67)$$

and

$$i\hbar\frac{\partial}{\partial t}\tilde{\chi}_{c,k_jl\lambda}(\boldsymbol{R},t) = \left[\hat{T}_N + V_c(\boldsymbol{R}) + \varepsilon_{k_j}\right]\tilde{\chi}_{c,k_jl\lambda}(\boldsymbol{R},t) + E_p(t)\sqrt{w_j}k_j C_{nc,k_jl\lambda}(\boldsymbol{R},\hat{\Omega})\chi_n(\boldsymbol{R},t). \quad (3.68)$$

Now we see that the coupled equations of motion may be written in a compact matrix form,

$$i\hbar\frac{\partial}{\partial t}\boldsymbol{\chi}(\boldsymbol{R},t) = [\mathbf{1}\hat{T}_N + \mathbf{V}_D(\boldsymbol{R}) + \mathbf{V}_O(\boldsymbol{R},t)]\boldsymbol{\chi}(\boldsymbol{R},t), \quad (3.69)$$

where $\boldsymbol{\chi}(\boldsymbol{R},t)$ is a column vector of the neutral state vibrational wavefunction and the ionized state vibrational partial-wave components,

$$\boldsymbol{\chi}(\boldsymbol{R},t) = \begin{pmatrix} \chi_n(\boldsymbol{R},t) \\ \tilde{\chi}_{c,k_jl\lambda}(\boldsymbol{R},t) \\ \vdots \end{pmatrix}, \quad (3.70)$$

$\mathbf{V}_D(\boldsymbol{R})$ is the diagonal part of the potential energy matrix comprised of the neutral state potential energy function and the potential energy function of the ionized states,

$$\mathbf{V}_D(\boldsymbol{R}) = \begin{pmatrix} V_n(\boldsymbol{R}) & 0 & \cdots \\ 0 & V_c(\boldsymbol{R}) + \varepsilon_{k_j} & 0 \\ \vdots & 0 & \ddots \end{pmatrix}, \quad (3.71)$$

and $\mathbf{V}_O(\boldsymbol{R},t)$ is the off-diagonal part of the potential energy function,

$$\mathbf{V}_O(\boldsymbol{R},t) = \begin{pmatrix} 0 & V^*_{1,2} & V^*_{1,3} & \cdots & V^*_{1,N} \\ V_{1,2} & 0 & \cdots & \cdots & 0 \\ V_{1,3} & \vdots & \ddots & & \vdots \\ \vdots & \vdots & & \ddots & \vdots \\ V_{1,N} & 0 & \cdots & \cdots & 0 \end{pmatrix}, \quad (3.72)$$

where each nonzero element of the matrix $V_{1,J}$ represents a term of form $E_p(t)\sqrt{w_j}k_jC_{nc,k_jl\lambda}(\boldsymbol{R},\hat{\Omega})$. Note that the order of the off-diagonal matrix N is $N_k \times N_{l\lambda}$, where $N_{l\lambda}$ is the number of (l,λ) used in the single-center expansion, which can be of the order 10^2 to 10^3 or more; however, the nonzero elements are only in one row and one column.

We may apply the split-operator method (Sec. 3.2.1) to the three matrices $\mathbf{1}\hat{T}_N$, $\mathbf{V}_D$, and $\mathbf{V}_O$ for a short-time numerical propagation scheme. Ionization is now described by population of the neutral state $\chi_n(\boldsymbol{R}, t)$ transferring to the ionized state partial-wave components $\chi_{c,k_jl\lambda}(\boldsymbol{R}, t)$ over time through the interaction represented by the matrix $\mathbf{V}_O(\boldsymbol{R}, t)$.

3.3.7 *Efficient diagonalization of the interaction matrix*

Diagonalization of the Hermitian matrix V_O (at each discretized geometry point) can be very time consuming using a general eigensolving technique, which usually takes on the order of N^3 operations. Since the number of elements in the matrix, N, can easily be on the order of 10^2 or 10^3 depending on the involved (l, λ) channels and the number of quadrature points k, the computation is rather impractical with such fast scaling.

This operation count can be reduced greatly considering the special form the interaction matrix takes in the present problem. Significant improvement of computation time comes from noting that nonzero elements of the matrix appear only in one row and one column, as given in Eq. (3.72). It can be shown that such a matrix is diagonalized analytically to result in just two nonzero eigenvalues, $\pm\lambda$, where

$$\lambda = \Big(\sum_{J=2}^{N} |V_{1,J}|^2 \Big)^{1/2}, \tag{3.73}$$

which means that after diagonalizing, only two of the resulting linear combinations of the wavefunctions of Eq. (3.70) are needed to apply the interaction operation. It turns out that only two of the eigenvectors of the matrix need to be calculated.

Short-time propagation with the interaction matrix may be written as

$$\exp\Big(- \frac{i\Delta t}{\hbar} \mathrm{V}_O \Big) = \exp\Big(- \frac{i\Delta t}{\hbar} \mathbf{U}\mathbf{D}\mathbf{U}^\dagger \Big) = \mathbf{U} \exp\Big(- \frac{i\Delta t}{\hbar} \mathbf{D} \Big) \mathbf{U}^\dagger, \tag{3.74}$$

where $\mathbf{U}$ is a unitary transformation matrix and $\mathbf{D}$ is the diagonalized interaction matrix. Analytical derivation shows that the transformation matrix has the form

$$\mathbf{U} = \frac{1}{\lambda\sqrt{2}} \begin{pmatrix} -\lambda & \lambda & X \cdots \\ V_{1,2}^* & V_{1,2}^* & \vdots \ \cdots \\ \vdots & \vdots & \vdots \ \cdots \\ V_{1,N}^* & V_{1,N}^* & X \cdots \end{pmatrix}, \tag{3.75}$$

where X elements here represent transformation matrix elements that we do not wish to compute. These elements arise as eigenvectors corresponding to zero eigenvalues, and obtaining their actual values amount to orthogonalizing the degenerate $N-2$ linear combinations of the wavefunctions, which is an $O(N^3)$ operation we would better avoid.

Instead we split the transformation matrix into two parts, one analytically known and the other part computationally nontrivial,

$$\mathbf{U} = \mathbf{U}_1 + \mathbf{U}_2, \tag{3.76}$$

where

$$\mathbf{U}_1 = \begin{pmatrix} U_{22} & 0 \\ U_{N2} & 0 \end{pmatrix}, \qquad \mathbf{U}_2 = \begin{pmatrix} 0 & U_{2N} \\ 0 & U_{NN} \end{pmatrix}, \tag{3.77}$$

with submatrices U_{22}, U_{N2}, U_{2N}, and U_{NN} of sizes 2×2, $(N-2) \times 2$, $2 \times (N-2)$, and $(N-2) \times (N-2)$, respectively. The diagonalized short time propagator can be split into two terms in a corresponding manner,

$$\exp\Big(-\frac{i\Delta t}{\hbar}\mathbf{D}\Big) = \begin{pmatrix} e^{i\Delta t\lambda/\hbar} & 0 & 0 \cdots & 0 \\ 0 & e^{-i\Delta t\lambda/\hbar} & 0 \cdots & 0 \\ 0 & 0 & 1 & 0 \\ \vdots & \vdots & & \ddots \\ 0 & 0 & 0 & 1 \end{pmatrix} = \mathbf{D}_1 + \mathbf{D}_2, \tag{3.78}$$

where

$$\mathbf{D}_1 = \begin{pmatrix} D_{22} & 0 \\ 0 & 0 \end{pmatrix}, \qquad \mathbf{D}_2 = \begin{pmatrix} 0 & 0 \\ 0 & D_{NN} \end{pmatrix}. \tag{3.79}$$

D_{22} is a 2×2 diagonal matrix of the exponentiated nonzero eigenvalues, and D_{NN} is the $(N-2) \times (N-2)$ unit matrix. Noting that

$$\mathbf{D}_2\mathbf{U}_2^\dagger = \mathbf{U}_2^\dagger, \tag{3.80}$$

and

$$\mathbf{U}_1\mathbf{U}_2^\dagger = \mathbf{U}_2\mathbf{U}_1^\dagger = \mathbf{D}_1\mathbf{U}_2^\dagger = \mathbf{D}_2\mathbf{U}_1^\dagger = \mathbf{U}_2\mathbf{D}_1 = 0, \tag{3.81}$$

and also that

$$\mathbf{1} = \mathbf{U}\mathbf{U}^\dagger = \mathbf{U}_1\mathbf{U}_1^\dagger + \mathbf{U}_2\mathbf{U}_2^\dagger, \tag{3.82}$$

Eq. (3.74) is reorganized as

$$\begin{aligned} \mathbf{U}\exp\Big(-\frac{i\Delta t}{\hbar}\mathbf{D}\Big)\mathbf{U}^\dagger &= (\mathbf{U}_1 + \mathbf{U}_2)(\mathbf{D}_1 + \mathbf{D}_2)(\mathbf{U}_1 + \mathbf{U}_2)^\dagger \\ &= \big[\mathbf{1} + \mathbf{U}_1(\mathbf{D}_1 - 1)\mathbf{U}_1^\dagger\big], \end{aligned} \tag{3.83}$$

where the last expression depends only on the analytically known terms. Both the computation of the matrix elements of $\mathbf{U}_1$ and $\mathbf{D}_1$ and the matrix-vector multiplication in the last expression of Eq. (3.83) scale linearly in N, so that the $O(N^3)$ diagonalization operation is now reduced to $O(N)$.

Note in passing that additional efficiency is obtained from the fact that diagonalization of the $N \times N$ matrix is computationally equivalent to the diagonalization of a smaller matrix of size $N_{\text{eff}} \times N_{\text{eff}}$, where N_{eff} is the number of elements after omitting the elements that do not significantly contribute at a particular time, geometry, or photoelectron energy or angular momentum. Varying the size of the interaction matrix at different time and geometry may result in a more efficient implementation.

3.3.8 *The external fields introduced*

The external linearly polarized electronic field is given as the time derivative of the vector potential [see Eqs. (3.45) and (3.46)],

$$\begin{aligned} E_p(t; E_p^{\max}, \omega_p, t_p, \phi_p) &= E_p^{\max} f_p(t-t_p)\cos\left(\omega_p(t-t_p)+\phi_p\right) \\ &\quad + \frac{E_p^{\max}}{\omega_p}\frac{df_p(t-t_p)}{dt}\sin\left(\omega_p(t-t_p)+\phi_p\right), \end{aligned} \tag{3.84}$$

where the second term is required when the frequency of the field, ω_p, is small compared to the time-dependence of the pulse envelope function, $f_p(t)$. The value of $f_p(t)$ is in the range $[0,1]$ so that $E_p^{\max}$ is the maximum amplitude of the field and t_p is at the center of the pulse. The intensity is $I_p = c\varepsilon_0(E_p^{\max})^2/2$ where c is the speed of light and ε_0 is the electric constant. Typically, $p = \text{pu}$ (the pump pulse to initiate the excited state dynamics of interest) or $p = \text{pr}$ (the probe, or the ionizing pulse) but others may be added (such as $p = \text{cntrl}$ for a control pulse to modify the excited state dynamics). We usually take $t_{\text{pu}} = 0$ so that t_{pr} is the delay time between the pump and the probe pulses. The phase ϕ_p looses significance for longer pulses.

One convenient form for the envelope function is a Gaussian function,

$$f_p(t) = e^{-a_p^2 t^2}, \tag{3.85}$$

which has a full-width at half-maximum (FWHM) of

$$\tau_p = \frac{2(\log 2)^{1/2}}{a_p}. \tag{3.86}$$

3.3.9 *Photoelectron spectra from the dynamics and their transient counterparts*

In the present framework, all the dynamical information, including that of electrons, is formally contained in the nuclear wavefunctions $\chi_n(\boldsymbol{R},t)$ and $\chi_{c,\boldsymbol{k}}(\boldsymbol{R},t)$, the coefficients of the stationary electronic wavefunctions in Eq. (3.51). After numerically solving Eq. (3.69) for the time-evolution of the nuclear wavefunctions, $\chi(\boldsymbol{R},t)$ and $\chi_{klm}(\boldsymbol{R},t)$, we may obtain the total ion population at any time for any given pump–probe delay time t_{pr},

$$\begin{aligned}P_{\text{ion}}(t;t_{\text{pr}}) &= \sum_c \int_0^{\infty} k^2 dk \int_0^{2\pi} d\phi_k \int_0^{\pi} \sin\theta_k d\theta_k \int d\boldsymbol{R} \\ &\quad \times \left| \sum_{l\lambda} \chi_{c,kl\lambda}(\boldsymbol{R},t;t_{\text{pr}}) Y_{l\lambda}(\theta_k,\phi_k) \right|^2 \\ &\approx \sum_c \sum_{j=1}^{N_k} w_j k_j^2 \int_0^{2\pi} d\phi_k \int_0^{\pi} \sin\theta_k d\theta_k \int d\boldsymbol{R} \\ &\quad \times \left| \sum_{l\lambda} \chi_{c,k_j l\lambda}(\boldsymbol{R},t;t_{\text{pr}}) Y_{l\lambda}(\theta_k,\phi_k) \right|^2 \end{aligned} \qquad (3.87)$$

To emphasize that $\chi_{c,\boldsymbol{k}}(\boldsymbol{R},t)$ is also a function of the parameter t_{pr}, the pump–probe delay time when the probe laser ionizes the system, we have explicitly included it in $\chi_{c,kl\lambda}(\boldsymbol{R},t;t_{\text{pr}})$.

The photoelectron kinetic energy (or wave number) dependent photoelectron angular distribution for ionization to state c, $P_{c,k_j}(\theta_k,\phi_k,t;t_{\text{pr}})$, and the photoelectron energy distribution for ionization to state c, $P_c(\varepsilon_k,t;t_{\text{pr}})$, may be defined as parts of Eq. (3.87),

$$\begin{aligned}P_{\text{ion}}(t;t_{\text{pr}}) &= \sum_c \sum_{j=1}^{N_k} w_j \int_0^{2\pi} d\phi_k \int_0^{\pi} \sin\theta_k d\theta_k P_{c,k_j}(\theta_k,\phi_k,t;t_{\text{pr}}) \\ &= \sum_c \sum_{j=1}^{N_k} w_j k_j P_c(\varepsilon_k,t;t_{\text{pr}}). \end{aligned} \qquad (3.88)$$

The energy-integrated photoelectron angular distribution for state c is given as

$$P_c(\theta_k,\phi_k,t;t_{\text{pr}}) = \sum_{j=1}^{N_k} w_j P_{c,k_j}(\theta_k,\phi_k,t;t_{\text{pr}}). \qquad (3.89)$$

At times it may be useful to obtain the angular momentum components of the population by integrating over the angles but not summing over l

and λ,

$$P_{c,kl\lambda}(t;t_{\rm pr}) = k^2 \int_0^{\pi} \sin\theta_k d\theta_k \int_0^{2\pi} d\phi_k \int d\boldsymbol{R} |\chi_{c,kl\lambda} Y_{l\lambda}|^2$$
$$= k^2 \int d\boldsymbol{R} |\chi_{c,kl\lambda}(\boldsymbol{R},t;t_{\rm pr})|^2. \tag{3.90}$$

All the above angular distributions are with respect to the molecular frame of a single oriented molecule. Closer to current experimental interest, we may consider the probing of molecules aligned along the single polarization axis of the pump field but averaged about the other orientations. When the pump and probe polarizations are aligned, rotation of the wavefunctions to the pump frame,

$$\tilde{\chi}_{c,kl\tilde{m}}(\boldsymbol{R},t;t_{\rm pr}) = \sum_{\lambda} \mathcal{D}^{l*}_{\lambda\tilde{m}}(\hat{\Omega}) \chi_{c,kl\lambda}(\boldsymbol{R},t;t_{\rm pr}), \tag{3.91}$$

and averaging over the angle $\tilde{\phi}_k$ yield photoelectron angular distributions for a set of molecules with their molecular axes aligned along the pump polarization axis,

$$P_{c,k}(\tilde{\theta}_k,t;t_{\rm pr}) = k^2 \int_0^{2\pi} d\tilde{\phi}_k \int d\boldsymbol{R} \left| \sum_{l\tilde{m}} \tilde{\chi}_{c,kl\tilde{m}}(\boldsymbol{R},t;t_{\rm pr}) Y_{l\tilde{m}}(\tilde{\theta}_k,\tilde{\phi}_k) \right|^2$$
$$= 2\pi k^2 \int d\boldsymbol{R} \sum_{\tilde{m}=-L}^{L} \left| \sum_{l=|\tilde{m}|}^{L} \tilde{\chi}_{c,kl\tilde{m}}(\boldsymbol{R},t;t_{\rm pr}) Y_{l\tilde{m}}(\tilde{\theta}_k,0) \right|^2, \tag{3.92}$$

where L is the largest value of l we consider and the tilde indicate values in a frame aligned with the pump. The energy-dependent photoelectron angular distribution is a sum over the ionized states,

$$P_k(\tilde{\theta}_k,t;t_{\rm pr}) = \sum_c P_{c,k}(\tilde{\theta}_k,t;t_{\rm pr}), \tag{3.93}$$

where we assume the orthogonality condition in Eq. (3.52).

The physical meaning of these quantities is perhaps obvious but not trivial. Recall that $\chi_{c,\boldsymbol{k}}(\boldsymbol{R},t;t_{\rm pr})$ represents the amplitude of the final ionized electronic (ion plus photoelectron) state $\Phi^{(-)}_{c,\boldsymbol{k}}(\boldsymbol{r};\boldsymbol{R})$ and $|\chi_{c,\boldsymbol{k}}(\boldsymbol{R},t;t_{\rm pr})|^2$ is the probability density that has accumulated (either increasing or decreasing) in this state up to the time t due to the probe laser with its pulse centered at time $t_{\rm pr}$. Hence, when the observation time t is far later than $t_{\rm pr}$,

$$t \gg t_{\rm pr} \tag{3.94}$$

$P_{c,k}(\theta_k, \phi_k, t; t_{\mathrm{pr}})$ represents the population (flux) of the photoelectrons in the direction (θ_k, ϕ_k) leaving the system in the electronic state c. This actually represents the photoelectron spectrum at the end of the entire optical interaction, and which are the results presented in Secs. 3.3.12, 5.2, and 5.4. On the other hand, if

$$t \sim t_{\mathrm{pr}}, \tag{3.95}$$

the electronic states are still interacting with the probe laser and are hence still evolving in time. Therefore $P_{c,k}(\theta_k, \phi_k, t; t_{\mathrm{pr}})$ may reflect the dynamics of the propagating electronic wavepackets driven by the laser fields for the intrinsic chemical process. If the probe pulse would be suddenly truncated at this time, the value of $P_{c,k}(\theta_k, \phi_k, t; t_{\mathrm{pr}})$ should be directly identified as the distribution of photoelectrons reaching an electron detector such as a CCD camera. However, the optical interaction continues and the scattering waves, which are still in the molecular site, may undergo interference among themselves, reducing or enhancing the intensity of photoelectrons that are indeed ejected from the molecule to a detector. We may talk of $P_{c,k}(\theta_k, \phi_k, t; t_{\mathrm{pr}})$ of $t \sim t_{\mathrm{pr}}$ as the *transient* photoelectron spectrum. This is not necessarily directly observed quantity but should require additional devices in experimental studies. We shall see an example of this in Sec. 5.6.

3.3.10 *History of dynamics in the transient photoelectron spectrum*

What may we expect to observe in the transient photoelectron spectroscopy? The dynamics of Eq. (3.47) as manifested in Eq. (3.51) is rewritten in the path integral form in such a way that

$$\begin{aligned}
\chi_{c,\boldsymbol{k}}(\boldsymbol{R}, t; t_{\mathrm{pr}}) \sim \lim_{n\to\infty} \sum_{n_k} \sum_{(n-1)_k} \cdots \sum_{1_k} \int d\boldsymbol{R}_n \cdots \int d\boldsymbol{R}_0 \\
\times \left\langle \boldsymbol{R}\Phi^{(-)}_{c,\boldsymbol{k}}(\boldsymbol{R}) \middle| U(t; t_n) \middle| \boldsymbol{R}_n \Phi_{n_k}(\boldsymbol{R}_n) \right\rangle \\
\times \left\langle \boldsymbol{R}_n \Phi_{n_k}(\boldsymbol{R}_n) \middle| U(t_n; t_{n-1}) \middle| \boldsymbol{R}_{n-1} \Phi_{(n-1)_k}(\boldsymbol{R}_{n-1}) \right\rangle \\
\ldots \times \left\langle \boldsymbol{R}_1 \Phi_{1_k}(\boldsymbol{R}_1) \middle| U(t_1; t_0) \middle| \boldsymbol{R}_0 \Phi_{n_0}(\boldsymbol{R}_0) \right\rangle \times \chi_0(\boldsymbol{R}_0),
\end{aligned} \tag{3.96}$$

in which time is sliced as

$$t_i = i \times (t - t_0)/(n + 1) + t_0 \tag{3.97}$$

with $i = 1, 2, \cdots n$ $(t = t_{n+1})$, and $\boldsymbol{R}_n \Phi_{n_k}(\boldsymbol{R}_n)$ is a basis in the joint space of (nuclear position) and (electronic state) such that, for instance,

$$\sum_{n_k} \int d\boldsymbol{R}_n \big|\boldsymbol{R}_n \Phi_{n_k}(\boldsymbol{R}_n)\big\rangle\big\langle \boldsymbol{R}_n \Phi_{n_k}(\boldsymbol{R}_n)\big| = 1. \tag{3.98}$$

Φ_{n_0} is an arbitrary initial electronic state and χ_0 is the vibrational wavepacket associated with it. $U(t_k; t_{k-1})$ is the propagator of Eq. (3.47) from time t_{k-1} to t_k. The longest value for t is the time when the probe laser field is completely switched off. (The time for the photoelectrons to reach a detector is not considered here.) As can be seen intuitively in the above expression, the molecular states can evolve along many different paths in the space of wavefunctions. The longer and more intense the applied laser field is, the more complicated this evolution (history of the state) can be. For example, Rabi oscillations and multiphoton absorption are the main causes of this behavior. Therefore the wavelength of the field is also a critical factor. Single photon ionization will generally give a monotonic spectrum with the passage of t. On the other hand, if $\chi_{c,\boldsymbol{k}}(\boldsymbol{R}, t; t_{\rm pr})$ is composed of many independent components traveling through different channels, an interference pattern should result. The electron dynamics within a molecule before ionization should also be manifested in the interference pattern of the transient photoionization spectrum. In particular, the different channels will contribute to the different l, λ components of $\chi_{c,kl\lambda}(\boldsymbol{R}, t; t_{\rm pr})$ in Eq. (3.87), and the interference among these will in turn bring about a time-dependent modulation or quantum beats in $P_{c,k}(\theta_k, \phi_k, t; t_{\rm pr})$ on the time-scale of electron dynamics.

3.3.11 *Velocity map imaging and its time derivative*

The information contained in the energy-dependent photoelectron angular distributions $P_k(\tilde{\theta}_k; t_{\rm pr})$ [Eq. (3.93)] may be conveniently visualized as velocity map images. Examples will be given in Sec. 5.4. Comparison of such images across different delay times $t_{\rm pr}$ reveal the time evolution of the system.

When the system undergoes fast (or somewhat sudden) evolution, such dynamics reflected in the velocity map image can be more sensitively extracted in terms of its time derivatives. Time derivatives of the image can be computed from $P_k(\tilde{\theta}_k; t_{\rm pr})$ using finite differences using a small time shift τ,

$$\Delta P_k(\tilde{\theta}_k, t_{\rm pr}) = \frac{P_k(\tilde{\theta}_k, t_{\rm pr} + \tau) - P_k(\tilde{\theta}_k, t_{\rm pr} - \tau)}{2\tau}. \tag{3.99}$$

Note that the time-derivative is to be taken with respect to the change of the time-delay between the pump and probe lasers, t_{pr}, and not with respect to t. An example will be shown in Secs. 5.4.2.4 and 5.4.3.3, illustrating that time-derivative photoelectron velocity map imaging can be quite helpful in uncovering and/or identifying useful characteristics of wavepacket motion, which may otherwise be overlooked in the original images.

3.3.12 *An example: The Na_2 double minimum state*

As an example we present the results of time-resolved photoelectron spectra of the wavepacket dynamics on the (2) Σ_u^+ double minimum state of the Na_2 molecule computed as outlined in the previous subsections.

3.3.12.1 *The system*

Figure 3.5 shows the potential energy curves for the ground $X\ ^1\Sigma_g^+$ (V_g) and double-minimum $^1\Sigma_u^+$ (V_e) states of neutral Na_2 and the ground $X\ ^2\Sigma_g^+$ state of the Na_2^+ ion and illustrates our pump–probe scheme. The double-minimum $^1\Sigma_u^+$ state has been studied extensively, and results from an avoided crossing of two adiabatic states. The inner well is a Rydberg state with a principal configuration of $\sigma_g(3s)\sigma_u(4s)$ and the outer well is a strong admixture of the $\sigma_g(4s)\sigma_u(3s)$ and $\sigma_g(3s)\sigma_u(3s)$ configurations. The $\sigma_g(3s)\sigma_u(3s)$ configuration lends significant ionic character to this state.

The electronic wavefunctions and energies for the ground states of Na_2 and Na_2^+ were obtained from extensive configuration-interaction CI studies using the GAMESS package. Calculations were done at 19 internuclear distances between 2 and 11 Å. For convenience in the calculation of the dipole matrix elements between the ground and excited states and for photoionization of the excited state, we employed the Hartree–Fock (HF) basis of the ion in these CI calculations. For a more accurate description of the excited state, we adopt the rotationless experimental potential of Cooper et al. [95] in the wavepacket calculations. All potential curves were adjusted to the known excitation energies and ionization potential.

Figure 3.5 shows the transition dipole moment between the ground and excited (2) Σ_u^+ states of Na_2. While around the inner well Rydberg region, the dipole moment function is very flat, indicating Franck–Condon behavior, and the dipole moment changes substantially in the outer well ionic region, reflecting the ionic character of the system.

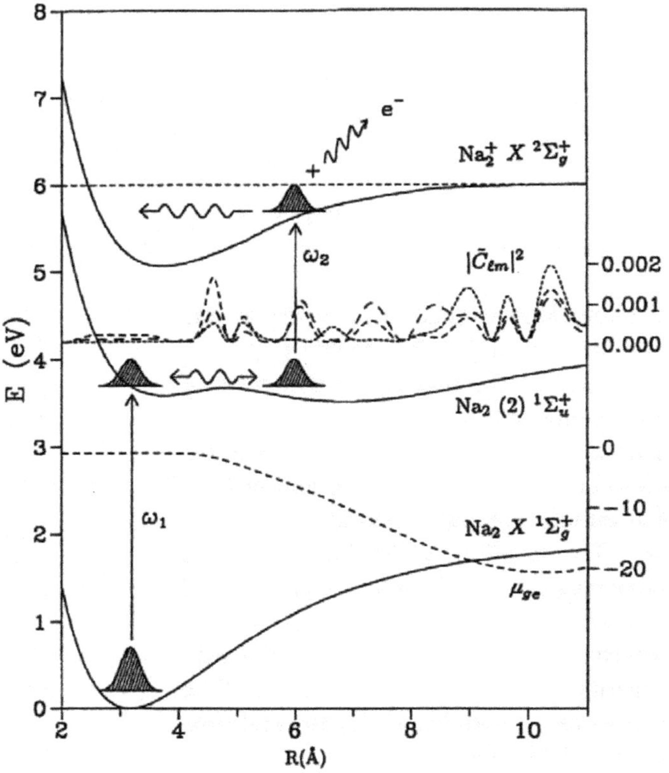

Fig. 3.5 Potential curves for the X $^1\Sigma_g^+$ and (2) $^1\Sigma_u^+$ states of Na_2 and the X $^2\Sigma_g^+$ state of Na_2^+. The transition dipole moment function μ_{ge} and photoionization matrix elements $\tilde{C}_{lm}$ for the molecule parallel to the pump and probe polarization are also shown for a kinetic energy of 0.60 eV. The partial waves $l = 0$, 2, and 4 with $m = 0$ are denoted by long, medium, and short dashed lines, respectively. (Reprinted with permission from Y. Arasaki *et al.*, J. Chem. Phys. 112, 8871 (2000)).

To obtain the photoionization amplitudes, we used our CI wavefunction for the double-minimum state and a frozen-core Hartree–Fock (FCHF) description of the wavefunction for the ionized state. For the FCHF model the wavefunction is taken to be an antisymmetrized product of HF ion orbitals and a photoelectron orbital that is a solution of a one-electron Schrödinger equation containing the Hartree–Fock potential of the ion core. The HF wavefunction provides a very adequate description of Na_2^+ over all internuclear distances of interest.

The $C_{l\lambda}$ coefficients of Eq. (3.59) incorporate the underlying dynamics into our formulation of time-resolved photoelectron spectroscopy. These

$C_{l\lambda}$ coefficients for polarization vectors of the pump and probe pulses parallel to the molecular axis and for a photoelectron energy of about 0.6 eV are shown as a function of internuclear distance in Fig. 3.5. For this state of Na_2 only even ls arise, while for this parallel arrangement only $m = 0$ terms are allowed. These $C_{l\lambda}$ display several important features. Among these, we note the striking change in the $C_{l\lambda}$ in the barrier region where the avoided crossing occurs. This behavior could be expected to be quite general in regions of nonadiabatic behavior. Furthermore, the large magnitude of these $C_{l\lambda}$ around the barrier will exert a strong influence on the ion signal as the wavepacket moves through this region. The pronounced oscillatory behavior of these coefficients for all ls ($l = 0, 2$, and 4 are shown) arises from the evolution of the electronic structure in the outer well where the wavefunction acquires significant ionic character. These features reflect non-Franck–Condon behavior of the underlying photoelectron matrix elements. Such behavior should arise quite generally in nonadiabatic regions of potential surfaces and, in fact, have been seen in the NaI system [17].

3.3.12.2 *The pump excited dynamics of Na_2*

We have used the formulation, potential curves (see Fig. 3.5), and photoionization matrix elements outlined in the previous sections to study the pump–probe photoelectron spectra in Na_2. Important parameters in these studies include,

(1) All potential energies are measured from the bottom of the ground potential energy curve, V_g. The initial wavepacket $\chi_g(0)$ is the vibrational ground state of V_g with energy $E_{\text{ground}} = 9.35 \times 10^{-3}$ eV.
(2) The characteristic points of the potential curves are as follows: The local minimum of the inner well of V_e lies at $R = 3.7$ Å and 3.589 eV, while that of the outer well is at 6.7 Å and 3.518 eV. The barrier between them lies at $R = 4.7$ Å and 3.675 eV, resulting in a barrier height of 0.0873 eV from the bottom of the inner well. The ion potential curve has its minimum at $R = 3.6$ Å and 5.064 eV.
(3) The laser parameters: We examine three different photon energies for the pump laser of 3.62, 3.68, and 3.70 eV resulting in a wavepacket on V_e below the barrier, at the top of the barrier, and above the barrier, respectively. The electric field strength and the full width at half maximum (FWHM) of the Gaussian pulses were taken to be 5.142×10^6 eV/m (0.001 au) and 100 fs, respectively.

(4) The probe laser is fixed for all the above cases at $\hbar\omega_2 = 2.28$ eV, $E_{02} = 1.029 \times 10^8$ eV/m, nd the FWHM is 40 fs. These values were chosen to give large gains in the photoionization signals.
(5) The molecular orientation is parallel to the pump laser. For each of these three ω_1, the polarization vector of the probe laser is set either parallel or perpendicular to the molecular axis.

The $|\chi_e(R,t)|^2$ for the wavepackets on the excited state are shown in Figs. 3.6(a)–3.6(c), and for pump photon energies below the barrier, at the barrier, and above the barrier, respectively. For a pump energy below the barrier, the $\chi_e(R,t)$ is clearly confined to the inner well [Fig. 3.6(a)]. For a pump photon energy at the barrier, the wavepacket bifurcates in the barrier region with one component proceeding to the outer well after some slowing down over the barrier, while the other component is reflected back into the inner well [Fig. 3.6(b)]. For a pump photon energy above the barrier, the wavepacket mostly passes over the barrier and moves into the outer well [Fig. 3.6(c)].

3.3.12.3 *Oscillatory total ion signal*

The non-Franck–Condon behavior of the photoionization amplitudes for this double-minimum state of Na_2 is evident in the oscillatory behavior of the $C_{l\lambda}$ coefficients with internuclear geometry. An unusual consequence of this behavior of the $C_{l\lambda}$ arises from their near-vanishing values for all $(k_j l\lambda)$ in a narrow range to the left ($R = 4.1$ Å) of the potential barrier on the excited state. One may thus predict that a wavepacket placed inside the inner potential well should exhibit a very small photoelectron signal whenever the wavepacket hits this "depleted" region. This should result in an oscillatory behavior of the total photoelectron signal as a function of delay time. This behavior is seen in the ion signal of Fig. 3.7 for a pump pulse of energy 3.62 eV which places the wavepacket in the inner well and for the polarization vectors of the pump and probe pulses parallel to the molecular axis. The rapidly oscillating solid line represents the total ion signal given by Eq. (3.87).

Upon excitation this wavepacket immediately begins an oscillatory motion of period about 347 fs, hitting the right-hand-side turning point in the inner well near 4.1 Å at $t = 158$, 505, 842, and 1180 fs, where the ion signal has minima. The signal at these minima gradually becomes larger due to dispersion of the wavepacket.

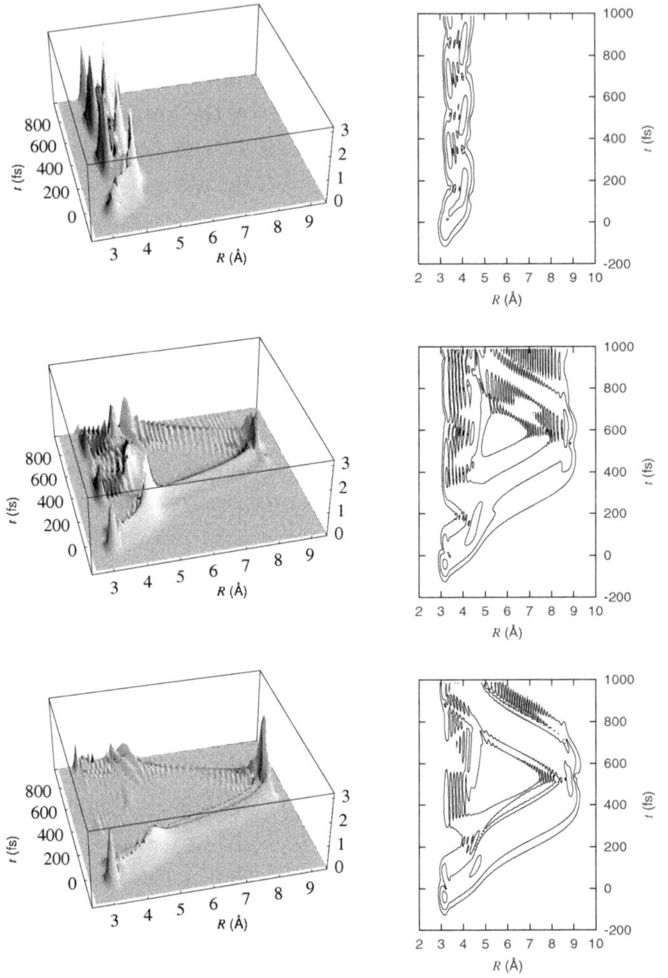

Fig. 3.6 Perspective view and contour plot of the absolute square of the wavepackets, $|\chi_e(R,t)|^2$ generated by the pump lasers with (a) $\hbar\omega_1 = 3.62$ eV (below the barrier), (b)3.68 eV (at the barrier), and (c)3.70 eV (above the barrier). (Reprinted with permission from Y. Arasaki *et al.*, J. Chem. Phys. 112, 8871 (2000)).

To further highlight this unusual behavior, we have calculated the total ion signal for ionization of these wavepackets, but with the Franck–Condon approximation, where $C_{20}(k_j, R)$, the largest of these coefficients, is simply replaced by its constant value at $R = 3.5$ Å at each photoelectron energy and all other $C_{l\lambda}(k_j, R)$ are set to zero. The dotted flat line in Fig. 3.7 shows

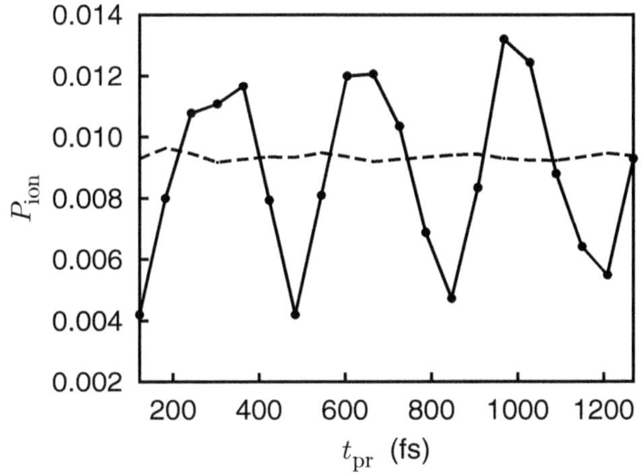

Fig. 3.7 Oscillatory behavior of the total ion signal solid line arising from the wave packet confined in the inner potential basin. Broken line: the total ion signal based on the Franck–Condon approximation. (Reprinted with permission from Y. Arasaki *et al.*, J. Chem. Phys. 112, 8871 (2000)).

the result. The difference is striking and illustrates the importance of the underlying photoionization dynamics to a mapping of the wavepacket via pump–probe photoelectron spectra.

3.3.12.4 *Energy-resolved photoelectron signals*

We next examine the kinetic energy distribution of the photoelectrons, $P(\varepsilon_k)$. As stated previously, three pump energies were studied with parallel and perpendicular polarizations. Figures 3.8(a)–3.8(c) show these $P(\varepsilon_k)$ as a function of delay time for pump photons that place the wavepacket in the inner well, on top of the barrier, and above the barrier for parallel polarizations of the pump and probe pulses. Figures 3.8(A)–3.8(C) show the spectra for perpendicular orientations of the pump and probe pulses. The only significant difference between these parallel and perpendicular cases is in the magnitudes of the signals. When the wavepacket is confined to the inner well [Fig. 3.6(a)], the major feature of the photoelectron spectra is a strong peak around 0.7 eV [Figs. 3.8(a) and 3.8(A)] which shows an oscillatory dependence on the pump–probe delay time, as discussed above. This feature is clearly a fingerprint of the localization of the wavepacket in the inner well.

When the pump photon places the wavepacket on top of the barrier, the photoelectron spectra [Figs. 3.8(b) and 3.8(B)] behave quite differently.

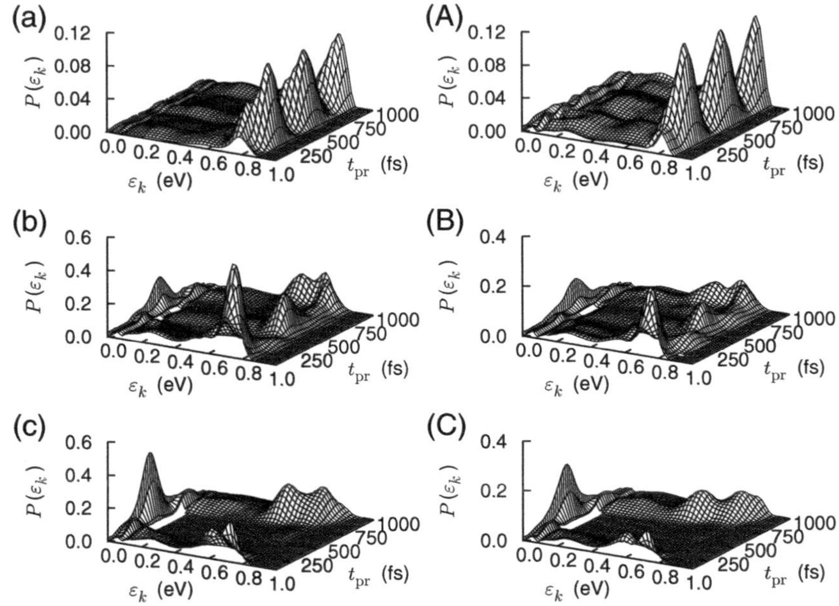

Fig. 3.8 Photoelectron signal $P(\varepsilon_k)$ vs kinetic energy and delay time. The photon energy of the pump laser is (a), (A) $\hbar\omega_1 = 3.62$ eV (below the barrier), (b), (B) 3.68 eV (at the barrier), and (c), (C) 3.70 eV (above the barrier). For each energy, the polarization vector of the probe laser is set parallel (a), (b), (c) and perpendicular (A), (B), (C) to that of the pump laser. Note that the vertical scales are different for different energies. (Reprinted with permission from Y. Arasaki *et al.*, J. Chem. Phys. 112, 8871 (2000)).

At about $t_{\rm pr} = 200$ fs, the wavepacket is moving slowly through the barrier region [Fig. 3.6(b)], where the photoionization amplitudes are large (see Fig. 3.5). This results in a strong peak at around 0.7 eV. In this region the wavepacket bifurcates into a component that moves into the outer well and another that is reflected back to the inner well. The period of recurrence of the wavepacket in the inner well happens to be about half of that of its outer-well counterpart. At about $t_{\rm pr} = 600$ fs, the outer-well component reaches its turning point, which results in a strong peak in the photoelectron signal at very low kinetic energy. This feature was first pointed out by Meier and Engel [270, 271] who also noted its value for mapping wavepackets at their turning points. On the other hand, the inner wavepacket, which becomes rather dispersed, leads to a peak at about $\varepsilon_k = 0.7$ eV at $t_{\rm pr} = 600$ fs. An interesting double-peak feature is seen around $\varepsilon_k = 0.7$ eV and $t_{\rm pr} = 1000$ fs.

When the pump pulse places the wavepacket above the barrier, the photoelectron spectra are simpler [Figs. 3.8(c) and 3.8(C), since no major bifurcation of the wavepacket occurs at the potential barrier. As in the other cases, the peak around $\varepsilon_k = 0.7$ eV arises from the wavepacket while it is in the inner well. Once over the barrier, this peak at 0.7 eV is no longer present, reflecting the absence of any major component in the inner well Fig. 3.6(c). A high peak at a very low photoelectron energy around $t_{\rm pr} = 500$ fs again arises at the outer turning point. The broad feature at $\varepsilon_k = 0.7$ eV and $t_{\rm pr} = 800$–1000 fs is due to the dispersed wave returning to the inner well.

3.3.12.5 *Angular distribution of photoelectrons*

In conventional photoelectron spectroscopy, angular distributions are known to provide valuable insight into the underlying dynamics. A few groups have, moreover, recently reported measurements of angular distributions in pump–probe photoelectron spectra [332, 406]. Althorpe and Seideman have also examined the angular distributions of photoelectrons in the pump–probe ionization for a rigid diatomic molecule NO [8].

These time-resolved photoelectron angular distributions are given by Eq. (3.89). Figure 3.9 shows a few such photoelectron spectra for pump photon energies below, on top of, and over the barrier in the excited state. In contrast to the kinetic energy distributions in Fig. 3.8, these angular distributions are significantly different for the parallel and perpendicular cases. For the parallel case, the angular distribution is generally of d_{z^2} type, as illustrated by the inset of Fig. 3.9, while it is basically of d_{yz} type for the perpendicular case. These shapes are expected on the basis of symmetry considerations for a dipole interaction. The oscillatory behavior of these angular distributions with time when the wavepacket is in the inner well simply reflect the oscillation of the ion signal seen in Fig. 3.7 and discussed previously. The main difference in the angular distributions for the cases when the pump photon places the wavepacket just at or above the barrier is the absence of the peak for a delay time of 600 fs in the latter case. This difference arises from the fact that in the latter case the wavepacket has virtually no inner-well component until the wavepacket returns there from the outer well at around 800 fs.

In conclusion, photoelectron angular distributions not only provide insightful fingerprints of vibrational wavepacket dynamics, but their strong

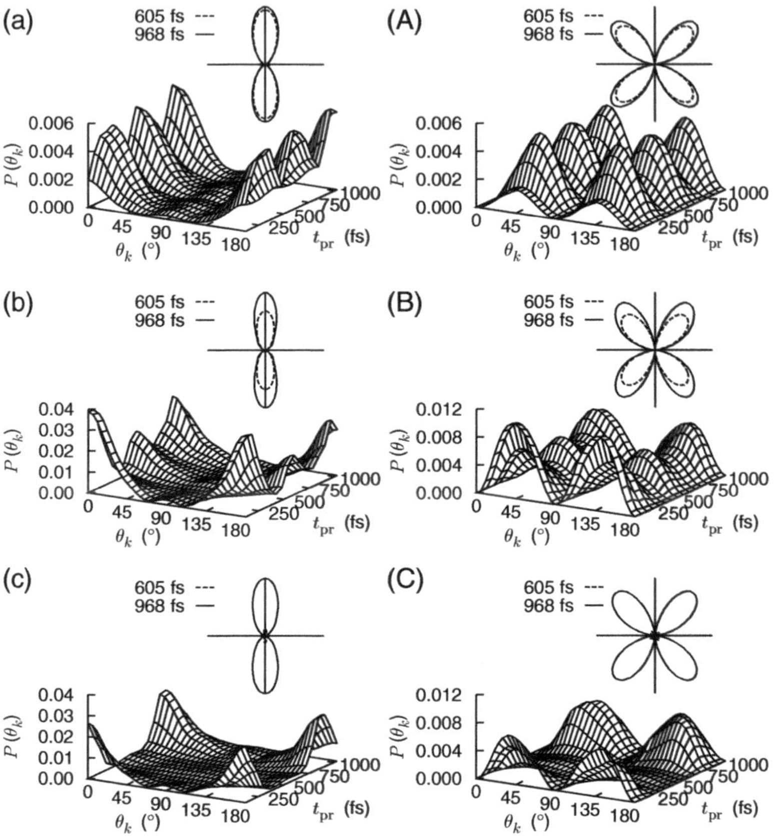

Fig. 3.9 The photoelectron angular distributions as a function of the delay time. As in Fig. 3.8, the photon energy of the pump laser is (a), (A) $\hbar\omega_1 = 3.62$ eV (below the barrier), (b), (B) 3.68 eV (at the barrier), and (c), (C) 3.70 eV (above the barrier). For each energy, the polarization vector of the probe laser is set parallel and perpendicular to that of the pump field. The insets show a polar coordinate representation of $P(\theta_k, \phi_k; t_{\rm pr})$ at $t_{\rm pr} = 605$ (broken curves) and 968 fs (solid curves) and $\phi_k = 0$. (Reprinted with permission from Y. Arasaki *et al.*, J. Chem. Phys. 112, 8871 (2000)).

dependence on the relative orientation of the pump and probe polarizations can also be exploited to monitor real-time molecular rotation. Those who are further interested should consult with Ref. [16].

Chapter 4

Breakdown of the Born–Oppenheimer Approximation: Classic Theories of Nonadiabatic Transitions and Ideas behind

Although this book attempts to guide readers to modern nonadiabatic electron wavepacket theory (both in the presence and absence of external laser fields), we should first review the traditional theories, which have led the study of nonadiabatic transitions for a long time. Our motive for reviewing time-honored theories is not a historical interest but to help readers to understand fundamental ideas in the nonadiabatic dynamics, and therefore the list of the theories presented here is far from perfect. More complete reviews are already available in Refs. [28, 84, 85, 117, 178, 198, 291, 295].

4.1 Theories for one-dimensional curve crossing problem

We begin the discussions of this section with the linear curve crossing model. In this model, though, the original problem is quite often radically simplified with small numbers of parameters so that analytical solutions are attained. This is in clear contrast to their modern counterparts, in which the properties (energies and matrix elements) of real molecules are faithfully reproduced by ab initio calculations. On the other hand, these time-honored theories have been successful in extracting the essential physics of nonadiabatic transitions, giving a vivid view of quantum transition dynamics. Therefore the solutions and the processes of deriving them directly lead us to the fundamental mechanism of the nonadiabatic transitions, which are often less clear in more elaborate numerical approaches. At the same time, we should always care about the validity range of the classic model theories. To make judicial use of the resulting formula, one should understand the assumptions and simplifications introduced in the course of model construction and in the solving processes.

4.1.1 *The Landau–Zener theory of curve crossing model*

We first visit the work of Landau and Zener on the linearized curve-crossing problem. Landau [229] and Zener [505] studied nonadiabatic transitions in atomic collisions and the (pre)dissociations of diatomic molecules. Although Landau also reached an equivalent conclusion independently, we will follow Zener's approach here. Zener introduced a significant simplification to the original problem so that it becomes analytically solvable yet it conserves an essential physics of the nonadiabatic transitions. The simplified model, which is now called "the linear curve crossing model", is a two-state model with following approximations: (1) The relevant nuclear motion near the curve crossing point is one-dimensional, which is referred to as the "reaction coordinate" in this context. The reaction coordinate is denoted as X, which is the distance from the curve crossing point. (2) Diabatic potentials near the crossing point, W_1 and W_2, are approximated as linear in X; $W_1 = -F_1X$ and $W_2 = -F_2X$ for state 1 and 2, respectively. (3) The off-diagonal term in the diabatic representation is approximated as the constant V throughout the region. (4) During the transition, the system is assumed to move with a constant velocity.

Although radical these approximations may seem, however, they may be justified in some limiting situations including atomic predissociation problem which Zener originally had in mind. In the region where the difference of the diabatic potentials $|W_1(X) - W_2(X)|$ is much larger than the off-diagonal term, the adiabatic theorem tells (note that in such situation, diabatic potentials are approximately equal to the adiabatic energies) that the nonadiabatic transitions are negligible. Hence the relevant dynamics is limited to the region near the diabatic crossing point where $|W_1(X) - W_2(X)| \gg V$ does not apply. The parameters in such region is then, in the simplest approximation, represented by the values at the curve crossing point, as in the (1) to (4) of the above approximation. Therefore the approximations are to be regarded as taking a single reference point, rather than literary assuming globally linear (constant) diagonal (off-diagonal) diabatic potential. The validity of approximation also depends on whether or not the nonadiabatic transitions can be represented by a single curve-crossing point. In practice, approximation (4) may further require that the nuclear motion be of high energy so as not to be affected by the electronic dynamics. The point will be modified in the generalized linear crossing model we shall encounter in the following section.

We assume that the electronic state is in the state 2 at the beginning. For simplicity, we also assume that $F_1 - F_2$ to be positive. We assume that the nuclear coordinate X has initially large positive value and propagates leftward (uphill) in constant velocity v, that is, $\dot{X} = -vt$. The wavefunctions of the two electronic states can be represented by $\Phi_1(\mathbf{r}; X)$ and $\Phi_2(\mathbf{r}; X)$, respectively. The total wavefunction is then represented as

$$\Psi(\mathbf{r}, X, t) = C_1(t)\Phi_1(\mathbf{r}; X) + C_2(t)\Phi_2(\mathbf{r}; X). \tag{4.1}$$

Considering X as a time-dependent external parameter, the Schrödinger equation is written in a matrix equation

$$i\hbar\frac{\partial}{\partial t}\begin{pmatrix} C_1(t) \\ C_2(t) \end{pmatrix} = \begin{pmatrix} W_1(X) & V \\ V & W_2(X) \end{pmatrix}\begin{pmatrix} C_1(t) \\ C_2(t) \end{pmatrix}. \tag{4.2}$$

Equation (4.2) is to be solved as an initial value problem under the conditions $C_1(-\infty) = 0$ and $|C_2(-\infty)| = 1$, so that the nonadiabatic transition probability (or, equivalently, the probability of the system remains as the diabatic state 2) is given by $P = |C_2(+\infty)|^2$. Eliminating C_2, the problem is reduced to a second order differential equation

$$\frac{d^2U_1}{dz^2} + \left(i\gamma + \frac{1}{2} - \frac{z^2}{4}\right)U_1 = 0. \tag{4.3}$$

Here the variables are scaled by an energy scale $\sqrt{\alpha} \equiv \sqrt{(F_1 - F_2)\,\hbar v}$ as $z \equiv (t/i\hbar)\sqrt{\alpha}e^{i\pi/4}$ and $\gamma \equiv V^2/\alpha$, whereas U_1 is related to the state amplitude by $C_1 = e^{-\frac{i}{2\hbar}\int^t (W_1+W_2)dt'}U_1(t)$. The solutions of Eq. (4.3) are known as Waber functions. Here the solution that vanishes in the limit of large negative value of t is $D_{-i\gamma-1}(-iz)$ hence we can write

$$U_1(t) = AD_{-i\gamma-1}(-iz), \tag{4.4}$$

where A is a constant that will be determined so as to satisfy another boundary condition. The asymptotic behavior of Eq. (4.4) in the limit of large negative value of t is known as

$$D_{-i\gamma-1}(ie^{-i\frac{\pi}{4}}\sqrt{\alpha}|t|/\hbar) \to e^{-i\frac{\pi}{4}(i\gamma+1)}e^{-i\frac{1}{4}\left(\sqrt{\alpha}|t|/\hbar\right)^2}\left(\sqrt{\alpha}|t|/\hbar\right)^{-i\gamma-1}. \tag{4.5}$$

Recalling that $C_2 = \left(i\hbar\dot{C}_1 - W_1C_1\right)/V$, the boundary condition $|C_2(-\infty)| = 1$ fixes the constant A as $A = \sqrt{\gamma}e^{-\pi\gamma/4}$. We can then consider the other asymptote for large positive t and obtain

$$|C_1| \to A\left|\frac{\sqrt{2\pi}}{\Gamma(i\gamma+1)}\right|. \tag{4.6}$$

This results in $|C_1|^2 \to 1 - e^{-2\pi\gamma}$ as $t \to \infty$, which is the probability of adiabatic passage. Cases with different signs of F_1 and F_2 are also obtained in the same manner. With a slight generalization, the transition probability is,

$$P = \exp\left[-\frac{2\pi}{\hbar v}\frac{V^2}{|F_1 - F_2|}\right], \tag{4.7}$$

which is now well-known as the Landau–Zener (LZ) formula.

Due to radical simplification in the model construction, the resultant estimate of probability, Eq. (4.7) is often not numerically accurate enough in the modern context of quantum chemical dynamics. There are also many important aspects in molecular nonadiabatic dynamics that are completely ignored; nuclear configuration dependence of matrix elements, multidimensionality of nuclear dynamics, entanglement of nuclear and electronic dynamics, etc.

The formula nevertheless reflect the essential physics in nonadiabatic transition. The exponential dependence on parameters reflects the non-perturbative nature of this problem. The non-dimensional exponent $\frac{V^2}{\hbar v|F_1-F_2|}$, which represents "adiabaticity," ensures the correct limiting behavior (the adiabatic limit $v \to 0$ and its opposite $v \to \infty$ correspond to $P = 0$ and $P = 1$, respectively). One also sees the correct "scale" in this problem; for example, whether the passage velocity v is fast or not should be measured in unit $V^2/(\hbar|F_1 - F_2|)$. Qualitative accuracy of this formula is known to be rather robust; see for example, Ref. [445], in which they compare exact result with LZ formula as well as their proposed (surface hopping) method. These virtues are favorable for making rough estimate as well as constructing new guiding principle of one's study; see, for example, Ref. [404] to see that the LZ estimate is still conceptually important even in the state-of-art studies.

On application of the LZ formula in analysis of the real system, one should keep in mind the limitation of LZ formula; its application should be limited to cases where (1) the kinetic energy of nuclear translational motion is large enough that change during the passage is negligible and (2) the slope of (approximately) linear diabatic potential is of the same sign. The second condition in fact indirectly arises from a requirement that the nuclear wavepacket of the both state should pass through the crossing point to reach some asymptotic region with a constant velocity, which is often unlikely in the situation where the sign of slope is opposite. However, it is known that LZ type formula applies even for the opposite sign of slope if the crossing point is in the classically allowed region far from the turning points so that the above requirement is satisfied.

4.1.2 *Quantum phase arising from nonadiabatic transitions*

4.1.2.1 *Phase in nonadiabatic interactions*

We will next review the Stueckelberg theory [99, 395]. While Landau–Zener treatment concentrated on single curve-crossing, realistic atomic/molecular dynamics often includes multiple passages of such points resulting in superpositions of quantum states each of which has undergone different time evolution. The phase gained in each of time evolution is crucial to describe the quantum interference. Such interference effect, which is now called as the Stueckelberg oscillation, is indeed experimentally observable with modern techniques [153].

Stueckelberg also introduced two-state model but adopted time-independent formulation and used semiclassical approach for solution. The latter is in contrast to constant velocity assumptions in the treatment of Landau–Zener, but is essential for analytical derivation of correct adiabatic phase factors. Semiclassical contour integral method and analysis of accompanying Stokes phenomena is used for deriving transition amplitude in time-independent formulation of this problem [395], which will be briefly mentioned in the next subsection (also see Ref. [99] for more details including corrections).

Child [83, 84] developed a uniform diagrammatic approach for one-dimensional problems; classical turning points, barrier penetration, and curve crossings are represented in terms of individual diagrams. These are further connected with link matrices of the semiclassical wavefunction coefficients. Child showed that a wide range of spectroscopic phenomena such as predissociation and restricted rotation are described as a combination of these elementary diagrams. The wavefunction is written in a semiclassical form as

$$\psi(r) = \sqrt{\frac{\mu}{2\pi\hbar^2 k(r)}} \left\{ C' \exp\left[i \int_{r_0}^{r} k(r)dr \right] + C'' \exp\left[-i \int_{r_0}^{r} k(r)dr \right] \right\}, \tag{4.8}$$

where μ is the reduced mass, $k(r)$ is the semiclassical wave number given as $\hbar k(r) \equiv \sqrt{2\mu(E - V(r))}$ with energy E and potential $V(r)$, and r_0 is the coordinate of the curve-crossing point.[1] The link matrix M then connects the coefficient vector $\mathbf{C} = (C', C'')^T$ before and after the passage in the manner $\mathbf{C}^{(\text{after})} = M\mathbf{C}^{(\text{before})}$. Summarizing Stueckelberg's results in a compact manner, the curve crossing, in particular, takes the following

[1] Note that the prefactor in Eq. (4.8) differs from the standard bound-state normalization $\sqrt{\mu\bar{\omega}/(2\pi\hbar k(r))}$ with effective angular frequency $\bar{\omega}$ [83].

form: The wavefunctions before and after the curve crossing are written as

(for $r \ll r_0$)

$$\psi_j(r) \sim k_{\pm}^{-\frac{1}{2}} \left\{ U'_j \exp\left[i \int_{r_0}^{r} k_{\pm}(r) dr \right] + U''_j \exp\left[-i \int_{r_0}^{r} k_{\pm}(r) dr \right] \right\}$$

(for $r \gg r_0$)

$$\psi_j(r) \sim k_{\mp}^{-\frac{1}{2}} \left\{ V'_j \exp\left[i \int_{r_0}^{r} k_{\mp}(r) dr \right] + V''_j \exp\left[-i \int_{r_0}^{r} k_{\mp}(r) dr \right] \right\}, \tag{4.9}$$

where (diabatic) state indices j are either 1 or 2. $\pm$ ($\mp$) for $r \ll r_0$ ($r \gg r_0$) takes $+$ ($-$) for $j = 1$ and $-$ ($+$) for $j = 2$. $k_{\pm}$ are semiclassical wave number on the adiabatic potential

$$V_{\pm} \equiv \frac{W_1 + W_2}{2} \pm \sqrt{\left(\frac{W_1 - W_2}{2} \right)^2 + V^2}. \tag{4.10}$$

Then the relation between the coefficients are given as

$$\begin{pmatrix} V'_1 \\ V'_2 \end{pmatrix} = \begin{pmatrix} \lambda & -(1-\lambda^2)^{\frac{1}{2}} e^{-i\chi} \\ (1-\lambda^2)^{\frac{1}{2}} e^{i\chi} & \lambda \end{pmatrix} \begin{pmatrix} U'_1 \\ U'_2 \end{pmatrix} \tag{4.11}$$

and

$$\begin{pmatrix} U''_1 \\ U''_2 \end{pmatrix} = \begin{pmatrix} \lambda & (1-\lambda^2)^{\frac{1}{2}} e^{i\chi} \\ -(1-\lambda^2)^{\frac{1}{2}} e^{-i\chi} & \lambda \end{pmatrix} \begin{pmatrix} V''_1 \\ V''_2 \end{pmatrix}, \tag{4.12}$$

where $\lambda \equiv e^{-\pi\nu}$ is the transition amplitude with $\nu = \mathrm{Im}\left[\int_{r_1}^{r_2} dr [k_-(r) - k_+(r)] \right] / (2\pi)$, with r_1 and r_2 being the (complex) crossing points that exist on the analytically continued functions of the adiabatic potential surfaces, whereas the phase factor χ is given as $\chi = \arg\Gamma(i\nu) - \nu \ln \nu + \nu + \pi/4$. Thus, a unified semiclassical theory has been established for a chemical reaction in which plural elementary processes may be involved. If one replaces ν by the LZ exponent $\nu = \frac{V^2}{\hbar v |F_1 - F_2|}$, Eqs. (4.11), (4.12) can be seen as an extension of LZ formula including the phase factor for "adiabatic" passages (off-diagonal in Eqs. (4.11) and (4.12)). For example, in a predissociation problem where U'_j and U''_j are related by some round-trip processes around the classical turning points, the solution of Eqs. (4.11) and (4.12) describes multiple passages of avoided crossings with appropriate quantum phase effects. An interesting analysis on the predissociation dynamics in an "intermediate coupling" molecule, IBr is found in Ref. [86], where simplest

LZ-like estimate breaks down but analysis using Eqs. (4.11), (4.12) gives reasonable results.

It should be noted that, however, the treatment here implicitly assumes each elementary process as independent to each other, which is not always correct. For example, in case of a predissociation problem, there exist a curve crossing and classical turning points, both break the applicability of the Wentzel-Kramers-Brillouin (WKB) approximation.

It then follows that, in order for independent treatment to be applicable, these singularities should be separated by a finite region where the system recovers the WKB applicability condition. The point will be further specified in the next subsection.

4.1.2.2 *Generalized linear curve-crossing problem*

Here we review a generalized form of linear curve-crossing problem [84], which was introduced in Stueckelberg's paper and has been developed by many researchers including Ref. [99]. In the time-dependent curve-crossing model, introduced by Landau [229] and Zener [505], nuclear dynamics is completely decoupled based on the assumption of constant velocity. This apparently limits the applicability to high-energy collisions. Moreover, as we have seen in the last section, it does not allow inclusion of the (state-dependent) phase factors arising from nuclear propagation. The generalized curve-crossing model partly overcomes these limitations by treating the nuclear dynamics within the one-dimensional model. The whole problem is formulated in the form of time-independent scattering problem. From here on, we do not have to restrict the initial state to be in state 2, nor the initial velocity to be directed in the leftward, hence we can set, without loss of generality, $F_1 \geq F_2$ and $F_1 \geq 0$.

The Schrödinger equation is given as

$$
\begin{aligned}
E\psi_1(x) &= -\frac{\hbar^2}{2m}\frac{d^2}{dx^2}\psi_1(x) + W_1\psi_1(x) + V_{12}\psi_2(x) \\
E\psi_2(x) &= -\frac{\hbar^2}{2m}\frac{d^2}{dx^2}\psi_2(x) + W_2\psi_2(x) + V_{12}\psi_1(x)
\end{aligned}
\tag{4.13}
$$

and the solution is sought for in momentum representation as

$$
\phi_i(x) = \int \frac{dk}{\sqrt{2\pi}} u_i(k) e^{ikx}. \tag{4.14}
$$

The corresponding reduced Schrödinger equation

$$[-k^2 + f_1 i\frac{d}{dk} + \epsilon]u_1(k) + \alpha u_2(k) = 0$$
$$[-k^2 + f_2 i\frac{d}{dk} + \epsilon]u_2(k) + \alpha u_1(k) = 0 \tag{4.15}$$

where matrix elements are scaled by $\hbar^2/2m$ as $\epsilon \equiv 2mE/\hbar^2$, $\alpha \equiv 2mV/\hbar^2$ and $f_i \equiv 2mF_i/\hbar^2$. It will later appear to be useful to introduce dimensionless parameters

$$a^2 \equiv \sqrt{f_1|f_2|}(f_1 - f_2)/(8\alpha^3), \tag{4.16}$$

and

$$b^2 \equiv \epsilon(f_1 - f_2)/(2\alpha\sqrt{f_1|f_2|}). \tag{4.17}$$

The solutions for $u_i(k)$ are investigated in the following form, which is compatible to the asymptotic behavior

$$u_i(k) = A_i(k)\exp\left(\frac{i}{f_i}\left[\epsilon k - \frac{k^3}{3}\right]\right). \tag{4.18}$$

The asymptotic behavior of solution in the limit of large x is then obtained by the steepest descent method as

$$\psi_i(x) = \int \frac{dk}{\sqrt{2\pi}} A_i(k)\exp\left(\frac{i}{f_i}\left[\epsilon k + f_i kx - \frac{k^3}{3}\right]\right), \tag{4.19}$$

whose stationary phase condition is $k^2 = \epsilon + f_i x$. It then follows that the behavior at the classically allowed asymptotic region (positive or negative large x, for positive or negative f_i, respectively) is given by a linear combination of positively or negatively large k. $A_i(k)$ at the corresponding values of k will be symbolically denoted as $A_i(\infty)$ and $A_i(-\infty)$. These four values are related by the reduced scattering matrix. For $f_2 > 0$, $A_1(-\infty)$ and $A_2(-\infty)$ are directed inward to the crossing point while $A_1(\infty)$ and $A_2(\infty)$ are directed outward. The reduced scattering matrix is then

$$\begin{pmatrix} A^{(1)}(\infty) \\ A^{(2)}(\infty) \end{pmatrix} = \begin{pmatrix} S'_{11} & S'_{12} \\ S'_{21} & S'_{22} \end{pmatrix} \begin{pmatrix} A^{(1)}(-\infty) \\ A^{(2)}(-\infty) \end{pmatrix}. \tag{4.20}$$

On the other hand, for $f_2 < 0$, $A_1(-\infty)$ and $A_2(\infty)$ are directed inward to the crossing point while $A_1(\infty)$ and $A_2(-\infty)$ are directed outward. The reduced scattering matrix is then

$$\begin{pmatrix} A^{(1)}(\infty) \\ A^{(2)}(-\infty) \end{pmatrix} = \begin{pmatrix} S'_{11} & S'_{12} \\ S'_{21} & S'_{22} \end{pmatrix} \begin{pmatrix} A^{(1)}(-\infty) \\ A^{(2)}(\infty) \end{pmatrix} \tag{4.21}$$

The time-dependent problem in Zener's paper is thus reformulated to a time-independent scattering problem. After this reformulation, naive assumption of constant v is no longer necessary and one can discuss low energy physics as well. Another thing to be noted is the large differences between positive and negative f_2; since the asymptotic behavior is different, one has to discuss it in different form. It is known, however, even in the case with negative f_2, the effect of a single passage through the avoided crossing is essentially the same to that in the LZ problem if the crossing is in classically allowed region and is well separated from the classical turning point. The validity condition of WKB analysis requires $\frac{d\lambda}{2\pi dx} \ll 1$ where $\frac{\lambda}{2\pi}$ is the de Brogile wavelength [230], whose sufficient condition[2] is $x \gg \Delta x_1$ and $x \gg \Delta x_2$, where two characteristic length scales are given as $\Delta x_1 \equiv \frac{2V}{F_1-F_2}$ and $\Delta x_2 \equiv \sqrt{\frac{\hbar v}{2(F_1-F_2)}}$, with v being the wavepacket velocity $v \equiv \sqrt{2E/m}$. In other words, validity condition of semiclassical analysis is broken over the range characterized by Δx_1 and Δx_2 due to the curve crossing. As we have mentioned before, in order for independent treatment to be valid, there should be a finite range of x where the WKB analysis is valid. We therefore have, approximating the turning point coordinate as $|x_{TP}| \approx E/F$, $E/F \gg \frac{2V}{F_1-F_2}$ and $E/F \gg \sqrt{\frac{\hbar v}{2(F_1-F_2)}}$. Under such condition, one can apply the Stueckelberg's treatment, Eqs. (4.11), (4.12). The (reduced) scattering matrix becomes, in case $F_2 > 0$,

$$S' = \begin{pmatrix} \lambda^2 + (1-\lambda^2)e^{-2i\phi} & 2i\lambda\sqrt{1-\lambda^2}\sin\phi \\ 2i\lambda\sqrt{1-\lambda^2}\sin\phi & \lambda^2 + (1-\lambda^2)e^{2i\phi} \end{pmatrix} \tag{4.22}$$

while in case $F_2 < 0$,

$$S' = \frac{1}{1+(1-\lambda^2)e^{2i\tilde{\phi}}} \begin{pmatrix} \lambda^2 & -2i\sqrt{1-\lambda^2}\cos\tilde{\phi} \\ -2i\sqrt{1-\lambda^2}\cos\tilde{\phi} & \lambda^2 \end{pmatrix} \tag{4.23}$$

where $\lambda = e^{-\pi\delta}$ with $\delta = \frac{V^2}{\hbar v(F_1-F_2)}$ is the LZ exponent while $\phi = -2\theta + \arg\Gamma(i\delta) + \pi/4$ and $\tilde{\phi} = \frac{2}{3}b^3/a + \arg\Gamma(i\delta) + \delta\ln 4\delta - 2\delta\ln(2b/a) + \pi/4$ are the phase factors arising from the avoided-crossing and subsequent propagation around the turning point [84]. One can verify that, in case $F_2 > 0$, successive application of Eqs. (4.11), (4.12) indeed gives Eq. (4.22). Whereas in case $F_2 < 0$, which is less obvious, we can consider the following decomposition to two passages; (1) incident waves from left (left movers of state 1 and the state 2, the latter being not the asymptotic channel whose

[2]These conditions follows from the effective single-component Schrödinger equation. See Ref. [84] for more details.

amplitude is temporarily given as X_2) is first scattered to right movers at right (state 1 being an intermediate state X_1) as

$$\begin{pmatrix} X_1 \\ A_2(-\infty) \end{pmatrix} = \begin{pmatrix} \lambda & -i\sqrt{1-\lambda^2}e^{i\tilde{\phi}} \\ -i\sqrt{1-\lambda^2}e^{-i\tilde{\phi}} & \lambda \end{pmatrix} \begin{pmatrix} A_1(-\infty) \\ X_2 \end{pmatrix}. \quad (4.24)$$

(2) the incident waves from right, X_1 and $A_2(+\infty)$ are scattered to left movers at left as

$$\begin{pmatrix} A_1(\infty) \\ X_2 \end{pmatrix} = \begin{pmatrix} \lambda & i\sqrt{1-\lambda^2}e^{-i\tilde{\phi}} \\ i\sqrt{1-\lambda^2}e^{-i\tilde{\phi}} & \lambda \end{pmatrix} \begin{pmatrix} X_1 \\ A_2(\infty) \end{pmatrix}. \quad (4.25)$$

Elimination of intermediate state amplitudes, X_1 and X_2 from Eqs. (4.24) and (4.25) gives the scattering matrix in Eq. (4.23). Although the role of X_1 and X_2 are implicit, one can consider them as the amplitude at the turning points of the state 1 and 2, respectively.

4.1.3 *Zhu–Nakamura theory*

The WKB treatment, as we have seen in the previous subsection, starts with sets of asymptotic wavefunctions in the form $c_1u_1(x) + c_2u_2(x)$, valid for each separate domain. These solutions are then "connected" by contour integrals in the complex plane around some criticalities that separate domains. Such connection scheme, however, does not work if either of u_1 or u_2 becomes exponentially small during the process[3]. The difficulty, which is nothing but Stokes phenomena, arises from the fact that the behaviors of (usually multi-valued) functions $u_1(x)$ and $u_2(x)$ around the critical point does not coincide with that of true solution [274]. To approximate the true solution one has to redefine the expansion coefficients at different domains. The prescription is to add to a discontinuous change on the coefficient of subdominant component at the Stokes line [274, 514], where the relative scale of the subdominant component takes minimum. The added change should be proportional to the coefficient of the dominant component and the factor is called the Stokes constant.

Stokes phenomenon plays a key role in the theory of Zhu and Nakamura [510–512, 514], which is the only complete solution of the linear curve

[3]A simplest example is found in the WKB solution of one-dimensional barrier reflection problem where the contour integral around the classical turning point passing through the upper half plane connects the damping function inside the barrier to one (but *not both*) of the oscillating components outside the barrier. The other component cannot be obtained because it becomes exponentially small during the round trip. In this example, however, the latter can be easily obtained by changing the direction of the contour to pass through the lower half plane. See Ref. [230] §47. pp167

crossing problem. Although the details of their solution is beyond the scope of this textbook, we here describe the essence of their discussion. The original two-component problem is first reduced to the second order differential equation for a single component of the form $\frac{d^2B_1(t)}{dt^2} + q(t)B_1(t) = 0$ where t is a reduced variable proportional to wavenumber k, and $q(t)$ is a quartic function of variable t. The first analysis of Zhu and Nakamura was the parameter-dependent distribution of critical points of the function $q(t)$. They next analyze the Stokes phenomenon arising for the asymptotic solution $q^{-\frac{1}{4}} \exp\left[\pm i \int^z q^{\frac{1}{2}}(z')dz'\right]$ to obtain the expression of connection matrix in terms of Stokes constant(s). The reduced scattering matrix S' is obtained from the behavior of the asymptotic solution. For $F_2 > 0$, it gives

$$S' = \begin{pmatrix} 1 + U_1U_2 & -2a^2U_2 \\ -2a^2U_2 & 1 - U_1^*U_2 \end{pmatrix} \tag{4.26}$$

where $U_2 = \frac{2i\mathrm{Im}U_1}{4a^2+|U_1|^2}$[4]. Comparison to the reduced scattering matrix we found in the previous subsection, Eq. (4.22) identifies $\lambda = \frac{2a^2}{\sqrt{4a^4+|U_1|^2}}$ and $\phi = \arg U_1$, In case of opposite sign of slopes,

$$S' = \frac{1}{1 + U_1U_2} \begin{pmatrix} 1 & \frac{1}{2a^2}U_2 \\ \frac{1}{2a^2}U_2 & 1 \end{pmatrix} \tag{4.27}$$

where $U_2 = \frac{2i\mathrm{Im}U_1}{|U_1|^2-1/(4a^4)}$. Comparison to Eq. (4.23) gives $\lambda = \sqrt{1 - 4a^4|U_1|^2}$ and $\phi = \arg U_1 - \pi/2$, where in this case, LZ-like single passage interpretation applies only for the parameter range $b^2 \geq 1$ or $E \geq 2V\sqrt{F_1|F_2|}/(F_1 + |F_2|)$, which essentially means that the upper *adiabatic* potential is classically accessible. In both cases, the nonadiabatic transition (or tunneling, in case $F_2 < 0$) probability is given by $|S'_{12}|^2$, which appears to be much oscillating with parameters a^2 and b^2. Zhu and Nakamura give an analytic expression of Stokes constant, then it formally solves the problem in a general way.

The second [510] and the third [511] of their series of papers, Zhu and Nakamura discuss approximate expression of U_1 in a parameter-dependent manner, while in the fourth paper [512], they also propose more handy (empirical) formulae for LZ-like probability $p_{\mathrm{LZ}} \equiv \lambda^2$. To quote one of the latter type of formulae, suggested approximation to p_{LZ} in case $F_2 < 0$ is given as

$$p_{\mathrm{LZ}} = \exp\left[-\frac{\pi}{4ab}\left(\frac{2}{1 + \sqrt{1 - b^{-4}(0.72 - 0.62a^{1.43})}}\right)^{1/2}\right], \tag{4.28}$$

[4]The definition of U_1 and U_2 here follows those of Ref. [514], which is by a factor different from those in Ref. [512]

which corrects the original LZ formula $p_{\mathrm{LZ}}^{(0)} = \exp\left(-\frac{\pi}{4ab}\right)$. The validity of this formula is again limited by the condition $b^2 \geq 1$.

The Zhu–Nakamura theory also covers other cases such as low-energy collision dynamics, where the incident energy is much lower than the barrier of the lower adiabatic potential (see Ref. [512] for the explicit form appropriate for this parameter range). That is, it can handle correctly and conveniently a system in which nonadiabatic dynamics couples with tunnelling phenomenon.

One great advantage of the Zhu–Nakamura theory is that its transition amplitude can be represented in terms of the parameters related to the adiabatic potentials and incident energy alone. Although the derivation started from the diabatic representation of the system, transformation to the diabatic parameters from the adiabatic counterparts is in fact not needed, once the effective parameters as a and b in Eq. (4.28) are obtained. Thus it is applicable solely within the adiabatic representation [291].

4.2 Mixed quantum–classical formulation of electron-nucleus coupled nonadiabatic dynamics

4.2.1 *Pechukas path integrals*

Since nonadiabatic transitions are of highly quantum-mechanical nature, it is natural to treat nuclear dynamics in a quantum mechanical or semiclassical manner. However, due to large computational cost of (multidimensional) full-quantum or semiclassical calculation, it is a common practice to apply classical (or point-like) approximation on the nuclear degrees of freedom. The treatment is called mixed quantum–classical (MQC) method, and is an indispensable tool in multidimensional molecular dynamics. Justification for the classical point-like approximation comes from the shortness of the wavelength of nuclear matter-wave compared to a typical electronic length scale. In this treatment, the number of variables needed for nuclear dynamical calculation is greatly reduced by the use of point-like representation. Another numerical advantage of this treatment is drastic reduction of the required number of electronic state calculations by "on-the-fly" treatment; one has to carry out only a limited number of electronic state calculations along a nuclear path. These are in clear contrast to the full-quantum calculation, where one represent the nuclear wavefunction as a large size vector on some basis set (such as spatial grid) and has to calculate the relevant PESs and matrix elements over such basis set in advance of actual dynamical calculations.

Despite of its usefulness, however, incorporating quantum and classical mechanics in a consistent manner is a difficult problem. In the simplest case where nonadiabatic interactions are negligible, the electronic state evolves adiabatically (i.e. keeps on a single PES) and the nuclear dynamics reduces to a Newtonian dynamics driven by the gradient of PES. This is what is called ab initio molecular dynamics (AIMD) approach (see Refs. [450, 451] for examples of chemical use of AIMD).

However, in more general situations where there exists finite nonadiabatic state mixing, contradiction between quantum and classical logic arises; the quantum time-evolution bifurcates the electronic state to a coherent superposition of multiple states whereas the classical time-evolution does not allow a bifurcation. Even if one derives some effective bifurcating nuclear paths beyond the classical dynamics, there remains a difficult problem of coherence. At one extreme, one keeps a single nuclear path then the electronic states remain perfectly coherent, while at another extreme, the nuclear path splits into multiple branches but the coherence between them are lost. In the reality, coherence should somehow decay as the nuclear coordinate separation becomes larger, but there is no general methodology to realize this except for the exact full-quantum simulation.

To clarify the central problem, we first outline a formal theory of mixed quantum classical nonadiabatic dynamics. One of the clearest formulation of mixed quantum–classical approximation is given by Pechukas [318] using the path-integral approach. His work is distinguished by the following two virtues; (1) the derivation formally starts from the quantum-mechanically exact path integral and (2) reduction to an effective classical mechanics is performed only through the stationary phase approximation with no additional assumptions. In his formalism, the "classical" equation of motion of nuclei is derived from the stationary phase approximation of quantum mechanics. Only in the adiabatic limit it reduces to the Newtonian dynamics whereas in general case, it is very different from purely classical dynamics as will be shown below.

On the other hand, there have been established no direct implementation of his formulation that is shown to be practical as well as of wide applicability (see also discussions below). Nevertheless it remains as an important theoretical reference or a start point for discussing further practical methods.

4.2.1.1 *Pechukas formulation*

We first divide the dynamical variables into two groups; those of "heavy particle", which are treated in the coordinate representation $\mathbf{R}$, and those of "internal degrees of freedom," which are represented by state labels $\alpha, \beta, \ldots$. The former are to be treated in the classical approximation. In his original work, Pechukas considered atomic collision problems. Here we consider a general electron-nucleus coupled dynamics and set the former as the nuclear degrees of freedom and the latter as the electronic degrees of freedom.

We then consider the transition amplitude of the system from a state $\{\mathbf{R}', \alpha\}$ at time t' to another $\{\mathbf{R}'', \beta\}$ at time t''. The kernel connecting the two end points is given by

$$\mathcal{K}_{\beta,\alpha}(\mathbf{R}'', t''; \mathbf{R}', t') = \int_{\mathbf{R}'}^{\mathbf{R}''} \mathcal{D}\mathbf{R}(t) \exp(\frac{i}{\hbar} S[\mathbf{R}(t)]) \mathcal{K}^{\text{el}}_{\beta\alpha}[\mathbf{R}(t)], \tag{4.29}$$

where $S[\mathbf{R}(t)]$ is the action for the heavy particle and $\mathcal{K}^{\text{el}}_{\beta\alpha}[\mathbf{R}(t)]$ is the transition amplitude of the electronic state for the nuclear position $\mathbf{R}(t)$, defined as

$$\mathcal{K}^{\text{el}}_{\beta\alpha}[\mathbf{R}(t)] \equiv \lim_{N\to\infty} \langle \beta | e^{-\frac{i}{\hbar}\epsilon H^{\text{el}}(\mathbf{R}(t_{N-1}))} e^{-\frac{i}{\hbar}\epsilon H^{\text{el}}(\mathbf{R}(t_{N-2}))} \cdots e^{-\frac{i}{\hbar}\epsilon H^{\text{el}}(\mathbf{R}(t_0))} | \alpha \rangle, \tag{4.30}$$

where $\epsilon \equiv \frac{t''-t'}{N}$ is the time interval and $t_n \equiv t' + n\epsilon$ is the nth time point with the time partition number being N. The integration of the electronic degrees of freedom is performed for each given $\mathbf{R}(t)$, which is then followed by the integration over the path $\mathbf{R}(t)$. Here we assume that the path-integration of the electronic part is performed in a quantum mechanical manner. Although Pechukas did not show how to perform the integrations, recent reports, Refs. [164, 223], have figured out the relevant method using the coherent-state path integrals. To quote their conclusion, the electronic path integrals (assumed to be in fixed number bound state) are representative using an arbitrary basis set expansion. In practice we just have to integrate the Schrödinger equation (including possible derivative couplings) using the Hamiltonian at the time-dependent position $\mathbf{R}(t)$. We then apply the stationary phase approximation to the integral over $\mathbf{R}$. The stationary phase condition gives rise to

$$M_k \ddot{\mathbf{R}}_k = -\text{Re} \frac{\langle \beta | \hat{U}(t'', t) \frac{\partial H^{\text{el}}(\mathbf{R})}{\partial \mathbf{R}_k} \hat{U}(t, t') | \alpha \rangle}{\langle \beta | \hat{U}(t'', t') | \alpha \rangle}, \tag{4.31}$$

where

$$\hat{U}(t'', t') \equiv \lim_{N\to\infty} e^{-\frac{i}{\hbar}\epsilon H^{\text{el}}(\mathbf{R}(t_{N-1}))} e^{-\frac{i}{\hbar}\epsilon H^{\text{el}}(\mathbf{R}(t_{N-2}))} \cdots e^{-\frac{i}{\hbar}\epsilon H^{\text{el}}(\mathbf{R}(t_0))} \tag{4.32}$$

is the time evolution operator for the electronic degrees of freedom (dependent on $\mathbf{R}(t)$). The force in Eq. (4.31), which we refer to as Pechukas force, is dependent on the initial and final states α, β as well as the path $\mathbf{R}(t)$, which is to be obtained as a solution of the dynamics Eq. (4.31). Therefore the nuclear equation of motion in Eq. (4.31) does not have explicit solution but has to be solved in some self-consistent method using the both Eq. (4.31) as well as Eq. (4.32). As we mentioned before, a numerical algorithm to achieve such self-consistency, which ranges over quantum and classical dynamics, has not been established yet. Rare exception is found in Refs. [273, 392], where they map the original problem to an effective classical double-boundary problem to successfully achieve the self-consistency. Furthermore, there is no general proof that such solution exist for an arbitrary set of end-points. These difficulties reflect the quantum mechanical entanglement between the nuclear and electronic motions. Despite of practical difficulty, however, the Pechukas dynamics, composed of Eqs. (4.31) (4.32), satisfies special conservation laws that are relevant in the MQC representation of the nonadiabatic dynamics as we will see below.

4.2.1.2 *Properties of Pechukas formulation*

We first consider an energy-like quantity defined as

$$\mathcal{E}(t) \equiv \frac{1}{2}\sum_k M_k \dot{\mathbf{R}}_k^2 + \mathrm{Re}\frac{\langle\beta|\hat{U}(t'',t)H^{\mathrm{el}}\hat{U}(t,t')|\alpha\rangle}{\langle\beta|\hat{U}(t'',t')|\alpha\rangle}. \tag{4.33}$$

Assuming no explicit time-dependence in H^{el}, the quantity $\mathcal{E}(t)$ can be easily proven to be conserved under the Pechukas dynamics. Although $\mathcal{E}(t)$ is not an Energy in the strict sense and there is no reason to believe that the quantity should be conserved during the time-evolution, it becomes a true energy (in the sense of mixed QC) in the asymptotic region where we assume that α and β are the energy eigenstates of eigenvalue $E_\alpha(\mathbf{R}(t'))$ and $E_\beta(\mathbf{R}(t''))$, respectively[5];

$$\mathcal{E}(t'') = \frac{1}{2}\sum_k M_k \dot{\mathbf{R}}_k^2(t'') + E_\beta(\mathbf{R}(t'')) \tag{4.34}$$

$$\mathcal{E}(t') = \frac{1}{2}\sum_k M_k \dot{\mathbf{R}}_k^2(t') + E_\alpha(\mathbf{R}(t')), \tag{4.35}$$

[5]Here, for simplicity, we assume the situation where the nonadiabatic transitions asymptotes to zero in the limit $t \to \pm\infty$.

where t'' and the minus of t' are assumed to be positive large number. The conservation of $\mathcal{E}(t)$ then lead to a physically meaningful relation

$$\frac{1}{2}\sum_k M_k \dot{\mathbf{R}}_k^2(t'') + E_\beta(\mathbf{R}(t'')) = \frac{1}{2}\sum_k M_k \dot{\mathbf{R}}_k^2(t') + E_\alpha(\mathbf{R}(t')). \quad (4.36)$$

Considering the same type of relations for all possible final states from the same initial state, we see that the energy is conserved for each asymptotic channel. This conservation law is physically correct and also crucial in some experimental techniques including time of flight measurements [7].

In a central symmetric system, conservation of the angular momenta can also be discussed in the same manner using the quantity defined as

$$J(t) \equiv \sum_k M_k \mathbf{R}_k \times \dot{\mathbf{R}}_k + \mathrm{Re}\frac{\langle\beta|\hat{U}(t'',t)\mathcal{J}^{\mathrm{el}}\hat{U}(t,t')|\alpha\rangle}{\langle\beta|\hat{U}(t'',t')|\alpha\rangle} \quad (4.37)$$

where $\mathcal{J}^{\mathrm{el}} \equiv \mathcal{J}^{\mathrm{tot}} - \sum_k M_k \mathbf{R}_k \times \dot{\mathbf{R}}_k$ is the angular momentum operator for the internal degrees of freedom.

These simple conservation laws are nevertheless not assured, in fact they are sometimes even severely violated, in many of existing calculation methods including mean-field methods like the semiclassical Ehrenfest theory (SET). The surface hopping method circumvents the problem by adjusting the linear momenta of nuclei so as to satisfy the asymptotic energy conservation law.

4.2.2 *Mean-field path representation: Semiclassical Ehrenfest theory*

We next turn to much simpler and easily calculable methods. The semiclassical Ehrenfest theory (SET) is based on rather an intuitive combination of the electronic dynamics on time-dependent nuclear configuration and the Newtonian dynamics of point-like nuclei using the wavepacket average of the Hellmann–Feynman force.

4.2.2.1 *An intuitive formulation of SET*

One of the most powerful methods to study the nonadiabatic transition starts from a time-dependent electronic Schrödinger equation

$$i\hbar\frac{\partial}{\partial t}\Phi_I(\mathbf{r};\mathbf{R}(t)) = H^{(\mathrm{el})}(\mathbf{r};\mathbf{R}(t))\Phi_I(\mathbf{r};\mathbf{R}(t)), \quad (4.38)$$

which "seemingly" appears to be a natural generalization of Eq. (2.11) to a case of time-dependent nuclear positions. This method is generally called the semiclassical Ehrenfest theory. The associated total wavefunction is assumed to have the form (unnormalized)

$$\Psi(\mathbf{r},\mathbf{R},t)=\sum_{path}\Phi(\mathbf{r};\mathbf{R}_{path}(t))\delta\left(\mathbf{R}-\mathbf{R}_{path}(t)\right), \tag{4.39}$$

where the electronic wavepacket propagated along a path $\mathbf{R}_{path}(t)$ is expanded as

$$\Phi(\mathbf{r};\mathbf{R}_{path}(t))=\sum_{I}C_I(t)\Phi_I(\mathbf{r};\mathbf{R}_{path}), \tag{4.40}$$

with the coefficients being supposed to satisfy the electronic time-dependent Schrödinger equation

$$i\hbar\frac{\partial}{\partial t}C_I=\sum_J\left[H_{IJ}^{(\mathrm{el})}(\mathbf{R}(t))-i\hbar\sum_k\dot{R}_kX_{IJ}^k(\mathbf{R}(t))\right]C_J, \tag{4.41}$$

where

$$X_{JK}^k\equiv\left\langle\Phi_J(\mathbf{R}(t))\left|\frac{\partial}{\partial R_k}\right|\Phi_K(\mathbf{R}(t))\right\rangle, \tag{4.42}$$

and we have rewritten $\mathbf{R}_{path}(t)$ as $\mathbf{R}(t)$ for simplicity. The nuclear "classical" path carrying the electronic wavepacket is assumed to be driven by the mean-force (as a mathematical analog of the Hellmann–Feynman force), that is,

$$\begin{aligned}\ddot{\mathbf{R}}(t)&=-\left\langle\Phi(\mathbf{R}(t))\left|\frac{\partial H^{(\mathrm{el})}}{\partial\mathbf{R}}\right|\Phi(\mathbf{R}(t))\right\rangle\\&=-\sum_I C_I(t)^*C_J(t)\left\langle\Phi_I(\mathbf{R})\left|\frac{\partial H^{(\mathrm{el})}}{\partial\mathbf{R}}\right|\Phi_J(\mathbf{R})\right\rangle,\end{aligned} \tag{4.43}$$

where the nuclear masses are scaled to unity, if the basis set used is complete. If the electronic basis is not complete,

$$\ddot{\mathbf{R}}^k=-\sum_{I,J,K}C_I^*\left(X_{IK}^kH_{KJ}^{(\mathrm{el})}-H_{IK}^{(\mathrm{el})}X_{KJ}^k\right)C_J-\sum_{IJ}C_I^*\frac{\partial H_{IJ}^{(\mathrm{el})}}{\partial R_k}C_J \tag{4.44}$$

is the correct formula to be used more accurately [10].

SET is known to give very accurate nonadiabatic transition amplitude up to the quantum phases as long as the path, which is driven by the averaged electronic state as in Eq. (4.43), resides in the strong nonadiabatic region. However, this averaged path remains running on a thus averaged potential energy surface even after it gets out of the nonadiabatic interaction region. As seen in the preceding section, the nonadiabatic interactions let the full quantum nuclear wavepackets branch into those pieces that eventually run on the individual adiabatic potential surfaces. But an SET mean-field path can never branch in itself but runs on an averaged potential surface.

In conclusion, SET is quite useful and promising as a practical means to describe nonadiabatic transitions from the view point of electron dynamics. However the theoretical foundation is weak and therefore it is not easy to improve this theory so as to represent the phenomena of nuclear wavepacket bifurcation.

4.2.2.2 *Further properties of SET*

SET is sometimes categorized as "coherent limit" in the sense we mentioned in the previous subsection; the nuclear motion keeps on a single trajectory and the electronic state evolves coherently as superposition of various electronic state, which makes no sense in the asymptotic region after wavepacket bifurcation. (We will discuss what the "coherence" is all about later in this book.) Although being derived in rather intuitive manner, SET can also be derived from Pechukas formulation with an (often not correct) additional assumption.

In contrast to the Pechukas formulation, Eq. (4.41) and the nuclear dynamics driven by Eq. (4.43) can be solved in an explicit manner since it has no future time dependence (although both have to be solved stepwise in a self-consistent manner). Despite of computational easiness, however, Eq. (4.43) leads to unphysical dynamics when the electronic wavepacket bifurcates into multiple states each of which asymptotes to a distinct channel; the nuclear motion is driven by an unphysical superposition of these distinct channels. In order to see what is wrong, we recall the path-integral formulation. First observe that the mean-field force Eq. (4.43) can be derived from the Pechukas force in Eq. (4.31) by a special assumption that the final state $|\beta\rangle$ is given by a unitary transformation

$$|\tilde{\beta}\rangle = \hat{U}(t'', t)|\alpha\rangle \times \text{ (phase factor)} ,$$

which is, in general, not a single asymptotic state but some superposition of them. Substitution of this $|\tilde{\beta}\rangle$ as the final state for the Pechukas force in

Eq. (4.31) gives an effective force

$$-\langle\alpha|\hat{U}(t',t)\frac{\partial H^{\mathrm{el}}(\mathbf{R})}{\partial\mathbf{R}}\hat{U}(t,t')|\alpha\rangle, \tag{4.45}$$

which is nothing but the mean-field force Eq. (4.43) (note that we dropped *Re* as the expression is manifestly real). The unphysical nature of this treatment is obvious when the wavepacket is in the asymptotic region where the assumed final state $|\tilde{\beta}\rangle$ is some superposition of different asymptotic states.

We further observe that the Energy-like conserved quantity in SET becomes

$$\frac{d}{dt}\left(\frac{1}{2}\sum_k M_k\dot{\mathbf{R}}_k^2 + \frac{\langle\alpha|\hat{U}(t',t)H^{\mathrm{el}}\hat{U}(t,t')|\alpha\rangle}{\langle\alpha|\hat{U}(t',t')|\alpha\rangle}\right) = 0, \tag{4.46}$$

which is different from Eq. (4.33). This does not result in the correct asymptotic energy conservation but the conservation of state-averaged energy.

These apparent defects, however, arises only when the wavepacket bifurcation occur. There is no obvious fault in using mean-field force in the region where quantum state superposition makes sense. Within such limitations, SET is known as a good approximation before possible occurrence of a wavepacket bifurcation. There have hence been proposed numbers of improved implementation of SET as we shall see in the next section.

4.2.3 *Quantum variables mapped to classical ones: Meyer–Miller method*

We next proceed to an interesting variant of the SET. Suppose a chemical reaction on a single PES, whose initial and final vibrational quantum numbers are denoted as n_i and n_f, respectively. In semiclassical theory, these quantized states are specified in terms of the action integral satisfying the quantization condition such as

$$\oint_{C_\alpha}\sum_j p_j dq_j = \left(n_\alpha + \frac{1}{2}\right)\hbar, \tag{4.47}$$

where $\{q_j, p_j\}$ are positions and the associated momenta, respectively, in the initial or final states. (We here have assumed the simplest case of the Maslov index.) The computation of the reaction S-matrix from n_i to n_f in the classical S-matrix theory [278, 391] is reduced to a problem to find a classical trajectory connecting those having the relevant action variables. Thus the quantum numbers are replaced with the classical action integral.

What if then we consider a chemical reaction $n_i \to n_f$, which passes across an avoided crossing on the way? It should be certainly nice, if the electronic quantum numbers are also represented in terms of variables having classical analog. Meyer and Miller showed that this is indeed possible as follows [273]. In the expansion of Eq. (4.40), they set

$$C_K = \sqrt{n_K} e^{-iq_K}, \tag{4.48}$$

where n_K and q_K are classical analog of the action and angle (or phase) variables, respectively. With this variable transformation, one can rewrite the averaged electronic Hamiltonian

$$\langle \Psi(t)|H^{el}(\mathbf{R}(t))|\Psi(t)\rangle = \sum_{J,K} C_J^* C_K H_{JK}^{el}, \tag{4.49}$$

as

$$\sum_{J,K} \sqrt{n_J(t) n_K(t)} e^{i(q_J(t)-q_K(t))} H_{JK}^{el} \equiv H^{el}(\mathbf{n}, \mathbf{q}; \mathbf{R}(t)), \tag{4.50}$$

which is combined with the nuclear kinetic term as

$$H(\mathbf{n}, \mathbf{q}; \mathbf{R}(t)) = \sum_k \frac{\mathbf{P}_k^2}{2M_k} + H^{el}(\mathbf{n}, \mathbf{q}; \mathbf{R}(t)), \tag{4.51}$$

to give the total effective Hamiltonian. The variational principle shows that the new variables now satisfy the Hamilton canonical equations of motion such that

$$\dot{n}_K = -\frac{\partial H(\mathbf{n}, \mathbf{q}; \mathbf{R}(t))}{\partial q_K} \quad \text{and} \quad \dot{q}_K = \frac{\partial H(\mathbf{n}, \mathbf{q}; \mathbf{R}(t))}{\partial n_K}, \tag{4.52}$$

which makes it possible to analyze the electronic dynamics with the semiclassical propagator.

In practical applications of the above "classical" dynamics of electrons to the classical S-matrix theory, the choice of the initial phase degrees of freedom, $\{q_K\}$ is crucial for a path thus initiated to be able to reach the desired final condition on $\{n_K\}$. In order to realize the special condition $n_K(t_1) = \delta_{K,\alpha}$ and $n_K(t_2) = \delta_{K,\beta}$, the electronic Hamiltonian of Eq. (4.50) has to be further converted to Langer-modified form,

$$H^{el}(\mathbf{n}, \mathbf{q}; \mathbf{R}(t)) = \sum_{J,K} \sqrt{\left(n_J(t) + \frac{1}{2}\right)\left(n_K(t) + \frac{1}{2}\right)} e^{i(q_J(t)-q_K(t))} H_{JK}^{el} - \frac{1}{2}\sum_K H_{KK}^{el}. \tag{4.53}$$

If one can find such a path, the transition probability is then obtained by

$$P_{\beta\alpha} = \left[(2\pi)^{N_s} \left| \frac{\partial \mathbf{n}(t_F)}{\partial \mathbf{q}(t_I)} \right|^{-1} \right], \tag{4.54}$$

where N_s is the number of states to be taken into account. On the basis of the above reformulation, Sun and Miller demonstrated the success of this nonadiabatic theory in typical three models offered by Tully by using their algorithm combined with the initial value representation in path integration [400, 401]. Following this work, Sun, Wang and Miller also showed that a linear approximation of the time correlation function qualitatively well captures the nature of nonadiabatic transition dynamics.

A similar type of semiclassical approach has been developed by Stock and Thoss [392, 393], where again electronic states and operators are "mapped" into c-number variables and treated classically: Quantum mechanical states $\{|\Phi_I\rangle\}$ and the associated operators are mapped into the Fock states $\{|n_1, ..., n_N\rangle\}$ and to the associated creation/annihilation operators $\{a_I^\dagger,\, a_I\}$ in a manner,

$$\begin{aligned} |\Phi_I\rangle &\to |0_1, ..., 1_I, ..., 0_N\rangle \\ |\Phi_I\rangle\langle\Phi_J| &\to a_I^\dagger a_J. \end{aligned} \tag{4.55}$$

Then the creation/annihilation operators are transformed into pair of operators $q_I \equiv (a_I + a_I^\dagger)/\sqrt{2}$ and $p_I \equiv i(a_I^\dagger - a_I)/\sqrt{2}$, which satisfy $[q_I, p_J] = i\hbar\delta_{nm}$. The total Hamiltonian (in diabatic basis) becomes,

$$H = \sum_k \frac{\mathbf{P}_k^2}{2M_k} + \frac{1}{2}\sum_{I,J} H_{IJ}^{el}(q_I q_J + p_I p_J) - \frac{1}{2}\sum_I H_{II}^{el}. \tag{4.56}$$

From the transformed Hamiltonian Eq. (4.56), classical approximation is derived by the standard procedure of replacing operators q_I, p_I to classical variables, and commutators to the Poisson brackets. Time evolution of these variables, together with the nuclear degrees of freedom $\mathbf{R}$ and $\mathbf{P}$ are to be calculated with the semiclassical initial value representation (IVR) propagator [400, 401].

By construction, the method resembles closely to the mean-field method. The electronic part of Hamiltonian Eq. (4.56) is actually equivalent to the expectation value of Eq. (4.49) apart from the last term in Eq. (4.56). However, an interesting characteristic particular to this mapping technique was claimed by Stock and Thoss [393] that the mapping method reproduces the correct density of states in a double well model while the mean-field approach does not.

4.2.4 *Initial value representation of semiclassical estimate of nonadiabatic transition amplitudes*

The semiclassical Feynman kernel $K(q_f, t_f; q_i, t_i)$ to classically connect the two space-time points involved herewith is defined as

$$\begin{aligned} K(q_f, t_f; q_i, t_i) &= \left\langle q_f \left| \exp\left[\frac{i}{\hbar} H\left(t_f - t_i\right)\right] \right| q_i \right\rangle \\ &\simeq (2\pi i\hbar)^{N/2} \left|\frac{\partial q_f}{\partial p_i}\right|^{-1/2} \exp\left[\frac{i}{\hbar} S(q_f, t_f; q_i, t_i) - i\pi\nu/2\right], \end{aligned} \tag{4.57}$$

where the path is launched from an initial phase-space point (q_i, p_i) and ν is the Maslov index, which is the number of zeros that appear in the Jacobian determinant $\partial q_f/\partial p_i$ during the travel from q_i to q_f. By definition, Eq. (4.57) sets a boundary value problem for two end points (q_f, q_i). Applying this equation, one can propagate a wavefunction at any place q in such a way that

$$\psi(q, t_f) = \int \delta(q_f - q) K(q, t_f; q_i, t_i) \psi(q_i, t_i) dq_i dq_f. \tag{4.58}$$

This expression is not easy to use, since infinitely many boundary points (q_f, q_i) connected by classical trajectories have to be searched. In place of this direct evaluation, one can transform the integral variable q_f in such away that

$$\begin{aligned} \psi(q, t_f) &= \int \delta(q_f - q) K(q, t_f; q_i, t_i) \psi(q_i, t_i) dq_i \left|\frac{\partial q_f}{\partial p_i}\right| dp_i \\ &\simeq (2\pi i\hbar)^{N/2} \left|\frac{\partial q_f}{\partial p_i}\right|^{1/2} \exp\left[\frac{i}{\hbar} S(q_f, t_f; q_i, t_i) - i\pi\nu/2\right] dq_i dp_i. \end{aligned} \tag{4.59}$$

There are two obvious advantages in Eq. (4.59) over Eq. (4.58): The root-search problem to find pairs of (q_f, q_i) is avoided, since one can let a path start with a natural condition (q_i, p_i). The other one is the singularities inherent in Eq. (4.58) due to the zero of $\partial q_f/\partial p_i$ are all removed. This is called the Initial Value Representation (IVR).

The semiclassical transition amplitude from N dimensional electronic state Φ_{k_1} and F dimensional nuclear state χ_1 to a counterpart $\chi_2\Phi_{k_2}$ in the IVR is written as

$$
\begin{aligned}
S_{21}(t) &\equiv \langle \chi_2 \Phi_{k_2} | e^{-i\hat{H}t/\hbar} | \chi_1 \Phi_{k_1} \rangle \\
&= \int dx_1 \int dR_1 \int dx_2 \int dR_2 \\
&\quad \times \chi_2^*(R_2)\Phi_{k_2}^*(x_2)\langle x_2 R_2 | e^{-i\hat{H}t/\hbar} | x_1 R_1\rangle \chi_1(R_1)\Phi_{k_1}(x_1) \\
&= \int dx_1 \int dR_1 \int dx_2 \int dR_2 \sqrt{\frac{\partial(x_t, R_t)}{\partial(p_1, P_1)}/(2\pi i\hbar)^{F+N}} \\
&\quad \times \chi_2^*(R_t)\Phi_{k_2}^*(x_t) e^{iS_t(x_1,p_1,R_1,P_1)/\hbar - i\pi\nu/2}\chi_1(R_1)\Phi_{k_1}(x_1),
\end{aligned} \tag{4.60}
$$

In the last equation, the integral transformation to initial variable representation is carried out. $\hat{H}$ denotes the total electronic and nuclear Hamiltonian and again ν is the Maslov index at time t. Similarly, the transition amplitude from the initial total quantum state $\Phi_{k_1}\chi_1$ to the final nuclear position R at state k_2, $\Phi_{k_2}R$ is described as

$$
\begin{aligned}
\chi_{k_2 \leftarrow k_1}(t) &\equiv \langle R\Phi_{k_2} | e^{-i\hat{H}t/\hbar} | \Phi_{k_1}\chi_1\rangle \\
&= \int dx_1 \int dR_1 \int dp_1 \int dP_1 \sqrt{\frac{\partial(x_t, R_t)}{\partial(p_1, P_1)}/(2\pi i\hbar)^{F+N}} \\
&\quad \times \delta(R - R_t)\Phi_{k_2}^*(x_t) \cdot \Phi_{k_1}(x_1)\chi_1(R_1) \cdot e^{iS_t(x_1,p_1,R_1,P_1)/\hbar - i\pi\nu/2}.
\end{aligned} \tag{4.61}
$$

The outcome from these expressions are essentially the same as the results in Eq. (4.54) and nothing is new scientifically.

The so-called coherent state representation is useful also in semiclassical calculations. One choice of the coherent states is due to Herman and Kluk type, provide that the relevant trajectories do not undergo chaotic behavior. The transition amplitude is expressed by

$$
\langle \Psi_2 | e^{-i\hat{H}t} | \Psi_1 \rangle = \int \frac{dq_1 dp_1}{(2\pi\hbar)^F} C_{qpt} \Psi_1^g(q_1, p_1)\Psi_2^{g*}(q_t, p_t) e^{iS_t(q_1,p_1)/\hbar}, \tag{4.62}
$$

where $\{\Psi_n^g\}$ are the coherent state transformations of wavefunctions. Here, $\Psi^g(q,p)$ is expressed by

$$
\Psi^g(q,p) = \langle g|\Psi\rangle = \int dq' \Psi(q') \left(\frac{\gamma}{\pi}\right)^{F/4} e^{-(\gamma/2)(q-q')^2 + ip(q-q')/\hbar}. \tag{4.63}
$$

The prefactor has a form of

$$
C_{qpt} = \sqrt{\det\left[\frac{1}{2}\left(\frac{q_t}{q_1} + \frac{p_t}{p_1} - i\gamma\hbar\frac{q_t}{p_1} + \frac{i}{\gamma\hbar}\frac{p_t}{q_1}\right)\right]}. \tag{4.64}
$$

Here, q and p are composite classical position and momentum variables of total system.

The nuclear quantum amplitude at nuclear position R and the electronic state k starting from the initial electronic state 1 with associated nuclear wavefunction is described in terms of

$$\begin{aligned}
&\chi_{k\leftarrow 1}(R,t) \\
&= \int dR_1 \int dP_1 \int dq_1 \int dp_1 \int dq_2 \int dp_2 \; C_{qpt} \\
&\times \chi_1^g(R_1,P_1)\Phi_1^g(q_1,p_2,q_2,p_2)g^*(R;R_t,P_t)\Phi_k^{g*}(q_{1t},p_{1t},q_{2t},p_{2t}) \\
&\times e^{iS_t(R_1,P_1,q_1,p_1,q_2,p_2)/\hbar - i\pi\nu/2}, \qquad (4.65)
\end{aligned}$$

with

$$g(R;R_t,P_t) = \langle R|g\rangle = \left(\frac{\gamma}{\pi}\right)^{F/4} e^{-(\gamma/2)(R-R_t)^2 + iP_t(R-R_t)/\hbar}. \qquad (4.66)$$

4.3 Surface hopping scheme and beyond

4.3.1 *Surface hopping model*

Tully and Preston [444] proposed a drastically simplified computational scheme to treat nonadiabatic transition, in which a multidimensional classical trajectory is allowed to suddenly hop at a selected place(s) on an adiabatic potential energy surface (PES) with a transition probability that is to be numerically provided by other methods like the Zhu–Nakamura theory. The first version of surface hopping is based on an assumption that nonadiabatic transition takes place only in a limited region of the potential energy surfaces such as a crossing point (or crossing seam in multidimensional systems) on the corresponding diabatic surfaces. On hopping, the velocity in the perpendicular direction to the crossing seam is corrected so as to satisfy the energy and momentum conservation. As an idea on where and how to let a trajectory hop between two adiabatic PES, Shenvi proposed the so-called phase-space hopping [368] and Martínez et al. their method of optimized hopping [489]. The statistical hopping is achieved through Monte Carlo algorithm, and basically the phase information is destroyed in the hopping procedure. For the technical issues, consult Ref. [444].

In this approach the nonadiabatic transition probability is estimated in terms of the number of classical trajectories reaching the goal areas of the individual PES in asymptotic regions. Yet, another stochastic computation of the total transition probability is possible by using both daughter (one hops to another PES) and mother (the other remaining on the original PES) trajectories. Let them have the individual transition amplitudes estimated with, for instance, the Zhu–Nakamura formula. These pieces of transition-probability information carried by the individual trajectories are to be summed up asymptotically. Besides, the transition-amplitude information containing the quantum phase, along with the phases coming from the action integral with each path, may be summed up asymptotically. An appropriate numerical method may make it possible to partly take account of the quantum interference among the paths in this way (see the comment in Ref. [279]).

Thanks to the inherent simplicity and tractability even in multidimensional systems, the surface hopping method has been successfully applied to a wide range of phenomena, and thereby has been making a tremendous contribution to the progress of chemistry. On the other hand, such a simplified treatment sacrifices some important features of nonadiabatic dynamics which are vital in modern experimental studies. One such feature is the interference effect arising from multiple transitions. In real systems, nonadiabatic transitions are not limited on a single "crossing seam", but occurs everywhere as long as nonadiabatic coupling exists. The true transition amplitude therefore becomes a superposition of such multiple transitions. In the first version of the surface-hopping scheme we have discussed so far, multiple transitions and their interference effects are missing, though the phase arising from a *single* transition can be calculated using the Zhu–Nakamura theory.

To overcome such severe limitation of the naive surface-hopping scheme, many improved versions have been proposed, including the important one due to Tully himself [445], which will be discussed in the next subsection.

4.3.2 *Surface hopping driven by several types of state couplings*

4.3.2.1 *Surface hopping in the general adiabatic representations*

Surface hopping can be caused not only by the standard nonadiabatic interactions like that in Eq. (4.42) but others such as the spin-orbit interactions. Likewise one can apply the spirit of the surface hopping model to other coupled-state systems. For instance, one can incorporate the electron dynamics such as that in Eq. (4.41) to calculate the transition probability, with which a path can jump from one adiabatic state to another.

The greatest advantage of the surface hopping model is its simplicity, which can be applied to any adiabatic representations. For instance, we showed in Sec. 2.4 there have been proposed other definitions of adiabatic representation besides the standard one. Therefore, let us consider the surface hopping among the general adiabatic states. The transition probabilities are obtained in terms of a time derivative coupling among general adiabatic states [335]. Let $\Psi(\mathbf{r}, t; \mathbf{R}(t))$ be an electronic wavepacket state on a nuclear position $\mathbf{R}(t)$ at time t in a generalized adiabatic representation, which is to be expanded in a basis set of a generalized adiabatic function $\phi_m^{SA}(\mathbf{r}, t; \mathbf{R}(t))$ such as the super adiabatic function as

$$\Psi(\mathbf{r}, t; \mathbf{R}(t)) = \sum_m c_m^{SA}(t)\phi_m^{SA}(\mathbf{r}, t; \mathbf{R}(t)), \tag{4.67}$$

where a super adiabatic basis are defined by

$$\hat{h}^{el}(\mathbf{R}(t))\,\phi_m^{SA}(\mathbf{r}, t; \mathbf{R}(t)) = \epsilon_m^{SA}\phi_m^{SA}(\mathbf{r}, t; \mathbf{R}(t)), \tag{4.68}$$

with the effective electronic Hamiltonian as eigenfunctions of effective Hamiltonian $\hat{h}^{el}(\mathbf{R}(t))$. The coupling coefficients c_m^{SA} are to be determined by

$$i\hbar\frac{\partial c_m^{SA}(t)}{\partial t} = \sum_n \left\{\epsilon_m^{SA}[R(t), t]\delta_{mn} + K_{mn}^{SA}[R(t), t]\right\} c_n^{SA}(t), \tag{4.69}$$

where the state couplings are given as

$$K_{mn}^{SA}[R(t), t] = \langle\phi_m^{SA}(\mathbf{r}, t; \mathbf{R}(t))\,|\frac{d}{dt}|\phi_n^{SA}(\mathbf{r}, t; \mathbf{R}(t))\rangle. \tag{4.70}$$

$K_{mn}^{SA}[R(t), t]$ in turn is decomposed into two pieces;

$$K_{mn}^{SA}[R(t), t] = K_{mn}^{\phi}[R(t), t] + K_{mn}^{U}[R(t), t], \tag{4.71}$$

where

$$K_{mn}^{U}[R(t), t] = \sum_{\alpha\beta} U_{m\alpha}^{*}\frac{\partial U_{n\beta}}{\partial t} \tag{4.72}$$

and

$$K_{mn}^{\phi}[R(t), t] = \sum_{\alpha\beta} U_{m\alpha}^{*}U_{n\beta}K_{\alpha\beta}. \tag{4.73}$$

The unitary matrix $U_{n\beta}$ operated at each time is used to transform an original (arbitrary) basis set $|\phi_m(\mathbf{r}, t; \mathbf{R}(t))\rangle$ to the (super) adiabatic states $|\phi_m^{SA}\mathbf{r}, t; \mathbf{R}(t)\rangle$, and

$$K_{\alpha\beta}[R(t), t] = \langle\phi_\alpha(\mathbf{r}, t; \mathbf{R}(t))\,|\frac{d}{dt}|\phi_\beta(\mathbf{r}, t; \mathbf{R}(t))\rangle. \tag{4.74}$$

while the transformation matrix from original basis, to super adiabatic basis, are related with

$$U_{m\alpha} = \langle \phi_m (\mathbf{r}, t; \mathbf{R}(t)) \, | \phi_\alpha^{SA} (\mathbf{r}, t; \mathbf{R}(t)) \rangle. \tag{4.75}$$

Then, the hopping probability among super adiabatic states in a time region $[t, t + dt]$ is given as

$$P_{m\to n}^{SA} = \frac{2\mathrm{Re}\left\{ c_m^{SA*}(t) c_n^{SA}(t) \left[\frac{i}{\hbar} \epsilon_m^{SA} \delta_{mn} + K_{mn}^{SA} \right] \right\}}{|c_m^{SA}(t)|^2}. \tag{4.76}$$

Practically, this scheme can well treat the state coupling having relatively slow transition in time compared to nuclear motion, since in such a case the superadiabatic basis can be generated well at each time step.

4.3.2.2 *Laser field-induced surface-hopping method for the simulation and control of ultrafast photodynamics*

We next introduce a surface hopping scheme devised by R. Mitric, J. Oetersen and V. Bonačić-Koutecký for treating optical transition under an external strong laser field having non resonant photon energy. They call this scheme as stochastic field induced surface-hopping (FISH) method [280]. They prepare an initial nuclear wavepacket state in the Wigner phase-space distribution function, from which classical trajectories are launched. They have carried out surface hopping trajectory calculations without energy scaling at any optical transition. In these calculations the intrinsic nonadiabatic transitions due to the nuclear derivative coupling have been neglected since the field induced transition is dominant in this particular system. With good description of the coherent Rabi oscillations, they successfully applied this scheme to typical photochemical systems such as nonlinear absorption dynamics of zinc porphyrin in water system with a combination of ab initio on-the-fly calculations [331]. Another application of their method was made to a quantum control based on electronic coherence of K_2 molecule [321].

4.3.3 *Tully's fewest switch surface hopping method and its variants*

As explained above, the naive surface hopping model lets a classical trajectory running on an adiabatic potential surface jumps to another at a specific point with a transition probability borrowed from other theories. Although this sometimes provides a good description of population dynamics depending on the geometrical condition of nonadiabatic dynamics under study, the electronic coherence is totally neglected. Or, dynamical self-consistency between electrons and nuclei is totally violated. To overcome, Tully proposed an algorithm to take account of electronic coherence along a path, whose hopping points is in turn modified by the electronic transition [445].

The basic idea of the algorithm is to allow a path to jump with the probability estimated internally (in a self-contained manner) with the electronic-state mixing as in the semiclassical Ehrenfest theory but the number of transitions (hops) should be minimized. Suppose we have an electronic wavepacket $|\Phi(t)\rangle = \sum_I C_I(t)|\Phi_I\rangle$ as in the SET, the dynamics of the corresponding density matrix, $\rho_{IJ}(t) \equiv C_I^*(t)C_J(t)$, is written as

$$\dot{\rho}_{IJ} = \frac{1}{i\hbar}\sum_K \left(H_{IK} - i\hbar\dot{\mathbf{R}}\mathbf{X}_{IK}\right)\rho_{KJ} - \frac{1}{i\hbar}\sum_K \left(H_{JK}^* + i\hbar\dot{\mathbf{R}}\mathbf{X}_{JK}\right)\rho_{KI}^*. \tag{4.77}$$

A main interest exists in its diagonal part, which is compactly expressed as

$$\dot{\rho}_{II} = \sum_K B_{IK}, \tag{4.78}$$

where

$$B_{IK} \equiv \frac{2}{\hbar}\mathrm{Im}\left\{H_{IK}\,\rho_{KI}\right\} - \mathrm{Re}\{\dot{\mathbf{R}}\cdot\mathbf{d}_{IK}\,\rho_{KI}\}. \tag{4.79}$$

With this representation, one can estimate a transition probability from state K to I during one time step, $\tau \to \tau + d\tau$, as

$$G_{IK} \equiv \begin{cases} 0 & (B_{IK} \le 0) \\ \dfrac{B_{IK}(\tau + d\tau)\cdot d\tau}{\rho_{KK}(\tau + d\tau)} & (B_{IK} > 0) \end{cases}, \tag{4.80}$$

which in turn can be used to determine stochastically the places of trajectory hopping in the dynamics without any external parameter. One can make a compromise in determining the number of hops and quality of the transition probability to be obtained. This fewest switch surface hopping method has been shown to be quite successful by Tully and his coworkers.

Recently, Shenvi [368] proposed to use the scheme of the fewest switching surface hopping algorithm along with the so-called phase-space adiabatic basis (Sec. 2.4), which diagonalize the Hamiltonian

$$H_{JK}(\mathbf{R},\mathbf{P}) = \frac{1}{2M}\left(\delta_{JK}\mathbf{P} - i\hbar\mathbf{X}_{JK}(\mathbf{R})\right)^2 + H_{JJ}^{(\mathrm{el})}(\mathbf{R})\,\delta_{JK}, \tag{4.81}$$

where the nonadiabatic coupling is included explicitly, and the adiabatic representation is to be used for the electronic Hamiltonian matrix $H_{JJ}^{(\mathrm{el})}(\mathbf{R})$.

(See [28] for an extensive discussion and algebra arising from this gauge-field type Hamiltonian.) Then one can diagonalize this matrix defined at each nuclear phase-space point, attaining $E_n(\mathbf{R}, \mathbf{P})$ as its eigenvalues. Further, regarding $E_n(\mathbf{R}, \mathbf{P})$ as classical Hamiltonians, one can generate classical paths, which are in turn used in the fewest surface hopping method. Shenvi illustrated that this method, terms as phase-space surface hopping method, gives an excellent nonadiabatic transition probability even in a low energy case.

Another rather obvious aspect to be improved is classically forbidden transition. Truhlar et al. proposed a time-energy uncertainty approach for the remedies of jumping positions [452].

Yet, some theoretical problems are left to be discussed to seek for the ultimate and idealistic features as a nonadiabatic-transition theory: Although a trajectory thus hopping plural times converges to run on an adiabatic potential surface asymptotically, the off-diagonal density matrix element $\rho_{IJ}(t)$ does not vanish practically, as in the original SET. This is ascribed to an incomplete treatment of the nuclear-electronic entanglement. This issue, often referred to as the problem of decoherence, is originated from the nuclear wavepacket bifurcation due to different slopes of potential surfaces, which will be discussed more precisely below.

4.3.4 *Spawning method of Martínez*

Suppose a surface hopping model, in which a mother path gives birth to a daughter path on a single hopping, leaving two paths behind. The daughter path is also a classical one. It is the idea of Martínez et al. [155, 261–263, 459] that each nuclear path can be partly quantized by placing a frozen Gaussian function (Gaussian functions with a fixed exponent totally irrespective of the underlying dynamics [177]) individually. Actually their entire method is more sophisticated in that they use the electronic-state mixing to identify the hopping points. Since the these paths are classical (running on an average potential though), the action integrals can be readily evaluated. Besides, the Heller thawed Gaussian approximation or the WKB theory may be applied in this context. Hence all the necessary quantities are available now to build a quantum wavepacket of the Gaussian type along each path. Many extensive applications of this so-called spawning method have been made in the analyses of nonadiabatic dynamics [155, 261–263, 459, 489].

The greatest advantage of the frozen Gaussian is its simplicity and therefore can be applied to many generalizations of the naive surface hopping method and so on. For instance, it is not difficult to imagine that the spawning method can be generalized to the fewest switch surface hopping method [445]. It can be also applied to the method of the natural decay of mixing of Truhlar et al. [197, 452, 509, 515] and to other methods.

On the other hand, the obvious concern about this method is inaccuracy arising from the frozen Gaussian exponents (see below in Sec. 4.3.5 though). It is desirable to seek for a theory more satisfactory to quantize *non-classical* path in the quality level similar to or better than that of the WKB theory.

4.3.5 *Remixing of electronic states to incorporate the quantum nature of nuclear dynamics and interference among the paths*

What we have surveyed thus far in this section for the mixed quantum–classical dynamics is to see (i) how the electronic state mixing is performed along non-Born–Oppenheimer paths or the locally classical paths in the hopping scheme and then (ii) how those resultant nuclear paths are quantized. Consequently, the feedback effect of quantization of nuclear paths on the electronic state mixing has not been yet considered. Also, in most of the theories discussed above, the electronic-state mixing is performed independently along each path. However, paths co-residing in a strong nonadiabatic region can still interfere after hopping or branching, as in the final state interaction. (This procedure reminds us of the self-consistent procedure of Pechukas as discussed in Sec. 4.2.1.) These effects are also missing at least theoretically in those theories. It is again Martínez and his coworkers who have made a major step to this end [261].

We resume with the total Hamiltonian

$$\hat{H} = -\frac{\hbar^2}{2}\sum_{\alpha}^{nuc}\frac{1}{M_\alpha}\nabla_\alpha^2 + \hat{H}^{el}, \tag{4.82}$$

where α denotes a nuclear degree of freedom and M_α means its corresponding mass. A total electronic and nuclear wavefunction given by the on-the-fly scheme is generally written as

$$\Psi(\mathbf{r},\mathbf{R},t) = \sum_{I}^{el.state}\sum_{i}^{N_I(t)} C_I^i(t)\chi_I^i(\mathbf{R};\mathbf{R}_I^i(t),\mathbf{P}_I^i(t),\gamma_I^i(t))\Phi_I(\mathbf{r};\mathbf{R}_I^i(t),\mathbf{P}_I^i(t)). \tag{4.83}$$

In the multiple spawning method of Martínez at al., the time-dependent Gaussian nuclear wavefunction basis $\chi_I^i(t)$ are supposed to be parametrized with their guiding centers in phase space, $\{R_I^i(t), P_I^i(t)\}$, and width matrix, $\gamma_I^i(t)$. The classical phase is included in nuclear basis implicitly. I denotes an electronic state and i labels a guiding center (path) running on it. The electronic basis function, $\Phi_I(\mathbf{r};\mathbf{R}_I^i(t),\mathbf{P}_I^i(t))$, is carried along $(\mathbf{R}_I^i(t),\mathbf{P}_I^i(t))$, satisfying $\langle\Phi_I|\Phi_J\rangle = \delta_{IJ}$ at each given nuclear position. The idea itself is rather straightforward: Substituting the wavefunction of Eq. (4.83) into the total Schrödinger equation, $i\hbar|\dot{\Psi}\rangle = \hat{H}|\Psi\rangle$, we can determine the electronic mixing coefficients $C_I^i(t)$. The functions $\chi_I^i(\mathbf{R};\mathbf{R}_I^i(t),\mathbf{P}_I^i(t),\gamma_I^i(t))\Phi_I(\mathbf{r};\mathbf{R}_I^i(t),\mathbf{P}_I^i(t))$ are therefore regarded as basis functions to expand $\Psi(\mathbf{r},\mathbf{R},t)$. The determining coupled equations obtained after some appropriate approximations are not simple but can be derived in a seemingly compact form as

$$i\hbar\sum_i\left[S_{JJ}^{ji}\dot{c}_J^i + \dot{\bar{S}}_{JJ}^{ji}c_J^i\right] = \sum_{\alpha i} T_{\alpha JJ}^{ji}c_J^i + \sum_{Ii}\bar{H}_{JI}^{ji}S_{JI}^{ji}c_I^i - \hbar^2\sum_{\alpha Ii}\frac{1}{M_\alpha}\mathbf{F}_{\alpha JI}^{ji}\bar{\mathbf{f}}_{\alpha JI}^{ji}c_I^i, \qquad (4.84)$$

with a generalized nuclear kinetic operator super matrix being $T_{\alpha JI}^{ji} \equiv -\frac{\hbar^2}{2M_\alpha}\langle\chi_J^j|\nabla_\alpha^2|\chi_I^i\rangle$. $S_{JI}^{ji} \equiv \langle\chi_J^j|\chi_I^i\rangle$ and $\dot{\bar{S}}_{JI}^{ji} \equiv \langle\chi_J^j|\dot{\chi}_I^i\rangle$ appeared in this equation of motion are nuclear wavepacket transition coupling matrix elements induced from the zeroth and first-order time derivative, respectively. The nuclear derivative terms consist of $\mathbf{F}_{\alpha JI}^{ji} \equiv \langle\chi_J^j|\nabla_\alpha|\chi_I^i\rangle$, $\mathbf{f}_{\alpha JI}^{ji} \equiv \langle\Phi_J^j|\nabla_\alpha|\Phi_I^i\rangle$. The electronic potential matrix elements are described in a form of $H_{JI}^{ji} \equiv \langle\Phi_J^j|\hat{H}^{el}|\Phi_I^i\rangle$. Practically the two center electronic integrals, $f_{\alpha JI}^{ji}$ and H_{JI}^{ji}, often demand much computational effort and need practical approximations. An overbar in the equation denotes an approximation of one center electron integral. The details of which along with other technical matters should be referred to in Ref. [44].

4.4 Coherence and decoherence before and after nonadiabatic interaction

4.4.1 *Decay of mixing with coherence switching*

In their studies of nonadiabatic dynamics in condensed phase, Rossky and his colleagues emphasized the role of stochastic perturbations that actually decohere the surrounded quantum states (solute state) [327, 468]. This is one of the most significant phenomena in the dynamics of open systems.

On the other hand, in nonadiabatic dynamics within a single molecule or isolated molecular reaction system, it is Hack, Jasper and Truhlar [197, 452, 509, 515] who first devised how to implement the decay of electronic state coherence, or decay of mixing, into the nuclear dynamics of individual path, thus bringing about the notion of non-Born–Oppenheimer path in an explicit manner.

As in the work of Neria and Nitzan [297], Truhlar et al. also studied how the off-diagonal elements of the electronic density matrix can vanish after the nonadiabatic coupling is switched off. Then they have introduced an external force into the semiclassical Ehrenfest nuclear dynamics such that it can cancel the persisting coherence. The construction of this force is rather artificial but it is quite well designed so as to fulfill the conservation laws of energy and linear momentum, as summarized as follows. Let us consider the semiclassical Ehrenfest dynamics of the off-diagonal electronic density matrix $\dot{\rho}_{IJ}^{SE}$, to which an additional term $\dot{\rho}_{IJ}^{D}$ is added in the form of $\dot{\rho}_{IJ} = \dot{\rho}_{IJ}^{SE} + \dot{\rho}_{IJ}^{D}$. ρ_{IJ}^{D} is responsible for canceling the coherence remaining in ρ_{IJ}^{SE}, thus making the off-diagonal elements of ρ_{IJ} vanish. To realize such dynamics, the Hamilton canonical equations of motion (in the SET dynamics) are modified by adding externally velocity and force in such a way that $\dot{\mathbf{R}} = \dot{\mathbf{R}}^{SE} + \dot{\mathbf{R}}^{D}$ and $\dot{\mathbf{P}} = \dot{\mathbf{P}}^{SE} + \dot{\mathbf{P}}^{D}$, respectively, to decohere in the electronic state mixing. These additional terms $(\dot{\mathbf{R}}^{D}, \dot{\mathbf{P}}^{D})$ are designed so that they are switched off after a decoherence time and asymptotically and thereby the path should run on one of the desired adiabatic potentials. Various ways to determine the key quantities, the decoherence time and $(\dot{\mathbf{R}}^{D}, \dot{\mathbf{P}}^{D})$, have been proposed by the original authors [197, 452, 509, 515]. With this intelligent but artificial treatment, a nonadiabatic path is continuously evolved in time without surface hopping and with correct momentum and energy conserved. The electronic-state mixing is carried out consistently along every non-Born–Oppenheimer path. Thus the transition probability may be determined in terms of the ratio of the number of asymptotic paths accumulated on the individual potential surfaces. On the other, the electronic state mixing due to the semiclassical Ehrenfest dynamics on each path should give a nonadiabatic transition amplitude along with an associated phase. It is not clear though whether these pieces of information about transition probability are always consistent with each other.

It is interesting to compare the Truhlar non-Born–Oppenheimer path with those expected to emerge from the mechanics of Pechukas, Eq. (4.31). In the latter, a non-Born–Oppenheimer path, if it indeed exists, starts from one adiabatic potential energy surface to another continuously without hopping yet being driven by the nonadiabatic coupling. Recall though it can be determined only through the self-consistent manner, which requires essentially an iterative computation. This is the way of decoherence in the theory of Pechukas theory. On the other hand, Truhlar's theory produces smooth path connecting two different potential energy surfaces without any self-consistent procedure. It is therefore an interesting issue to comprehend how these two paths are theoretically related. For instance, one may be led to a conjecture that if the Truhlar paths are geometrically akin to the Pechukas paths, Pechukas' self-consistent procedure can be avoided only by introducing an external force. This may further lead to a consequence that a global solution in the mechanics of Eq. (4.31) does not always exit. This is not an absurd guess, if we recall that the essential feature of nonadiabatic dynamics is wavepacket branching, which seems to be hard to be represented directly in terms of a set of continuous non-branching paths.

4.4.2 *Notion of decoherence in quantum subsystems by contact with classical subsystems and decoherence time*

Decoherence is an essential concept appearing in a system in which a quantum subsystem contacts classical subsystem(s) in one way or another. As is widely recognized, the SET cannot describe this dynamics since there is no mechanism in it to switch off the electronic coherence along the nuclear path. The decoherence problem is critically important not only in our nonadiabatic dynamics but in other contemporary science such as spin-Boson dynamics in quantum computation theory and more extensively a quantum theory in open (dissipative) systems [147]. The decoherence problem is also critical to chaos induced by nonadiabatic dynamics [136, 137, 182, 453, 454]. Therefore, in the rest of this section, we pay deeper attention to the aspect of the effect of electronic state decoherence strongly coupled with the relevant nuclear motion. A review about the notion of decoherence related to quantum mechanical measurement theory is found in the papers by Rossky et al. [53].

Neria and Nitzan [297] proposed an idea to remove "artificially" the electronic coherence by damping the off-diagonal terms of the electronic density matrix. To do so, they define a damping time, or the decoherent time, in terms of the characteristic time for a wavepacket $\chi(t)$ running on a reference potential energy surface to lose its overlap with its bifurcating counterpart into the Kth surface $\chi_K(t)$. Assume such an overlap integral and approximate it in a Gaussian process as

$$|\langle \chi_K(t)|\chi(t)\rangle| \approx \exp\left(-\sum_n \frac{(F^n(0) - F_K^n(0))^2}{4a_n\hbar^2}t^2\right), \tag{4.85}$$

where a_n is a Gaussian width for nuclear nth degree of freedom. Here, $F^n(0)$ and $F_K^n(0)$ are nth component of mean-force and Kth adiabatic force vectors, respectively. This suggests a decoherence time τ_K as

$$\tau_K = \left(\sum_n \frac{(F^n(0) - F_K^n(0))^2}{4a_n\hbar^2}\right)^{-1/2}. \tag{4.86}$$

This property is expected to reflect the basic feature of damping of coherence. Subtonik et al. [398, 399] have proposed more sophisticated formulae to remove the artificial parameter a_n.

4.5 Some specific methods recently proposed for nonadiabatic dynamics

We here introduce a couple of modern theories to treat nonadiabatic transitions, which are somewhat characteristic and contain something to drive us to consider what a nonadiabatic transitions is. These theories are not necessarily tractable or powerful in actual applications though.

For a given potential energy surface, a nuclear wavefunction in the Bohm formalism is written as

$$\psi(\mathbf{R},t) = A(\mathbf{R},t)\exp\left(\frac{i}{\hbar}S(\mathbf{R},t)\right), \tag{4.87}$$

where both $A(\mathbf{R},t)$ and $S(\mathbf{R},t)$ are assumed to be real-valued. Then $S(\mathbf{R},t)$ satisfies

$$\frac{\partial}{\partial t}S(\mathbf{R},t) + \sum_I \frac{1}{2M_I}(\nabla_I S)^2 + V(\mathbf{R}) - \sum_I \frac{\hbar^2}{2M_I}\frac{\nabla_I^2 A(\mathbf{R},t)}{A(\mathbf{R},t)} = 0, \tag{4.88}$$

and

$$\frac{\partial}{\partial t}A(\mathbf{R},t) + \sum_I \frac{1}{2M_I}(\nabla_I S(\mathbf{R},t))\cdot(\nabla_I A(\mathbf{R},t)) + \sum_I \frac{1}{2M_I}\left(\nabla_I^2 S(\mathbf{R},t)\right)A(\mathbf{R},t) = 0. \tag{4.89}$$

The equation of motion for $A(\mathbf{R},t)$ is equivalent to the equation of continuity, representing the conservation of the quantum probability. On the other hand, $S(\mathbf{R},t)$ satisfies the generalized Hamilton-Jacobi equation except that the so-called quantum potential $-\sum_I \frac{\hbar^2}{2M_I}\frac{\nabla_I^2 A(\mathbf{R},t)}{A(\mathbf{R},t)}$ is included. This suggests that to determine the path solutions in Eq. (4.88), $A(\mathbf{R},t)$ should be known beforehand. This is a nonlinear process. However, the Bohmian dynamics is known to be a very important reformulation of the Schrödinger equation with respect to the interpretation of quantum mechanics. Much study has been devoted to practical methods to numerically realize the Bohmian dynamics in the context of chemical dynamics (see [481] and references therein).

The Bohmian formalism has been recently generalized to cases of multistate systems that couple nonadiabatically [100]. Readers are suggested to consult the original papers for the explicit form of the coupled equations of motion. Since the Bohmian dynamics is very difficult to solve even for a single potential energy surface, its extension to nonadiabatic dynamics requires a number of specific approximations and assumptions.

Tannor has developed a quantum dynamics by writing a wavefunction in an exponential form such that

$$\psi(\mathbf{R},t) = \exp[iS_T(\mathbf{R},t)/\hbar], \tag{4.90}$$

$S_T(\mathbf{R},t)$ of which is complex and obviously different from $S(\mathbf{R},t)$ in Eq. (4.87). He has derived an equation of motion for $S_T(\mathbf{R},t)$ and developed a method to solve it numerically. Furthermore he has extended his formalism to treat nonadiabatic dynamics [501, 502]. Though its numerical implementation is difficult, the formalism is interesting and deserves being studied for deeper understanding of nonadiabatic dynamics.

4.6 Hybrid methods for nonadiabatic dynamics in large molecular systems

The electron dynamics coupled with nuclear motion, whether coherent or incoherent, principally drives any chemical process in a moderately large molecular system. Dynamical chemical properties of material strongly depend on the time-dependent behavior of the electronic wavefunction, and many calculation schemes have been proposed for treating nonadiabatic electron dynamics in large scale systems.

One possible way to treat chemical dynamics in a large molecular system involved with nonadiabatic electron dynamics is to divide the system into the electron mixing part and the rest. Methodologies in multiscale physics may help in treating more complicated situations involving multi-layer or different sizes of dynamical objects.

Fragmenting the molecular system for nonadiabatic dynamics may allow a compact form of the Hamiltonian. According to the approximation level, interactions among fragments are included in the Hamiltonian through an additional (one step treatment) or self-consistent treatise (multi step treatment).

The typical molecular systems requiring a large scale treatise include (a) an exciton transfer system such as a photosynthesis system as a molecular aggregation of a reaction center and antenna molecules with classical bath, (b) aggregation system consisting of Rydberg atoms which may be expected with a small perturbation of vibrational motion, (c) bio-molecular system involved with drastic charge transfer accompanied with nonadiabatic electron dynamics, such as nonadiabatic dynamics of solvated DNA and photoisomerization of retinal in rhodopsin via a conical intersection, and (d) large molecular system involved with highly excited electrons such as Auger electrons where the classical picture of electrons works.

(a) and (b) are often treated as follows: nonadiabatic subsystems with their interactions structured by the dipole-dipole transition interaction in the off-diagonal parts and electronically static classical forces in the diagonal parts [464].

With regard to example (c), Elstner et al. developed an efficient algorithm of nonadiabatic dynamics for the hole transfer dynamics of solvated DNA, with a combination of the fragment orbital DFT method and the surface hopping scheme in the spirit of a QM/MM approach [225].

Goddard et al. studied a case of type (d) in materials science. All particles including electrons and nuclei are treated classically. This may be efficient for a liquid like situation where the system is at a high electron temperature. The antisymmetry effect and the exchange repulsion, related to the fermion nature of electrons is included through an effective electron potential called the Pauli potential. They especially explored the Auger induced chemical process caused by drastic rearrangement of the electronic structure in a materials system after deep orbital ionization [396, 397].

As another nonadiabatic QM/MM approach, Doltsinis et al. introduced a further layer in the MM part. This type of multiscale dynamics includes nonadiabatic QM parts, interacting MM part, and course-grained interaction with the environment. This approach intends to treat a drastic structure change in a macro scale originated from a microscopic phenomena associated with nonadiabaticity of electrons after shining a light [55].

Fragmentation methods with use of proper reduction of calculation level [150, 225] will play an important role in future research on electron dynamics in large molecular systems. The states are constructed tightly within the electron active space in terms of fragment orbitals separately obtained in each fragment monomer. For treating the electronic transfer among the monomers, one needs to enlarge the description of the monomer (where the truncation and re-coupling scheme work well computationally) with aid of massive computational schemes. In the part where a classical Coulomb interaction from another monomer is dominant, the separation of the monomer provides a convenient description. Reduction of the computational cost of the non-quantum mechanical part allows calculation of spatially delocalized electron dynamics in a large system.

Chapter 5

Direct Observation of the Wavepacket Bifurcation due to Nonadiabatic Transitions

The Born–Oppenheimer (BO) approximation based on an infinitesimal limit of electron mass against nuclear one provides a solid notion as well as powerful theoretical tools for a description of molecular static properties. However, there exist many situations where this picture breaks down when a nuclear kinematic coupling magnitude sufficiently exceeds an adiabatic BO energy gap. Because the energy gaps in excited state manifold tend to become small compared to that against the ground electronic state, the BO break down is usual in an excited state dynamics in molecule. The typical examples are the photochemical reaction dynamics accompanied with the crossing of different electronic characters and electron fluctuation induced by structural deformation in electronically unstable system. This BO breakdown assists many important chemical phenomena involved with an electronic quantum nature, for instance, a long range energy transfer in an excited state manifold, flexible chemical bond deformation, and so on. The molecular model should be constructed according to the characteristics of its constituents in as possible as natural manner. Otherwise, the theory would not work on prediction of molecular properties. In this chapter, we look at its physical origin and overview the several theoretical formalisms intended for describing this non-BO chemistry.

5.1 How does the Born–Oppenheimer approximation break down?

In cases where the Born–Oppenheimer perturbation theory of Eq. (2.36) is not valid, mainly as a result of the degenerate situation among the static electronic states, quantum mechanical mixing among them through the nuclear kinetic operator becomes significant. Consequently nonadiabatic transition takes place in molecular configuration at which the nonadiabatic coupling elements, X^k_{IJ} in Eq. (2.7), are large enough and/or the potential energy surfaces "avoid-cross" each other. [See Ref. [91] for a latest review on the faithful approaches to solve the coupled equations.] In such a nonadiabatic situation, a nuclear wavepacket starting from an electronic state $\Phi_1(\mathbf{r};\mathbf{R})$ is bifurcated into two by a single passage across a nonadiabatic region, which is mathematically written as follows;

$$\chi_1(\mathbf{R},t_b)\Phi_1(\mathbf{r};\mathbf{R}) \rightarrow \chi_1(\mathbf{R},t_a)\Phi_1(\mathbf{r};\mathbf{R}) + \chi_2(\mathbf{R},t_a)\Phi_2(\mathbf{r};\mathbf{R}) + \cdots . \quad (5.1)$$

Here, t_b and t_a, respectively, indicate a time before and after passing through the nonadiabatic region. The right hand side of this equation represents a typical situation of quantum entanglement, in which it is only after the relevant measurement that one can realize which one of $|\chi_1(\mathbf{R},t_a)\Phi_1(\mathbf{r};\mathbf{R})|^2$, $|\chi_2(\mathbf{R},t_a)\Phi_2(\mathbf{r};\mathbf{R})|^2$, and so on actually appears. This kind of quantum entanglement generated by the wavepacket bifurcation is one of the key features of nonadiabatic dynamics (see below for a more thorough argument about entanglement in the BO approximation).

To be a little more precise about the notion of quantum entanglement, it should be noted that even a single term in the Born–Huang expansion such as $\chi(\mathbf{R},t)\Phi(\mathbf{r};\mathbf{R})$ in Eq. (5.1) has already, at least to some extent, taken into account quantum entanglement through the functional form of the electronic wavefunction. This is made clearer if we formally expand it into "independent" basis functions as

$$\chi(\mathbf{R},t)\Phi(\mathbf{r};\mathbf{R}) = \sum_a \check{\chi}_a(\mathbf{R},t)\check{\Phi}_a(\mathbf{r}), \quad (5.2)$$

where the electronic function is free from the nuclear coordinates and thereby does not track the nuclear motion. This is in contrast to the spirit of Born–Oppenheimer. The basis set $\{\check{\Phi}_a(\mathbf{r})\}$ is sometimes referred to as a completely diabatic basis, and in fact it is physically poor and slowly converging. Moreover, it has been shown [198] that such a complete diabatic electronic basis does not globally "exist" (we will also show this fact later in this book, see Sec. 6.9). Nevertheless, $\{\check{\Phi}_a(\mathbf{r})\}$ is mathematically simple to handle. Hence some of these basis functions might prove useful depending on how they are exploited. In fact, the simultaneous quantization of electronic and nuclear degrees of freedom is a somewhat classic problem in theoretical chemistry and consequently has a long history. A more recent development has been the direct determination of self-consistent nuclear orbitals along with electronic orbitals (molecular orbitals) in the functional form of Eq. (5.2) [73, 74, 160, 161, 288, 289]. These methods are expected to work for estimating the vibronic coupling energy, but only if a large amplitude displacement of nuclear positions does not take place. We will not consider these basis functions further in this book, not only because of the slow convergence expected in the expansion of Eq. (5.2), but more importantly because they obscure the essential characteristics of nonadiabaticity.

5.2 Nuclear wavepacket bifurcation as observed with time-resolved photoelectron spectroscopy

In Sec. 3.3 we have shown how the nuclear wavepacket dynamics on a single electronic state can be tracked in real-time with use of the time-resolved photoelectron spectroscopy. Here in this section we extend the theory to cope with nonadiabatic dynamics and explore how the associated wavepacket bifurcation can be observed in this spectroscopy.

5.2.1 *Coupled nuclear dynamics on diabatic potential energy surfaces*

When the neutral states to be probed are coupled through nonadiabatic interaction, we need some way of including them in our formalism. In the adiabatic picture, the time-dependent Schrödinger equation for the nuclear wavefunctions $\chi(\boldsymbol{R}, t)$, which is a column vector

$$\chi(\boldsymbol{R}, t) = \begin{pmatrix} \chi_1(\boldsymbol{R}, t) \\ \chi_2(\boldsymbol{R}, t) \\ \vdots \end{pmatrix}, \tag{5.3}$$

is given by

$$i\hbar \frac{\partial}{\partial t} \chi(\boldsymbol{R}, t) = [\hat{\mathbf{T}}_N + \mathbf{V}(\boldsymbol{R})] \chi(\boldsymbol{R}, t), \tag{5.4}$$

where $\mathbf{V}(\boldsymbol{R})$ is diagonal, but the kinetic energy operator $\hat{\mathbf{T}}_N$ has off-diagonal elements, termed nonadiabatic couplings (recall Eq. (2.6)). Then, every electronic state $\Phi_j(\boldsymbol{r}; \boldsymbol{R})$ associated with $\chi_j(\boldsymbol{R}, t)$ is an adiabatic electronic state with a potential energy surface (potential energy *curve* in one dimension) $V_j(\boldsymbol{R})$, on which motion of $\chi_j(\boldsymbol{R}, t)$ takes place. The nonadiabatic couplings get the nuclear dynamics of different adiabatic states mixed. At the heart of nonadiabatic couplings are the derivative couplings,

$$\boldsymbol{F}_{jk} = \langle \Phi_j | \nabla_{\boldsymbol{R}} | \Phi_k \rangle. \tag{5.5}$$

A straightforward way of putting Eq. (5.4) into a form better suited for numerical computation is to make an attempt to put away the nonadiabatic off-diagonal elements through a unitary transformation,

$$\tilde{\mathbf{T}}_N = \mathbf{U}(\boldsymbol{R})\hat{\mathbf{T}}_N\mathbf{U}^\dagger(\boldsymbol{R}), \tag{5.6}$$

$$\tilde{\mathbf{V}}(\boldsymbol{R}) = \mathbf{U}(\boldsymbol{R})\mathbf{V}(\boldsymbol{R})\mathbf{U}^\dagger(\boldsymbol{R}), \tag{5.7}$$

and

$$\tilde{\boldsymbol{\chi}}(\boldsymbol{R},t) = \mathbf{U}(\boldsymbol{R})\boldsymbol{\chi}(\boldsymbol{R},t). \tag{5.8}$$

In one dimension and with a complete set of electronic states, such a transformation is always possible. $\tilde{\mathbf{T}}_N$ becomes diagonal, and $\tilde{V}(\boldsymbol{R})$ now has off-diagonal elements. The off-diagonal elements are simple functions and not operators so that they are much easier to make use of. This is the diabatic representation, in which we now have diabatic electronic states. In the case of more than one dimension or with an incomplete set of electronic states, this transformation is not exactly possible. But frequently (or we are hoping that) a transformation which makes the off-diagonal kinetic energy operator matrix elements almost zero is possible. This is also named the diabatic representation (or the *quasi-diabatic* representation when we have need to make clear that the transformation is not exact).

When a group of adiabatic electronic states have (almost) no couplings with all the other adiabatic electronic states, we may keep only the equations for the group of states in the matrix of Eq. (5.4), and make a diabatic representation from the group. This is different from making diabatic states from the complete set of adiabatic states and then keeping only a group of them, but it is in fact a better approximation, and is used more frequently.

In the two state case, the adiabatic to diabatic transformation is defined by the adiabatic-to-diabatic transformation angle (ADT angle) $\alpha(\boldsymbol{R})$,

$$\mathbf{U}(\boldsymbol{R}) = \begin{pmatrix} \cos\alpha(\boldsymbol{R}) & \sin\alpha(\boldsymbol{R}) \\ -\sin\alpha(\boldsymbol{R}) & \cos\alpha(\boldsymbol{R}) \end{pmatrix}, \tag{5.9}$$

which goes from 0 (diabatic the same as adiabatic) to $\pi/2$ (diabatic the same as adiabatic with lower and upper state exchanged). $\alpha(\boldsymbol{R})$ taking values between 0 and $\pi/2$ say how much the adiabatic states are mixed in forming the diabatic states, with the diabatic states becoming half the sum of adiabatic states at $\alpha(\boldsymbol{R}) = \pi/4$. In the one dimension case, the adiabatic-to-diabatic transformation angle comes directly from the derivative coupling [173]:

$$\alpha(R) = \int_R^\infty F_{12}(R')dR'. \tag{5.10}$$

When the system is put into the diabatic representation, time evolution of the nuclear wavefunction including nonadiabatic transitions is simple in a split-operator context, which amounts to diagonalizing the potential matrix at each time step [9]. In fact, the coding would be the same as for two states coupled though an external field (Sec. 3.3.2), except that the coupling potential would be a time-independent, intrinsic function of the system.

Photoionization out of several neutral states to several cation states may be simply included as separate terms of a split-operator scheme, assuming the ionized states do not strongly couple. Alternately, when the different ionization channels interact weakly, ionization out of each channel may be computed separately and the results combined. Examples of nonadiabatic dynamics seen through time-resolved photoelectron spectroscopy will be seen in Secs. 5.2 and 5.4. In the rest of this section, we will see examples of the diabatic representation for two classes of nonadiabatic interaction, the avoided crossing (Sec. 5.2.1.1) and the conical intersection (Sec. 5.2.1.2), and then discuss the necessary extensions to the standard vibrational wavefunction time-propagating schemes.

5.2.1.1 *An example: The LiF avoided crossing*

In the adiabatic representation, it is not possible for the two electronic states of a diatomic molecule in the same irreducible representation of a point group to have the same potential energy. This is the noncrossing rule. (When the irreducible representation is doubly degenerate, for example, two states do have the same potential energy.) Because of the noncrossing rule, the adiabatic potential energy curves in one dimension have avoided crossings, where the potential curves come near each other but do not touch each other. At an avoided crossing, the derivative coupling is not small.

See Fig. 5.1(a) for an example of an avoided crossing and the derivative coupling. Alkali halides, including the LiF molecule shown, are among the most noted systems with avoided crossings (see, for example, [128]). Computation for the potential energy curves was done with how it was done in [469] in mind, at the multiconfiguration self-consistent field (MC-SCF) level. We get the derivative coupling by taking the difference of the electronic wavefunction at different nuclear coordinates. We see good agreement between the derivative coupling curve of Fig. 5.1(a) and that given in Fig. 9 of Ref. [456], a computation at a much higher level of multireference configuration interaction (MR-CI) at the complete basis set (CBS) limit.

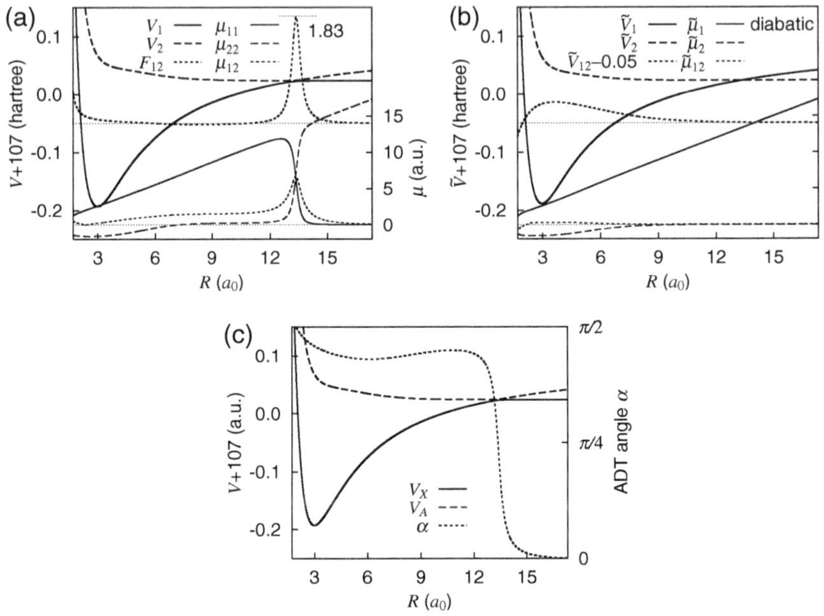

Fig. 5.1 (a) LiF MC-SCF adiabatic curves. (b) MC-SCF diabatic curves. (c) ADT angle.

The adiabatic-to-diabatic transformation angle $\alpha(R)$ is obtained by numerical integration of Eq. (5.10). We see this curve in Fig. 5.1(c). It goes near $\pi/2$ for R smaller than the avoided crossing position, but does not quite get to $\pi/2$. From this angle we get the diabatic curves seen in Fig. 5.1(b). Because α does not go to $\pi/2$, $V_{12} \neq 0$ for small R. The curves (V_{12} also) agree in shape with those given in Refs. [469] and [146], from diagonalizing the dipole moment matrix [251].

Because $V_{12} \neq 0$ for small R, there is some transition between diabatic states at small R, but this only implies that the dynamics there is adiabatic.

5.2.1.2 *An example: The NO_2 conical intersection*

We use a diabatic representation to describe the two electronic states coupled with the conical intersection. This aids in time propagation of the vibrational wavefunctions as the interaction between the states appear in the potential function term rather than in the kinetic energy operator. The adiabatic kinetic energy coupling diverges at the conical intersection. Although there is no rigorous diabatic representation for the multidimensional case, an approximate representation that makes the kinetic coupling "small enough" can often be constructed, either by introducing constraints to compute diabatic states directly, or by transforming adiabatic states by some criterion. Transformation of adiabatic to quasi-diabatic states by *ab initio* procedures have recently seen great advances [28, 218]. Here instead we use a simple phenomenological method previously used by Hirsch et al. successfully on the NO_2 system [183].

We want to obtain the coordinate-wise adiabatic-to-diabatic transformation matrix $U(\boldsymbol{R})$ [Eq. (5.9)] and the adiabatic-to-diabatic transformation angle $\alpha(\boldsymbol{R})$. The NO_2 $X\,^2A'/A\,^2A'$ conical intersection is located at C_{2v} symmetry configuration, where the adiabatic dipole moment perpendicular to the bond angle bisector ($\mu^y_X(\boldsymbol{R})$ for the ground state and $\mu^y_A(\boldsymbol{R})$ for the excited state) is zero. Assuming the C_{2v} states as the diabatic states, $\mu^y_X(\boldsymbol{R})$ and $\mu^y_A(\boldsymbol{R})$ can be identified as the deviation of the C_s states from the diabatic C_{2v} states. The adiabatic-to-diabatic transformation angle is then obtained as the angle that maximizes the off-diagonal terms and eliminates the difference between diagonal terms of the dipole moment matrix,

$$\alpha(\boldsymbol{R}) = \frac{1}{2}\tan^{-1}\left(\frac{\mu^y_X(\boldsymbol{R}) - \mu^y_A(\boldsymbol{R})}{2\mu^y_{XA}(\boldsymbol{R})}\right), \tag{5.11}$$

where $\mu^y_{XA}(\boldsymbol{R})$ is the adiabatic transition dipole moment perpendicular to the bond bisector. The dipole moments cannot be made zero as in the C_{2v} case, but are both made to be as small as they can simultaneously. Hirsch et al. computed spatial derivatives of the transformed electronic wavefunctions and have shown that the states thus made are indeed diabatic for NO_2 [183].

We use the transformation matrix $U(\boldsymbol{R})$ to represent the potential energy matrix, dipole moment matrix, and photoionization amplitudes in the diabatic representation. Then the vibrational wavefunctions are computed in the diabatic representation, but can also be transformed with $U(\boldsymbol{R})$ when we wish to see the adiabatic functions. The vibrational wavepacket dynamics and time-resolved photoelectron spectroscopy of the system is treated in Sec. 5.4.

5.2.2 *Wavepacket bifurcation in the NaI system*

In this Subsection we apply the above theory of time-resolved photoelectron spectroscopy to the study of wavepacket motion on the nonadiabatically coupled ionic (Na^+I^-) and covalent (NaI) states of sodium iodide. Here we show results of calculations of the energy and angular distributions of photoelectrons from wavepackets on these coupled covalent and ionic states of NaI. A key objective of these studies is to explore the potential of energy- and angle-resolved pump-probe photoelectron spectra for probing the intramolecular electron transfer occurring between the covalent and ionic states (NaI $\rightarrow$ Na^+I^-) in the nonadiabatic region. Our pump (3.73 eV) and probe (4.89 eV) photon energies, which differ from the pump (3.97 eV) and probe (4.71 eV) energies of Refs. [201] and [78], are well suited for probing the wavepacket motion across the avoided crossing and are chosen with this in mind.

Pump-probe studies of this system have demonstrated the decrease of wavepacket population on the excited adiabatic surface due to dissociation at the avoided crossing [94, 340]. Oscillations in the population of the dissociative products (Na and I) due to interference of wavepackets on the covalent and ionic potentials merging at the avoided crossing have also been observed [340]. Direct observation of wavepacket bifurcation is interesting in itself, but also quite significant to studies of electron transfer, wavepacket engineering, and reaction control through wavepacket splitting and mixing. Also, from the perspective of quantum measurement, wavepacket bifurcation corresponding to an intramolecular double-slit experiment will shed light on the evolution of quantum entanglement between electronic and nuclear motion [431]. Bifurcation and merging of wavepackets are important as an intrinsic mechanism of "quantum chaos," which has no simple classical counterpart [136, 137, 182, 217].

5.2.2.1 *The NaI system*

The seminal studies of Zewail and coworkers established NaI as a benchmark system for monitoring wavepacket motion [94, 340, 506]. In these experiments a femtosecond pump laser launches a wavepacket on the covalent excited state which is coupled nonadiabatically to the ground state (Fig. 5.2). The wavepacket oscillates across the adiabatic potential well formed by the avoided crossing of the covalent (NaI) state and ionic (Na^+I^-) states. In the region of the avoided crossing an electron is transferred from Na to I and the bond converts from covalent to ionic. In these studies [94, 340, 506], the time evolution of the wavepacket was monitored by excitation to a higher state and fluorescence from this state. More recently, Jouvet et al. [201] have followed wavepacket evolution on the same state of NaI by time-resolved detection of photoelectron energy distributions and photoions. These photoelectron spectra and ion energy distributions provided valuable insight into

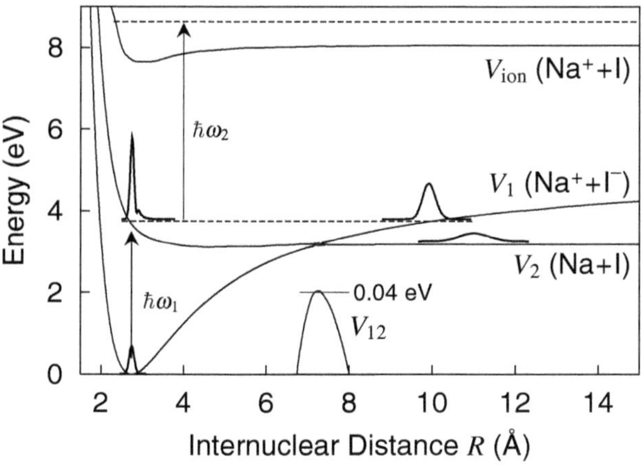

Fig. 5.2 NaI pump–probe photoelectron spectroscopy. (Reprinted with permission from Y. Arasaki *et al.*, J. Chem. Phys. 119, 7913 (2003)).

the nuclear dynamics of this system. There have also been several theoretical studies of the pump-probe photoelectron energy distributions [63, 78] and of dissociative ionization [78] of this same state of NaI. While the early studies of this system by Engel and coworkers [63] did not include any dependence of the photoionization amplitude on geometry, Charron and Suzor-Weiner [78] employed an empirical form to account for the change in the transition dipole as the wavepacket moved from the covalent to ionic regions and obtained photoelectron and photoion spectra in good agreement with the measurements of Ref. [201]. Such R-dependence of the photoionization amplitude has also been observed to be essential in accounting for the unexpected inward-outward asymmetry in femtosecond two-photon ionization of wavepackets on the A state of NaI [152].

5.2.2.2 *Potential curves and photoionization matrix elements*

The adiabatic potential energy curves for the ground, excited, and ion states were obtained from high-level electronic structure calculations using the MOLPRO code [471, 472]. It was still necessary, however, to shift the neutral potential curves slightly to adjust the excitation energy to the experimental values and the final value of the nonadiabatic coupling matrix element [see Fig. 5.2]. It is worth noting that it was also essential to include spin-orbit interaction to correctly describe the covalent potential curve in the region where the wavepacket is launched.

To account for the dependence of the photoionization amplitude on internuclear distance in this system, particularly in the crossing region, geometry-dependent amplitudes are employed in these calculations. Here we comment briefly on a few salient and relevant features. The cross sections for ionization of the wavepacket on the ionic branch of the excited state are much larger than on the inner (covalent) well and depend quite strongly on energy and distance. They range in value between 15 Mb and 30 Mb and continue to show molecular-like behavior out to the turning point (10 Å), reflecting the ion-pair character of the state. In contrast, the cross section on the covalent branch can be of the order of 1 to 3 Mb, which is significantly larger than the cross section for the sodium atom (∼ 0.1 Mb). This difference reflects the molecular nature of the covalent (NaI) state.

5.2.2.3 *Excited wavepacket motion*

Figure 5.3 shows the wavepackets on the ionic (V_1, lower panel) and covalent (V_2, upper panel) diabatic states for a linearly polarized pump pulse of 3.73 eV with a FWHM of 40 fs and centered at $t = 0$. These diabatic potentials provide a very pictorial view of the wavepacket motion on these nonadiabatically coupled surfaces. After launching, the wavepacket moves to the right and undergoes bifurcation at the crossing point between V_1 and V_2. The component of the wavepacket that continues along V_2 represents dissociation into neutral atoms. The component moving on the ionic surface reaches its right-hand-side turning point near 10 Å at around 475 fs. This component returns to the crossing point at around 725 fs and bifurcates. The component on the V_2 (covalent) surface reaches the left-hand-side turning point at ∼925 fs.

5.2.3 *Photoelectron signals arising from the NaI dynamics*

We compute the photoelectron spectra using a quantum wavepacket method described in [15] with enhancements to handle the nonadiabatic interactions. Briefly, the system is expanded in terms of the three relevant electronic states: Φ_1 (V_1 state), Φ_2 (V_2 state), and $\Phi_{\mathbf{k}}^{(-)}$ (ion state), as

$$\Psi(\mathbf{r}, R, t) = \chi_1(R,t)\Phi_1(\mathbf{r}; R) + \chi_2(R,t)\Phi_2(\mathbf{r}; R) + \int d\mathbf{k}\chi_{\mathbf{k}}(R,t)\Phi_{\mathbf{k}}^{(-)}(\mathbf{r}; R), \tag{5.12}$$

where $\mathbf{k}$ labels the wave vector of the photoelectron. Substitution into the time-dependent Schrödinger equation results in coupled dynamical

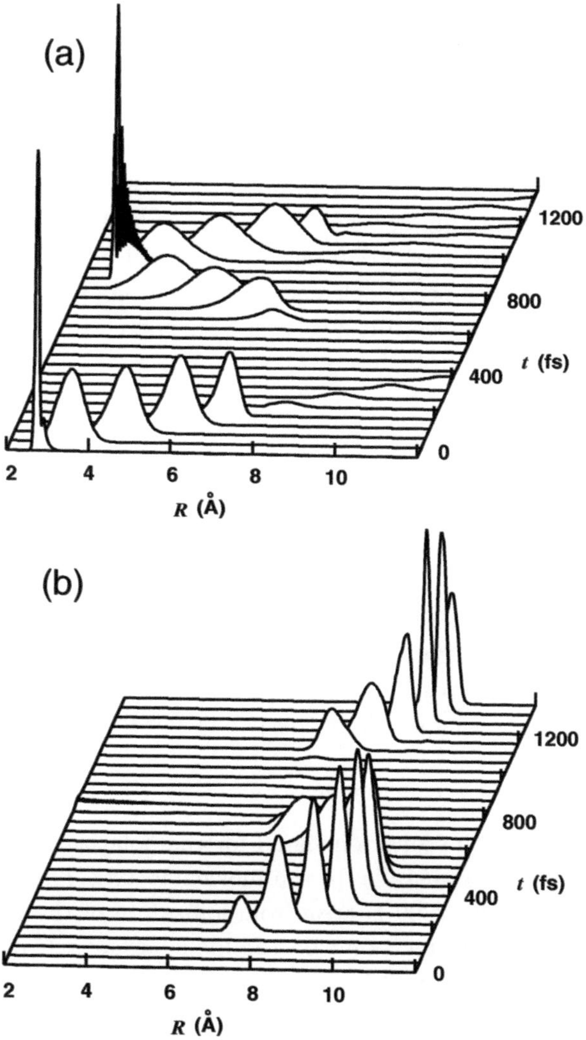

Fig. 5.3 Pump-pulse excited wavepacket motion on the diabatic states, (a) V_1 (ionic) and (b) V_2 (covalent). (Reprinted with permission from Y. Arasaki *et al.*, 119, 7913 (2003)).

equations for the vibrational wavepackets χ_1, χ_2, and $\chi_{\mathbf{k}}$ of the form

$$i\hbar\frac{\partial}{\partial t}\chi_1(R,t) = [T_{\rm N} + V_1(R)]\,\chi_1 + V_{12}(R)\chi_2 + \int d\mathbf{k}V_{1,\mathbf{k}}(\mathbf{R},\theta_P,\mathbf{k},t)\chi_{\mathbf{k}}, \tag{5.13}$$

$$i\hbar\frac{\partial}{\partial t}\chi_{\mathbf{k}}(R,t) = \left[T_{\mathrm{N}} + V_{\mathrm{ion}}(R) + \frac{\hbar^2 k_j^2}{2m_e}\right]\chi_{\mathbf{k}} + \sum_{d=1,2} V_{\mathbf{k},d}(\mathbf{R},\theta_P,\mathbf{k},t)\chi_d, \tag{5.14}$$

where T_{N} is the kinetic energy operator; $V_{\mathbf{k},d}$ is the ionization interaction matrix element, including dependence on molecular geometry ($\mathbf{R}$), photoelectron energy and angle, probe pulse shape, and polarization (θ_P); and molecular rotation has been neglected. The equation for χ_2 is analogous to that for χ_1. The photoelectron spectra are calculated from the vibrational wavepackets after the probe interaction as

$$P(\varepsilon_k) = k\int dRd\hat{k}|\chi_{\mathbf{k}}(R)|^2. \tag{5.15}$$

The photoionization matrix elements employed in these studies are obtained using high-quality wavefunctions for the bound states and photoelectron continuum. Preliminary results for this calculation were presented earlier [421].

For ionization of these wavepackets we choose a probe pulse with a photon energy of 4.89 eV which is just sufficient for ionization of the wavepacket on V_2 (covalent) beyond the crossing point. This pulse has a FWHM of 40 fs and its polarization can be either parallel or perpendicular to the molecular axis.

5.2.3.1 *Total photoelectron signal*

Figure 5.4(a) shows the total ion signal for the polarization vectors of the pump and probe lasers parallel to the molecular axis. The peaks centered around 400, 1350, and 2300 fs arise from ionization of the wavepackets at the right-hand-side turning point on the ionic surface while those around 900, 1850, and 2800 fs are due to wavepackets located at the left-hand-side turning point on the V_2 covalent surface. The shape of the ion signal at the right-hand-side turning point (400, 1350, and 2300 fs) reflects the slowing down of the wavepacket there and the dependence of $C_{1,lm}$ on R. The valleys between peaks for ionization at the right- and left-hand-side turning points, e.g., $t_{\mathrm{pr}} = 800$ fs, indicate passage of the wavepacket through the avoided crossing region. The decreasing height of these peaks with delay time results from loss of NaI population to dissociation.

The ion signal with the probe polarization perpendicular to the molecular axis [Fig. 5.4(b)] is significantly larger than for the parallel alignment. The shapes of these ion signals for parallel and perpendicular orientation of the polarization of probe laser to the molecular axis reflect differences in

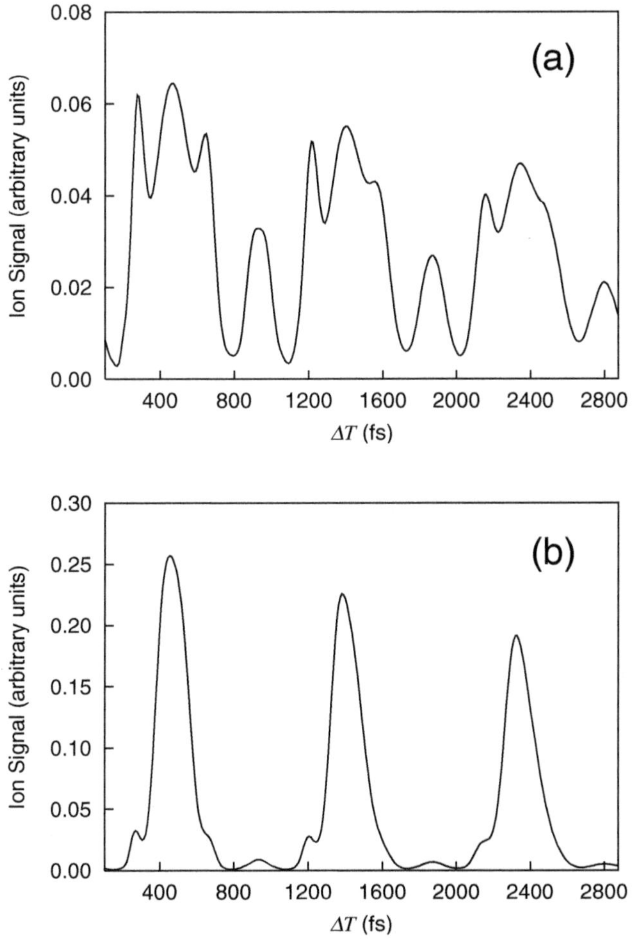

Fig. 5.4 Total ion signals vs. pump-probe delay time. Probe polarization (a) parallel and (b) perpendicular to the pump polarization. (Reprinted with permission from Y. Arasaki *et al.*, J. Chem. Phys. 119, 7913 (2003)).

the behavior of the photoionization amplitudes with R. The perpendicular signal is dominated by ionization of the wavepacket from the region around the right-hand turning point.

5.2.3.2 *Photoelectron kinetic energy distributions*

Fig. 5.5 shows the photoelectron energy distributions as a function of pump-probe delay time with the probe polarization parallel to the molecular axis. As is expected from previous model studies of this system [63, 78], the kinetic

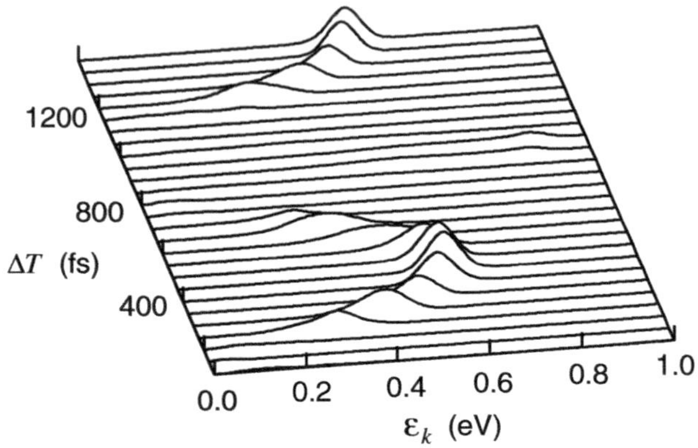

Fig. 5.5 Delay time dependence of photoelectron kinetic energy distributions. Probe polarization parallel to that of the pump. (Reprinted with permission from Y. Arasaki *et al.*, J. Chem. Phys. 119, 7913 (2003)).

energy distributions map the vibrational wavepacket motion quite nicely: the photoelectron energy increases as the wavepacket moves to the right on the covalent state, bifurcates at the crossing point, climbs up the ionic potential, and reaches its turning point around 475 fs where the photoelectron energy is about 0.6 eV. The much weaker peak at 0.8 eV and 900 fs maps the wavepacket near the left-hand turning point on the covalent state. The very low photo-

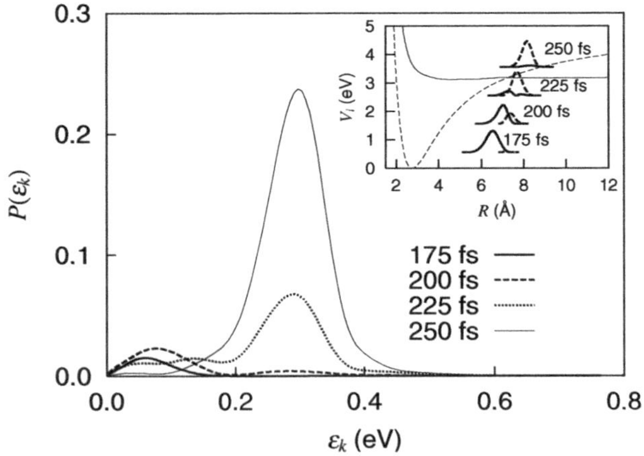

Fig. 5.6 Photoelectron energy spectrum as the wavepacket bifurcates at the avoided crossing (first crossing). The probe polarization is perpendicular to that of the pump in Figs. 5.6 and 5.7. (Reprinted with permission from Y. Arasaki *et al.*, J. Chem. Phys. 119, 7913 (2003)).

electron signal between pump-probe delays of $\sim$ 750 fs and 1150 fs, when the wavepacket is on the covalent state, reflects the vast difference in the amplitudes for ionization on the covalent and ionic states.

We now look more closely at how these photoelectron energy distributions map the motion of the wavepacket in the crossing region. Figure 5.6 shows the photoelectron spectra for pump-probe delays of 175, 200, 225, and 250 fs. Snapshots of the wavepackets on the diabatic potentials are also shown as insets in this figure. At 175 fs the single peak around 0.1 eV in the photoelectron spectrum and its low intensity are signatures of the wavepacket on the covalent potential. By 200 fs the wavepacket has bifurcated, resulting in components on the covalent and ionic potentials which give rise to peaks in the photoelectron spectrum $\sim$0.1 eV and 0.3 eV, respectively. The peak at 0.3 eV reflects the greater kinetic energy available to photoelectrons from the wavepacket on the ionic potential than from the wavepacket on the covalent potential. By 225 fs continued bifurcation onto the ionic potential leads to a dominant peak at 0.3 eV and a weak shoulder to lower energy in the photoelectron spectrum. At 250 fs the spectrum has evolved into a strong feature around 0.3 eV in the spectrum arising from the wavepacket on the ionic potential. The near 15-fold larger intensity of the peak at 0.3 eV (250 fs) over the peak at 0.1 eV (175 fs) is a result of the much greater magnitude of the photoionization amplitude for the ionic state than for the covalent state. The photoelectron signal from the wavepacket on the covalent potential beyond the crossing point is expected to be negligible.

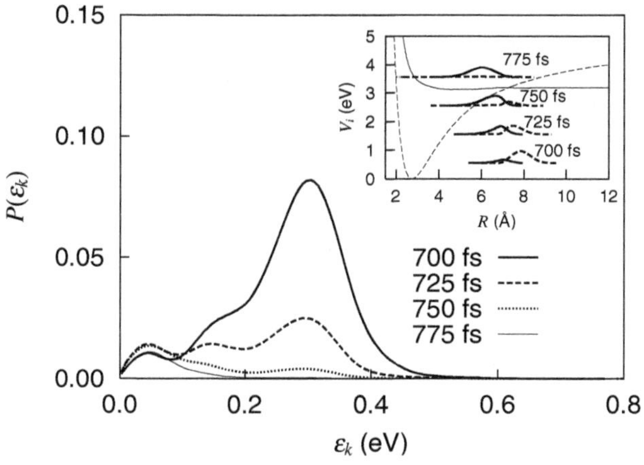

Fig. 5.7 Photoelectron energy spectrum as the wavepacket transfers from the ionic to the covalent curve (second crossing). (Reprinted with permission from Y. Arasaki *et al.*, J. Chem. Phys. 119, 7913 (2003)).

As a second example, we look at the photoelectron spectra for the wavepacket as it moves back down on the ionic potential and begins to reenter the crossing region. Fig. 5.7 shows these spectra for pump-probe delays of 700, 725, 750, and 775 fs. The evolution of these spectra with pump-probe delay again illustrates how the photoelectron energy distributions map the motion of the wavepacket through the crossing region. In contrast to the previous example where the wavepacket was bifurcating onto the ionic potential, here the photoelectron signal decreases with delay time as the wavepacket moves onto the covalent potential where the photoionization amplitude is smaller.

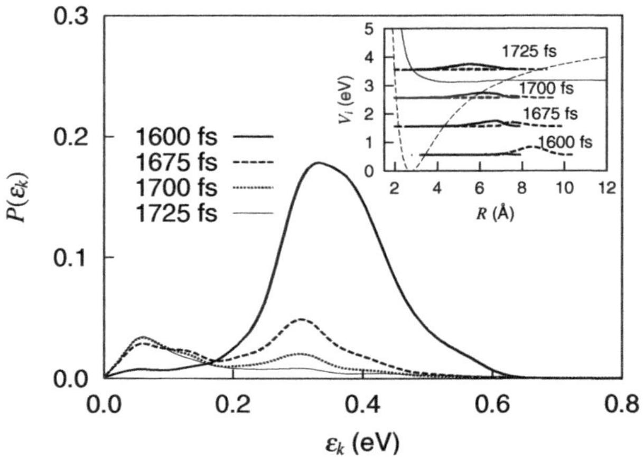

Fig. 5.8 Photoelectron energy spectrum as the wavepacket moves from the ionic onto the covalent curve (third crossing). (Reprinted with permission from Y. Arasaki *et al.*, J. Chem. Phys. 119, 7913 (2003).)

As a final example, Fig. 5.8 shows photoelectron spectra with the probe polarization parallel to the molecular axis for pump-probe delays of 1600, 1675, 1700, and 1725 fs when the wavepacket is reentering the crossing region for the second time from the ionic potential. The evolution of the lower and higher energy peaks in the spectra with time are seen to monitor the bifurcation of the wavepacket through the avoided crossing region. The broadening of the energy distribution at 1700 and 1725 fs to higher energy is due to greater spreading of the wavepacket on the covalent potential at these times than at the first crossing, e.g., 775 fs.

5.2.3.3 *Photoelectron angular distributions*

We now explore a few examples of molecular frame photoelectron angular distributions to see how they may reflect intramolecular electron transfer between the covalent and ionic states, i.e., NaI $\rightarrow$ $Na^+ + I^-$, in the crossing region. Figure 5.9 (upper panel) shows the angular distribution of photoelectrons with 0.05 eV, the peak energy in the spectrum, for photoionization of the wavepacket at a pump-probe delay of 125 fs. In this figure the sodium and iodine ends of the molecule are at 0 and 180 degrees, respectively. The peaking of the distribution at lower angles ($\sim 20°$) implies that ionization of the wavepacket on the covalent surface to the left of the crossing (see lower right panel) favors ejection of electrons from the sodium side. This behavior is also observed at other photoelectron energies in the spectrum, e.g., 0.13 and 0.3 eV. This can also be seen in the polar plot of the angular distribution, integrated over kinetic energy, shown in the lower left frame of Fig. 5.9.

The upper panel of Fig. 5.10 shows the angular distribution of photoelectrons at 0.08 eV, the peak energy in the spectrum, for photoionization at a pump-probe delay of 175 fs when the wavepacket has moved into the crossing region. Electrons are now preferentially ejected from the iodine side of the molecule, reflecting the increasing ionic character of the wavepacket as the electron transfers from sodium to iodine, i.e., NaI $\rightarrow$ $Na^+ + I^-$. This same behavior is seen in the kinetic energy integrated angular distributions in the lower panels of Fig. 5.10. By 225 fs the wavepacket has moved to the right of the crossing (Fig. 5.10, inset) and the angular distribution continues to favor electron ejection from the iodine (I^-) side of the molecule. The evolution of these angular distributions as the wavepacket moves from the covalent (NaI) to ionic (Na^+I^-) potentials suggests that such spectra can be a useful window on the real-time dynamics of electron transfer.

Figure 5.11 tracks the evolution of the angular distribution of photoelectrons as the wavepacket enters the crossing region on its way back from the right-hand turning point. The frames at 700, 750, and 800 fs show the energy-integrated angular distributions as they evolve from two dominant lobes towards the iodine end of the molecule at 700 fs, where the wavepacket is essentially on the ionic state (Na^+I^-), to two dominant lobes towards the sodium end of the molecule at 800 fs when the wavepacket is on the covalent state (NaI). The angular distribution at 750 fs displays features transitional between those at 700 and 800 fs. The 10-fold decrease in intensity of the distributions from 700 to 800 fs again reflects the much larger magnitude

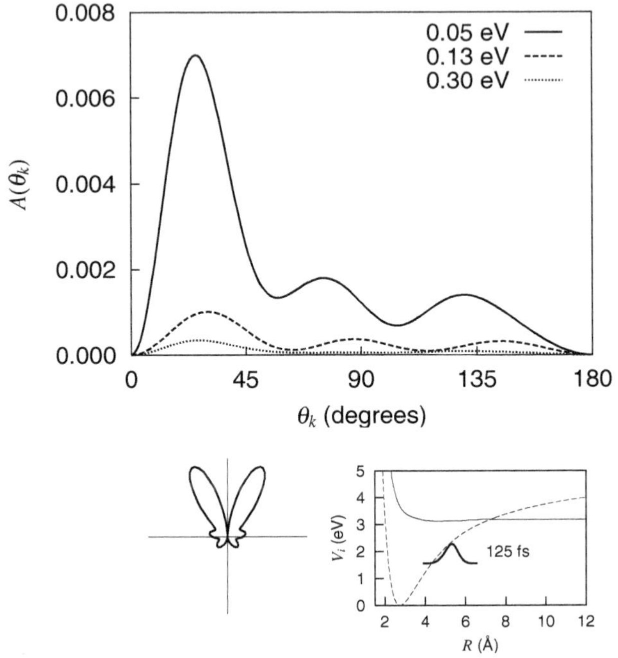

Fig. 5.9 Photoelectron angular distributions at delay time $t_{\rm pr} = 125$ fs (upper panel). Polar plot of the energy-integrated photoelectron angular distribution for the same delay time (lower left panel). Sodium lies at the top of the vertical axis in the polar plots in Figs. 5.9 through 5.12. Wavepacket snapshot (lower right panel). Probe polarization perpendicular to that of the pump. (Reprinted with permission from Y. Arasaki *et al.*, J. Chem. Phys. 119, 7913 (2003)).

of the photoionization matrix elements on the ionic potential. For further comparison of these angular distributions as the wavepacket moves through the crossing, the lower left panel of Fig. 5.11 shows the angular distribution at the peak energy of each spectrum.

The energy-integrated angular distributions of photoelectrons for pump-probe delays of 1050, 1125, and 1200 fs and with the probe polarization parallel to the molecular axis, Fig. 5.12, serve as a final example. As the wavepacket moves into and through the crossing, the characteristic d_z^2 angular distribution of the photoelectron evolves from one that is more symmetrical with respect to ejection from the Na and I sides of the molecule at 1050 fs to one that strongly favors ejection from the I side.

In this section we have discussed the requisite extension of our theory of energy- and angle-resolved photoelectron spectra for pump-probe ionization

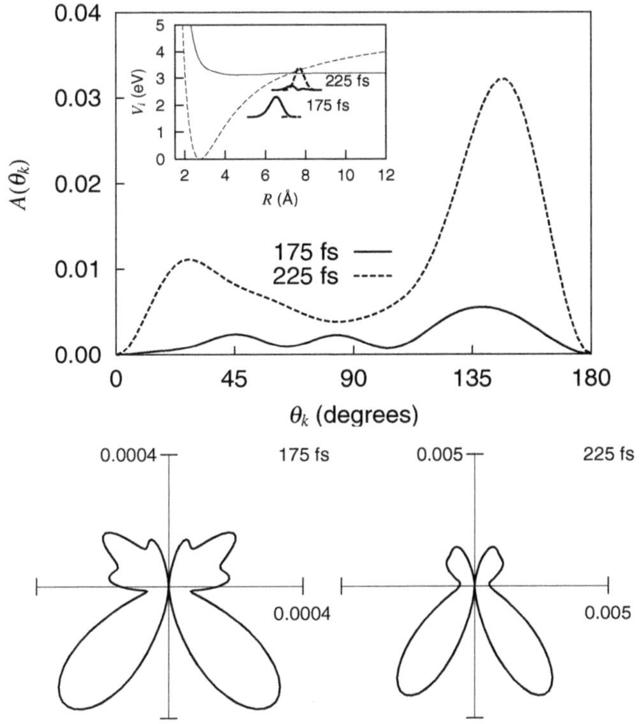

Fig. 5.10 Photoelectron angular distributions at peak energy (0.08 eV for 175 fs and 0.30 eV for 225 fs) of each photoelectron spectrum (upper panel) and wavepacket snapshots at corresponding delay times (inset). Energy integrated photoelectron angular distributions (lower panel). Probe polarization perpendicular to that of the pump. (Reprinted with permission from Y. Arasaki *et al.*, J. Chem. Phys. 119, 7913 (2003)).

of molecules [15] to study wavepacket motion on the nonadiabatically coupled ionic (Na^+I^-) and covalent (NaI) states of sodium iodide. A principal objective of these studies has been to explore the potential of such pump-probe photoelectron spectra for direct observation of wavepacket bifurcation due to a nonadiabatic transition and for probing the intramolecular electron transfer that occurs in the crossing region between the covalent and ionic states. The results of these studies suggest that the energy and angular distributions of such photoelectron spectra do indeed map the motion and bifurcation of the wavepacket through the crossing region and should provide a useful window on the intramolecular electron transfer occurring between the covalent and ionic states (NaI $\rightarrow$ Na^+I^-) in the nonadiabatic region. Use of robust values of the photoionization amplitudes and their

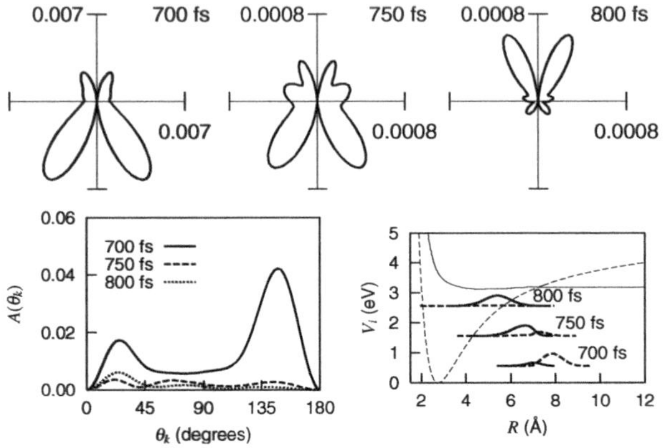

Fig. 5.11 Energy-integrated photoelectron angular distributions as wavepacket transfers from the ionic to the covalent curve (upper panel). Photoelectron angular distributions at peak energy (0.30 eV for 700 fs, 0.05 eV for 750 and 800 fs) of each photoelectron spectrum (lower left panel). Wavepacket snapshots at corresponding delay times (lower right panel). Probe polarization perpendicular to that of the pump. (Reprinted with permission from Y. Arasaki *et al.*, J. Chem. Phys. 119, 7913 (2003)).

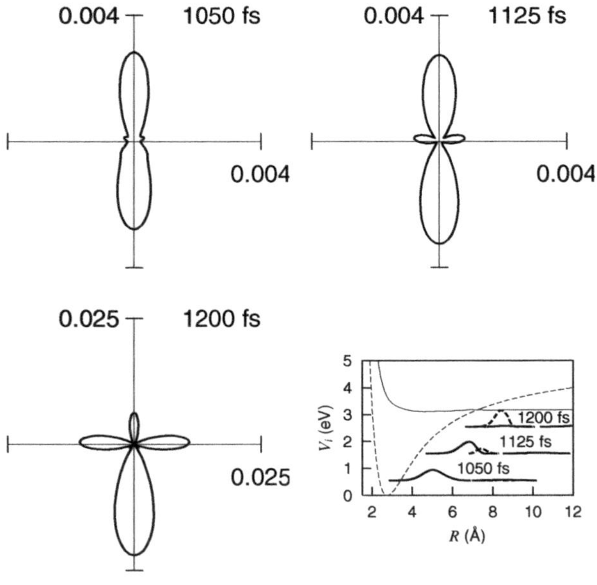

Fig. 5.12 Polar plots of the energy-integrated angular distributions and snapshots of the wavepacket moving through the crossing. Probe polarization parallel to that of the pump. (Reprinted with permission from Y. Arasaki *et al.*, J. Chem. Phys. 119, 7913 (2003)).

evolution with internuclear distance through the crossing region, however, is essential in such studies. While in this section we have focused on the behavior of pump–probe photoelectron spectra in the crossing region, these spectra provide useful insight on the wavepacket dynamics in other regions of these potential surfaces, e.g., near the right- and left-turning points on the ionic and covalent potentials respectively.

5.3 Control of nonadiabatic chemical dynamics

The recent advances in the technology of generating pulsed fields have inspired studies into not only ways to observe so far unobserved molecular states and dynamics but at the same time ways of controlling the forming or the outcome of them. The intense interest and rapid progress in the field of coherent control can be witnessed as almost yearly publications of review articles and monographs over the last decade [64, 105, 304, 306, 309, 364, 365, 372, 436, 475, 476]. In this section, instead of going into the details that are amply given in earlier excellent works, we wish to briefly overview the various techniques and put them in perspective from a dynamical point of view.

5.3.1 *Various time domains of external field control*

Equation (3.49) describing the effect of external fields on the molecular system is suggestive in that for a good approximation over a wide range of parameters the interaction between the molecular system and the external fields may be thought of as a time-dependent potential energy function. We may focus on the time scale of the external fields and recognize different domains. Each such domain has brought forth control schemes applicable in the time scale of interest.

When the frequency of the external field is compatible with the electronic excitation energy and its duration long compared to the nuclear dynamics, we may regard the individual electronic excitation channels as units and exploit coherent interaction among the channels [65].

Intense external fields may be regarded better as modifying the potential surfaces in forming "dressed states" [460], with the surfaces time evolving in the time scale of the envelope function of the external field; the vibrational wavefunctions evolving on the dressed states.

For pulses of shorter durations, we may regard the results as the formation and time evolution of vibrational wavepackets on (the not dressed) electronic states and exploit the timing of electronic excitation/deexcitation [434], directing the wavepacket's destination [324] or avoiding certain potential energy surface features [281]. Alternately, several wavepackets may be formed and the interference among them exploited for control [306].

For dynamics of even shorter durations, when the frequency of the external field is such that the time scale of the oscillation of the field is comparable to the time scale of nuclear dynamics, it may be better to regard the potential surfaces as becoming time-dependent through including the interaction potential within them [355]. The potential surfaces time-evolve not only with the envelope function of the external field but follows the oscillation of the external field itself, enabling dynamic Stark control [404].

A robust experimental concept potentially applicable to all time domains is the feedback-optimized control [21] which requires no knowledge of the underlying mechanism in effecting control. Optimal control theory [187] provides theoretical and computational description of the process to help understand the control mechanisms in effect.

Current technology is enabling pulses shorter in time scale than the nuclear dynamics [61, 76], opening up the field of attosecond dynamics for control; the current focus is mostly on controlling the electron (recollision) dynamics [222].

5.3.2 *An example: Fluctuating potential curves*

The coupled time-dependent Schrödinger equation for the vibrational wavepackets $\Psi_1(R,t)$ and $\Psi_2(R,t)$ on the electronic states 1 and 2, respectively, in a diabatic representation is given as

$$i\hbar\frac{\partial}{\partial t}\begin{pmatrix}\Psi_1(R,t)\\ \Psi_2(R,t)\end{pmatrix} = \hat{H}\begin{pmatrix}\Psi_1(R,t)\\ \Psi_2(R,t)\end{pmatrix}, \tag{5.16}$$

where the Hamiltonian $\hat{H}$ is given by

$$\hat{H} = \hat{T}_{\mathrm{N}}\mathbf{1} + \mathbf{V}(R) + \mathbf{V}_F(R,t). \tag{5.17}$$

with $\hat{T}_{\mathrm{N}}$ the nuclear kinetic energy operator. The potential energy matrix $\mathbf{V}(R)$ is

$$\mathbf{V}(R) = \begin{pmatrix}V_{11}(R) & V_{12}(R)\\ V_{12}(R) & V_{22}(R)\end{pmatrix}, \tag{5.18}$$

where $V_{11}(R)$ and $V_{22}(R)$ are the diabatic potential energy curves and $V_{12}(R)$ is the diabatic coupling potential between them. An avoided crossing in the adiabatic representation becomes literally a crossing of the curves $V_{11}(R)$ and $V_{22}(R)$ in the diabatic representation. The magnitude of $V_{12}(R)$ near the crossing point $R_{\rm cross}$ indicates the strength of state mixing at the avoided crossing. A larger $V_{12}(R)$ there induces more population transfer between the diabatic states, thereby making the dynamics more adiabatic. A smaller $V_{12}(R)$ there induces less population transfer between the states.

For a linearly polarized external field in the dipole approximation, the time-dependent interaction $\mathbf{V}_F(R,t)$ is given by

$$\mathbf{V}_F(R,t) = \begin{pmatrix} \mu_{11}(R) & \mu_{12}(R) \\ \mu_{12}(R) & \mu_{22}(R) \end{pmatrix} F(t), \tag{5.19}$$

where $\mu_{ij}(R)$ is a dipole moment matrix element between the states i and j in the direction of field polarization, and $F(t)$ represents the time-dependent amplitude of the external field. For the case of significant dipole moment values we neglect the less contributing higher order terms for simplicity. In the present work, we consistently choose a diabatic representation such that the off-diagonal dipole matrix elements vanish at any R, that is, $\mu_{12}(R) = 0$ by diagonalization of the dipole matrix. (We discuss the practical diabatization of the LiF molecule more in Sec. 5.2.1.1.)

The electric field generated by the classical vector potential of the electromagnetic field of a pulse laser to be used in Eq. (5.23) is of the standard form

$$\begin{aligned} F(t) &= -Ef(t-t_0)\sin(\omega(t-t_0)+\phi) \\ &\quad + \frac{E}{\omega}\frac{df}{dt}\cos(\omega(t-t_0)+\phi), \end{aligned} \tag{5.20}$$

where we use a Gaussian envelope function $f(t-t_0) = e^{-a^2(t-t_0)^2}$ with a full width at half maximum (FWHM) $\tau = 2\sqrt{\log 2}/a$. The field energy $\hbar\omega$, center of the pulse t_0, field amplitude E (in terms of intensity $I = c\epsilon_0 E^2/2$, where c is the speed of light and ϵ_0 is the electric constant), and phase ϕ are the parameters of the control field. The time integral of the electric field associated with the pulse is equal to zero [118].

It will prove convenient to rewrite the Hamiltonian, Eq. (5.17), as

$$\hat{H} = \hat{T}_{\rm N}\mathbf{1} + \mathbf{V}^{\rm eff}(R,t), \tag{5.21}$$

where the effective potential matrix $\mathbf{V}^{\rm eff}$ is given by

$$\mathbf{V}^{\rm eff}(R,t) = \mathbf{V}(R) + \mathbf{V}_F(R,t). \tag{5.22}$$

This representation is useful since we are considering an external field oscillating with a time-scale comparable to that of the nuclear motion. The diagonal matrix elements of Eq. (5.22) are defined as

$$V_i^{\rm eff}(R,t) = V_{ii}(R) + \mu_{ii}(R)F(t), \tag{5.23}$$

and the off-diagonal elements as

$$V_{12}^{\text{eff}}(R,t) = V_{12}(R) + \mu_{12}(R)F(t), \tag{5.24}$$

that is, they are obtained by a time- and coordinate-dependent energy shift of the original potential energy curves by the control field through the intrinsic dipole moments. We may therefore regard the nuclear wavepacket dynamics as that taking place on dynamically fluctuating (on the same time scale as the wavepacket dynamics) effective potentials $V_i^{\text{eff}}(R,t)$ and $V_{12}^{\text{eff}}(R,t)$ rather than a dynamics on the static potential energy curves $V_{ij}(R)$. This implies that not just the potential energy curves, but also the crossings between potential energy curves change in position as a function of time (or the crossing may even disappear for certain times). In other words the static crossing, located at R_{cross}, is replaced by a *dynamical* crossing, and at the dynamically crossing position $R_{\text{X}}(t)$ (driven by the control field), a propagating wavepacket on one state may split and be transferred to the other state, mediated by $V_{12}^{\text{eff}}(R,t)$ there.

5.3.2.1 *Fluctuating potential curves in the LiF molecule*

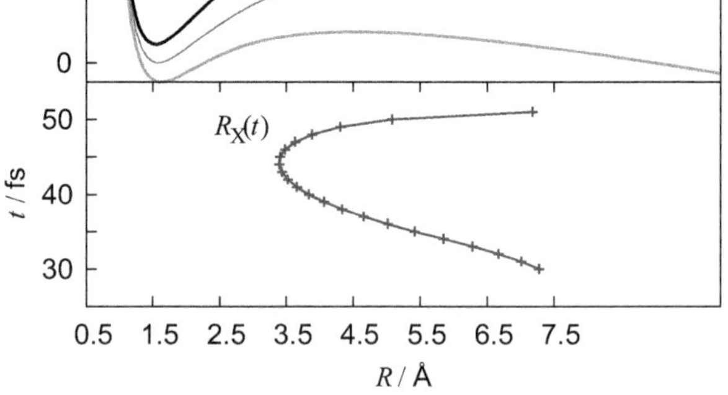

Fig. 5.13 Dynamically fluctuating potential curves of the two low lying states of LiF under control laser. Upper panel: the extreme positions reached by the effective potential energy curves $V_1^{\text{eff}}(R,t)$ and $V_2^{\text{eff}}(R,t)$. Lower panel: time evolution of the dynamical crossing point $R_{\text{X}}(t)$ during the first half of the control pulse. (Reprinted with permission from S. Scheit *et al.*, J. Phys. Chem. A 116, 2644 (2012)).

The two lowest electronic states of the LiF molecule will serve as a concrete example (Fig. 5.13). In the diabatic representation, state 1, which supports a bound state, is of ionic character, while state 2 is of neutral and globally repulsive character, leading to dissociation into the ground state of Li and F atoms. The crossing between the diabatic potential energy curves is located at $R_{\mathrm{cross}} = 7.18$ Å. The intrinsic dipole moments of electronic state 2, $\mu_{22}(R)$ is very small and approaches zero asymptotically as $R \to \infty$. $\mu_{11}(R)$ is much larger, always positive, and moreover increases almost linearly with R. $\mu_{12}(R)$ is zero by definition (see Sec. 5.2.1.1). Therefore, when considering the effective time-dependent potential curves defined by Eq. (5.23), the control pulse affects $V_1^{\mathrm{eff}}(R,t)$ very strongly while it affects $V_2^{\mathrm{eff}}(R,t)$ only minimally. Since the shift of $V_1^{\mathrm{eff}}(R,t)$ increases with R and $V_2^{\mathrm{eff}}(R,t)$ remain almost constant, the crossing position $R_{\mathrm{X}}(t)$ moves in a wide range according to its amplitude at each instant.

The control is exercised by means of a few-cycle control pulse field with a cycle duration comparable to the time-scale of the nuclear dynamics and a short pulse width such that the control can be switched on/off at the right moment with respect to the position of the evolving wavepacket. First we survey the effects the moving crossing has on the nuclear wavepacket dynamics by looking at the results of a simplified single-cycle control pulse of the form

$$F(t, t_0) = -Ef(t - t_0)\sin(\omega(t - t_0) + \phi), \tag{5.25}$$

with $\phi = 0$. The time integral of the electric field associated with the pulse is approximately zero for these particular phases. The pulse intensity chosen is $I = 1.69 \times 10^{13}$ W cm^{-2}, pulse energy is $\hbar\omega = 0.1$ eV, and the pulse width is FWHM $\tau = 20$ fs.

The field shape consists of two symmetric half-cycles with zero field amplitude at the pulse center. The upper panel of Fig. 5.13 shows the effective potential curves $V_i^{\mathrm{eff}}(R,t)$ obtained under the control field ($t_0 = 51$ fs). The plotted curves at $t = 44$ fs represent the maximum shifts of the PECs in the first half-cycle of the control pulse, and those at $t = 58$ fs represent the maximum shifts in the later half-cycle of the control pulse. In the first half-cycle of the pulse, the control field, through the positive transition amplitude function $\mu_{11}(R)$, acts to shift the $V_{11}(R)$ potential curve upwards, and thus the dynamical crossing position $R_{\mathrm{X}}(t)$ is first shifted to the left. It reaches the leftmost position R_{left} at the height of the control field, and then is shifted right and restored to R_{cross} by $t = t_0$. The lower panel of Fig. 5.13 shows the resulting time-dependence of $R_{\mathrm{X}}(t)$. The fact that the dynamical crossing is slower when going from R_{cross} to R_{left} (taking 15 fs) than when returning from R_{left} to R_{cross} (taking 5 fs) has important consequences on the controlled wavepacket dynamics. In the later half-cycle of the control pulse, the field acts to lower $V_{11}(R)$, and thus $R_X(R,t)$ is shifted to the right. For a certain time range, $V_1^{\text{eff}}(R,t)$ is shifted so much that it is below $V_2^{\text{eff}}(R,t)$ for all R and the crossing disappears. After disappearing for 15 fs, the crossing is reestablished and is restored at R_{cross} when the control field is turned off.

5.3.2.2 *Population dynamics in the controlled LiF system*

Figure 5.14(a) shows the time dependence of the diabatic state 1 population $P_1(t)$ for various delay times t_0 of the control pulse between 30 and 120 fs. In each panel, the black curve shows $P_1(t)$ computed without application of the control pulse. The curve shows a drop in population between -20 to 20 fs; this is the pump pulse excitation. The diabatic representation and the significant $V_{12}(R)$ around the initial wavefunction position results in the oscillatory features on top of the population drop seen in this time range (these oscillations are not seen in the adiabatic representation). Then the pump-excited state decrease by 13% (from $P_2(t) = 0.31$ to 0.27) around $t = 110$ fs; this is due to the static crossing at R_{cross}. Without the control pulse, this is the only population transfer seen.

The curves in the upper panels of Fig. 5.14(a) show $P_1(t)$ under control pulses with various t_0. The pump pulse excitation ($t < 20$ fs) is common to all the cases, and the control pulse affects the dynamics after the pump. We see the final population ($P_1(t)$ at $t = 250$ fs) increase from $t_0 = 30$ to 60 fs, reach the maximum around 70 fs, then decrease as t_0 is made larger, and become even less than the final population found for the case without control by $t_0 > 100$ fs. For $t_0 > 120$ fs, the effect of the control pulse becomes smaller, and is completely without effect for $t_0 > 140$ fs (not shown in Figure). The initial rise in $P_1(t)$ increases as t_0 is increased from 30 to 65 fs, and then decreases as t_0 is increased from 65 to 120 fs. For $t_0 < 80$ fs, we see an immediate drop in $P_1(t)$ following the initial rise. Then after some time ($t \geq 100$ fs), we see the population increase at R_{cross} that is also present without the control pulse. For $80 \leq t_0 < 100$ fs, we instead see a slow but lasting decrease some time after the initial rise.

To see the geometrical relation between the dynamical crossing point and the wavepacket propagation, we show in Fig. 5.14(b) the time-dependence of $R_X(t)$ during the first half-cycle of the control pulse, for $t_0 = 30, 51, 65,$

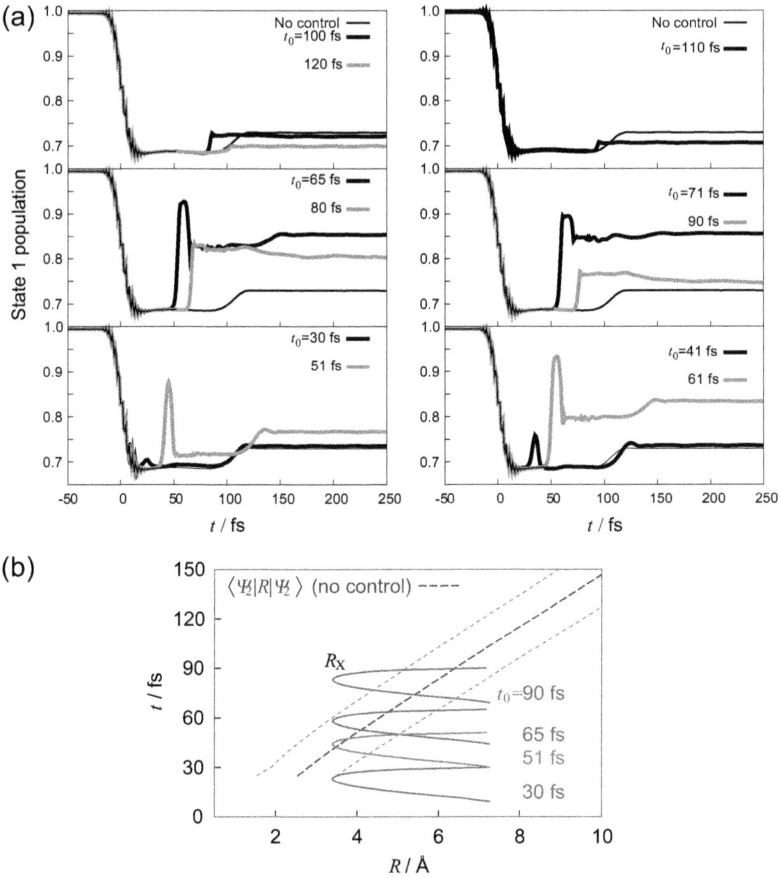

Fig. 5.14 (a) Time-evolution of the state 1 population $P_1(t)$ under the action of the pump and control pulses. The population in absence of control is shown in black for reference. (b) $R_X(t)$ for several t_0 seen in panel (a) and position of $\Psi_2(R,t)$ in the absence of control. Thick dashed curve indicates the center of wavepacket; thin dotted curves indicate the width of the wavepacket. $R_X(t)$ is shown only for the first half-cycle of the control pulse. (Reprinted with permission from S. Scheit *et al.*, J. Phys. Chem. A 116, 2644 (2012)).

and 90 fs, together with an indication of the position of $\Psi_2(R,t)$ computed for the case without application of a control pulse. The thick dashed curve is the expectation value $\langle\Psi_2(R,t)|R|\Psi_2(R,t)\rangle$. The thin dotted curves indicate the range of $|\Psi_2(R,t)| > 0.006$, which is 1/10 of the maximum height of $\Psi_1^{\text{new}}(R,t)$ created at R_{cross} (without the control pulse). The wavepacket is roughly Gaussian-shaped and the indicated range is meant to suggest where the dynamical crossing may start to interact with the wavepacket.

When $R_X(t)$ first being shifted towards the left by the control pulse meets the rightmost tail of $\Psi_2(R,t)$, the transfer of the wavepacket from electronic state 2 to state 1 starts, resulting in a new component $\Psi_1^{\rm new}(R,t)$. This is the initial rise in $P_1(t)$. For $t_0 = 30$ fs, $\Psi_2(R,t)$ is located mostly towards the left of $R_X(t)$ when $R_X(t)$ reaches $R_{\rm left}$, and there is little population transfer. As t_0 is increased from 30 fs, the amount of the wavepacket contacting $R_X(t)$ increases until $t_0 = 65$ where the whole wavepacket is involved in the initial transfer, and the initial rise in $P_1(t)$ increases through this range of t_0. For $t_0 \geq 65$ fs, $\Psi_2(R,t)$ and the dynamical crossing interact where $R_X(t)$ is moving faster and faster, and $V_{12}(R)$ smaller and smaller, causing the initial rise to diminish.

The created $\Psi_1^{\rm new}(R,t)$ initially inherits momentum from $\Psi_2(R,t)$ but slows down because of the steeper slope of $V_1^{\rm eff}(R,t)$ as compared to $V_2^{\rm eff}(R,t)$. The difference in position of $\Psi_1^{\rm new}(R,t)$ and $\Psi_2(R,t)$ widens as time passes. The slowing down of $\Psi_1^{\rm new}(R,t)$ is larger when the control field amplitude is larger (and positive), because of the larger deformation of the potential energy curve.

After the initial transfer, $R_X(t)$ keeps going to the left and reaches $R_{\rm left}$ and then starts to come back towards the right. When $R_X(t)$ now moving towards the right overtakes $\Psi_1^{\rm new}(R,t)$, a transfer from state 1 to state 2 is caused. This is the drop in $P_1(t)$ following the initial rise. We call this process the back-transfer. Since $\Psi_2(R,t)$ on the dissociating curve moves faster than $\Psi_1^{\rm new}(R,t)$ on the bound curve, the two wavepacket components do not mix during the back-transfer. Nevertheless, when the interaction between the states continue for a sufficiently long time by $R_X(t)$ taking some time to overtake $\Psi_1^{\rm new}(R,t)$, interference between the wavepacket components on the two states is observed as small oscillations in $P_1(t)$ following the back-transfer. The ratio of population transferred in the back-transfer to that in the initial transfer becomes less as t_0 is increased. As t_0 is made larger, $R_X(t)$ meets $\Psi_1^{\rm new}(R,t)$ at larger speed and smaller $V_{12}(R)$. For $t_0 = 30$ fs, the back-transfer cancels out the initial transfer, but by $t_0 = 80$ fs, the back-transfer is hardly seen.

During the later half-cycle of the control pulse, either $R_X(t) > R_{\rm cross}$ or the crossing has vanished and the wavepackets propagate undisturbed towards larger R. For $t_0 < 70$ fs, by the time $\Psi_2(R,t)$ reaches $R_{\rm cross}$, the crossing position has already been restored there and $\Psi_2(R,t)$ is partially transferred to state 1 with the same efficiency as in the case without the control pulse. This transfer starts at a time in between 100 and 130 fs depending on the value of t_0, gradually shifting towards later times as t_0

is increased. The delayed onset is due to $\Psi_2(R,t)$ slowing down during the population transfer between states. This deceleration is accentuated the longer the transfer lasts.

The transfer at R_{cross} after the original crossing is restored is no longer very effective at $t_0 = 70$ fs, and for $t_0 > 70$ fs is substituted by a slow but continuous population transfer in the opposite direction, from state 1 to state 2. For $t_0 = 71$ fs, there is less time between the end of the initial-transfer–back-transfer process and the wavepackets reaching R_{cross}, and both $\Psi_2(R,t)$ and $\Psi_1^{\text{new}}(R,t)$ go through the restored crossing at the same time. Transfer takes place in both directions and we do not see a net population change. For $71 < t_0 \leq 90$ fs, $\Psi_1^{\text{new}}(R,t)$ remains partly localized around R_{cross} for a long time while $\Psi_2(R,t)$ goes through quickly to the right, leading to the slow population transfer from state 1 to state 2.

For $90 < t_0 \leq 120$ fs, $\Psi_2(R,t)$ reaches R_{cross} just before the control field is switched on. The transfer at R_{cross} thus lasts for a much shorter time than in the control-free case, and the generated $\Psi_1^{\text{new}}(R,t)$ is very small. Furthermore, when $R_{\text{X}}(t)$ is returning from R_{left} to R_{cross} and overtakes the wavepackets, $R_{\text{X}}(t)$ is moving too quickly for any relevant population transfer to take place.

For $t_0 > 120$ fs, $\Psi_2(R,t)$ will have already reached the original crossing before the control pulse is switched on so that the effect of the control pulse is small, and completely without effect for $t_0 > 140$ fs.

5.3.3 *Shift of conical intersection and replacement by avoided crossing*

Nonadiabatic electronic transitions are of fundamental importance in chemistry. In particular, because a conical intersection (conical intersection) between two electronic states provides a very fast and efficient pathway for radiationless relaxation [117], there has been much interest in controlling transitions through a conical intersection. Indeed, several methods have already been proposed to control the dynamical processes associated with a conical intersection. One of these concerns the modification of electronic states involved in the conical intersection by environmental effects of polar solvents on the PES (potential energy hypersurface) through orientational fluctuations [6, 67, 68]. Another strategy is to apply a static electric field to shift the energy of a state of ionic character as in the Stark effect [384, 482] (see Ref. [403, 404] for the non-resonant dynamical Stark effect). More dynamical methods, which aim to suppress the transition either by preparing wavepackets that do not reach the conical intersection [281] or that destructively interfere there [1], have also been proposed. Furthermore, de Vivie-Riedle et al. have developed an optimal control theory for the dynamics associated with a conical intersection [143, 144, 461, 462]. Recently, Lim et al. [241] have used conformational changes of an excited state PES with chemical substitution to modify the pathway of the relevant nuclear (vibrational) wavepacket with respect to the location of the conical intersection manifold.

In this section we show how a conical intersection can be transformed into an avoided crossing at the geometry of the conical intersection by externally breaking its symmetry. Although symmetry breaking can be caused not only by optical control but also chemically by nearby substituent groups and/or by solvent effects [108], a significant advantage of optical control is that the laser used has a number of parameters such as wavelength, intensity, pulse shape, polarization, phase, and so on, that can be externally controlled. Moreover, optical control can be applied in conjunction with the above methods based on chemical modification. The mechanism we propose here may hence have potential for external control of *gating* of the branching pathways for nuclear wavepackets. This study suggests that by varying the laser parameters, one can control the rate of passage of wavepackets through a conical intersection.

The description of our control scheme for transformation of a symmetry-allowed conical intersection to an avoided crossing in the region of the conical intersection begins with a two-state model. The Hamiltonian for n nuclei in a molecular system with a nonadiabatic interaction between two electronic states can be written in the diabatic representation as

$$\hat{H} = \hat{T}_{\boldsymbol{R}}\mathbf{1} + \mathbf{V}(\boldsymbol{R}), \tag{5.26}$$

where $\hat{T}_{\boldsymbol{R}}$ is the kinetic energy operator for the nuclear coordinates $\boldsymbol{R}$, $\mathbf{1}$ the unit matrix, and the diabatic potential energy matrix

$$\mathbf{V}(\boldsymbol{R}) = \begin{pmatrix} V_1(\boldsymbol{R}) & V_{12}(\boldsymbol{R}) \\ V_{12}(\boldsymbol{R}) & V_2(\boldsymbol{R}) \end{pmatrix}, \tag{5.27}$$

contains the potential energy surfaces $V_1(\boldsymbol{R})$ and $V_2(\boldsymbol{R})$ for the two diabatic electronic states 1 and 2, and the diabatic coupling element $V_{12}(\boldsymbol{R})$ between them. For a conical intersection, $V_1(\boldsymbol{R}) = V_2(\boldsymbol{R})$ and $V_{12}(\boldsymbol{R}) = 0$, thereby forming a $(3n-8)$-dimensional intersection space. Applying a perturbation $\mathbf{V}_E(\boldsymbol{R}, t)$, which can also be represented as a two dimensional matrix and may be time-dependent, one can modify the property and geometry of the conical intersection manifold. $\mathbf{V}_E(\boldsymbol{R}, t)$ might be due to environmental effects of solvents or a functional group adjacent to the location of the conical intersection which would change the force field at the conical intersection. The perturbation may also be more conveniently introduced externally in the form of a linearly polarized pulsed optical field, which is the case treated here.

Consider the dynamics of a molecule in which two coupled electronic states belong to the same irreducible representation in an initial molecular geometry (an asymmetric triangular geometry for instance), thus resulting in level repulsion. As the molecule changes its geometry and acquires a higher symmetry (say, an isosceles triangle), the two electronic states can come to belong to different irreducible representations, resulting in degeneracy. It is around such a geometry that we can prevent the raising of the symmetry by applying an external optical field and inducing an additional coupling between the two electronic states of different irreducible representations (i.e. breaking local symmetry). The two states will then continue to undergo level repulsion as in an ordinary avoided crossing. To be more concrete, $\mathbf{V}_E(\boldsymbol{R}, t)$ might be

$$\mathbf{V}_E(\boldsymbol{R}, t) = -\begin{pmatrix} \mu_{11}(\boldsymbol{R}) & \mu_{12}(\boldsymbol{R}) \\ \mu_{12}(\boldsymbol{R}) & \mu_{22}(\boldsymbol{R}) \end{pmatrix} E(t), \tag{5.28}$$

where $E(t)$ is the electric field in a particular direction, $\mu_{11}(\boldsymbol{R})$ and $\mu_{22}(\boldsymbol{R})$ the diabatic dipole moments along this direction, and $\mu_{12}(\boldsymbol{R})$ the transition dipole moment in the same direction. We may think of $\mathbf{V}_E(\boldsymbol{R}, t)$ as having the effect of coupling of the two states rather than shifting the individual potential surfaces given by $\mathbf{V}(\boldsymbol{R})$. At the geometry of the conical intersection, $\boldsymbol{R}_X$, the diabatic coupling potential $V_{12}(\boldsymbol{R}_X)$ is by definition zero. On the other hand, one may be able to choose a direction of the applied laser field so that $\mu_{12}(\boldsymbol{R}_X)$ is nonzero, and this is generally possible by a simple symmetry consideration. Diagonalization of the Hamiltonian of Eq. (5.26) with the potential term replaced by $\mathbf{V}(\boldsymbol{R}) + \mathbf{V}_E(\boldsymbol{R}, t)$ results in the instantaneous "adiabatic" potential energy matrix, $\tilde{\mathbf{V}}(\boldsymbol{R}, t)$, where degeneracy at the geometry of the conical intersection is lifted. Specifically, the potential energy surfaces at the conical intersection geometry will become separated by a gap of $\Delta\tilde{V}(\boldsymbol{R}_X) = 2|\mu_{12}(\boldsymbol{R}_X)E(t)|$. The overall effect of the external field is, in fact, to shift the conical intersection away from its original location. Nevertheless, in the very short time of passage for a single wavepacket through the conical intersection region, the trajectory cannot adjust to the changing shape of the potential energy surfaces and the wavepacket continues to go through the original conical intersection region. Thus, an avoided crossing is, in effect, formed at the original geometry of the conical intersection and nonadiabatic population transfer is expected to be greatly suppressed there.

In Sec. 5.4.3, conversion of the conical intersection between the lowest two $^2A'$ states of the NO_2 will be taken up as a concrete example.

5.4 Conical intersection and wavepacket dynamics there

Conical intersections are ubiquitous in polyatomic molecules and are among the most important of nonadiabatic processes in these systems. They play a fundamental role in the excited-state dynamics of simple polyatomic molecules and are also believed to be responsible for the underlying photostability of DNA under ultraviolet radiation [47, 98, 447]. Femtosecond time-resolved photoelectron spectroscopy can be expected to be a versatile probe of wavepacket dynamics in and around conical intersections. In fact, internal conversion in polyatomic molecules was among the earliest suggested application of this technique [363], which was subsequently realized experimentally [54]. Recently, Bisgaard et al. [52] and Horio et al. [190] could also track the change in electronic character with vibrational motion in an excited state from measured molecular-frame and laboratory-frame photoelectron angular distributions, respectively. Advances in ultrashort pulse shaping technology may also well enable the observation of wavepacket dynamics through a conical intersection on the actual time scale of the nonadiabatic transition.

In this section we extend these earlier studies by employing robust values of the photoionization matrix elements in calculations of the time-resolved photoelectron spectra for ionization into the singlet and triplet ion states. We also report the photoelectron angular distributions for these time-resolved spectra.

5.4.1 *The NO_2 system*

The conical intersection between the first two $^2A'$ states of the NO_2 molecule is known to lead to an extremely complex absorption spectra [110–112], and has been the subject of numerous studies of nonadiabatic dynamics [125, 157, 183, 226, 252, 256, 348]. For C_{2v} geometries the two surfaces ($^2A_1$ and $^2B_2$) intersect at a bond angle that depends on the bond length and form a one-dimensional conical intersection seam [183]. The seam is located close to the bottom of the excited state and is readily accessible by a vibrational wavepacket launched onto the excited electronic surface from the Franck–Condon region of the ground state. In Ref. [13] we investigated the application of femtosecond pump–probe photoelectron spectroscopy to explore the ultrafast dynamics near this conical intersection in NO_2, and showed that the energy-resolved photoelectron spectra could identify the passage of wavepackets through the conical intersection. These studies, however, accounted only for ionization into the triplet cation. There [13] we also noted the need to incorporate the geometry- and energy-dependence of the photoionization matrix elements and to include ionization into the singlet ground state of the cation in order to achieve a more realistic and useful description of the time-resolved photoelectron spectra of these wavepackets.

The scheme for our studies of the pump–probe photoelectron spectra in NO_2 is illustrated in Fig. 5.15. The lower panel shows the two neutral surfaces $^2A_1$ and $^2B_2$ in the diabatic representation ($r_1 = r_2 = 1.22$ Å, C_{2v} geometry) and adiabatic representations of the two ion surfaces $^1A_1$ and $^3B_2$ used in these studies. The four possible ionization channels between these ion surfaces and the neutral states are indicated with arrows. The dominant configurations are shown along each potential curve. Dotted horizontal lines indicate the energy reached by the pump ($v_0 + \hbar\omega_{\mathrm{pu}} = 3.3$) and probe pulses ($v_0 + \hbar\omega_{\mathrm{pu}} + \hbar\omega_{\mathrm{pr}} = 16.8$ eV). The upper panel shows the classical photoelectron kinetic energy expected ($\bar{\varepsilon}_k = \hbar\omega_{\mathrm{pr}} - \Delta V$, where ΔV is the difference between the ion and neutral state) as a function of geometry ($r_1 = r_2 = 1.22$ Å) for each ionization channel.

NO_2 molecules in the ground vibrational level are first transiently aligned using short laser pulses [52, 338]. This initial state is then pumped to an excited electronic state by a femtosecond pulse. Because of the ultrafast time-scale of the associated dynamics, we employ pulses of a full-width at half-maximum (FWHM) of 8 fs in these studies. Wavepacket motion on the excited state, as well as on the ground electronic state that is coupled to the excited state by the conical intersection, is probed with a time-delayed femtosecond pulse that directly ionizes the molecule. The photoelectrons are then energy- and angle-resolved for signatures of the wavepacket motion. To simulate these photoelectron spectra we numerically time-evolve the wavepackets on the relevant electronic surfaces in all three dimensions (neglecting rotation), employing the diabatic representation to handle the

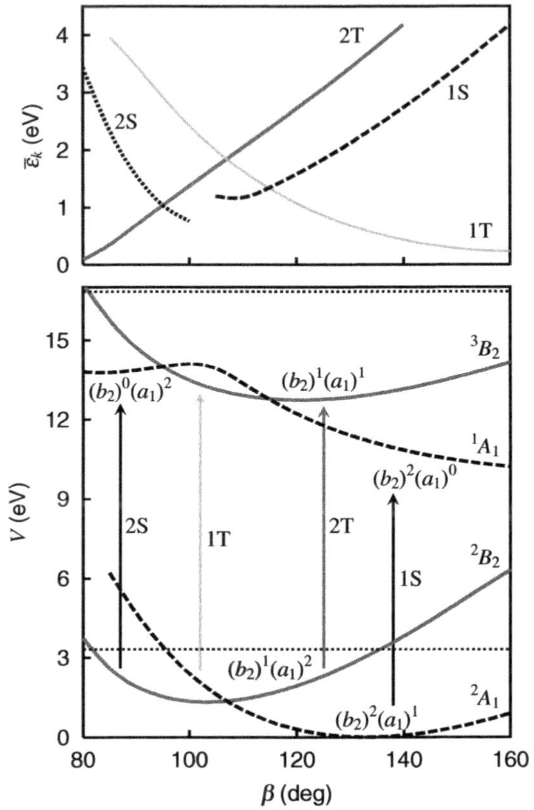

Fig. 5.15 Scheme to observe vibrational wavepacket dynamics through the $^2A_1/^2B_2$ conical intersection in NO_2 (lower panel). Upper panel shows expected photoelectron energies from the four photoionization channels. (Reprinted with permission from Y. Arasaki *et al.*, J. Chem. Phys. 132, 124307 (2010)).

nonadiabatic interaction at the conical intersection. The coupling of the electronic surfaces due to the pump and probe pulses is explicitly included in the Hamiltonian and geometry- and energy-dependent photoionization matrix elements are employed throughout.

The photoelectron spectrum of NO_2 has been well characterized [31, 124, 196]. The ground state of the cation is linear, and ionization from the bent neutral ground state results in a long vibrational progression. The first excited state of the ion, $^3A'$ (or $^3B_2$ in C_{2v} notation), is bent at its equilibrium geometry [185, 429]. We consider this state to be a better candidate for probing the dynamics at the conical intersection because of the relative proximity of its equilibrium geometry to the geometry of the conical intersection. Although there are several excited states of the ion lying close in energy to this state at the geometry of the conical intersection [185], most of these are dissociative [124] and we expect photoelectrons associated with this ion surface to be discernible from others. In our previous study [13] employing model (constant) photoionization amplitudes, the time-resolved photoelectron energy spectra to the triplet state were seen to track the vibrational wavepacket through the conical intersection. The singlet surface of the ion lies close in energy to the triplet state at the geometry of the conical intersection [185] and ionization to the singlet state can be expected to make it more difficult to unravel the photoelectron signal from the wavepackets.

5.4.1.1 *Potential energy surfaces*

For computation of pump–probe spectra, we need global potential energy surfaces for all the relevant states at roughly the same level of theory. The NO_2 $X\,^2A'$ and $A\,^2A'$ potential energy surfaces have been studied by many authors. Because the pump pulse excites a part of the wavefunction to energies above the dissociation threshold, a global description (including dissociation) as well as a good description of the conical intersection region is necessary for our purpose.

The most recent and high level computation satisfying both these requirements is the potential energy surfaces by Kurkal et al. [226], at the multireference configuration interaction (MRCI) level with a quadruple zeta basis set. 5000 points were calculated (up to bond length of 8.0 a.u.) and joined with cubic splines, and an ab initio quasi-diabatization method [373] was used to obtain diabatic surfaces. Wavepacket dynamics studies employing this surface has not been published however.

Other global surfaces include works by Santoro et al. [350, 351] and Salzgeber et al. [347] (these are modified forms of the surfaces by Hirsch et al. [184]), and double many-body expansion surfaces of Varandas [455] and Reignier et al. [333]. The last do not take nonadiabatic coupling into account.

Surfaces with good description of the conical intersection include work by Mahapatra et al. [252], computed at the MRCI level with a triple zeta basis set, but dissociation is not considered. There is also an effective Hamiltonian by Joyeux et al. [202] to describe the conical intersection region.

Compared to the neutral surfaces, there have been much fewer NO_2^+ global potential surface studies [185, 429].

Instead of trying to obtain each surface individually at a highly accurate level, we here resort to the much simpler scheme of complete active space self consistent field (CASSCF). Together with the phenomenological diabatization scheme of Sec. 5.2.1.2, the potential surfaces are computed at fairly low cost. Several accurate potential surfaces have been computed for the NO_2 system by other groups as stated above. By comparison we conclude that features relevant to the conical intersection dynamics are well represented already at the CASSCF level.

The $X\,^2A'$ and $A\,^2A'$ NO_2 and the $1\,^3A'$ NO_2^+ surfaces were computed using the algorithms of Werner and Knowles [214, 470] as incorporated in the Molpro quantum chemistry package [471, 472] as points on a dense grid over the bond lengths $1.7 \leq r_2 \leq r_1 \leq 4.2a_0$, and bond angle $70 \leq \beta \leq 180°$, with Dunning's correlation consistent polarized triple (cc-pVTZ) basis set [121] at full-valence CASSCF level of theory.

The two coupled neutral states were computed together in a state-averaged calculation. The cation triplet ground state also has interaction with higher states at small bond angles and at linear conformation, so that a single reference method cannot be used to compute the surface globally. Therefore CASSCF was used, but only the single surface was computed.

Diabatic surfaces were obtained for the neutral according to the scheme outlined in Sec. 5.2.1.2. For $\beta \leq 80°$, the $2\,^2A'$ surface approaches very close to the $3\,^2A'$ (both $^2A_1$ in C_{2v} notation) and the diabatization can no longer be described by a two-state scheme. Also near $\beta = 160°$ the $2\,^2A'$ and $3\,^2A'$ surfaces cross and again the simple phenomenological two-state diabatization scheme fails. We employ Gaussian functions to damp the computed coupling potential surface $V_{12}(\boldsymbol{R})$ in these regions to make the diabatic curves coincide with the adiabatic.

The computed points were interpolated by a local polynomial interpolation scheme [5] for generation of the Jacobi coordinate grid used in the wavepacket propagation.

The CASSCF surfaces do not go to the correct dissociation limits. Long bond distance regions of the potential were corrected with the switching function form previously employed by Salzgeber et al. [347],

$$\begin{aligned} V_i(\boldsymbol{R}) = (1 - Q(a, b(r_1 - r_{\text{long}})))V_i^0(\boldsymbol{R}) \\ + Q(a, b(r_1 - r_{\text{long}}))(V_{\text{diss}} + V_{\text{NO}}(r_2)), \end{aligned} \qquad (5.29)$$

where $V_i^0(\boldsymbol{R})$ is the uncorrected CASSCF potential surface, V_{diss} is the (experimental) dissociation energy 3.23 eV [142, 191, 200], $V_{\text{NO}}(r)$ is a Morse potential form of the NO ($X^2\Pi$) ground state curve [235], $Q(a, b)$

the incomplete gamma function [326], and a, b, r_{long} arbitrary parameters. $r_{\text{long}} = 1.7$ Å. The parameters $a = 4.0$ a.u. and $b = 7.0$ a.u. were chosen to bring the dissociation curve of the ground state in rough agreement with the surface given by Kurkal et al. [226]. Both diabatic $V_1(\boldsymbol{R})$, $V_2(\boldsymbol{R})$, and adiabatic $V_X(\boldsymbol{R})$, $V_A(\boldsymbol{R})$ were adjusted this way and diabatic coupling $V_{12}(\boldsymbol{R})$ was reevaluated for consistency. The excited state also shares the same set of parameters as the ground state (otherwise it was difficult to consistently adjust the coupling $V_{12}(\boldsymbol{R})$) and does not reproduce the barrier to dissociation in the excited state reported by Kurkal et al. [226].

The $1\,^3A'$ cation surface was shifted vertically to match the experimental ionization energy [31] but was otherwise used without modification as it is only used for the short pulse duration of the probe and global accuracy is of less concern.

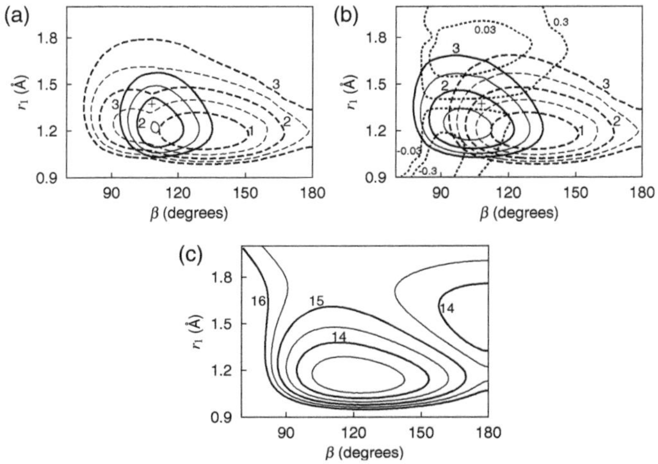

Fig. 5.16 Potential energy surfaces ($r_2 = 1.73$ Å). (a) NO_2 adiabatic $A\,^2A'$ (solid) and $X\,^2A'$ (dashed). (b) Diabatic V_2 (solid), V_1 (dashed), and V_{12} (dotted). (c) NO_2^+ $1\,^3A'$. Contours in 0.5 eV steps. Thick contours are labeled in eV. Energy measured from bottom of $X\,^2A'$. Conical intersection on the neutral surface is marked with a cross (+). (Reprinted with permission from Y. Arasaki et al., Chem. Phys. 338, 175 (2007)).

Figure 5.16 shows the computed potential surfaces as contour plots in the (β, r_1) plane at $r_2 = 1.37$ Å. Location (bond angle) and energy of the $X\,^2A_1/A\,^2B_2$ conical intersection seam is shown in Fig. 5.17. Table 5.1 lists structure and energy parameters of the computed potential energy surfaces. The neutral ground state has an equilibrium bond length of 1.20 Å and bond angle of 133° (1.196 Å and 134.3° in the multireference configuration

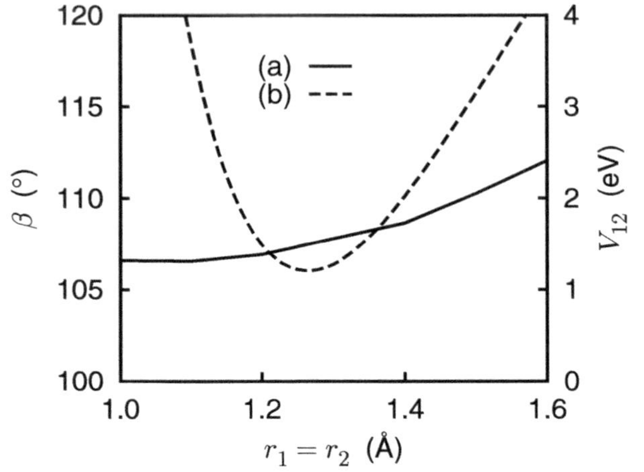

Fig. 5.17 (a) Bond angle (scales to the left) and (b) energy (scales to the right) of the conical intersection seam. (Reprinted with permission from Y. Arasaki et al., Chem. Phys. 338, 175 (2007)).

Table 5.1 Structure parameters of $X\,^2A'$ and $A\,^2A'$ NO_2 and $1\,^3A'$ NO_2^+ potential energy surfaces in this study.

	$^2A_1$	$^2B_2$	Cusp	$^3B_2$
r_e (Å)	1.20	1.27	1.26	1.22
β_e (°)	133	102	108	121
T_e (eV)	—	1.15	1.21	—
ω_1 (cm^{-1})	1283			
ω_2 (cm^{-1})	747			
ω_3 (cm^{-1})	1547			

r_e, β_e, T_e, ω_1, ω_2, ω_3 represents the equilibrium bond length, equilibrium bond angle, energy difference with the global minimum, symmetric stretch, bending, and asymmetric stretch, respectively. $^2A_1$, $^2B_2$, $^3B_2$ refer to equilibrium geometry of the $X\,^2A'$, $A\,^2A'$, and $1\,^3A'$ states. Cusp refers to the minimum along the $X\,^2A_1/A\,^2B_2$ conical intersection seam.

interaction calculation (MRCI) of Ref. [226]) while the first excited $^2A'$ state has a minimum at a smaller bond angle of 102° (101.4° in Ref. [226]). The conical intersection is located near the excited state minimum, with the cusp for example located at bond length 1.26 Å, bond angle 107.5°, and energy 1.21 eV. These can be compared with the values of 1.248 Å, 106.6°, and 1.28 eV, respectively, in the MRCI calculation of Ref. [226].

There have been far fewer studies of the global surfaces for the NO_2^+ $^1A'$ ground state ($^1\Sigma_g^+$ at the equilibrium geometry) and for the $^3A'$ triplet ground state ($^3B_2$ at C_{2v} geometry) [185, 429] than for the neutral system. The ion surfaces used in the present study were obtained with the same method and basis set as the neutral surfaces. However, each surface was computed separately without any state-averaging. The resulting singlet and triplet ion surfaces were shifted vertically up by 1.9 and 1.7 eV, respectively, to bring them into agreement with experiment [31]. The singlet ground state of the ion has its minimum at a linear geometry and a rather steep slope in the Franck–Condon region for ionization from the bottom of the neutral ground state (see Fig. 5.15). The first excited state ($^3B_2$ in C_{2v} geometry) is bent and has an equilibrium geometry much closer to the conical intersection region of interest. In the same region the two surfaces approach each other and are both between 13–14 eV from the bottom of the neutral state 1 (Fig. 5.15).

The four ionization channels for the coupled neutral states included in the present studies are shown in the lower panel of Fig. 5.15 by vertical arrows. In C_{2v} geometry the dominant configurations of the diabatic neutral states 1 and 2 are $\cdots(4b_2)^2(6a_1)^1$ and $\cdots(4b_2)^1(6a_1)^2$, respectively. Ionization from either the $4b_2$ orbital of state 1 or the $6a_1$ orbital of state 2 leads to the triplet ion. We hereafter refer to these channels as 1T (ionization of state 1 to the triplet ion) and 2T (ionization of state 2 to the triplet ion). The singlet ion state has a dominant configuration of $\cdots(4b_2)^2(6a_1)^0$ for bond angles larger than 100°, which changes to $\cdots(4b_2)^0(6a_1)^2$ for smaller bond angles. Therefore for bond angles larger than 100°, only the neutral state 1 can ionize to the singlet ion, and we refer to this channel as 1S. For bond angles less than 100°, only the neutral state 2 can ionize to the singlet ion, and we refer to this channel as 2S. Note that for ionization to the triplet ion, the 1T and 2T channels are open for all bond angles, in contrast to the case of ionization to the singlet ion.

The upper panel of Fig. 5.15 shows the photoelectron kinetic energy expected classically in the Condon approximation,

$$\begin{aligned}\bar{\varepsilon}_k(\boldsymbol{R}) &= (E_{\mathrm{pr}} - V_{\mathrm{ion},j}(\boldsymbol{R})) - (E_{\mathrm{pu}} - V_i(\boldsymbol{R})) \\ &= \hbar\omega_{\mathrm{pr}} - (V_{\mathrm{ion},j}(\boldsymbol{R}) - V_i(\boldsymbol{R})),\end{aligned} \tag{5.30}$$

where $E_{\mathrm{pu}} = v_0 + \hbar\omega_{\mathrm{pu}}$ and $E_{\mathrm{pr}} = v_0 + \hbar\omega_{\mathrm{pu}} + \hbar\omega_{\mathrm{pr}}$ are the energies reached by the pump and probe pulses, respectively, v_0 the zero point energy, and the indices $i = 1, 2$ and $j = \mathrm{S}, \mathrm{T}$ refer to the neutral state probed and the final ion state, respectively. The first passage of the wavepacket through the conical intersection is nearly one-dimensional along the bond angle coordinate but with a small amount of symmetric stretch, so a one dimensional plot as in Fig. 5.15 is useful. This figure shows the photoelectron kinetic energies only in the regions where each ionization channel is open. The open ionization channels change with bond angles and each channel produces different photoelectron energies. Changes in the ionization channels with bond angle can be used to monitor the evolution of the wavepacket on the potential surfaces.

5.4.1.2 *Wavepacket dynamics near the conical intersection*

For time propagation of vibrational wavefunctions in three dimensions, we employ the Jacobi coordinates (r, R, θ), respectively the bond distance for one NO moiety, the distance between the center of mass of this NO moiety and the other O, and the associated angle. Numerical time propagation of the vibrational wavefunction within this coordinate system was discussed in Sec. 3.2.2. $\theta = \pi$ places the second O on the same side as N with respect to the center of mass of NO. Although this choice does not reflect the symmetry of the NO_2 molecule, it is convenient as the molecule can dissociate into NO+O in the energy range of interest.

Potential surfaces obtained as described in the previous section was used to time propagate coupled vibrational wavefunctions.

Table 5.2 Grid parameters.

	min	max	N
r	0.9	2.6	64
R	0.8	4.0	144
θ	80	180	143

Grid spans for each coordinate from the coordinate under the heading min to the coordinate under the heading max. N the number of grid points. Lengths in Å and angles in degrees.

Table 5.2 tabulates the parameters of the grid used. The grid parameters were chosen to represent up to 4.0 eV in kinetic energy. Numbers in the table indicate the grid to have more than 1.3 million points, but the wavefunction has nonzero amplitude in only about half of the grid points, and for efficiency only the nonzero region was used in the time propagation.

To handle dissociation, an optical potential, $V_{\text{opt}}(\boldsymbol{R})$,

$$V_{\text{opt}}(\boldsymbol{R}) = V_{\text{opt}}(r_1) + V_{\text{opt}}(r_2), \tag{5.31}$$

$$V_{\text{opt}}(r_i) = \begin{cases} -i\alpha_{\text{opt}}(r_i - r_{\text{opt}}) & r_i > r_{\text{opt}} \\ 0 & r_i \le r_{\text{opt}}, \end{cases} \tag{5.32}$$

defined through bond lengths r_1 and r_2 rather than the Jacobi coordinates was added to the diagonal parts of the potential. Wave function beyond $r_{\text{opt}} = 2.1$ Å is considered dissociated and is eliminated by the optical potential.

The molecule is taken to lie in the xy-plane, and for C_{2v} geometry, the two O atoms lie parallel to the x-axis, and the N atom lies on the negative y axis. The pump pulse is polarized along the x-axis. The system is propagated from the lowest vibrational level of the ground electronic state with a time step Δt of 0.1 fs. The excited state wavepacket is generated with a pump pulse of intensity $I = 3.2 \times 10^{13}$ W cm^{-2}, $\hbar\omega_{\text{pu}} = 3.1$ eV and a Gaussian envelope with an FWHM of 8 fs. The center of the pump pulse is taken as time $t = 0$ fs.

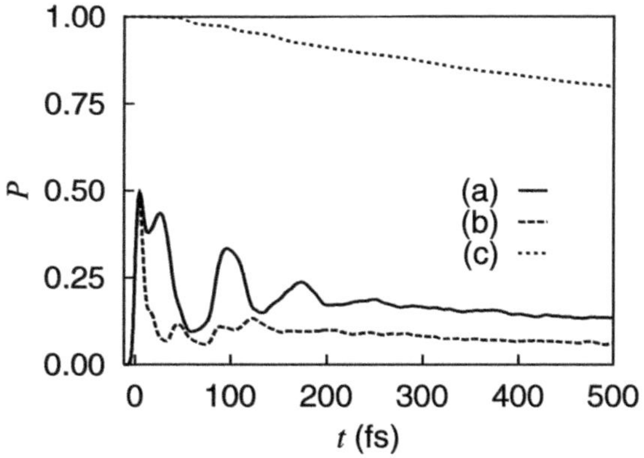

Fig. 5.18 Time dependence of (a) diabatic bound population P_2, (b) adiabatic bound population P_A, and (c) total bound population P_{bound}. (Reprinted with permission from Y. Arasaki et al., Chem. Phys. 338, 175 (2007)).

Time dependence of the excited state population by the pump pulse is shown in Fig. 5.18. Initial wavefunction is the vibrational ground state of the electronic ground state. The diabatic excited state population P_2 is shown as solid curve (a), and the adiabatic excited state population P_A is shown as dashed (b). Also shown is the bound population $P_{\text{bound}} = P_1 + P_2 = P_X + P_A$. Part of the wavepacket damped by the optical potential

Eq. (5.32) is considered dissociated, whose population is $P_{\text{dissoc}} = 1 - P_{\text{bound}}$ (by definition).

The center of the pump pulse is taken as time $t = 0$. The pump pulse excites almost half of the initial ground state wavefunction on to the excited electronic state. The excited wavepacket immediately starts to move towards the conical intersection/bottom of the excited state, even during the pumping, to first reach the conical intersection by $t = 5$ fs. By $t = 50$ fs, the first dissociating components reach the bond distance of 2.1 Å, beyond which we consider the molecule dissociated.

The diabatic excited state population P_2 shows rather regular oscillation for 4 times before becoming steadily decreasing after 300 fs. We note that the steady decrease in population beyond 300 fs is due to the wavefunction dissociating from the excited state (which is unphysical, see Sec. 5.4.1.1) and not from population transfer to the V_1 surface.

In the adiabatic representation, the excited state population P_A quickly damps from 50% to 10% by $t = 150$ fs.

The overall feature of initially fast population transfer and subsequent stability (with small fluctuations of seemingly short and random periods) for the *adiabatic* state seems to disagree with the results of previous studies [252, 350], where such behavior was found for the *diabatic* state. Our results indicate a much more adiabatic behavior [256]. This is understandable by noting the wavefunction energy is much higher than that of Mahapatra et al. [252]. Our wavefunction extends to larger bond lengths, away from the conical intersection, where it is necessarily more adiabatic. Santoro et al.'s surface [350] does consider longer bond lengths, but lets $V_{12} = 0$ there, resulting in a more diabatic dynamics.

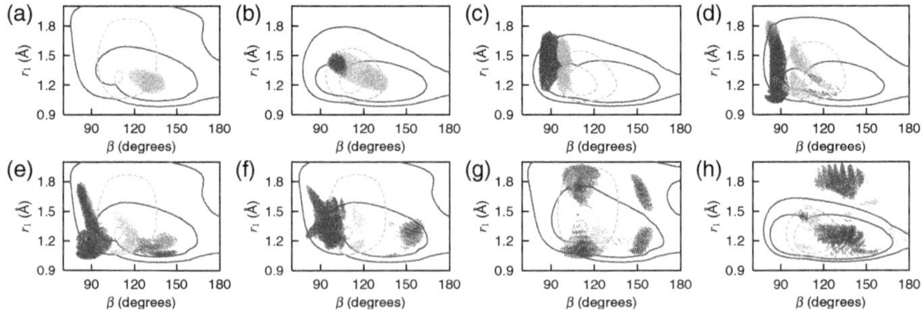

Fig. 5.19 Wavepacket snapshots in the adiabatic representation. (a) $t = 0$, (b) 6, (c) 16, (d) 20, (e) 26, (f) 32, (g) 42, and (h) 52 fs. (Reprinted with permission from Y. Arasaki et al., Chem. Phys. 338, 175 (2007)).

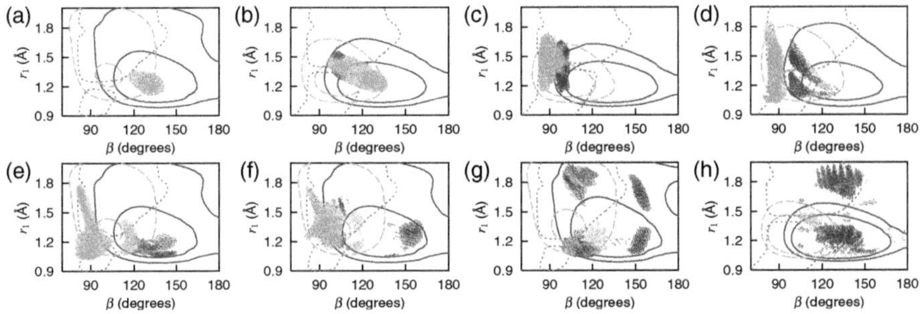

Fig. 5.20 Wavepacket snapshots in the diabatic representation. (a) $t = 0$, (b) 6, (c) 16, (d) 20, (e) 26, (f) 32, (g) 42, and (h) 52 fs. (Reprinted with permission from Y. Arasaki et al., Chem. Phys. 338, 175 (2007)).

Time evolution of the wavefunction is shown in Figs. 5.19 and 5.20, for the adiabatic and diabatic representation, respectively. Each panel shows the wavepacket and potential surfaces for a different time [t = 0, 6, 16, 20, 26, 32, 42, and 52 fs, in panels (a)–(h), respectively]. Contours of the potential energy surfaces in the (β, r_1) plane at a representative r_2 for the wavefunction at that time (outer contour 3.0 eV, inner contour energy at the conical intersection for that bond length) are shown in each panel [r_2 = 1.22, 1.37, 1.40, 1.32, 1.22, 1.22, 1.19, and 1.43 Å, in panels (a)–(h), respectively]. Significant presence of wavefunction amplitude $(\sin\theta|\chi(r, R, \theta)|^2)$ for each grid point used in the time propagation is indicated by a point plotted in the (β, r_1) plane (without regard to the value of the r_2 coordinate). For Fig. 5.19, dark shade indicate V_X and χ_X, and light shade indicate V_A and χ_A. For Fig. 5.20, dark shade indicate V_1 and χ_1, and light shade indicate V_2 and χ_2. Contours for $|V_{12}| = 0.03$ eV (dotted) are also shown in Fig. 5.20. Because of the mapping from Jacobi coordinates (r, R, θ) used in the computation to the bond length bond angle coordinates (r_1, r_2, β) used in the plots, plotting the wavefunction onto the (β, r_1) plane together for several different values of r_2 makes the "centers" of wavepackets appear more dense than the tails of the wavepacket, giving some impression of depth in the two-dimensional plots.

The pump pulse leaves about half the population still in the ground electronic state. This population mostly remains in the initial ground vibrational state, but some low vibrational states (up to 2 quanta for bending, 3 for symmetric stretch) are excited due to the Rabi oscillation. These low energy (< 0.7 eV) components were projected out of the wavefunction when plotting for clarity of presentation.

At $t = 0$ [Fig. 5.19(a)], the excited state wavepacket is formed directly above the initial wavefunction at the bottom of the ground electronic state. Only about 25% of the initial wavefunction has been excited by this time. The excited state wavepacket moves downhill to smaller bond angles and longer bond lengths as the wavefunction is still being formed by the pump pulse.

At $t = 6$ fs [Fig. 5.19(b)], the leading edge of the excited state wavepacket has reached the conical intersection. The center of the wavepacket moves through the conical intersection where the diabatic coupling V_{12} is zero. There is no transfer between the diabatic states there, or alternately, in the adiabatic representation the nonadiabatic transfer there is complete. Nonzero diabatic coupling away from the conical intersection causes parts of the wavepacket spread in the asymmetric stretch direction to transfer between diabatic states, or prevents interaction between adiabatic states. Thus in the adiabatic representation [Fig. 5.19(b)], the (nonadiabatic) transfer is seen to be from the center of the leading part of the wavepacket, while in the diabatic [Figs. 5.20(b) and (c)], the transfer (between diabatic states) occurs where r_1 is larger or smaller than r_2 of the center of wavepacket, resulting in a short bond length component and a long bond length component for χ_1. After the first passage through the conical intersection between $t = 5$ and 12 fs, roughly 80% of the excited state wavefunction remain in the diabatic V_2 state.

Wave function component that transferred to the V_1 state reach the turning point earlier than the V_2 component. By $t = 20$ fs [Fig. 5.20(d)], the V_1 component reflected at the turning point has approached the conical intersection for the second time. Transfer from V_1 to V_2 starting here is responsible for the second peak at $t = 26$ fs for P_2 in Fig. 5.18. By $t = 26$ fs [Fig. 5.20(e)], this second crossing of the conical intersection has finished.

The diabatic χ_2 component that did not transfer at the first crossing has spread out in the bond length direction by $t = 20$ fs [Fig. 5.20(d)]. At $t = 32$ fs [Fig. 5.20(f)], the χ_2 component has again reached the conical intersection angle. But this time most of the wavefunction is not passing through the conical intersection but around it in an asymmetric conformation (one of the bond lengths longer than the other), so that V_{12} is large and transfer between diabatic states large. This is responsible for the large dip in P_2 between $t = 26$–60 fs. In Fig. 5.20(g), the wavefunctions passing around the conical intersection is seen as two components, one for short r_1 (and long r_2) and the other for long r_1 (and short r_2). A part of this χ_2 to χ_1 transfer becomes the dissociating component.

By $t = 52$ fs [Fig. 5.20(h)], the fast moving wavefunction component on V_1 from the "second" passage through the conical intersection at $t = 20$–26 fs, and those from the "third" passage between $t = 30$–50 fs have collided. Interference pattern from the collision is seen between the bond angles 120 to 150°. Repeated splitting and rejoining of the wavefunction is partly responsible for its rapid increase in complexity and we can see here the beginning of chaos [136, 182].

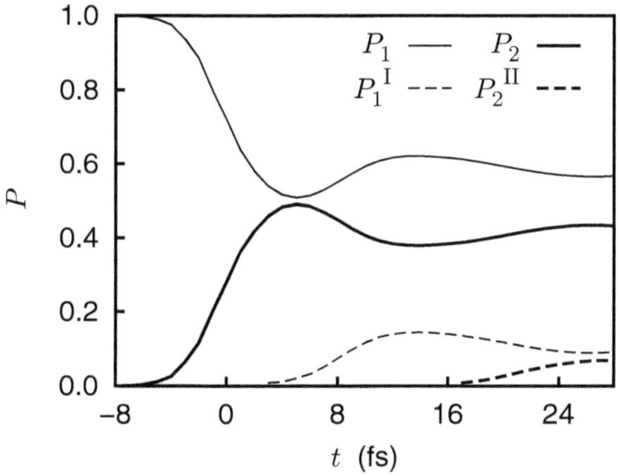

Fig. 5.21 Time evolution of (component) populations following the pump pulse. Populations of the two diabatic electronic states (P_1, P_2), and contributions from the first two passages through the conical intersection (P_1^{I}, P_2^{II}) are shown. (Reprinted with permission from Y. Arasaki *et al.*, J. Chem. Phys. 132, 124307 (2010)).

Focusing on the early part of the excited state dynamics, Fig. 5.21 shows the time evolution of the population of diabatic states 1 and 2 as well as contributions from the first two passages of the wavepacket through the conical intersection. The population of state 1, P_1, (thin solid curve), is seen to decrease to ~ 0.5 during the pump pulse with an FWHM of 8 fs centered at $t = 0$ fs. Conversely, population builds up in state 2, P_2 (thick solid curve). The wavepacket formed on state 2 by the pump pulse immediately begins to move towards the conical intersection, and the population on state 2 is subsequently affected by the conical intersection. It first reaches the conical intersection by $t = 4$ fs. Most of the population passing through the conical intersection remains on the same diabatic state, but some adiabatically transfers to state 1, as shown by P_1^{I} (thin dashed curve). This component of the population on state 1 is obtained by eliminating the contribution of low vibrational levels with energy < 0.7 eV using calculated vibrational eigenfunctions. Population of the low vibrational levels results from the intense pump pulse used here whereas high vibrational levels are populated through the conical intersection.

The first passage through the conical intersection is complete by around 12 fs. The wavepacket components on the two states then evolve separately with the component on state 1 returning for a second passage through the conical intersection around $t = 20$ fs, where some of it adiabatically transfers to state 2 (thick dashed curve, P_2^{II}). The newly formed component on state 2 is spatially well separated from the component on state 2 that remained diabatically on the state during the first crossing of conical intersection for the time span shown, so its population, P_2^{II}, can be determined. The increase in P_2^{II} in Fig. 5.21 near $t = 20$ fs indicates the second passage of this component of the wavepacket through the conical intersection.

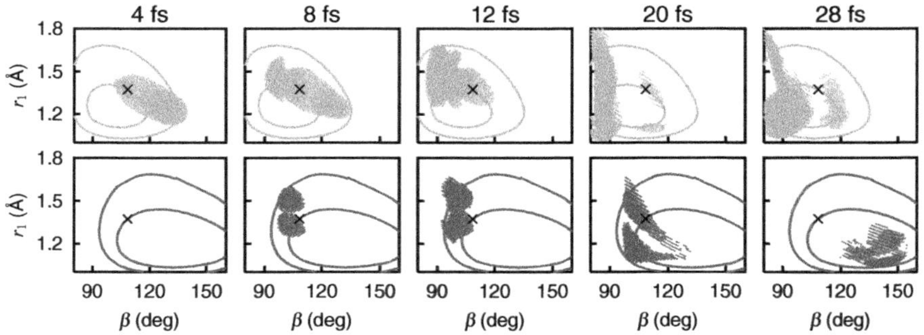

Fig. 5.22 Time evolution of the diabatic wavepacket near the conical intersection. (Reprinted with permission from Y. Arasaki *et al.*, J. Chem. Phys. 132, 124307 (2010)).

Figure 5.22 shows the time evolution of the wavepackets. Each column shows a projection of the amplitude of the wavefunction onto the (β, r_1) plane at times indicated at the top of the panels. The upper row shows the amplitudes of the wavepacket on state 2, and the lower row the amplitude for state 1. Potential energy contours are shown for 3.00 eV (outer contour) and 1.77 eV (inner contour). In each panel the conical intersection at $r_2 = 1.37$ Å is indicated with a cross ($\times$). The initial wavefunction left unexcited on state 1 by the pump pulse has been removed for clarity in plotting Fig. 5.22.

5.4.2 *Time-resolved photoelectron spectroscopy of the conical intersection dynamics in the NO_2 system*

Time-resolved photoelectron spectra of the excited state dynamics of the NO_2 molecule presented in Sec. 5.4.1.2 are computed and the results are given in the following sections, demonstrating the use of the technique in tracking wavepacket dynamics across a conical intersection. We assume that the molecules were initially transiently aligned beforehand (such as by using short laser pulses [52, 338]), and the polarization of the probe is taken parallel to that of the pump. The propagation of the vibrational wavefunction is done entirely in internal (Jacobi) coordinates, and the molecular orientation enters into the calculation only through the matrix elements of the pump and probe couplings. Photoelectron spectra of the pump excited wavepackets are obtained for various delay times, t_{pr}, for a probe pulse with $\omega_{\mathrm{pr}} = 13.5$ eV and an FWHM of 8 fs. Photoelectron spectra were obtained separately for ionization to the singlet and triplet ion states.

The Hamiltonian matrix for the ionization run includes coupling between either diabatic neutral state and all ion partial wavefunctions $\{\chi_{k_j l \lambda}\}$. The number of quadrature points for the photoelectron kinetic energy N_k was 50 for a maximum kinetic energy of 5.0 eV. Partial waves up to $l = 9$ were included in these calculations. An analysis of the resulting spectra shows that waves with $l \leq 5$ accounted for over 99 % of the ion population.

5.4.2.1 *Time-resolved photoelectron kinetic energy spectra*

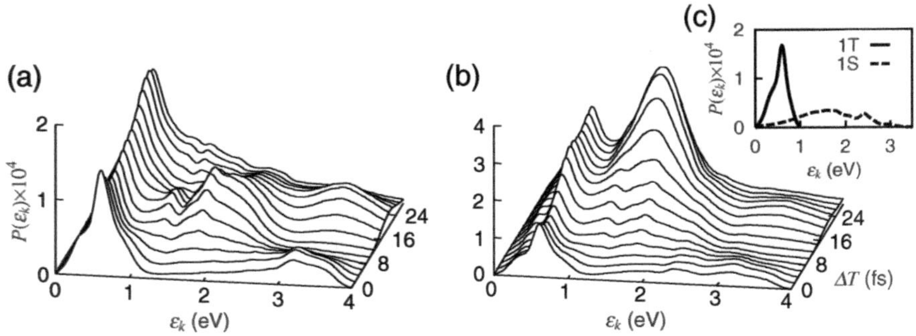

Fig. 5.23 Time evolution of photoelectron kinetic energy distribution in the NO_2 system. (a) Ionization to triplet state only. (b) Ionization to both triplet and singlet states. (c) Ionization from the initial wavefunction only (no pump pulse). (Reprinted with permission from Y. Arasaki *et al.*, J. Chem. Phys. 132, 124307 (2010)).

Figure 5.23 shows the time-resolved photoelectron kinetic energy spectra for a probe pulse polarized parallel to the pump polarization (x-axis) and for molecules that have been transiently aligned. The spectra for ionization to the triplet state only are shown in Fig. 5.23(a). The evolution of the photoelectron spectrum is plotted in steps of 2 fs.

Figure 5.23(a) shows the result for photoelectrons from the triplet state only and the excited state dynamics is seen most clearly in these signals. The wavepacket formed on the excited surface directly above the ground state global minimum by the pump pulse should result in photoelectrons with energy more than 3 eV by the probe (Fig. 5.15). As the wavepacket moves downhill towards the conical intersection region, the likely photoelectron energy becomes lower, down to 1.5 eV at the intersection. Then after going through the conical intersection, wavefunction components that transferred to the ground state, moving away from the conical intersection, should cause photoelectron energy to quickly lower for shorter bond distances (less than $\sim$ 1.5 Å), or result in approximately constant photoelectron energy for larger bond distances. Components remaining on the excited state beyond the conical intersection region (at smaller bond angles) can cause photoelectron energy to rise again, up to about 2.5 eV, but is not expected to go as high as it was for larger bond angles.

These estimates should also depend on the bond length coordinates (not shown in Fig. 5.15); the conical intersection is located lowest at $r_2 = 1.26$ Å, and higher in energy for r_2 less than or greater than that (Fig. 5.17). Also to be considered is that some of the energy transferred by the probe pulse may be converted to vibrational energy in the ion, thus may result in lower energy photoelectrons than the simple estimate.

The time resolved photoelectron kinetic energy distributions shown in Fig. 5.23(a) overall show the expected behavior outlined above. High energy (> 3 eV) photoelectrons are found at $t_{\mathrm{pr}} = 0$ fs that indicate that a wavepacket is formed in the excited electronic state, then wavepacket motion to the conical intersection region can be tracked in time by the lowering photoelectron energy. At all times there is a high peak at 0.5 eV; this is from the initial wavefunction that was left not excited by the pump pulse. First passage through the conical intersection, 6–12 fs, appears as a peak in the photoelectron signal about 1.9 eV (slightly higher than the 1.5 eV mentioned above, indicating the wavepacket does not pass through the cusp). Immediately after passing through the conical intersection, the peak shifts to a slightly higher energy as the wavepacket remaining on the excited state surface moves away from the intersection. The wavefunction transferred to the ground electronic state may cause a peak lower in energy than that at the conical intersection region, but is not clearly seen in the present case where less than 1 eV part of the spectrum is dominated by signal from the large amount of initial ground state wavefunction remaining unexcited by the pump pulse. At later times the spectra features become much less pronounced due to the wavefunction spreading.

Photoelectron kinetic energy spectra for ionization into both the triplet and singlet states are shown in Fig. 5.23(b). Compared to Fig. 5.23(a), a broad and intense peak centered at 1.7 eV now appears after about 14 fs and obscures the photoelectron spectra from the triplet ion, making it more difficult to infer the wavepacket dynamics from these signals. For comparison Fig. 5.23(c) shows the photoelectron kinetic energy spectrum for ionization of the initial vibrational eigenstate to the triplet (1T) and singlet (1S) ions in the absence of the pump field.

5.4.2.2 *Interpretation of spectra: Wavepacket components and ionization channels*

The excited wavepacket formed by the pump pulse evolves into several components as seen in Figs. 5.21 and 5.22: (1) part of the initial wavefunction that was not electronically excited by the pump pulse, (2) the bulk of the excited wavepacket formed by the pump pulse, (3) two symmetrically equivalent components formed on electronic state 1 at the time of the first passage through the conical intersection (P_1^{I} in Fig. 5.21, snapshot shown in the lower panel at 8 fs in Fig. 5.22), and (4) another set of symmetrically equivalent components formed on electronic state 2 at the time of the second passage through the conical intersection (P_2^{II} in Fig. 5.21, or the additional components shown in the upper panel at 20 fs in Fig. 5.22). These wavepacket components are probed by the four ionization channels depicted in Fig. 5.15.

Some insight into these photoelectron energy spectra can be obtained with the help of the classical picture of ionization from the centers of the components of the wavepacket. Figure 5.24 shows a top view of the same time-resolved photoelectron kinetic energy spectrum as in Fig. 5.23(b), with the signal strength indicated by color. This chart exposes the qualitative changes in the dynamics more vividly. Curves in this chart show the photoelectron kinetic energy expected at the centers of the various wavepacket components, $\bar{\varepsilon}_k(\boldsymbol{R}_{\mathrm{c}})$, where $\boldsymbol{R}_{\mathrm{c}}$ is the expectation value of the coordinate for each wavepacket component. The curves are labeled by ionization channel (1S, 1T, 2S, or 2T) and a superscript indicating the wavepacket component

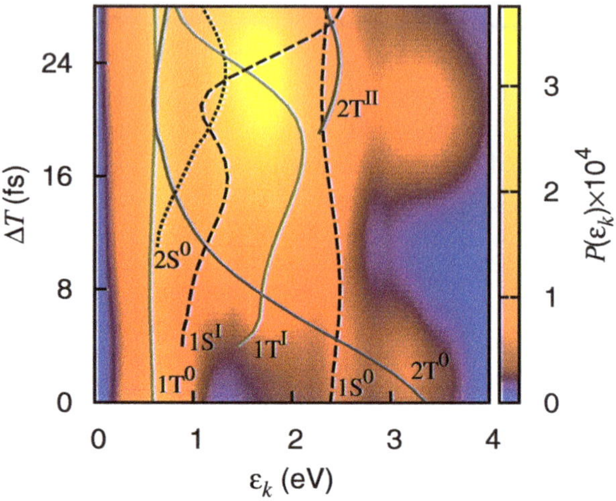

Fig. 5.24 Expected ionization energies for components given on top of the time-resolved photoelectron kinetic energy spectrum of Fig. 5.23(b). (Reprinted with permission from Y. Arasaki *et al.*, J. Chem. Phys. 132, 124307 (2010)).

from which ionization occurs (I, the component formed at the first passage through the conical intersection, II, the component formed at the second passage through the conical intersection, and 0, the bulk of the wavefunction on state 1 that was electronically unexcited by the pump pulse, or the component on state 2, formed by the pump pulse but excluding the portions formed at the conical intersection). Triplet ionization channels are shown as solid curves, and singlet ionization channels as dashed (ionization from state 1) or dotted (ionization from state 2) curves. The classical photoelectron energies can be read off a figure such as the upper panel of Fig. 5.15 using the coordinates of the center of a wavepacket component.

The $1S^0$ (dashed) curve in Fig. 5.24 represents ionization to the singlet state out of the component of the ground state wavepacket that is electronically unexcited ($\sim$ 50%) by the pump pulse and corresponds to the broad 1S distribution of Fig. 5.23(c). The $1T^0$ (solid) curve, representing ionization of the same component of the ground state wavepacket to the triplet ion, corresponds to the sharper 1T channel peak of Fig. 5.23(c). These features do not change with time after the pump pulse is over. The 1T peak in Figs. 5.23(a) and (b) is the dominant feature below 1 eV but lies well below features arising from ionization of excited wavepackets. In contrast, the broad 1S photoelectron distribution obscures features in the time-resolved spectrum above 1 eV, making it more difficult to infer excited-state wavepacket dynamics.

As it moves towards its first passage through the conical intersection, the wavepacket formed on state 2 by the pump pulse can be tracked by the 2T ionization channel ($2T^0$ blue curve in Fig. 5.24), while after passage through the conical intersection the bulk of the wavepacket can be tracked by the 2S channel signal ($2S^0$ green dotted curve). The 2S channel becomes active only after the wavepacket has passed through the conical intersection and is enhanced as it approaches the turning point in the bending coordinate. The 2T channel signal is evident in Fig. 5.23(a) as a peak above 3 eV for delay times up to 8 fs and another near 1 eV between 6 and 16 fs. The pronounced peak below 2 eV in Fig. 5.23(b) after $t_{\rm pr} = 14$ fs is due to the 2S signal. Though the population probed by the 2S channel is smaller than that initially probed by the 2T channel due to some population loss near the conical intersection, the 2S signal is much stronger because of the large photoelectron matrix elements for singlet ionization.

$P_1^{\rm I}$ of Fig. 5.21 is the population on state 1 due to the first passage of the wavepacket through the conical intersection. This wavepacket component is formed symmetrically on either side of the conical intersection ($r_1 > r_2$ or $r_1 < r_2$; see Fig. 5.22 at 8 fs). Only one of these two components is used to obtain the expectation value of the wavepacket coordinates. Once formed, this wavepacket component can be probed by either the 1T channel ($1T^{\rm I}$ curve in Fig. 5.24), or the 1S channel ($1S^{\rm I}$ dashed curve). The 1S channel signal is too weak to be seen due to the background signal from the unexcited portion of the ground state wavepacket. The 1T signal from this wavepacket, seen as the peak near 2 eV in Fig. 5.23(a) between $t_{\rm pr} = 6$ and 20 fs, is, however, obscured by the large 2S signal when ionization to the singlet ion is included [Fig. 5.23(b)].

The curve $P_2^{\rm II}$ in Fig. 5.21 represents the population of the wavepacket formed on state 2 from the second passage through the conical intersection. While ionization of this wavepacket to the singlet ion is not allowed, ionization to the triplet ion is, and the corresponding classical photoelectron energy is shown by the curve $2T^{\rm II}$ in Fig. 5.24. Only one of the two symmetrically formed components is used in obtaining the average coordinate, as in the case for $1T^{\rm I}$. Though this component can be tracked by the 2T signal in principle, the signal, expected around $\varepsilon_k = 3.4$ eV, is too weak to be seen in Fig. 5.23. Instead, the peak seen around this energy at $\Delta T = 22$ fs, actually arises from ionization to the triplet ion (1T) of the wavepacket component near the turning point on state 1. The lower panel of Fig. 5.22

shows that around $t - 20$ fs the leading edge of the wavepacket on state 1 is at a longer bond length, while the center of the wavepacket is at shorter bond length and moving towards larger bond angles. The average value of the coordinate of the wavepacket for this delay time cannot fully describe its multidimensional character.

Although, in principle, the time-resolved photoelectron spectra of the wavepackets on the potential surfaces should make it possible to monitor their associated dynamics, broad overlapping peaks in the calculated photoelectron spectra make such an analysis here difficult for short pump and probe pulses.

5.4.2.3 *Time-resolved photoelectron angular distributions*

While overlapping signals from different ionization channels make it difficult to infer the wavepacket dynamics from the photoelectron kinetic energy spectra in Secs. 5.4.2.1 and 5.4.2.2, the angular distributions provide a clearer and more useful picture of the underlying ionization channels and wavepacket dynamics around the conical intersection. Figure 5.25 shows our calculated time- and energy-resolved photoelectron angular distributions from spatially aligned molecules [$P_k(\tilde{\theta}_k; t_{\rm pr})$ of Eq. (3.93)] for the same pump and probe pulse parameters as in the photoelectron kinetic energy spectra of Sec. 5.4.2.1. Each row shows, from left to right, the time evolution of the angular distributions for the photoelectron kinetic energy ε_k indicated on the left. The delay time $t_{\rm pr}$ for each column is shown on the bottom. The probe polarization lies along the vertical axis of each figure and the distributions are symmetric with respect to rotation around this axis as a result of averaging in the laboratory frame [Eq. (3.92)]. Each figure shows the angular distributions from the individual channels while the outer contour for each plot shows the angular distribution summed over all four ionization channels. The upper half of each figure shows the angular distributions associated with the 2T (dark solid curve) and 2S (dotted) ionization channels, and the lower half the distributions for the 1T (light gray) and 1S (dashed) channels. All distributions are symmetric with respect to the horizontal (molecular) axis, reflecting the overall symmetry of the molecule, *i.e.*, bond lengths $r_1 > r_2$ and $r_1 < r_2$ for the same pair of (r_1, r_2) occur with equal probability. Only the upper or lower half of the distributions are drawn for individual components. For clarity the plots are normalized separately for each photoelectron kinetic energy and hence they do not represent relative strengths of the signal at different energies.

Relative strengths of the total signal can be seen in Fig. 5.23. For a given energy the plots in Fig. 5.25 show the time evolution of the strength of the signal as well as its shape. For delay times with a weak signal, the plots are magnified by factors indicated at the top right. The contribution of each component is shown to scale with the total signal to illustrate the relative importance of each channel.

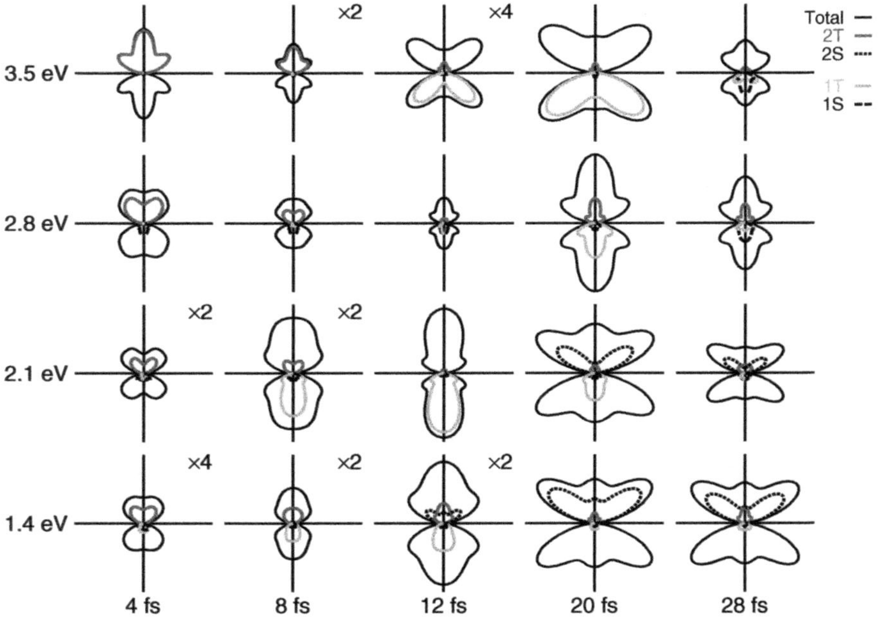

Fig. 5.25 Time evolution of the energy-resolved photoelectron angular distributions. (Reprinted with permission from Y. Arasaki *et al.*, J. Chem. Phys. 132, 124307 (2010)).

These time- and energy-resolved laboratory-frame angular distributions are seen to change with delay time and photoelectron kinetic energy, reflecting the dependence of the photoionization amplitudes on geometry and energy. The photoelectron angular distributions for each ionization channel could be used to track the transfer of population among electronic states as the vibrational wavepacket moves through the nonadiabatic region. The signal for $\varepsilon_k = 3.5$ eV illustrates this well. As discussed in Sec. 5.4.2.1, the wavepacket formed by the pump pulse on state 2 is at first probed through the 2T ionization channel (3.5 eV and 4 fs in Fig. 5.25). As this wavepacket goes through the conical intersection region at 8 fs, a part of the wavepacket is adiabatically transferred to state 1, leading to a signal in the 1T ionization channel (3.5 eV and 12 fs in Fig. 5.25). After passage through the conical intersection region, the photoelectron energy expected classically in the 2T channel is less than 1.5 eV (see upper panel of Fig. 5.15) and the signal at 3.5 eV is hence much weaker than the 1T signal at this time. This results in a significant change in the overall photoelectron angular distribution as the wavepacket passes through the conical intersection region. When this portion of the wavepacket on state 1 passes through the conical intersection ($\sim$ 20 fs) again, the 1S channel becomes dominant, leading to a significant change in the overall angular distribution (3.5 eV and 28 fs in Fig. 5.25). These changes in the overall photoelectron angular distributions should make it possible to track the evolution of the wavepacket as population is transferred from one region to another through nonadiabatic interaction.

At 1.4 eV the evolution of the aggregate photoelectron angular distributions in Fig. 5.25 reflects changes in the distributions within a given channel in addition to changes in active channels. Whereas the angular distributions for each ionization channel have similar shapes throughout at $\varepsilon_k = 3.5$ eV, at 1.4 eV the distributions for the 2T channel (between 4 and 12 fs) and 1T channel (between 12 and 28 fs) are seen to change shape (Fig. 5.25). These changes arise from the geometry dependence of the $C_{l\lambda}$ matrix elements of Eq. (3.59), with the relative contribution of (l, λ) partial waves changing on either side of the conical intersection. The evolution of the angular distributions in the 2T channel between 4 and 12 fs at 1.4 eV reflects the passage of the wavepacket through the conical intersection. Removal of the 2T contribution would result in aggregate distributions of similar shapes over this time interval. During this time the shape of the angular distributions in the 1T channel remains about the same. These angular distributions in the 1T channel probe the component of the wavepacket formed adiabatically by transfer from diabatic state 2 to state 1 near the conical intersection and, because this component lies only on one side of the conical intersection, no large dependence on geometry is expected. During the second passage of the wavepacket through the conical intersection (12 to 28 fs), the 1T signal tracks the diabatic state 1 population through the conical intersection, and its angular distribution shows a strong dependence on geometry. At the same time, however, there is a strong 2S channel signal due to the arrival of wavepacket population from the first conical intersection passage to the small bond-angle turning point. This obscures the 1T signal from the second passage through the conical intersection. Nevertheless, these results suggest that evolution of the energy-resolved angular distribution with time, even when only one ionization channel is effective, can be useful in tracking the wavepacket through a conical intersection. This is in contrast to the angular distributions at 3.5 eV which essentially monitor the build-up of population, due to nonadiabatic interactions, in each region probed by different ionization channels.

The evolution of the aggregate photoelectron angular distributions (4 fs to 12 fs) at 2.1 and 2.8 eV in Fig. 5.25 are seen to track the wavepacket as it moves through the conical intersection. Though this behavior is similar to that seen at 3.5 eV, it is important to note the significant dependence of these distributions on energy, e.g., 2.8 eV versus 3.5 eV. The angular distributions at 2.1 and 2.8 eV also show a dependence on geometry similar to those at 1.4 eV, with changes in the 2T channel reflecting the first passage of the wavepacket through the conical intersection and changes in the 1T channel its second passage. Changes in the shape of these distributions in the 2T channel at 2.1 eV, however, occur later than at 1.4 eV, while the changes in the distributions at 2.8 eV occur even later. Figure 5.24 shows the photoelectron kinetic energy expected from a classical view. The wavepacket is composed of a distribution of components with some higher and others lower in nuclear kinetic energy than the average energy. Wavepacket components with higher nuclear kinetic energy should give rise to lower-energy photoelectrons, and vice versa for lower-energy wavepacket components. Hence, around the conical intersection where the angular distributions depend strongly on energy, the dynamics at a lower photoelectron energy should reflect the faster components of the wavepacket, while the higher-photoelectron energy signal reflects the dynamics of the slower part of the wavepacket. Ionization giving rise to 3.5 eV photoelectrons probes a region too far away from the conical intersection compared to the energy spread of the wavepacket for this effect of conical intersection to be observed. Likewise, the 2S channel angular distributions, away from nonadiabatic regions, do not change its shape significantly over time.

5.4.2.4 *Velocity map image and its time derivative*

From the time- and energy-resolved photoelectron angular distributions of the previous section, we can construct femtosecond time-resolved photoelectron images of these spectra, which are shown in Fig. 5.26(a). The vertical axis indicates the photoelectron momentum along the direction of the pump and probe polarizations in atomic units, and the horizontal axis the momentum perpendicular to it. The intensities [$P_k(\tilde{\theta}_k) \times 10^3$ of Eq. (3.93)] are indicated by color and brightness (key given as the rightmost scale box). These images offer a compact representation of the angular distribution plots of Fig. 5.25 over the entire range of photoelectron energies and reaffirm how these distributions track the wavepacket dynamics.

The bright innermost ring at $k = 0.2$, mostly in the direction parallel to the probe polarization and seen for all delay times, corresponds to the 1T peak in Fig. 5.23(c) (photoionization from the unexcited initial state). The $k = 0.3$ and $k = 0.4$ rings correspond to the 1.4 and 2.1 eV plots of Fig. 5.25, and we again see the build-up in intensity in the direction parallel to the probe as the wavepacket passes through the conical intersection between 4 and 12 fs causing the 1T channel to become active for the excited wavepacket. At later times (20 to 24 fs), the peak intensity lies away from the probe polarization axis, which comes from the 2S and 1T channels in Fig. 5.25 (wavepacket remaining in state 2 reaching the small bond angle turning point region). For the outermost ring ($k = 0.5$), the intensity along the diagonal direction from the 1T channel weakens relative to the 1S channel signal in the parallel direction. This shift indicates the second passage of the wavepacket through the conical intersection, as discussed for the $\varepsilon_k = 3.5$ eV case in Fig. 5.25.

The time-derivative velocity map images (Sec. 3.3.11) were also computed from the obtained $P_k(\tilde{\theta}; t_{\mathrm{pr}})$ through finite differences. A small time shift of $\tau = 1.0$, 0.4, and 0.2 fs were tried and all gave the same results within the resolution of the figures shown.

Figure 5.26(b) shows the time evolution of $\Delta P_k(\tilde{\theta}_k, t) \times 10^4$. The axes are the same as in Fig. 5.26(a). Red through yellow signify positive values (increasing photoelectron signal) and blue through white signify negative values (decreasing photoelectron signal). Black signifies no change in the photoelectron signal at that instant. Since the image time derivatives are sensitive to the time evolution of the photoelectron angular distribution, but not necessarily to the absolute strength of components of the signal, the rapid but delicate changes in the dynamics near the conical intersection are better represented by these derivatives than by the velocity map images.

One obvious example where the derivative image better represents the dynamics is seen in the inner ring which is bright in the direction parallel to the probe polarization axis at $k = 0.2$ in the velocity map image, but is missing in the image time derivative. In the velocity map image, this ring comes from the initial wavefunction that was left unexcited by the pump pulse through the 1T channel. Although bright, it does not change in time and does not reflect the excited state wavepacket dynamics.

For delay times $t_{pr} = 4$ and 8 fs, the signal is seen to decrease for $k > 0.4$ and increase in the range $0.2 < k < 0.4$. We may interpret this as the lowering of the photoelectron kinetic energy over time. The photoelectron kinetic energy from the 2T channel follows the rapid progress of the wavepacket from an initial location at a larger bond angle to a smaller bond angle during 4–8 fs (see upper panel of Fig. 5.15), and this variation in time is reflected in the image time derivative for a single delay time. We also see a similar case at later times $t > 20$ fs as the state 1 wavepacket rapidly moves towards the opposite direction. The 1T channel photoelectrons provide a signal in the direction parallel to the probe polarization axis that tracks the wavepacket motion through its decreasing kinetic energy. This is seen in the time derivative image for a single delay time ($t_{pr} = 20$ or 24 fs) where the higher energy signal is decreasing and lower energy signal increasing.

Comparison of the time derivative images for different delay times also provides additional insight into the vibrational dynamics. Between $t_{pr} = 12$ and 16 fs, the signal in the direction parallel to the probe polarization axis at $k = 0.4$ changes from increasing to decreasing. This corresponds to the (relatively) slow build up of population at the turning point and the population leaving the turning point. The 1T channel monitoring the state 1 component reaching its small angle turning point is reflected as a change in the sign of the time derivative between $t_{pr} = 12$ and 16 fs. A similar case for the state 2 component is seen for the 2S signal at $t_{pr} = 20$ and 24 fs.

5.4.3 *Monitoring the effect of a control pulse on a conical intersection by time-resolved photoelectron spectroscopy*

In this section we show with full quantum wavepacket dynamics on coupled *ab initio* potential energy surfaces how a conical intersection can be transformed into an avoided crossing at the geometry of the conical intersection by externally breaking its symmetry, the idea developed in Sec. 5.3.3. The population transfer through the conical intersection in NO_2 can be significantly suppressed using a phase-controlled far-infrared pulse.

We also show with full quantum mechanical calculations that such optical control of a conical intersection can actually be monitored in real time with femtosecond angle- and energy-resolved photoelectron spectroscopy. This suggests that one can optimally control the above gating process through the conical intersection by a feedback loop based on varying the control parameters and monitoring the photoelectron velocity map images.

This strategy should provide a far more efficient and rapid means of optimal control than measuring the ratio of products.

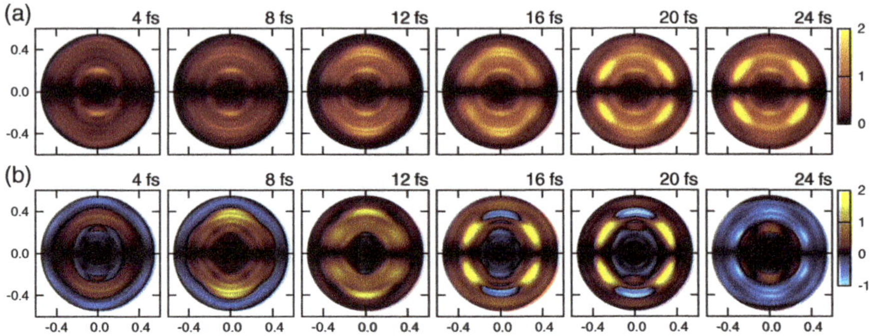

Fig. 5.26 (a) Computed femtosecond time-resolved photoelectron velocity map images near the conical intersection of the NO_2 molecule. (b) Corresponding time derivative images. (Reprinted with permission from Y. Arasaki *et al.*, J. Phys. B 45, 194006 (2012)).

5.4.3.1 *Transformation of a symmetry-allowed conical intersection to an avoided crossing*

To transform the conical intersection into an avoided crossing, we use a control pulse of the form of Eq. 3.84, where the field intensity is 3.2×10^{13} W cm^{-2}, the energy is $\hbar\omega = 0.1$ eV, the center of pulse is at $t_0 = 8$ fs, the phase is $\phi = 0$, and the FWHM is 16 fs. We note that the second term in Eq. (3.84) is necessary because the time derivative of the pulse envelope is not small compared to the frequency of the control pulse. The pulse width is short compared to its oscillation period and the phase is such that the maximum of the wave matches that of the pulse envelope, and the shape of the resulting field is very similar to the pulse envelope. Polarizability terms that may be quantitatively significant [34] are neglected in this study. Figure 5.27 illustrates the control pulse and its effect on the potential energy surfaces. The solid curve in Fig. 5.27(b) shows the shape of the control pulse field. Off diagonal element of $\mathbf{V}_E(\boldsymbol{R}, t)$ at the conical intersection geometry is shown in units of eV. For comparison, the pump pulse shape computed in the same way at the global minimum geometry is also shown as the dashed curve. The fields are polarized along the direction perpendicular to the bond angle bisector for C_{2v} geometry and belongs to the irreducible representation B_2. Upon application of the control field the

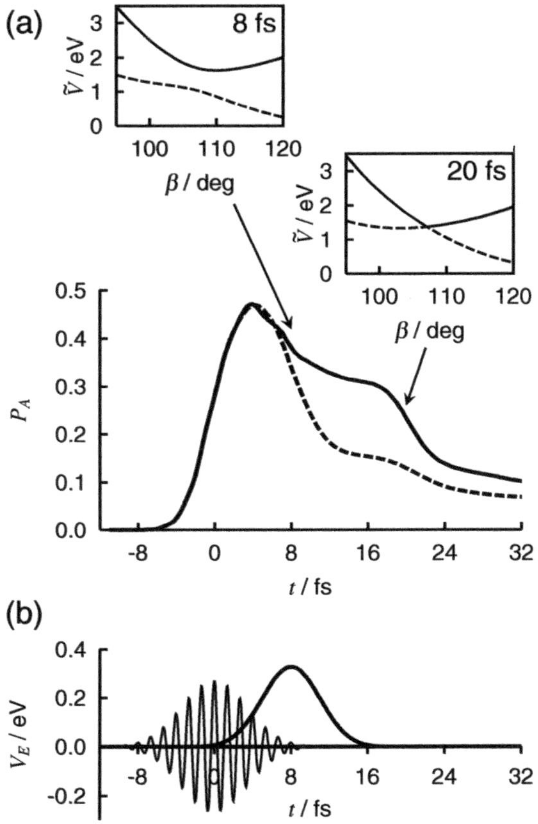

Fig. 5.27 (a) Time evolution of adiabatic excited state population formed by the pump pulse, under influence of control pulse (solid) or without control (dashed). Level repulsion of the potential surfaces by the control pulse ($t = 8$ fs), and restoration of the conical intersection after control ends ($t = 20$ fs) are shown as insets. (b) Control pulse shape (thick curve) and pump pulse shape (thin curve). (Reprinted with permission from Y. Arasaki *et al.*, Phys. Chem. Chem. Phys. 12, 1239 (2010)).

two potential surfaces of A_1 and B_2 character belong to the same irreducible representation (A'). At its peak ($t = 8$ fs) the control pulse splits the levels at the conical intersection by ~ 0.6 eV, as shown in the inset of Fig. 5.27(a).

5.4.3.2 *Wavepacket dynamics under effect of the control pulse*

We now see the effect of the control pulse on the wavepacket dynamics. The center of the pulse are taken to be $t_0 = 8$ fs, about the time the center of excited state wavepacket is at the conical intersection. Field shape is shown as thick curve of Fig. 5.27(b), as the value of off diagonal of $\mathbf{V}_{\mathrm{E}}(\boldsymbol{R}_X, t)$ in units of eV. Its maximum height is over 0.3 eV. This is an order of magnitude larger than the value of V_{12} slightly ($|r_2 - r_1| \sim 0.1$ Å) away from the conical intersection. As a result, the potential surfaces at the conical intersection separate by over 0.6 eV at the center of the pulse. The oscillation period of the field (41.4 fs for $\hbar\omega = 0.1$ eV) is chosen long enough compared to the time scale of population transfer through the conical intersection and the phase explicitly controlled so that the field amplitude is positive during its influence. Otherwise the deformation of conical intersection cancels out by either the rapid oscillation or small amplitude during the time the wavepacket is near the conical intersection. Field shape of the 3.1 eV pump pulse is also shown in Fig. 5.27(b) for comparison. Note that the pump and control pulses do not significantly overlap, and the vibrational wavepacket initially excited is not affected by the control pulse.

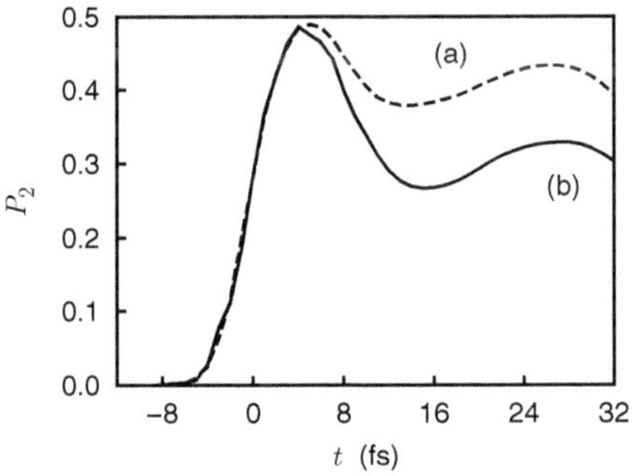

Fig. 5.28 Time evolution of diabatic state 2 population following the pump pulse, (a) without and (b) with the control pulse. (Reprinted with permission from Y. Arasaki *et al.*, Phys. Chem. Chem. Phys. 13, 8681 (2011)).

Figure 5.28 shows the time evolution of the population of the diabatic state 2. The dashed curve (a) shows the population without the control pulse and the solid curve (b) shows the population when the control pulse is applied. During the FWHM 8 fs pump pulse, the state 2 population builds up to ~ 0.5 and then decreases as the wavepacket passes through the conical intersection region around 8 fs. The state 1 population created in the first passage through the conical intersection returns to the conical intersection region for the second time around 20 fs and results in a population transfer in the reverse direction. Comparison of the curves in Fig. 5.28 shows that application of the control pulse centered at $t = 8$ fs enhances population transfer from diabatic state 2 to state 1. At later times when the control pulse becomes insignificant, the populations for the two cases evolve in the same way with time.

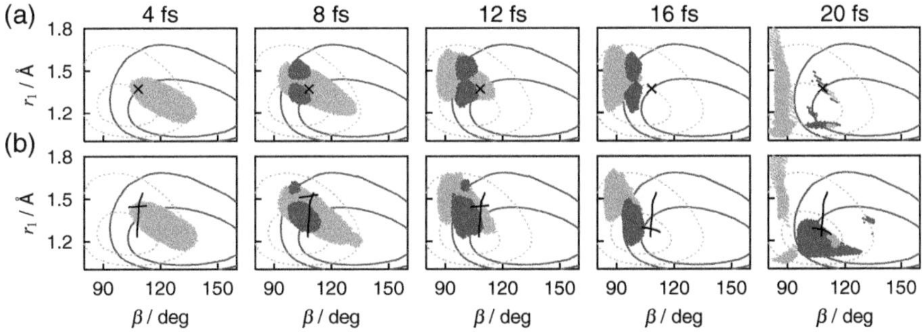

Fig. 5.29 Time evolution of the wavepacket formed by the pump pulse, (a) without and (b) with application of the control pulse. (Reprinted with permission from Y. Arasaki *et al.*, Phys. Chem. Chem. Phys. 13, 8681 (2011)).

Figure 5.29 shows the evolution of the wavepackets with time increasing from left to right. The top frames (a) show the results without application of the control pulse, and the bottom frames (b) the results with the control pulse on. Each column shows a projection of the wavefunction onto the (β, r_1) plane at times indicated at the top of the panels. Light shade shows the wavepacket component on diabatic state 2, and dark shade shows the component on diabatic state 1. Potential energy contours for $r_2 = 1.37$ Å are shown for 3.0 (outer contour) and 1.8 eV (inner contours) with the same shades. A cross ($\times$) indicates the position of the conical intersection for the same bond length in the upper row panels; in the lower row, the intersection of the $V_1 = V_2$ and $V_{12} - \mu_{12}E(t) = 0$ surfaces (black horizontal and vertical curves, respectively) is shown for the same bond length and represents the position of the conical intersection under the effect of the control pulse. The portion of the initial wavefunction left unexcited on state 1 by the pump pulse has been removed for clarity in Fig. 5.29. Similar figures (for the case without control) were shown in Sec. 5.4.1.2; the current figures have higher thresholds for plotting compared to previous figures to emphasize differences in heights for different spatial regions.

Without the control pulse [Fig. 5.29(a)], there is no transfer between diabatic states at the center of the wavepacket ($r_1 = r_2$) and the center of the wavepacket goes through the conical intersection, where the interaction potential between diabatic states is zero ($V_{12} = 0$ at the conical intersection). Thus, population transfer from the initially excited state 2 to state 1 occurs away from the position of the conical intersection and symmetrically for $r_1 < r_2$ and $r_1 > r_2$. There are thus two regions of state 1 population at 8 fs. After passage through the region of conical intersection, the state 2 wavepacket component reaches its small angle turning point by 20 fs, while the state 1 wavepacket is reflected earlier by its steeper inner well compared to state 2. By 20 fs, the state 1 component is moving quickly through the conical intersection region for the second time, resulting in some transfer of population from state 1 to state 2. The state 1 component in the plot at 20 fs is shown much smaller compared to that at 16 fs because the wavepacket at 20 fs is much flatter and is spread over a large region of space compared to when the wavepacket peaks at its turning point at 16 fs.

When the control pulse is applied [Fig. 5.29(b)], the conical intersection is effectively lifted and there is population transfer between diabatic states even at the center of the wavepacket. The fastest component of the wavepacket reaches the conical intersection before the surfaces are shifted by the control pulse and behaves in a similar fashion to the case without the control pulse. However, the bulk of the population that is transferred no longer splits into two components, and because of the asymmetry introduced by the control pulse, the resulting wavepacket is also asymmetric with respect to the two bond lengths. The behavior of the conical intersection under the control pulse results in greater state 1 and less state 2 population after the first passage through the region of the conical intersection, as is evident from comparison of the two cases at 20 fs (rightmost panel).

5.4.3.3 *Photoelectron angular distributions reflecting the controlled dynamics*

Energy-resolved photoelectron angular distributions [Eq. (3.92)] were computed for probing of the wavepackets in Fig. 5.29: (a) without the control pulse, $P_{k_j}^{(a)}(\tilde{\theta}_k)$, and (b) with the control pulse, $P_{k_j}^{(b)}(\tilde{\theta}_k)$. The probe pulse parameters are the same as it those in Sec. 5.4.2.3. Figure 5.30(a) shows the time evolution of $P_{k_j}^{(a)}(\tilde{\theta}_k)$ as photoelectron velocity map images. The delay time between the pump and probe increases from left to right for $t_{\rm pr} = 4$, 12, and 20 fs, representing times before reaching the conical intersection region, after the first passage through the conical intersection region, and when the state 2 component reaches the small angle turning point, respectively. The axes in each frame represent photoelectron momenta in atomic units and the momentum along the polarization of the pump and probe is shown on the vertical axis (parallel to O–O for C_{2v} geometry). Brightness indicates the signal strength $P_{k_j}^{(a)}(\tilde{\theta}_k)$ and the numbers on the rightmost scale box are given in units of 10^{-4}. A similar figure was given in Sec. 5.4.2.4. As the wavepacket passes through the conical intersection region for the first time (between 4 and 12 fs) a signal in the direction parallel to the pump and probe polarization develops. At later times when the wavepacket on diabatic state 2 reaches the turning point at small angles, the 2S channel becomes active and a strong signal in the diagonal direction develops.

The difference between the energy-resolved photoelectron angular distributions with and without the control pulse, $\Delta P_{k_j}(\tilde{\theta}_k) = P^{(b)}_{k_j}(\tilde{\theta}_k) - P^{(a)}_{k_j}(\tilde{\theta}_k)$, is plotted in Fig. 5.30(b) as velocity map images. Red brightness indicates a more intense signal with the control pulse while blue brightness indicates a less intense signal with the control pulse. At 4 fs (leftmost), before the control pulse has significant amplitude, the difference between the two cases is negligible. After the first passage through the conical intersection region (12 fs), the relative populations of the two diabatic states have been modified by the control pulse and an increase in signal for the region around $k = 0.35$ (1.7 eV) and a decrease for higher and lower momenta are seen in the difference images [middle frame of Fig. 5.30(b)]. At later times [rightmost frame of Fig. 5.30(b)], the signal from the 2S channel is much reduced by the control pulse because there is less population in state 2 reaching the small angle region for this ionization channel.

Figure 5.31(b) shows the time derivative of the velocity map images in the same way as in Fig. 5.26(a). Numbers on the rightmost scale box are given in units of 10^{-3} in (a) and 10^{-4} in (b). Comparing Fig. 5.31 with Fig. 5.26, we expect signal associated with state 2 to decrease and signal associated with state 1 to increase reflecting the effect the control pulse had on the wavepacket dynamics.

For $t_{pr} = 4$ fs, before the control pulse has significant amplitude, the difference between the two cases is negligible. After the first passage through the conical intersection region (12 fs), in the middle panel of Fig. 5.31(a), we see an additional signal in the direction parallel to the probe axis at $k = 0.2$ for the controlled case, in between the existing signals at higher and lower momenta for the uncontrolled case. The existing signal at $k = 0.35$ is also enhanced under control, and the signal at higher momenta decreases. For Fig. 5.30, corresponding changes were identified with the decreasing signal associated with the 2T channel and the increasing signal associated with the 1T channel, reflecting the change in the relative population due to the control pulse. Comparison of the time derivative of the images shows a more striking change, where the region of increasing signal in the middle panel of Fig. 5.26(b) is less pronounced with the control pulse in Fig. 5.31(b), and additional regions of increased signal clearly appear in the controlled case.

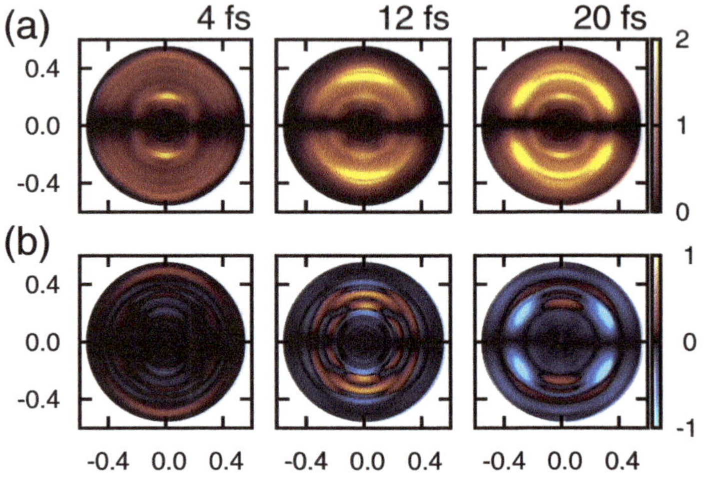

Fig. 5.30 (a) Femtosecond time-resolved photoelectron velocity map images without the control pulse. (b) Velocity map image difference between the results with or without the control pulse. (Reprinted with permission from Y. Arasaki *et al.*, Phys. Chem. Chem. Phys. 13, 8681 (2001)).

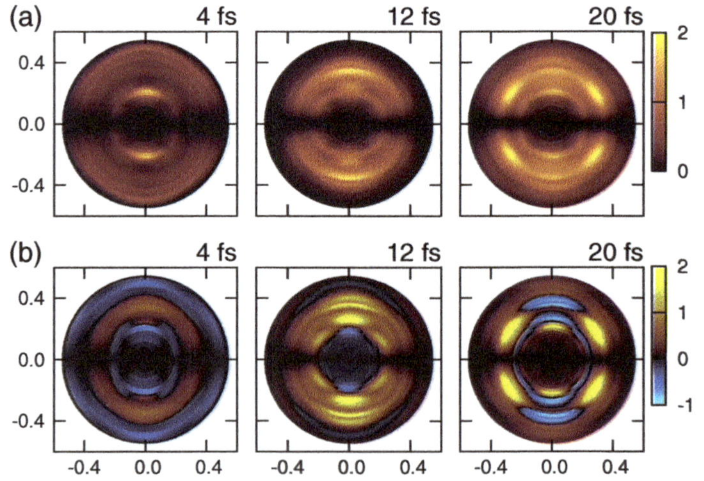

Fig. 5.31 Time evolution of (a) velocity map images and (b) their time derivatives for the controlled NO_2 dynamics. (Reprinted with permission from Y. Arasaki *et al.*, J. Phys. B 45, 194006 (2012)).

At $t_{\rm pr} = 20$ fs, the strongest signal comes from the 2S channel when the state 2 wavepacket is at its small angle turning point. Though we expect the signal associated with state 2 to decrease and the signal associated with state 1 to be enhanced under control, the difference is not easily seen in the velocity map images because of the relative strength of the 2S channel signal. In the time derivative image we see that under the control pulse a decrease in the $k = 0.3$ region and increase in the $k = 0.2$ region parallel to the probe polarization axis are much more prominent. This behavior was associated with the lowering in energy of the 1T channel signal reflecting motion of the state 1 wavepacket to larger bond angles in the previous subsection. The region where the signal is increasing is the same region where the signal is bright in the velocity map image for $t_{\rm pr} = 4$ fs; at $t_{\rm pr} = 20$ fs, we see the increase in the state 1 population approaching the bond angle region of the initial wavefunction because of the control pulse.

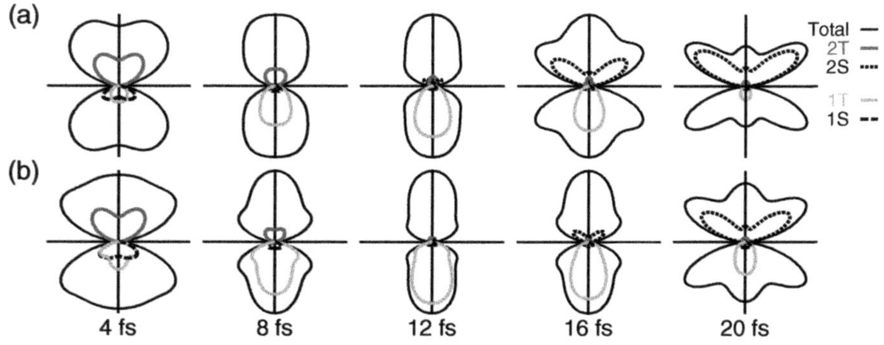

Fig. 5.32 Time evolution of the photoelectron angular distributions for a kinetic energy of $\varepsilon_k = 1.7$ eV: (a) without and (b) with the control pulse. (Reprinted with permission from Y. Arasaki *et al.*, Phys. Chem. Chem. Phys. 13, 8681 (2011)).

To provide some insight into the velocity maps of Fig. 5.30, Fig. 5.32 shows the time evolution of the photoelectron angular distribution for $\varepsilon_k = 1.7$ eV. This energy corresponds to a value of $k = 0.35$ a.u. in the 12 fs frame of Fig. 5.30 where the signal with the control pulse on is larger than the signal without the control pulse. The top row (a) in Fig. 5.32 shows the angular distributions obtained without the control pulse, $P^{(\rm a)}_{k_j}(\tilde{\theta}_k)$, while the bottom row (b) shows the angular distributions with the control pulse on, $P^{(\rm b)}_{k_j}(\tilde{\theta}_k)$. The delay time between the pump and probe increases from left to right. For each frame, the polarization of the pump and probe lies along the vertical axis and the distributions are cylindrically symmetric around this axis because of the averaging of Eq. (3.92). The angular distribution for each frame is normalized separately for each delay time for clearer presentation of the overall shape; information on the relative signal strengths between different delay times is omitted. Inside the solid black contour of the photoelectron angular distribution for each frame, the contribution from each ionization channel is plotted in the same way as in Fig. 5.25.

Figure 5.32 shows that the overall time evolution of the photoelectron images in Fig. 5.30 is mainly due to the change in the relative composition of the components of the signal. During the first passage through the conical intersection region (between 4 and 12 fs), the 2T component signal shrinks while the 1T component signal grows, reflecting the transfer of population between the diabatic states. For later times (16 and 20 fs), a strong 2S component signal develops as the wavepacket on diabatic state 2 reaches the turning point at small angles. These overall changes in the photoelectron angular distributions were noted in Sec. 5.4.2.3 without a control pulse.

The difference between the photoelectron angular distributions with the control pulse turned on [Fig. 5.32(b)] and without the control pulse [Fig. 5.32(a)] mainly reflects the change in the composition of the photoelectron signal due to the control pulse. With the control pulse on, there is somewhat more state 1 population from the first passage through the conical intersection region centered at 8 fs, resulting in a larger 1T component and smaller 2T component for subsequent times (12 fs), and a smaller 2S component at later times (20 fs). This is the main reason for the increase/decrease of signal strength seen in the time evolution of the difference photoelectron velocity images in Fig. 5.30(b).

Most of the component signals themselves do not change shape over time. The exceptions are the 2T channel signal during passage through the conical intersection region, and the 1T channel signal during the same time, but only with the control pulse on. In Sec. 5.4.2.3 we noted for the 2T channel signal that, when the wavepacket has amplitude on both sides of the conical intersection, the signal resulting from that region changes shape during passage through the region. In contrast, the state 1 component of the wavepacket appears only after the wavepacket has gone through the conical intersection region. Because it exists only on one side of the conical intersection, the angular distribution for the 1T channel did not change during the time of the first conical intersection passage.

When affected by the control pulse, there is much greater transfer between the diabatic states before the wavepacket reaches the conical intersection geometry. Thus the wavepacket component on state 1 is formed both just before and after first passage through the conical intersection region, and hence the 1T channel component signal changes shape during this time. The photoelectron angular distribution from the state 1 component before reaching the conical intersection position has protrusions out to the sides [4 and 8 fs frames of row (b) of Fig. 5.32] in contrast to the distributions parallel to the polarization axis for the state 1 component on the other side of the conical intersection (12 fs and later). These protrusions out to the sides survive even in the difference map images at 12 fs in Fig. 5.30(b). This is an example where the time evolution of the photoelectron angular distribution reflects not only the composition of the evolving component signals, but also the change in shape of the angular distribution of a single component signal from one side of the conical intersection to the other. This behavior can hardly be seen in the aggregate signal without the application of the control pulse.

Time-resolved photoelectron kinetic energy spectra are seen to track the vibrational wavepacket dynamics near the conical intersection. However, overlapping signals from several ionization channels make it difficult to infer the wavepacket dynamics from the photoelectron energy spectra alone. Nevertheless, the evolution of the photoelectron angular distributions make it possible to track the evolution of the wavepackets as population is transferred in nonadiabatic regions. The results suggest that this behavior might arise quite generally in polyatomic molecules.

We have also shown that a symmetry allowed conical intersection can be transformed to an avoided crossing by externally breaking its symmetry. With the NO_2 molecule system under a phase controlled far infrared pulse field as an example, we have shown that the population transfer through the conical intersection is significantly suppressed. Symmetry breaking can be caused not only by optical control but chemically by the presence of substituent groups nearby and/or by the solvent effect. In fact, the pioneering work by Hamaguchi et al. [108] for retinal isomerization may be comprehended in this view. However, an important difference between our optical means and these chemical modification including the new method of Lim et al. [241] is that the optical control contains a number of parameters that can be externally adjusted easily such as the wavelength, intensity, shape of the pulse, electronic polarization, phase, and so on. Besides, the optical control can be applied simultaneously along with the above methods of chemical modification. Therefore, the present mechanism unique to the conical intersection may have potential as a means of "control" of *gating* to the branching pathways for nuclear wavepacket to proceed.

For the symmetry allowed conical intersection and the symmetry required conical intersection as well (the definition due to Yarkony, see Ref. [117]), the mechanism of converting conical intersection to avoided crossing and the role of the laser have been thus evident. Yet, even for other types of conical intersection, the conversion of conical intersection by an appropriate choice of $\mathbf{V}_{\mathrm{E}}(\boldsymbol{R}, t)$ may be possible depending on the electronic-state origin [316] of the appearance of the degeneracy. Therefore the idea of laser conversion of conical intersection is worth further investigation both experimentally and theoretically.

A combination of the methodologies of such optical control at conical intersections and the photoionization-spectroscopic observation of the wavepacket in these regions can enable studies of the control and monitoring of the gating process through a conical intersection. The present study suggests that time-resolved photoelectron spectroscopy may be utilized as a sensitive tool to explore how nature might control conical intersections in optical systems.

5.5 High-harmonic spectroscopy to monitor nonadiabatic transition

In the previous section, femtosecond photoelectron spectroscopy of the vibrational dynamics near a conical intersection was discussed. Progress in attosecond light shaping technology has enabled complementary approaches using the recombination of the photoionized electron as the probe. It was recently demonstrated that high-harmonic interferometry [375, 465] is a viable means of tracking electronic and nuclear coupled dynamics. Especially in the case of interference of ionization channels out of the same orbital, the signals were found to track the diabatic population dynamics near the conical intersection [221, 466]. In this section we shall outline this emerging experimental method.

5.5.1 *High-harmonic generation and associated phases*

When an external field ionizes an electron and subsequently drives it back to the parent hole-state for recollision, high-order harmonics are generated. This high-harmonic generation, as detailed in recent reviews and tutorials [101, 212, 222], has been the focus of current intense studies, both as a source of ultrashort, attosecond pulses [76] and as a means of extracting information out of molecular systems at the electronic temporal and spatial resolutions [346].

Briefly, the process can be understood by a three process model [96, 237], where (1) the external field tunnel ionizes an electron, (2) the ionized electron is accelerated by the external field, gaining energy, and (3) the electron recollides with the parent ion, emitting light at multiples of the external frequency corresponding to the energy gained during free electron propagation. The process imparts a phase difference between the recolliding electron and the parent ion wavefunctions. The ionization phase, embodied in the photoionization matrix elements of Sec. 5.2, the phase gained through the continuum evolution, and the recombination phase enter into the phase gained by the electron. Phase of the parent ion also evolves between the time of ionization and recollision. Simultaneous measurement of amplitude and phase of high harmonics provides detailed information into the concerned orbitals and dynamics of the process.

5.5.2 *Transient grating and interferometry*

To obtain the relative phase of high-order harmonics, transient grating spectroscopy has been applied to high-harmonic spectroscopy [253]. Prior to the probe pulse that generate the high-harmonics, the pump pulse is configured so that a transient grating of excitation to be probed is formed. The pump pulse is split in two, offset spatially, and recombined at the sample molecular jet, creating a spatial alternation of excitation in the sample.

When the probe pulse generates high-harmonics from this mixed sample, the signals from the excited and unexcited population interfere, causing diffracted signals at the detector. Thus the relative phase difference between the excited and unexcited populations is be found in the diffraction pattern. In particular, when the pump pulse creates a vibrational wavepacket in both the ground and excited electronic states [221], the zero-order diffraction signal,

$$I_{m=0}(\Omega,t) = \left|[1-r(t)]\tilde{d}_g(\Omega) + \sum_{i=X,A} r_i(t)\tilde{d}_i(\Omega,t)\right|^2, \tag{5.33}$$

is spatially resolved from the first-order diffraction signal

$$I_{m=\pm 1}(\Omega,t) = \frac{1}{4}\left|\sum_{i=X,A} r_i(t)[\tilde{d}_i(\Omega,t) - \tilde{d}_g(\Omega)]\right|^2, \tag{5.34}$$

and can both be measured. Here, Ω is the photon frequency, r_i ($i = X$ and A) is the vibrationally excited population in the ground X and excited A electronic states, $r(t) = \sum r_i(t)$, $\tilde{d}_i(\Omega, t)$ represents the time-dependent complex spectral component of high-harmonic emission from state i, and $\tilde{d}_g(\Omega)$ is that from the unexcited component.

$\tilde{d}_i(\Omega, t)$ can be represented as an amplitude and phase,

$$\tilde{d}_i(\Omega, t) = d_i(\Omega) e^{i\phi_i(\Omega, t)}, \tag{5.35}$$

or in the three-step model of Sec. 5.5.1,

$$\tilde{d}_i(\Omega, t) = \int d\boldsymbol{R} |\chi_i(\boldsymbol{R}, t)|^2 \sum_f \sqrt{I^{if}(\boldsymbol{R})} e^{-iI_p^{if}(\boldsymbol{R})\tau} \tilde{d}_r^{if}(\boldsymbol{R}), \tag{5.36}$$

where f represents the states of the cation, I^{if} is the rate of strong-field ionization from state i to the cation state f, I_p^{if} the ionization potential, $\tilde{d}_r^{if}$ the photorecombination matrix element, and τ the transit time.

5.5.3 *Transient grating interferometry of the conical intersection dynamics in NO_2*

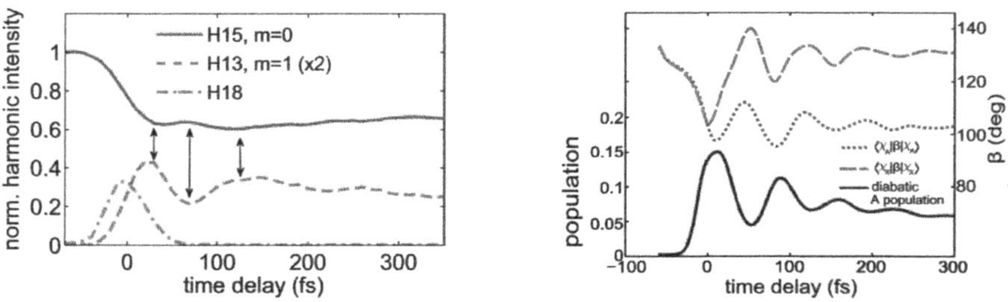

Fig. 5.33 Measured signal from high-harmonic transient grating spectroscopy of NO_2 (left) and computed diabatic state population (right). Reprinted form [221]. (Reprinted with permission from P. M. Kraus *et al.*, Phys. Rev. A 85, 043409 (2012)).

In Refs. [466] and [221], the transient grating interferometry signal from the conical intersection dynamics in NO_2 (see Sec. 5.4.1.2) was measured for a 401 nm (3.09 eV), FWHM 40 fs pump pulse. The left panel of Fig. 5.33 shows the time-dependence of $I_{m=1}(\Omega, t)$ for the 13th harmonic, which reflects the time evolution of the diabatic excited state population (Fig. 5.33, right). In the case of ionization out of the same orbital and spatially spread vibrational wavefunction as in the present case, the dependence of $\tilde{d}_i(\Omega, t)$ on the molecular geometry (vibrational dynamics) was found to be small compared to the signal's dependence on population dynamics, thus demonstrating the method to be a potentially powerful method of probing the nonadiabatic population dynamics.

5.6 Electron and nucleus dynamics tracked with pulse train in time-resolved photoelectron spectroscopy

With use of single attosecond pulses as pump and/or probe [222], the dynamics of electron wavepackets in the valence state is anticipated to be tracked in real time [476]. Theoreticians are currently involved in numerical calculations of ionization dynamics of electron(s) in very simple molecular systems like hydrogen molecules and its cation in restricted electronic and nuclear configurations [205, 498]. However, there are a number of reasons that cast questions to such naive anticipation: (i) Electronic wavefunctions are generally not localized in space, in a keen contrast to the nuclear wavefunctions. Therefore it is hard to identify the motion of electrons with respect to the frame (geometrical structure) of a molecule. (ii) Attosecond pulses usually contain very many high harmonics, which may result in a huge energy uncertainty beyond the energy range of chemical interest. (iii) Typical time scale of chemical reaction is of the order of ten femto- to subpico-seconds. How can we make use of attosecond pulses in tracking a chemical reaction? In other words, we ask ourselves whether we can track *electron–nucleus simultaneous dynamics* with femtosecond lasers.

It seems almost intuitively obvious on one hand that nonadiabatic electron wavepacket theory should offer direct methods to study dynamics of electrons in chemical reactions. On the other hand, we recently began to explore possible situations and relevant physical observables that can be detected as a manifestation of electron dynamics in femto- and subfemtosecond time-resolved photoelectron spectroscopy [20]. Most of our previous studies have aimed at direct photoionization with a single pulse from the Hartree–Fock level electronic state, which is a half-collision process represented by a scattering orbital and a Hartree–Fock cation core. (See Refs. [418, 419] for a more general scattering theory to take account of the electronic polarization of a target molecule in electron scattering.) Within the direct photoionization experiments, however, it is generally hard to track complicated electron dynamics. A straightforward idea to detect an electronic-state propagation with photoelectrons is to apply the method of multiphoton ionization of high pulse intensity, for which the Rabi oscillation is expected before ionization. However, this is not easy either, since the Rabi oscillation takes place much faster than the time scale of the femtosecond time-resolved photoelectron spectroscopy.

In this section we try to practically realize the idea of Ref. [20] by applying a train of subfemtosecond pulses composed of frequencies appropriate to photoelectron spectroscopy. Just as the vibrational progression is clearly recorded in conventional (time-independent) photoelectron spectra, time-resolved photoelectron spectroscopy with a well designed pulse train is anticipated to be able to monitor the electronic state progression along with signatures of the nuclear vibrational dynamics.

The present work was stimulated by the idea of HHG spectroscopy as a kind of inverse process: In HHG spectroscopy ATI is first induced and modulation in the HHG spectrum is probed as a change of the molecular state [465], whereas here we shine a designed pulse train to ionize the molecule to detect electronic-nuclear simultaneous dynamics. In this conjunction, it should be recalled that feedback control experiments of chemical reactions based on the pulse engineering of the shapes and phases of the applied laser [64, 187, 473] has become a fertile field in chemical dynamics. Among others a very important experimental work was conducted in the group of Wöste, in which they tried to optimize the photoionization yield from NaK with use of the genetic algorithm [37]. This was soon followed by a theoretical analysis by the group of Bonačić-Koutecký [353], which clarified that the optimized process has been achieved in the series of resonant electronic excitation from the ground state ($1^1\Sigma^+$) to intermediate excited states sequentially ($\rightarrow 2^1\Sigma^+ \rightarrow 3^1\Pi$) and finally leading to the ionized state NaK^+ ($1^2\Sigma^+$). Besides, in each intermediate electronic state the nuclear wavepacket running on it takes some time to move to a position where the succeeding electronic excitation has a large quantum mechanical probability. The laser pulses that materialize this optimization have been found to be a train of pulses whose frequencies and time-delay between the successive pulses are determined as designed. Such a sequence of pulses results from optimization not only for the diatomic molecule but for more complex cases [81, 463] as well. Thus, a pulse sequence of a given separation, or a pulse train is seen as a potentially good optical source to control the excited state dynamics [320]. Here, on the other hand, we show how the regularized pulse train can be utilized to track the electronic-state (not only excitation but deexcitation as in the Rabi oscillation) and nuclear dynamics before ionization and illustrate numerically that the photoelectron spectroscopy is a practical means of obtaining a record of the electron nucleus simultaneous dynamics traveling among several excited states.

5.6.1 *Generation of pulse train*

A train of ultrashort pulses are naturally generated as high harmonic generation from atoms and molecules, which may be used as an optical source to excite another target. In our consideration here, however, a pulse train is meant to be a series of pulses formed from well designed combination of multiples of a phase controlled continuous external field. To be precise, we assume a field formed of multiples of the base frequency ω,

$$E_x(t) = \frac{1}{N} E_x^{\max} f(t - t_x) \sum_{n=n_1}^{n_2} s_n \cos(n\omega t + \phi_n), \tag{5.37}$$

where $s_n = 0$ or 1 control which individual components in the range given by $[n_1, n_2]$ is included, and N is the number of components included. An overall Gaussian envelope $f(t)$ is included for switching the pulse train on and off, and the temporal center of the pulse train is to be t_x [cf. Eq. (3.84)].

All the phases are taken $\phi_n = 0$ to generate a pulse train. Inclusion only of odd multiples of the base frequency generates a field with an envelope symmetric with respect to the sign of the field, with the separation between the pulses in the pulse train Δt_{p} given by the period of the lowest frequency component

$$\Delta t_p = \pi/(n_1 \omega \hbar). \tag{5.38}$$

The number of pulses within the envelope is given by $\tau_{\mathrm{e}}/\Delta t_p$, where the duration of the pulse train τ_{e} is

$$\tau_{\mathrm{e}} = -\frac{\ln 0.2}{2 \ln 2} \tau, \tag{5.39}$$

when considering subpulses greater than or equal to, say, one fifth (0.2) in height of the maximum subpulse as part of the pulse train.

5.6.2 *A case study on LiH molecule*

To investigate possible detection of the history of sequential transitions between electronic states by a pulse train with time-resolved femtosecond photoelectron spectroscopy, we compute the quantum wavepacket dynamics of the LiH system. The LiH system is suited for our purpose of studying time evolution of electronic interaction in that (1) the lowest excited state

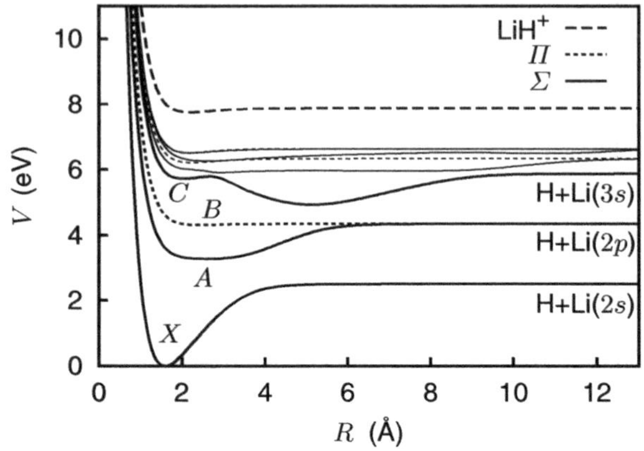

Fig. 5.34 Potential energy curves of the LiH system. (Reprinted with permission from Y. Arasaki *et al.*, ChemPhysChem 14, 1387 (2013)).

has a flat potential minimum allowing relatively long vibrational periods and (2) energy separation between the lowest three excited states are rather uniform among them and much larger than between states above, so that we may have multiple electronic transitions without considering a dense set of states.

Potential energy and dipole moment curves for the LiH system relevant to our study were computed with the complete active space self-consistent field method (CASSCF, 2 electrons in 31 orbitals) [214, 470] state averaged over all 13 singlet states up to those leading to dissociation to Li($3d$), followed by multireference configuration interaction (MRCI) [215] using aug-cc-pwCVQZ (Li) [325] and cc-pVQZ (H) [121] basis sets using the Molpro program package [471, 472]. The potential energy curve for the ground state of LiH^+ was computed with CASSCF (3 electrons in 5 orbitals) followed by singles and doubles configuration interaction (CISD) with the same basis set. Dissociation and ionization energies computed with the efficient weighted core-valence basis set [325] compare well with experiment and more extensive calculations [138] for the states computed. Figure 5.34 shows the computed potential energy curves.

The X, A, and C states belong to the Σ^+ irreducible representation and the B state belong to the Π irreducible representation. The energy differences between the B and A states and between C and B states are about 1 eV, suggesting an external field of that energy linearly polarized perpendicular to the bond axis should couple the excited states. We include up to the C state (and the cation ground state) in the wavepacket dynamics calculations. Nonadiabatic coupling between the A and C states that may become significant when the wavepacket reach the internuclear distance of 5 Å is neglected. Any other kinetic coupling between the electronic states are also neglected in this study. The molecule is assumed to lie in the laboratory z-axis and rotation is neglected for the short interaction time considered. Ionization energy from the A state is approximately 5 eV for the range of internuclear distances within the flat bottom of the A state.

In the present study, which is a semi-model or semi-realistic study dedicated to a survey on the "pure" effects of pulse trains on the photoionization spectrum, we will forego the computation of the much more expensive to compute photoelectron angular distributions, and neglect the dependence on the spatial coordinates θ_k and ϕ_k altogether. All the neutral states are taken to be coupled to the ion state in the form $-E_x(t)\tilde{\mu}$, $\tilde{\mu} = \sqrt{w_i}k_i\mu$ [see Eqs. (3.67) and (3.68)], where, for lack of extensive calculation of the photoionization coupling matrix elements, we take for ionization μ the same constant ($\mu = 1$) for all states. Although dependence on the molecular geometry and photoelectron momentum as well as on the neutral state to be ionized is important in obtaining detailed features of the photoelectron spectrum reflecting the dynamics of the system, our previous experience justifies the use of the simplified model for qualitative features of the photoelectron kinetic energy distribution [13, 18].

Quantum vibrational wavepacket dynamics is computed using repeated application of the split-operator short-time propagation method (Sec. 3.2.1) with the kinetic energy operator diagonalized by fast Fourier transform. The potential energy operator is numerically diagonalized at each short time step. The spatial grid was taken to be 512 points between 0.5 to 13.8 Å, and the short time step for time propagation is 0.01 fs. The initial state is easily obtained with the energy screening method (Sec. 3.2.5).

After determining the time evolution of the system, the population of any neutral state may be computed as Eq. (3.50), and the ionized population may be computed as Eq. (3.87).

5.6.3 *Pulse train induced dynamics*

Starting from the ground vibrational state on the ground (X) electronic state, we form a wavepacket on the first excited (A) state with a z-axis polarized pump pulse of the form of Eq. (3.84) with intensity $I_z = 4 \times 10^{12}$ W cm^{-2}, $\hbar\omega_z = 3.63$ eV, and a Gaussian envelope of FWHM $\tau_z = 8$ fs.

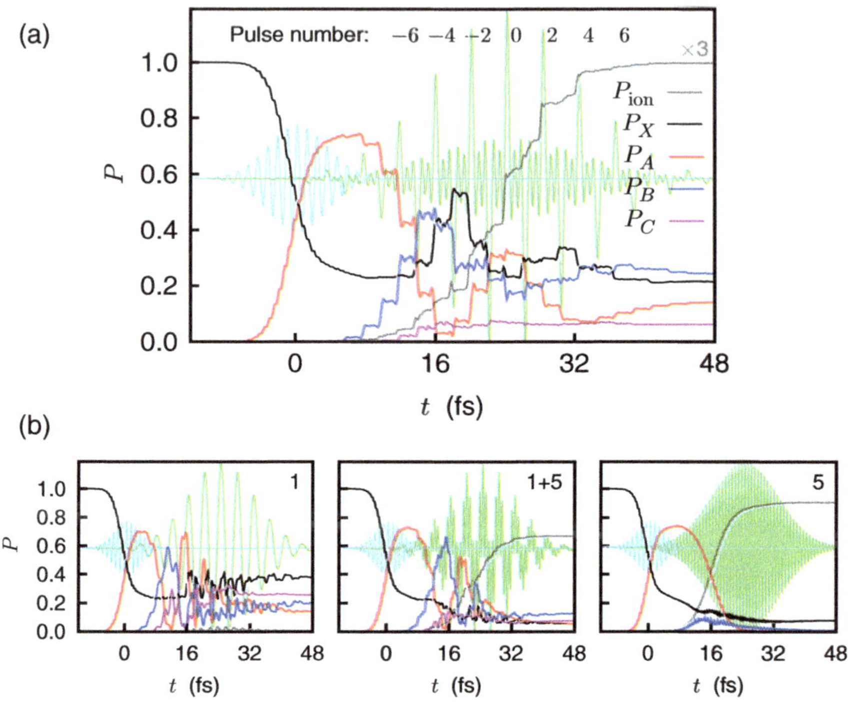

Fig. 5.35 Time evolution of the population of each state under a pulse train. In (a) the ion population is plotted magnified ×3. (Reprinted with permission from Y. Arasaki *et al.*, ChemPhysChem 14, 1387 (2013)).

A probe field composed of 1, 3, and 5th multiples of the base frequency $\hbar\omega = 1$ eV with intensity $I = 4 \times 10^{13}$ W cm^{-2}, $\tau = 20$ fs, $t_x = 24$ fs polarized in the x direction is applied ($n_1 = 1$, $n_2 = 5$, $s_n = 1$ for odd multiples, and $s_n = 0$ for even multiples in Eq. (5.37)). The FWHM of the subpulses is 0.3 fs and delay between the subpulses, or refractory period in analogy to nerve activity, is 2.1 fs.

Figure 5.35(a) shows the time evolution of the populations of the ground and excited electronic states and the ionized population during the pump/probe interaction, overlaid on the electric field strengths of the pump pulse (light blue) and the probe pulse train (light green) in an arbitrary scale (but with the same scale for the pump and probe). Each state population: P_X (black), P_A (red), P_B (blue), P_C (pink), and P_{ion} (gray) are shown in color. For later identification, the pulse at the center of the pulse train is numbered pulse number 0, and positive numbers are given to subsequent pulses and negative numbers are given to pulses prior to the center pulse. The even pulse numbers are indicated at the top of Fig. 5.35(a). [We discuss Fig. 5.35(b) later in Sec. 5.6.5.]

The pump pulse causes 74% excitation into the A state (between $t = -8$ and 4 fs). Afterwards the probe pulse train causes transfer of population between the A and B states, and between B and C (and X) states. Transfer of population occurs synchronous with the pulses in the pulse train, causing the step-like appearance seen in Fig. 5.35(a). The pulse train causes ionization simultaneously with population transfer among the neutral states. Time evolution of the ionized population is also seen to be step-like, with change synchronous with the pulses. Thus the electronic excitation and deexcitation seems to be a Rabi oscillation whose timing of transition is well controlled. On the other hand, the relevant nuclear vibrational motion is autonomously evolved in time during the refractory period.

Figure 5.36 shows the spatial distribution of vibrational wavefunction density on each electronic state at different times (solid curves). The panels are drawn for a time 1 fs after the center of the pulse whose number is indicated at the top right of each panel (see Fig. 5.35 for pulse numbers). The vertical scaling of the wavepacket density is arbitrary but consistent among all states and panels. Vertical placing of the wavepacket densities indicate each wavepacket's energy expectation value. The population for each state is indicated on the right for each panel. The dashed curves indicate the wavepacket density calculated without photoionization, suggesting what part of the wavepackets contributed to the photoelectron signal.

We see that by Pulse −6 significant population is transferred from the A state to the B state. The wavepacket density, initially located at 1.6 Å (equilibrium geometry of the ground electronic state) is moving towards longer bond lengths and is located at 2.5 Å by the time of Pulse −6. Pulses −5 and −4 causes further transfer of population to the B state with depletion of the A state, but simultaneously causes excitation into the C state as well as deexcitation from the B to the X state, resulting in a spatially split wavefunction on the X state. During subsequent pulses population again builds up on the A state, and transfer from the time evolving A state into the B state and from it to the C state at the various wavepacket positions is causing a wide spatial distribution by Pulse 6.

The transition from the B state to the X state takes place at a turning point of the latter, where a large value of the Franck–Condon overlap is expected. On the other hand, major transitions in $B \longleftrightarrow A$ and $B \longleftrightarrow C$ are realized in the regions where the involved potential functions are geometrically flat, in the range of 1.5–3.5 Å for the A state, 2.0 Å and larger for the

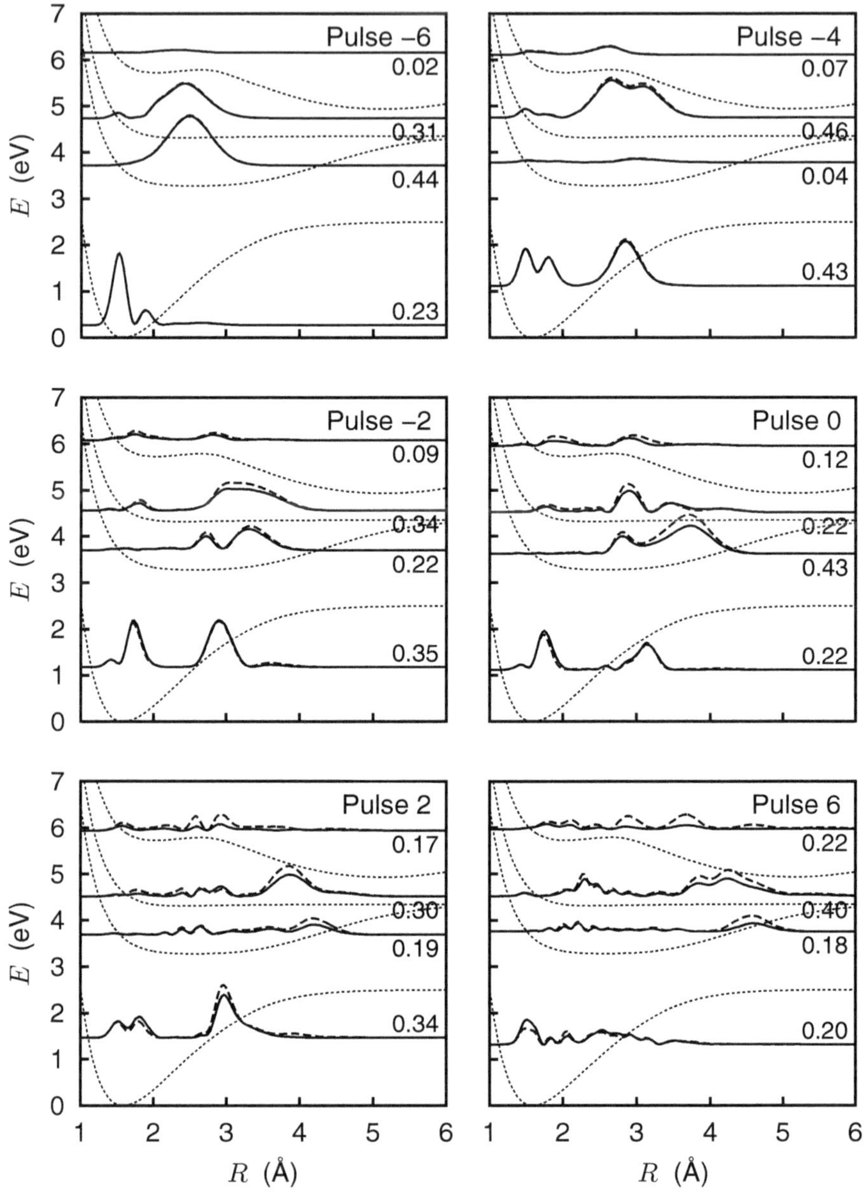

Fig. 5.36 Wavefunction density on each electronic state after each pulse. (Reprinted with permission from Y. Arasaki *et al.*, ChemPhysChem 14, 1387 (2013)).

B state, and 1.8–2.5 Å for the C state. In these ranges, electronic transition takes place as though it makes a copy of vibrational wavefunction in the common regions between the relevant electronic states. This is a reflection of the Condon principle. We note that there is counter-intuitive decrease of population in the potential well of the X state when omitting ionization, especially in the panels for Pulse 2 and Pulse 6, indicating there is some deexcitation from the excited electronic states via Rabi-oscillation like coupling with the ion continuum. Thus, complicated transfer, overlapping, and dispersion of the vibrational wavepackets proceed in each electronic state in a stepwise manner. Very fine information in the attosecond time scale is thus folded in the complicated structures and phases of the set of vibrational wavepackets [305].

5.6.4 *Transient photoelectron spectrum*

The insets of Fig. 5.37 show the transient photoelectron kinetic energy distribution at the times corresponding to those indicated in Fig. 5.36, during the pulse train interaction, with the photoelectron kinetic energy ε_k on the vertical axes and $P(\varepsilon_k)$ on the horizontal. We see the number of peaks increasing over time. Because of the energy separation between the excited states, ionization from different electronic states are separated in photoelectron kinetic energy. Ionization from the A state results in photoelectron energies below 1 eV, from the B state less than 2 eV, and from the C state above that. The dashed horizontal lines in Fig. 5.37 suggest separation of photoelectrons from different electronic states. The vibrational levels of the X state are too low in energy to be photoionized by a 5 eV external field.

Since the A state population is formed first by the pump pulse, then the B state starts to get populated by the pulse train, and then the C state is populated by excitation from the B state, we see the A state peak appearing first for Pulse −6, then B state peak appearing for Pulse −4, then the C state peak forming after then. Vibrational motion within the state and electronic transition between the states cause the wavepacket to change its spatial position over time and result in peaks splitting within the same electronic state.

Comparison of Fig. 5.37 with Fig. 5.36 suggests that the plural large components of vibrational wavepackets on the individual potential curves manifest themselves as the finer spectral peaks. Thus the electronic spectral assignment made in Figure 5.37 exposes in the chronological order the history of electronic and nuclear simultaneous dynamics in this molecule in the femtosecond time scale.

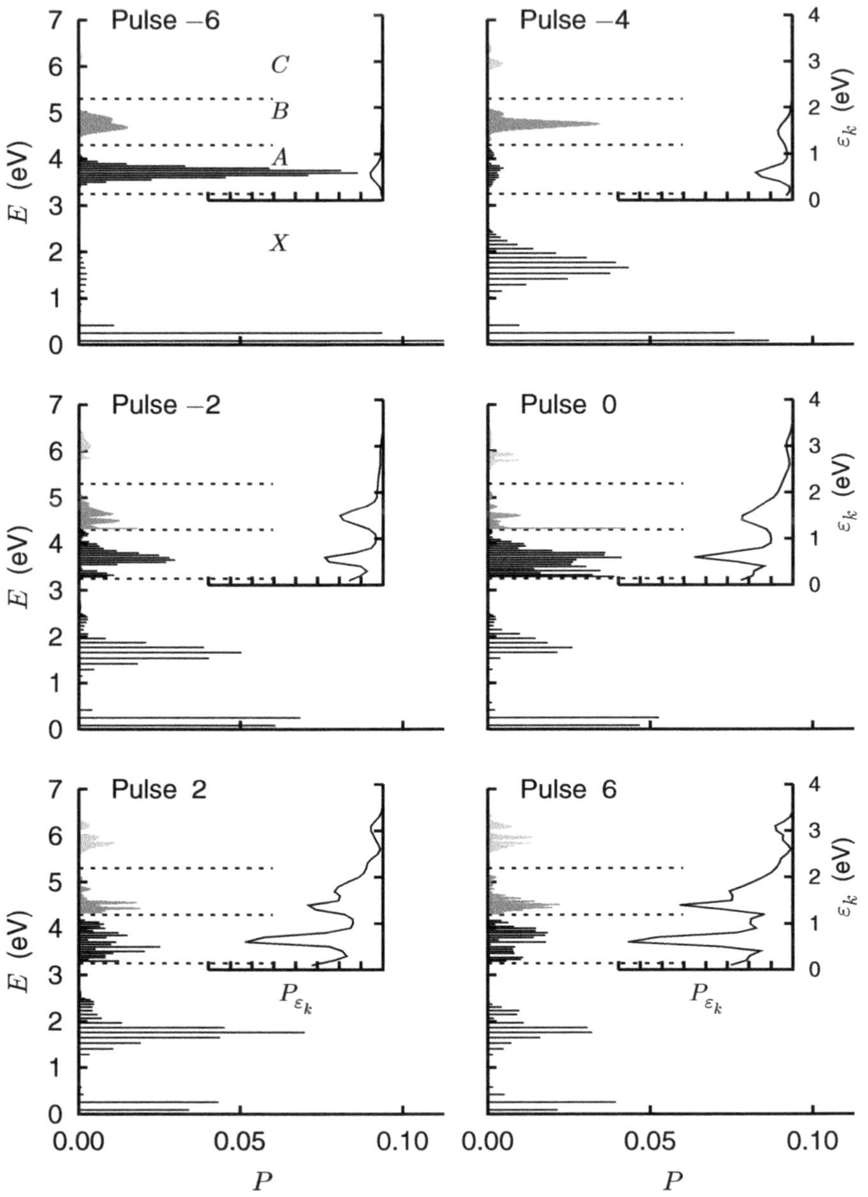

Fig. 5.37 Comparison of transient photoelectron kinetic energy distribution against time-evolving vibrational state distribution. (Reprinted with permission from Y. Arasaki *et al.*, ChemPhysChem 14, 1387 (2013)).

To further understand the photoelectron spectrum, we compare in Fig. 5.37 the spectrum with population in individual vibrational states of the electronic states (computed with the Fourier grid Hamiltonian method [260]). Population of individual vibrational states are shown as sticks during the pulse train interaction (time indicated by pulse number). The continuum of vibrational states above the dissociation threshold for each electronic state is discretized by the use of a finite grid representation of the wavefunction but is seen in the figure as approximately continuous. The vertical energy axis scaling is taken the same for the photoelectron kinetic energy and vibrational state distribution in Fig. 5.37, and the photoelectron kinetic energy inset is placed so that the peak positions for the A state population and photoelectron kinetic energy distribution match for Pulse −6. The B state is seen to hold a single bound state (seen for the times Pulse −2 and Pulse 0). We see a rather intuitive correspondence between the vibrational distribution and the photoelectron kinetic energy distribution. For Pulse 6, the small peak in the photoelectron kinetic energy distribution at $\varepsilon_k = 1$ eV, where there are no vibrational states, is a copy of the peak at $\varepsilon_k = 3$ eV due to photoionization with a probe field containing both a 5 eV (resulting in the higher energy peak) and a 3 eV (resulting in a peak 2 eV lower in energy) component.

5.6.5 *Roles of individual components*

The results in the previous two subsections considered a pulse train formed of three frequency components as the probe field. To identify the roles of the components, we compute the time evolution under a probe field formed of one or two components. Figure 5.35(b) shows the results. The probe field is indicated at the top right of each panel, for example, "5" for a probe field consisting of a single carrier frequency of 5ω [$n_1 = n_2 = 5$ and only $s_5 = 1$ in Eq. (5.37)] and "1+5" for a probe field consisting of mixtures of frequencies 1ω and 5ω ($n_1 = 1$, $n_2 = 5$, $s_1 = s_5 = 1$ with other $s_n = 0$). The curves are colored the same as in Fig. 5.35(a). Pulse train width $\tau = 20$ fs, central time $t_x = 24$ fs, and base frequency energy $\hbar\omega = 1$ eV as in Fig. 5.35(a). The case of "5" in this figure represents a typical and traditional pump–probe photoelectron arrangement using a single probe pulse, similar to those in Secs. 3.3.12, 5.2, and 5.4.2.

We see that the 1ω single carrier frequency field causes the coherent electronic transitions between the excited states (and also the ground state), which is consistent with the approximately 1 eV energy separation between the excited states. Population transfer alternates between transfer from the A to the B state (and through the B state, C state), and transfer from the B to the A state. There is little overall ionization for the 1ω probe field. The 5ω field mainly causes ionization from the first excited state, with some excitation directly from the ground state to the B state, whose population is subsequently ionized. These interactions are also consistent with the energy separation between the concerned states. The combined $1\omega + 5\omega$ field shows both the population alternating between the A and B states, and also the ionization. Therefore the 5ω component not only serves as photoionizing machinery but helps to shape the global feature of the probe pulse into the spike-like short pulses and refractory periods. We see that it is the inclusion of the 3ω component as well that results in the probe field becoming a pulse train, causing the step like population transfer in Fig. 5.35(a).

The effect of even multiples of the base frequency is briefly discussed in Ref. [14].

We have studied the application of pulse trains to probe some important aspects of the electronic excitation/deexcitation dynamics coupled with vibrational dynamics, with the LiH system as an example. A train of very short pulses well separated in time including frequency components suited for transfer between multiple electronic states and for photoionization resulted in step-like population transfers that may be recorded in the transient photoelectron signal.

The photoelectron angular distributions from the B state of LiH (Π-state) and all the others considered (X, A, C states having Σ symmetry) are qualitatively different from each other. Therefore future incorporation of geometry- and orientation-dependent photoionization matrix elements should enhance the understanding of information that may be obtained from such a study of the electron and nuclear dynamics.

Though we focused on a femtosecond means to monitor electron–nucleus simultaneous dynamics, we do never claim that attosecond investigation of electron dynamics is unnecessary in chemical applications. On the contrary, it would be extremely exciting to know how the electronic wavepacket is driven by the spike-like pulses of subfemtoseconds or shorter width. In addition, it would be interesting to apply additional (independent) attosecond pulses during the refractory period in a pulse train.

In conclusion, the realm for femtosecond time-resolved photoelectron spectroscopy to cover has been extended such that it is applicable to simultaneous electronic and vibrational dynamics in molecules.

5.7 Photoemission arising from electron transfer within a molecule

Recent progress in the studies of electron and nuclear dynamics in molecules sheds a new light on electron-photon interactions. For instance, the so-called high harmonic generation (HHG) technique [96, 101, 237] makes use of above-threshold ionization, in which electrons driven by an alternate electric field collide with the cation, emitting light consisting of many high harmonics of ultrashort pulses. HHG is now a viable optical source for chemical dynamics [49, 349, 375, 383, 465]. In this section we explore possible photoemission resulting from back-and-forth electron jumps induced by continuous modulation of a potential energy curve of ionic character that nonadiabatically couples with another potential curve of covalent nature. This type of nonadiabatic interaction is quite common to alkali halides. For instance the NaI molecule undergoes long-range quantum mechanical electron-jump $\mathrm{NaI} \rightarrow \mathrm{Na^{+}I^{-}}$ and $\mathrm{NaI} \leftarrow \mathrm{Na^{+}I^{-}}$, depending on the relative nuclear velocity at the nonadiabatic avoided crossing region [354]. It is therefore expected that quasi-periodical electron jumps forced by an external alternating electric field should result in an alternation of the large dipole field and lead to photoemission. As for ultrafast transitions, it has been shown theoretically [17–19] and experimentally [190] that the instant of the passage of nuclear wavepackets across the nonadiabatic region can be directly observed with femtosecond time-resolved photoelectron spectroscopy. Furthermore, recent experimental and theoretical studies have shown that such nonadiabatic transition probabilities can be controlled externally with an alternating electric field or laser fields by lifting up and down the ionic counterpart of the potential curves and thereby shifting the crossing point, which is known as the dynamical Stark effect [163, 355, 404].

If this expected photoemission really takes place, the resultant spectra should reflect the nonlinear dynamics of nonadiabatic vibrational motion under an external field, which is similar to classical driven oscillators such as a forced Duffing oscillator [156, 239]. Therefore various nonlinear phenomena such as limit cycle, frequency locking, and chaos (1.5-dimensional chaos) [156, 239] can be expected, which would be intrinsically originated from the quantum dynamics. Furthermore, one may be able to control the frequency and amplitude of the photoemission by varying the laser parameters applied. It may be possible to utilize the photoemission as a new optical source and also as finger-print signals to identify molecular species and/or molecular states. In this section we illustrate the appearance of such spectra with quantum wavepacket dynamics calculations on the ab initio potential curves of the LiF molecule.

As usual, the dynamics of one-dimensional (internuclear distance R) vibrational wavefunctions $\Psi_1(R,t)$ and $\Psi_2(R,t)$ associated with diabatic electronic states 1 and 2, respectively, is given by the Hamiltonian $\hat{H}$,

$$\hat{H} = \hat{T}_{\mathrm{N}}\mathbf{1} + \mathbf{V}(R) + \mathbf{V}_F(R,t), \tag{5.40}$$

where $\hat{T}_{\mathrm{N}}$ is the nuclear kinetic energy operator and the 2×2 potential energy matrix $\mathbf{V}(R)$ includes the diabatic potential energy curves $V_{11}(R)$ and $V_{22}(R)$ for the diabatic states 1 and 2, respectively, as well as the diabatic coupling potential $V_{12}(R)$ between them. The time-dependent interaction $\mathbf{V}_F(R,t)$ by a linearly polarized external field is given in the dipole approximation,

$$\mathbf{V}_F(R,t) = \begin{pmatrix} \mu_{11}(R) & \mu_{12}(R) \\ \mu_{12}(R) & \mu_{22}(R) \end{pmatrix} F(t), \tag{5.41}$$

where the external field $F(t)$ has the form $F(t) = -E_{\mathrm{d}} \sin(\omega_{\mathrm{d}} t) - E_{\mathrm{p}} f(t) \sin(\omega_{\mathrm{p}} t + \phi)$, consisting of a continuous driving field of frequency ω_{d} and amplitude E_{d} (or intensity $I_{\mathrm{d}} = c\varepsilon_0 E_{\mathrm{d}}^2/2$, where c is the speed of light and ε_0 is the electric constant) and a pump pulse with envelope function $f(t)$, frequency ω_{p}, and amplitude E_{p} (intensity I_{p}) acting through the dipole moments $\mu_{11}(R)$ and $\mu_{22}(R)$ for the respective states and the transition dipole moment $\mu_{12}(R)$ between them in the polarization direction of the external field. To see the lowest-order effects, second order and higher interactions with the field are neglected in this study. The center of the pump pulse defines time $t = 0$. The carrier phase difference ϕ between the driving and pump components is not important for the frequencies considered in this study. Numerically solving the coupled time-dependent Schrödinger equation [9, 220] we obtain the time evolution of the vibrational wavefunctions $\Psi_1(R,t)$ and $\Psi_2(R,t)$, and hence the time evolution of the molecular dipole $\mu(t)$,

$$\mu(t) = \int \begin{pmatrix} \Psi_1^* & \Psi_2^* \end{pmatrix} \begin{pmatrix} \mu_{11}(R) & \mu_{12}(R) \\ \mu_{12}(R) & \mu_{22}(R) \end{pmatrix} \begin{pmatrix} \Psi_1 \\ \Psi_2 \end{pmatrix} dR. \tag{5.42}$$

The power spectrum of this dipole moment

$$S(\omega) = \left| \int_{-\infty}^{\infty} \mu(t) e^{i\omega t} dt \right|^2 \tag{5.43}$$

is proportional to the spectral density of photoemission, though refined descriptions may be necessary, such as the Fourier transform of $\ddot{\mu}(t)$, when discussing the intensities of emission [32, 70, 114, 162, 402].

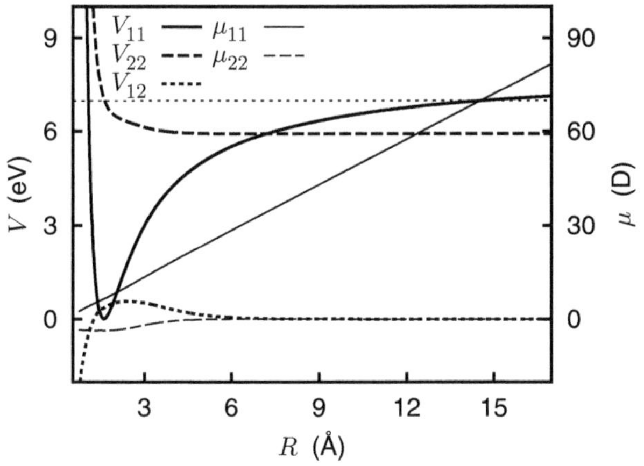

Fig. 5.38 Diabatic potential curves and dipole moment curves of the LiF molecule. (Reprinted with permission from Y. Arasaki *et al.*, J. Chem. Phys. 138, 161103 (2013)).

We computed the time-dependent molecular dipole function $\mu(t)$ [Eq. (5.42)] and associated emission spectra $S(\omega)$ [Eq. (5.43)] for the orientation-fixed LiF molecule under different driving fields. Potential energy and dipole moment curves were interpolated from the set of points computed by Giese and York with multireference configuration interaction at the complete basis set limit [146]. Diabatization is modeled by diagonalization of the dipole moment matrix [251, 469]. $\mu_{12} = 0$ for the diabatization model used. Although this transformation is not as rigorous as that proposed by Baer [28], obtained diabatic curves have been confirmed to be consistent with those obtained through computation of the nonadiabatic coupling matrix elements (see Sec. 5.2.1.1). Mathematical mechanisms of photoabsorption and nonadiabatic transition under the diabatic representation of $\mu_{12} = 0$ should be referred to Ref. [355]. The resulting curves are shown in Fig. 5.38. The system has a narrowly avoided crossing (in the adiabatic representation) of ionic (here, diabatic state 1) and covalent (state 2) potential energy curves at 7 Å, typical of an alkali halide.

Starting from the ground vibrational state on the ground electronic state, the time evolution of the system under a driving field of intensity $I_\mathrm{d} = 1.69 \times 10^{13}$ W cm^{-2} with energy $\hbar\omega_\mathrm{d} = 0.4$, 0.8, 1.6, 3.2, 6.4, or 12.8 eV was computed with or without a pump pulse of intensity $I_\mathrm{p} = 4.8 \times 10^{12}$ W cm^{-2}, energy $\hbar\omega_p = 6.94$ eV (dotted line in Fig. 5.38), and a Gaussian envelope $f(t)$ having a full width at half-maximum of 20 fs [355]. The dotted line in Fig. 5.38 indicates the pump pulse energy used. The spatial grid is taken as 2048 points between 0.7 and 26.3 Å. Time step of the split-operator short-time propagation [9] is 1.0 as. To eliminate abrupt turn-on effects of the continuous driving field, the system is evolved under a linearly ramping driving field for a sufficiently long time before sampling $\mu(t)$ for computation of $S(\omega)$ or applying a pump pulse.

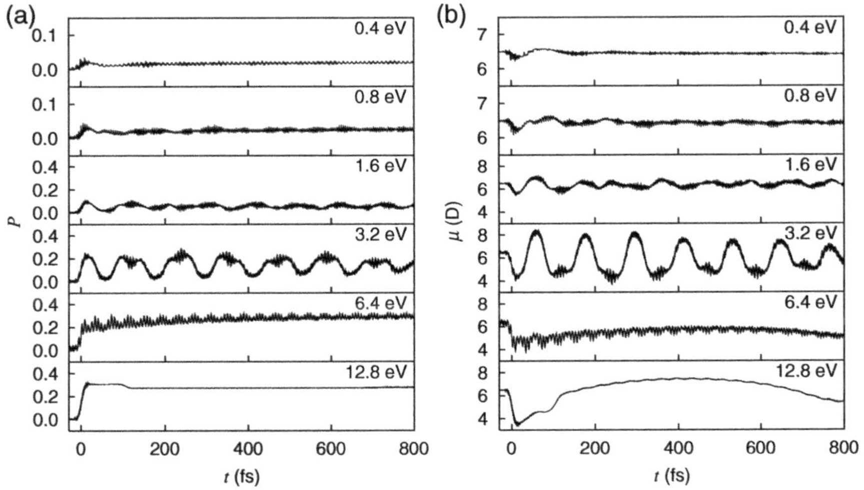

Fig. 5.39 Time evolution under driving field and pump pulse. (a) diabatic state 2 population $P(t) = \langle \Psi_2 | \Psi_2 \rangle$. (b) molecular dipole moment function $\mu(t)$ [Eq. (5.42)]. (Reprinted with permission from Y. Arasaki *et al.*, J. Chem. Phys. 138, 161103 (2013)).

Figure 5.39(a) shows the time evolution of the population on diabatic state 2 (covalent state) under the effect of a driving field with a pump pulse. Energy of the driving field is indicated at the top right of each panel. The time evolution of the dipole moment function $\mu(t)$ is shown in Fig. 5.39(b). Note the change in the vertical scale between smaller and larger $\hbar\omega_d$. State 2 dipole moment is zero at larger internuclear distances while state 1 dipole moment is linearly dependent on the internuclear distance. Thus the dipole moment function is large when there is more population in state 1 and when the wavepacket on state 1 is located at larger internuclear distances. The time evolution of the population of state 1 is seen to be directly reflected in that of the dipole moment function.

The $\hbar\omega_d = 12.8$ eV driving field is too far from resonant excitation energy to cause significant population transfer between the electronic states, so that the time evolution for this case is similar to what would be obtained without the driving field [355]. The pump pulse causes electronic excitation near $t = 0$ fs and the state 2 population is initially formed as a vibrationally excited wavepacket. Internuclear distance R increasing on the dissociative covalent state causes the wavepacket to reach the avoided crossing region at $t = 100$ fs (for the $\hbar\omega_{\mathrm{d}} = 12.8$ eV case), seen as a population decrease in Fig. 5.39(a). The ionic state population there would see a corresponding population increase. The bound ionic state population returns to the avoided crossing region near $t = 800$ fs; this motion on the bound ionic state is seen as the arching in Fig. 5.39(b).

At long times (> 400 fs) LiF on diabatic state 2 would dissociate, handled by an optical potential eliminating the wavepacket component going out of the spatial grid in the computation. The dissociating population is kept track of and the figures do not show the decrease due to dissociation.

The driving laser fields of other frequencies have significant effect on the population dynamics. For the $\hbar\omega_{\mathrm{d}} = 6.4$ eV case we see a rapid (subfemtosecond time scale) electronic state transfer corresponding to $\omega_{\mathrm{d}} = 6.4/\hbar$, which is superimposed with far longer time scale dynamics. For the cases of $\hbar\omega_{\mathrm{d}} = 3.2$ eV and less, we see a different behavior. The driving field perturbs more significantly the excited electronic state and causes the vibrational period on it to shorten by an order of magnitude to approximately 100 fs. When the excited vibrational wavepacket is located at the outer (inner) turning point, the excited population is mostly on the ionic (covalent) curve. At a moment when the external field effectively brings the avoided crossing to shorter distances where V_{12} is larger, there is much more transfer between the diabatic states through the interaction at the avoided crossing. Thus the location, speed, spatial distribution (shape) of the nuclear wavepackets, which are all affected by the laser frequency, intensity, and so on, create various patterns of the emission spectrum.

Figure 5.40 shows the spectra computed from the time-dependent dipole moments in Fig. 5.39. Energy of the driving field is indicated within each panel. Fast Fourier transform in the range from $t = 275$ to 800 fs, a time range long enough after the pump pulse, is used to obtain $S(\omega)$. ($t = 275$ was chosen so as to avoid unnecessary complication due to the interaction during the pump.) In each panel, the solid (upper) and dashed (lower) curves show the spectra obtained with and without the pump pulse, respectively. The lower curves are plotted shifted as $(\log S - 6)$ for clearer comparison with the upper curves. Note the difference in scales for the energy axes.

Without the pump pulse, the spectra simply show peaks at the fundamental frequency and at several harmonics of the respective driving fields

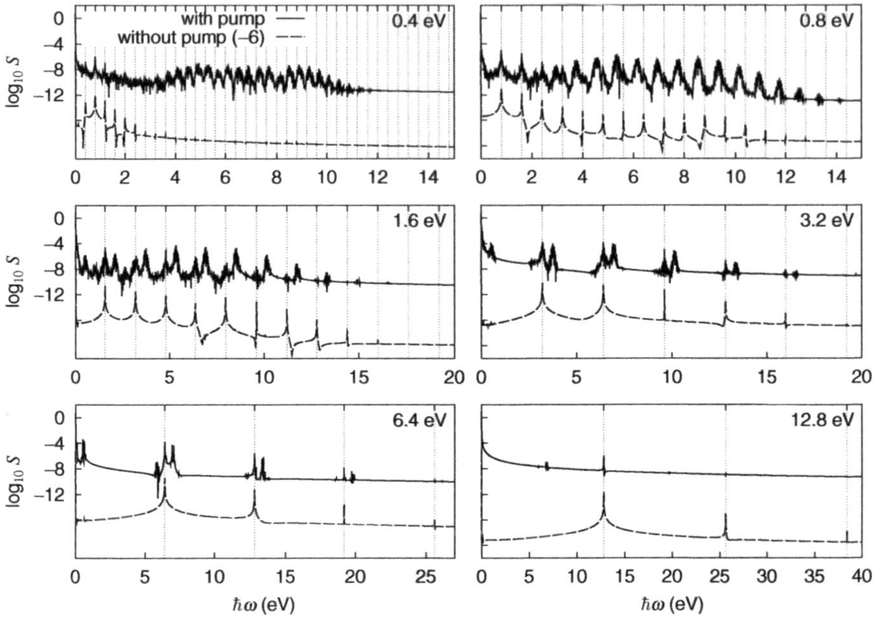

Fig. 5.40 Emission spectra computed from time evolutions of the molecular dipole moment function in Fig. 5.39(b). Solid (upper) curve: $\log S(\omega)$, computation with pump pulse. Dashed (lower) curve: $\log S(\omega) - 6$, computation without pump pulse. (Reprinted with permission from Y. Arasaki *et al.*, J. Chem. Phys. 138, 161103 (2013)).

(because the molecule is fixed in space, both even and odd harmonics appear). The nuclear wavepacket dynamics remains mainly on diabatic state 1 and virtually no electron transfer at the avoided crossing is involved in the generation of harmonics. Under application of the pump pulse, on the other hand, except for the $\hbar\omega_{\rm d} = 12.8$ eV case, we see significant change in the emission spectra. A second peak develops about 0.5 eV above each of the peaks at multiples of the fundamental frequency, which arises from a tentative state that has components both on V_1 and V_2 exchanging the populations in an oscillatory manner through the nonadiabatic and optical interactions at the region of the inner turning point. It would be interesting to compare this emission peak with the so-called "hyper-Raman" lines in harmonic generation spectrum from an atom in an intense laser field [141, 322]. Further, the present peak is associated with progressions of far finer structures. These fine structures have energy spacings of 0.1 eV (for the $\hbar\omega_{\rm d} = 6.4$ eV case) or smaller (for cases of 3.2 eV and less) which seem to be consistent with the vibrational periods of the excited wavepacket as modified by the driving fields. In the lower energy cases (0.4 eV to 1.6 eV), the signals truly arising from nonadiabatic coupling is more prominent.

We also note that the signals under application of the pump pulse exhibit continuous spectra of the order $S(\omega) \propto \omega^{-2}$ near the origin ($\hbar\omega = 0$).

We have observed with LiF as an example that nonadiabatic jump of electrons within a molecule can cause photoemission with characteristic harmonic generation. Since the nonadiabatic population transfer can be modified/controlled by an external field, the resultant photoemission is also controllable. Such emission spectra would provide signals of electronic and vibrational dynamics at femtosecond and subfemtosecond time scales as a function of tunable parameters of the external field. It may be utilized as a characteristic optical source, which is hard to be attained by other means.

The dynamics of molecular rotation and the effect of orientation have been neglected in the present study for simplicity. However, it should modulate the spectra in a way that would provide additional information about coupling between vibrational and rotational modes of nonrigid molecules in large amplitude motion.

Chapter 6

Nonadiabatic Electron Wavepacket Dynamics in Path-branching Representation

In this chapter we introduce a theory of electron dynamics involved with nonadiabatic path branching of nuclear motions. We have already emphasized in Chapter 1 an importance of the studies on real-time dynamics of electron wavepackets, which may undergo nonadiabatic transitions in the presence or absence of optical fields. The theory presented below can be regarded as one of the most general theories for dynamics in mixed quantum and classical representation. It will also give a complete theoretical foundation of the semiclassical Ehrenfest theory, which was derive rather intuitively without a rigorously theoretical basis. This section also serves as a preparation of the (nonadiabatic) electron wavepacket dynamics of molecules placed in intense and/or extremely short pulse lasers, which will be discussed in Chap. 8. Even without lasers, the electron wavepacket theory offers a promising method to analyze very fast chemical reactions, in which the dynamics of electrons are supposed to dominate chemical changes.

6.1 Path-branching representation for electron wavepacket propagation

6.1.1 *Theoretical background: A representation of the total wavefunctions*

6.1.1.1 *Summary of the standard nuclear wavepacket dynamics for nonadiabatic transitions*

For the convenience, we here briefly summarize the standard formalism of quantum dynamics of molecular dynamics of Chap. 2.

1) The quantum-mechanical molecular Hamiltonian is written generally as

$$H(\mathbf{r},\mathbf{R}) = \frac{1}{2}\sum_k \hat{P}_k^2 + H^{(\mathrm{el})}(\mathbf{r};\mathbf{R}), \tag{6.1}$$

where the many-body electronic Hamiltonian is defined by

$$H^{(\mathrm{el})}(\mathbf{r};\mathbf{R}) = \frac{1}{2}\sum_j \hat{p}_j^2 + V_c(\mathbf{r};\mathbf{R}). \tag{6.2}$$

Here, $\mathbf{r}$ and $\mathbf{R}$ denote the electronic and nuclear coordinates, respectively, and $\hat{p}_j$ and $\hat{P}_k$ are the operators of their conjugate momenta. $V_c(\mathbf{r};\mathbf{R})$ is the Coulombic interaction potential among electrons and nuclei.

2) The corresponding wavefunctions are usually represented in the Born–Huang expansion as

$$\Psi(\mathbf{r},\mathbf{R},t) = \sum_I \chi_I(\mathbf{R},t)\Phi_I(\mathbf{r};\mathbf{R}). \tag{6.3}$$

with the electronic basis that depends on the instantaneous positions of nuclei $\{\Phi_I(\mathbf{r};\mathbf{R})\}$. As usual we assume that electronic basis set $\{|\Phi_I\rangle\}$, either adiabatic or diabatic, are orthonormal

$$\langle\Phi_I(\mathbf{R})|\Phi_J(\mathbf{R})\rangle = \delta_{IJ}. \tag{6.4}$$

3) Coupled equations of motion for the nuclear wavepackets are

$$i\hbar\dot{\chi}_I = \frac{1}{2}\sum_k \hat{P}_k^2\chi_I - i\hbar\sum_k\sum_J X_{IJ}^k\hat{P}_k\chi_J - \frac{\hbar^2}{2}\sum_k\sum_J Y_{IJ}^k\chi_J + \sum_J H_{IJ}^{(\mathrm{el})}\chi_J, \tag{6.5}$$

or

$$i\hbar\dot{\chi}_I = \sum_J\left[\frac{1}{2}\sum_k\left[\left(\mathbf{I}\hat{P}_k - i\hbar\mathbf{X}^k\right)^2\right]_{IJ} + H_{IJ}^{(\mathrm{el})}\right]\chi_J, \tag{6.6}$$

where

$$X_{IJ}^k = \left\langle\Phi_I\middle|\frac{\partial\Phi_J}{\partial R_k}\right\rangle,\quad Y_{IJ}^k = \left\langle\Phi_I\middle|\frac{\partial^2\Phi_J}{\partial {R_k}^2}\right\rangle, \tag{6.7}$$

and

$$H_{IJ}^{(\mathrm{el})}(\mathbf{R}) = \left\langle\Phi_I\middle|H^{(\mathrm{el})}\middle|\Phi_J\right\rangle. \tag{6.8}$$

with $[\mathbf{I}]_{IJ} = \delta_{IJ}$ and $[\mathbf{X}^k]_{IJ} = X_{IJ}^k$.

4) When we solve Eq. (6.5) in a system, the phenomena of nuclear wavepacket bifurcation both in the passage of avoided crossing and conical intersection. It has been shown in Chap. 5 that this is not a mathematical consequence but can be experimentally observed as physical fact, which is a strong quantum entanglement between electronic and nuclear motions.

6.1.2 *Nonadiabatic electron wavepackets along branching paths*

It is actually very difficult to solve the entire scheme down to Eq. (6.5) for systems of chemical interest, even if a very good set of $\{|\Phi_I\rangle\}$ is available. (Note that electronic structure theory (quantum chemistry) can handle far larger molecular systems within the Born–Oppenheimer approximation) than the nuclear dynamics based on Eq. (6.5) can do.) This is because the short wavelength nature of nuclear matter wave blocks accurate computation and brings **classical nature** into the nuclear dynamics, in which path (trajectory) representation is quite often convenient and useful than sticking to the wave representation. Then what do the paths of nuclear dynamics look like on the occasion of nonadiabatic transitions, for which it is known that the nuclear wavepackets bifurcate, reflecting purely quantum nature.

When we talk about wavepacket bifurcation in nonadiabatic transitions, on the other hand, not only the nuclear wavepackets but the electronic counterparts should also undergo branching.

Therefore we explore below a theoretical framework in which electronic wavepackets propagate in time along bifurcating nuclear paths.

6.1.2.1 *The total Hamiltonian represented in nuclear-configuration and electronic-Hilbert spaces*

To formulate the electron wavepacket dynamics on a clear basis, we begin by representing the total Hamiltonian operator in a basis set $\{|\Phi_I(\mathbf{R})\rangle|\mathbf{R}\rangle\}$ such that [422, 423]

$$H(\mathbf{R}, elec) \equiv \frac{1}{2}\sum_k \left(\hat{P}_k - i\hbar\sum_{IJ}|\Phi_I\rangle X^k_{IJ}\langle\Phi_J|\right)^2 + \sum_{IJ}|\Phi_I\rangle H^{(\mathrm{el})}_{IJ}\langle\Phi_J|, \tag{6.9}$$

where $\hat{P}_k$ is the nuclear momentum operator). Equation (6.9) looks similar to Eq. (6.6) but is certainly different in that $H(\mathbf{R}, elec)$ is the total Hamiltonian. In this representation, the nuclear motion is described in coordinate space (usual $\mathbf{R}$-space, whereas the electron dynamics is tracked in the electronic Hilbert space. In the summation of $\sum_{IJ}$ in this Hamiltonian, one can include the continuum state in treating electronic scattering problem like photoionization. The electronic basis functions $\{\Phi_I\}$ can be either adiabatic or general diabatic-like states, provided that both X^k_{IJ} and

$H_{IJ}^{(\mathrm{el})}$ are included appropriately. The choice of the electronic representation depends on a system and situation under study.

An extension of this representation to more general form so as to include the vector potential arising from classical electromagnetic field will be made later in this book.

6.1.2.2 *Mixed quantum-classical mixed representation of the Hamiltonian and wavefunctions*

It is hard to find an advantage in directly trying to solve Eq. (6.9) compared to the standard method based on Eq. (6.5). We therefore proceed according to the general scenario that "electron wavepackets along nuclear paths".

To do so we replace the nuclear momentum operator $\hat{P}_k$ with its classical counterpart P_k such that

$$\tilde{H}(\mathbf{R},\mathbf{P},elec) \equiv \frac{1}{2}\sum_k \left(P_k - i\hbar \sum_{IJ} |\Phi_I\rangle X_{IJ}^k \langle\Phi_J|\right)^2 + \sum_{IJ} |\Phi_I\rangle H_{IJ}^{(\mathrm{el})} \langle\Phi_J|. \tag{6.10}$$

This is one of the most general expressions in the so-called mixed quantum-classical representation.

For this Hamiltonian, we choose the following representation of the total wavefunction: In contrast to the classic form of Eq. (6.3), we would rather expand the total wavefunction as

$$\Psi(\mathbf{r},\mathbf{R},t) = \sum_I^{path} c_I \Phi_I(\mathbf{r};\mathbf{R}_I(t))\chi_I(\mathbf{R}-\mathbf{R}_I(t),t), \tag{6.11}$$

where $\chi_I(\mathbf{R}-\mathbf{R}_I(t),t)$ is a (normalized) nuclear wavepacket localized nearby a path position $\mathbf{R}_I(t)$, while $\Phi_I(\mathbf{r};\mathbf{R}_I(t))$ is a (normalized) electronic wavepacket at this point. We will be back to this point later.

6.1.3 *Dynamics in the electron-nuclear quantum-classical mixed representation*

6.1.3.1 *Dynamics of quantum electron wavepackets*

The dynamics of electron wavepackets is determined by the time dependent variational principle

$$\delta \int dt \langle\Phi(\mathbf{R},t)| \left(i\hbar\frac{\partial}{\partial t} - \tilde{H}(\mathbf{R},\mathbf{P},elec)\right) |\Phi(\mathbf{R},t)\rangle = 0. \tag{6.12}$$

For an electron wavepacket expanded in a basis set as

$$|\Phi^{(el)}(\mathbf{R}(t))\rangle = \sum_I C_I(t)|\Phi_I\rangle|_{\mathbf{R}=\mathbf{R}(t)}, \tag{6.13}$$

the combination of the left and right variations gives rise to

$$i\hbar\frac{\partial}{\partial t}C_I = \sum_J \left[H_{IJ}^{(el)} - i\hbar\sum_k \dot{R}_k X_{IJ}^k - \frac{\hbar^2}{4}\sum_k \left(Y_{IJ}^k + Y_{JI}^{k*}\right)\right] C_J, \tag{6.14}$$

where the matrix in the right hand side has been Hermitized. This expression is similar to the standard representation of the semiclassical Ehrenfest theory (see Eq. (4.41)), except that Eq. (6.14) includes (nontrivial) corrections as the third term in its right hand side.

6.1.3.2 *"Classical-like" nuclear motions in nonadiabatic transitions—A generalization of classical mechanics*

With the above Hamiltonian, Eq. (6.10), an analogy to purely classical mechanics brings about the "canonical equations of motion" for nuclear classical variables $(\mathbf{R}, \mathbf{P})$ as [492]

$$\frac{d}{dt}R_k = \frac{\partial \tilde{H}}{\partial P_k} \tag{6.15}$$

$$\frac{d}{dt}P_k = -\frac{\partial \tilde{H}}{\partial R_k}. \tag{6.16}$$

These equations are also defined in the nuclear configuration space and electronic Hilbert space. Therefore the quantities $\frac{d}{dt}R_k$ and $\frac{d}{dt}P_k$ should work as transition operators for the electronic states. To emphasize this fact, we express our quantities with the calligraphic fonts as $\dot{\mathcal{R}}^k \equiv \frac{d}{dt}R_k$ and $\ddot{\mathcal{R}}^k \equiv \frac{d^2}{dt^2}R_k$. After some manipulation, [423, 492, 493] we find

$$\begin{aligned} -\frac{d^2}{dt^2}\mathcal{R}^k &= \sum_{IJ} \frac{\partial}{\partial R_k}\left(|\Phi_I\rangle H_{IJ}^{(el)}\langle\Phi_J|\right) \\ &\quad - i\hbar\sum_{IJ}\sum_l \dot{R}_l \left[\frac{\partial}{\partial R_k}\left(|\Phi_I\rangle X_{IJ}^l\langle\Phi_J|\right) - \frac{\partial}{\partial R_l}\left(|\Phi_I\rangle X_{IJ}^k\langle\Phi_J|\right)\right]. \end{aligned} \tag{6.17}$$

To use this operator in practice, we should transform it to a matrix form by sandwiching two electronic states, say, $\langle\Phi_I|$ and $|\Phi_J\rangle$ as

$$\begin{aligned}\mathcal{F}^k_{IJ} &\equiv \left\langle \Phi_I \left| \ddot{\mathcal{R}}^k \right| \Phi_J \right\rangle = \langle \Phi_I | \hat{\mathcal{F}}^k | \Phi_J \rangle \\ &= -\sum_K \left[X^k_{IK} H^{(el)}_{KJ} - H^{(el)}_{IK} X^k_{KJ} \right] - \frac{\partial H^{(el)}_{IJ}}{\partial R_k} + i\hbar \sum_l \dot{R}_l \left[\frac{\partial X^l_{IJ}}{\partial R_k} - \frac{\partial X^k_{IJ}}{\partial R_l} \right],\end{aligned} \tag{6.18}$$

which is equivalent to

$$\mathcal{F}^k_{IJ} = -\left\langle \Phi_I \left| \frac{\partial H^{(\mathrm{el})}}{\partial R_k} \right| \Phi_J \right\rangle + i\hbar \sum_l \dot{R}_l \left[\frac{\partial X^l_{IJ}}{\partial R_k} - \frac{\partial X^k_{IJ}}{\partial R_l} \right], \tag{6.19}$$

if the basis set used was complete. (Again, the masses are all scaled to unity.) We refer to $\mathcal{F}^k_{IJ}$ as the force matrix. Obviously, the first term in the right hand side of Eq. (6.19) comes from the electronic energy, while the second one represents the recoil from the delayed kinematic response of electrons to nuclear motions. This last term represents a force acting in the directions perpendicular to the velocity vector $\left\{\dot{R}_k\right\}$ just as the magnetic Lorentz force does. [423]

6.1.3.3 *The Born–Oppenheimer molecular dynamics*

In case of

$$X^k_{IJ} = 0, \tag{6.20}$$

it is most convenient to adopt the adiabatic wavefunctions $\{\psi_\alpha\}$, satisfying

$$H^{(\mathrm{el})}\psi_\alpha = E^{ad}_\alpha \psi_\alpha. \tag{6.21}$$

Then we have

$$\mathcal{F}^k_{\alpha\beta} = -\delta_{\alpha\beta}\frac{\partial E^{ad}_\beta}{\partial R_k}, \tag{6.22}$$

subject to the condition Eq. (6.20). Therefore the force matrix in the adiabatic limit reproduces the Born–Oppenheimer forces in its diagonal elements. Nuclear motion on each potential energy surface (PES) is governed by each force arising its belonging PES. This is the theoretical foundation of why one can make use of classical mechanics with use of the energy gradient of the adiabatic potential energy surfaces as a force working on nuclei.

On the other hand, when the force matrix $\mathcal{F}^k_{IJ}$ has the nonzero off-diagonal matrix elements, the classical dynamics should be changed accordingly. This is generalization of classical mechanics under the presence of nonadiabatic interactions.

6.1.4 *The semiclassical Ehrenfest theory as a special case*

Let us take an average of the force operator defined in Eq. (6.18) over the running total electron wavepacket of Eq. (6.13) such that

$$\begin{aligned}\mathcal{F}_{av}^k &= \langle \Phi^{(el)}(\mathbf{R}(t))|\hat{\mathcal{F}}^k|\Phi^{(el)}(\mathbf{R}(t))\rangle \\ &= -\sum_{I,J} C_I^* C_J \left[\frac{\partial H_{IJ}^{(el)}}{\partial R_k} + \sum_K \left(X_{IK}^k H_{KJ}^{(el)} - H_{IK}^{(el)} X_{KJ}^k \right) \right] \\ &\quad + i\hbar \sum_{I,J} C_I^* C_J \sum_l \dot{R}_l \left[\frac{\partial X_{IJ}^l}{\partial R_k} - \frac{\partial X_{IJ}^k}{\partial R_l} \right]. \end{aligned} \tag{6.23}$$

If the basis set was complete, $\mathbf{F}_{av}^k$ becomes dramatically simpler as

$$\mathcal{F}^k = \langle \Phi^{(el)}(\mathbf{R}(t))| \frac{\partial H^{(el)}}{\partial R_k} |\Phi^{(el)}(\mathbf{R}(t))\rangle. \tag{6.24}$$

Upon comparing this force with that in Eq. (4.43), $\mathcal{F}_{av}^k$ gives rise to the Ehrenfest (mean-field) force. Therefore, the mean-field force does not arise from a physical origin but from our "mathematical" or "artificial" operation of averaging. However, this averaging is not necessarily bad in case where the nonadiabatic interaction is so strong that the electronic states may largely fluctuate from time to time.

Note that the above derivation of SET is not intuitive one in contrast to that starting from Eq. (4.38). Also the equations of motion for electron dynamics, Eq. (6.14) are different from Eq. (4.41) in that the former includes the second order terms like $Y_{IJ}^k + Y_{JI}^{k*}$.

Thus the theoretical basis of the semiclassical Ehrenfest theory is found on a purely theoretical ground.

6.2 Methods of averaging and branching

6.2.1 *Electronic state mixing along branching paths*

6.2.1.1 *Eigenforce and branching paths*

The electronic wavepacket are to be carried along nuclear paths that are driven by the force matrix. Suppose we have an electronic wavepacket $\Phi(\mathbf{r};\mathbf{R}(t))$ at a phase space point $(\mathbf{R},\mathbf{P})$, where $\mathbf{R} = \mathbf{R}(t)$ and $\mathbf{P}$ is its conjugate momentum. To materialize an electronic-state mixing among given basis functions $\{\Phi_I(\mathbf{r};\mathbf{R})\}$ (either adiabatic or any diabatic basis), we first integrate Eq. (6.14) for a short time, say, Δt to give a new set of $\{C_I(t)\}$. Next we want to run a path using the force matrix $\mathcal{F}(\mathbf{R})$ again for a short time Δt. However, the presence of the off-diagonal elements in the force matrix can induce additional electronic-state mixing. To avoid this additional mixing, we diabatize the electronic basis for this time interval by diagonalizing the force matrix at $\mathbf{R}$ such that

$$\mathbf{U}(\mathbf{R})\,\mathcal{F}(\mathbf{R})\,\mathbf{U}(\mathbf{R})^{-1} = \begin{pmatrix} f_1(\mathbf{R}) & 0 & \cdots \\ 0 & f_2(\mathbf{R}) & \\ \vdots & & \ddots \end{pmatrix} \tag{6.25}$$

with the associated electronic basis-set transformation

$$\mathbf{U}(\mathbf{R}) \begin{pmatrix} \Phi_1(\mathbf{r};\mathbf{R}) \\ \Phi_2(\mathbf{r};\mathbf{R}) \\ \vdots \end{pmatrix} = \begin{pmatrix} \lambda_1(\mathbf{r};\mathbf{R}) \\ \lambda_2(\mathbf{r};\mathbf{R}) \\ \vdots \end{pmatrix}. \tag{6.26}$$

The electronic wavepacket obtained as above at $\mathbf{R}$ may be re-expanded in the eigenfunctions $\{\lambda_K(\mathbf{r};\mathbf{R})\}$ such that

$$\Phi(\mathbf{r};\mathbf{R}(t)) = \sum_K D_K(t)\lambda_K(\mathbf{r};\mathbf{R}). \tag{6.27}$$

Then, each electronic component $D_K(t)\lambda_K(\mathbf{r};\mathbf{R})$ should be carried by its own path that is to be driven by the eigenforce f_K, without the electronic mixing among $\{\lambda_L(\mathbf{r};\mathbf{R})\}$, to reach a new point after Δt. (Note however that mixing among $\{\lambda_L(\mathbf{r};\mathbf{R})\}$ can occur in the electron dynamics even in $\lambda_L(\mathbf{r};\mathbf{R})$-representation of Eq. (6.14).) Different eigenforces make different paths even if they start from a single phase-space point $(\mathbf{R},\mathbf{P})$ in such a way that

$$(\mathbf{R},\mathbf{P}) \rightarrow (\mathbf{R}_K,\mathbf{P}_K). \tag{6.28}$$

Therefore a path at $\mathbf{R}$ is branched to as many pieces as the number of electronic states involved in the nonadiabatic coupling. The electronic-state mixing should be considered again at the individual points $(\mathbf{R}_K,\mathbf{P}_K)$, regarding the corresponding component $D_K(t)\lambda_K(\mathbf{r};\mathbf{R})$ in Eq. (6.27) as the renewed condition to integrate Eq. (6.14). Again, the path branching is followed at each $(\mathbf{R}_K,\mathbf{P}_K)$. Hence, the cascade of path-branching should continue as long as the nonadiabatic coupling cannot be effectively ignored.

6.2.1.2 *Branching of electron wavepackets*

As discussed above, the nuclear paths can branch due to the presence of nonadiabatic interactions. Since the electronic wavepackets are to be carried over along those branching paths, the electronic packets should also branch accordingly. Such branchings can be viewed as follows.

An advantage of the semiclassical Ehrenfest theory (SET) is that it gives quite accurate transition amplitudes, as far as a single passage of the crossing region is concerned. [493] On the other hand, an obvious drawback is, as repeatedly claimed, that the path thus driven is forced to run on an artificial potential energy function built by average over the relevant adiabatic potential surfaces. More explicitly, one can write as

$$\begin{aligned}&\Phi(\mathbf{r};\mathbf{R}_I(t_{before}))\delta(\mathbf{R}-\mathbf{R}_I(t_{before}))\\&\rightarrow\delta(\mathbf{R}-\mathbf{R}(t_{after}))\sum_K C_K(t)\Phi_K(\mathbf{r};\mathbf{R}(t_{after})),\end{aligned}\tag{6.29}$$

where t_{before} and t_{after} are, respectively, a time before and after passing across the nonadiabatic region, and $\mathbf{R}(t_{after})$ is a path running on an averaged potential

$$\sum_{K,L}^{path} C_K(t)^* C_L(t)\left\langle\Phi_K(\mathbf{R}_{path})\left|H^{(\mathrm{el})}(\mathbf{R}_{path})\right|\Phi_L(\mathbf{R}_{path})\right\rangle.\tag{6.30}$$

It is obvious that the path is not branched by the nonadiabatic interaction, and therefore the important feature of quantum entanglement caused by nonadiabatic transition is not captured correctly in this formalism. Furthermore, in a more rigorous mathematical context, the above naive treatment of electron wavepackets along with classical paths is found to be not quite correct. Both of Eqs. (4.41) and (4.43) need corrections, which should be referred in [493].

To appropriately describe the entanglement situation induced by electronic nonadiabatic transition, we should impose the following condition

$$\begin{aligned}&\Phi(\mathbf{r};\mathbf{R}_I(t_{before}))\chi(\mathbf{R}-\mathbf{R}_I(t_{before}))\\&\rightarrow\sum_K\sum_k^{\text{on }K}\Phi_K(\mathbf{r};\mathbf{R}_{Kk}(t_{after}))\chi_{Kk}(\mathbf{R}-\mathbf{R}_{Kk}(t_{after})),\end{aligned}\tag{6.31}$$

where $\mathbf{R}_{Kk}(t_{after})$ represents the kth path running on the Kth adiabatic potential energy surfaces at t_{after}. This is equivalent to requiring that every path (or nuclear wavepacket associated with) should branch to plural pieces individually. path branching is in general not realized in a well-posed initial value problem such as the ordinary differential equations like the Newtonian equations of motion giving classical trajectories. Therefore a mathematical trick is necessary to materialize the branching. Below we exploit a natural way to achieve path branching in an efficient and accurate manner.

6.2.1.3 *Averaging over the paths to extract a (few) representative path(s) in the coupling region*

It is obvious that tracking every series of branching paths is extremely cumbersome and technically impossible, although the branching feature reflects the essential physics behind nonadiabatic interaction. However, it is anticipated in case where the coupling region is narrow enough that those branching paths should not geometrically deviate much from each other in phase space. In other words, they should localize along a representative path forming a tube-like structure. Therefore we extract such a representative path by taking an average of phase-space points in the following manner:

(i) Suppose we have a path ending at $(\langle \mathbf{R}(t)\rangle, \langle \mathbf{P}(t)\rangle)$ in phase space. At this point, diagonalize the force matrix,

$$\mathcal{F}(\langle \mathbf{R}\rangle)\,|\lambda_K(\langle \mathbf{R}\rangle)\rangle = |\lambda_K(\langle \mathbf{R}\rangle)\rangle f_K(\langle \mathbf{R}\rangle) \tag{6.32}$$

to obtain the eigenforces $\{f_K\}$ and its eigenstates $\{|\lambda_K\rangle\}$. The wavepacket $\Phi(\mathbf{r}; \langle \mathbf{R}(t)\rangle)$ is expanded in terms of these eigenstates as in Eq. (6.27).

(ii) The K-th eigenforce drives a path starting from $(\langle \mathbf{R}(t)\rangle, \langle \mathbf{P}(t)\rangle)$ for a short time Δt in terms of the Hamilton canonical equations of motion as

$$\mathbf{R}_K(t+\Delta t) = \langle \mathbf{R}(t)\rangle + \Delta \mathbf{R}_K \tag{6.33}$$

$$\mathbf{P}_K(t+\Delta t) = \langle \mathbf{P}(t)\rangle + \Delta \mathbf{P}_K. \tag{6.34}$$

(iii) Average them into the form

$$\langle \mathbf{R}(t+\Delta t)\rangle = \langle \mathbf{R}(t)\rangle + \sum_K |D_K(t)|^2 \Delta \mathbf{R}_K / \sum_L |D_L(t)|^2 \tag{6.35}$$

$$\langle \mathbf{P}(t+\Delta t)\rangle = \langle \mathbf{P}(t)\rangle + \sum_K |D_K(t)|^2 \Delta \mathbf{P}_K / \sum_L |D_L(t)|^2, \tag{6.36}$$

which makes the next point $(\langle \mathbf{R}(t+\Delta t)\rangle, \langle \mathbf{P}(t+\Delta t)\rangle)$ of the representative path.

(iv) With this averaged point, we calculate

$$\begin{aligned}&\mathcal{F}(\langle \mathbf{R}(t+\Delta t)\rangle)\,|\lambda_K(\langle \mathbf{R}(t+\Delta t)\rangle)\rangle \\ &= |\lambda_K(\langle \mathbf{R}(t+\Delta t)\rangle)\rangle f_K(\langle \mathbf{R}(t+\Delta t)\rangle)\end{aligned} \tag{6.37}$$

and turn anew to the step (ii).

The successive applications of the procedure (i)-(iii) gives a single finite path.

6.2.1.4 *Phase-space averaging vs. force averaging*

In the eigenforce representation, the averaged forces in the SET (recall Eq. (6.24)) are represented as

$$\ddot{\mathbf{R}}^k = \sum_K |D_K(t)|^2 f_K^k / \sum_L |D_L(t)|^2. \tag{6.38}$$

The first order approximation to the SET is

$$\begin{aligned}\langle \mathbf{P}(t+\Delta t)\rangle_{SET} &= \langle \mathbf{P}(t)\rangle + \ddot{\mathbf{R}}\Delta t \\ &= \langle \mathbf{P}(t)\rangle + \sum_K |D_K(t)|^2 \Delta \mathbf{P}_K / \sum_L |D_L(t)|^2 \end{aligned} \tag{6.39}$$

and

$$\begin{aligned}\langle \mathbf{R}(t+\Delta t)\rangle_{SET} &= \langle \mathbf{R}(t)\rangle + \langle \mathbf{P}(t)\rangle \Delta t \\ &= \langle \mathbf{R}(t)\rangle + \langle \mathbf{P}(t)\rangle \sum_K |D_K(t)|^2 \Delta t / \sum_L |D_L(t)|^2 \end{aligned} \tag{6.40}$$

Equations (6.39) and (6.36) are formally the same (to the first order) for the short interval. Also, it holds

$$\langle \mathbf{P}(t)\rangle \sum_K |D_K(t)|^2 \Delta t \simeq \sum_K |D_K(t)|^2 \Delta \mathbf{R}_K \tag{6.41}$$

in the first order approximation, and Eqs. (6.40) and (6.35) are also very close to each other in the first order of Δt. Thus, it turns out that phase-space averaging and SET are equivalent to each other in the first order approximation. Technically, however, the latter (SET) should be faster than the former because diagonalization of the force matrix is not necessary. Therefore we use the force averaging as an averaging procedure in what follows for its simplicity.

6.2.1.5 *Branching of the averaged path*

When a molecule gets into an effectively coupling-free region after passing through an avoided-crossing zone, the force matrix smoothly becomes a diagonal matrix, whose diagonal elements represent the forces arising from the individual adiabatic potential energy surfaces. Also, the adiabatic wavefunctions become the eigenfunctions of the force matrix. At the same time, the mixing of electronic states is switched off gradually. Therefore, we stop taking the average as in Eqs. (6.35) and (6.36), and simply let the every individual component

$$D_K(t)\lambda_K(\mathbf{r};\mathbf{R}_K(t)) \tag{6.42}$$

run being driven by its own force. Then the coefficients $D_K(t)$ coherently carry the information of transition amplitudes. In this way, a path can naturally branch as soon as the averaging is terminated.

After the first branching performed as above, the individual paths are being driven by their own eigenforces. Suppose we are tracking one of them, say, the Kth path, $(\mathbf{R}_K(t), \mathbf{P}_K(t))$. To emphasize that every force is generated along this path, we rewrite Eq. (6.32) in the matrix form explicitly as

$$\mathbf{U}(\mathbf{R}_K)\mathcal{F}(\mathbf{R}_K)\mathbf{U}(\mathbf{R}_K)^{-1} = \begin{pmatrix} f_1(\mathbf{R}_K) & 0 & \cdots \\ 0 & f_2(\mathbf{R}_K) & \\ \vdots & & \ddots \end{pmatrix}, \tag{6.43}$$

where the dependence of the force matrix on $\mathbf{R}_K$ has been stressed. The right hand side of this representation reminds that other eigenforces, say, f_L are also calculated along $\mathbf{R}_K(t)$. Therefore, at a point on the path $(\mathbf{R}_K(t), \mathbf{P}_K(t))$, one may switch the force from f_K to f_L to emanate *another* path, such that $(\mathbf{R}_K(t), \mathbf{P}_K(t)) \to (\mathbf{R}_L(t+\Delta t), \mathbf{P}_L(t+\Delta t))$ in terms of f_L. If one uses f_K at the same point, it follows that $(\mathbf{R}_K(t), \mathbf{P}_K(t)) \to (\mathbf{R}_K(t+\Delta t), \mathbf{P}_K(t+\Delta t))$. Thus this procedure allows further branching. These are the paths of the first generation of branching. Along the paths of the first generation, we carry out the additional electronic-state mixing.

The individual branched paths of the first generation may bifurcate again to just run toward the asymptotic region without further electronic mixing. Or, they can proceed to the second generation, taking another account of electronic-state mixing and the final branching of just releasing to the adiabatic surfaces. In the similar way, we can make a cascade of branching.

The method presented in this section, which is a combined use of phase-space averaging and natural branching, is referred to as Phase Space Averaging and Natural Branching (PSANB).

6.2.2 *The electronic wavepackets on the branching paths*

We have thus made up electron wavepacket dynamics nonadiabatically coupled with branching nuclear paths within the above quantum-classical mixed representation. The total wavefunction up to this point is explicitly written as follows. Suppose an electron wavepacket $\Phi_I(\mathbf{r};\mathbf{R}(t_{before}))$ begins to run on an adiabatic potential energy surface I being carried by the associated nuclei, whose initial condition is $(\mathbf{R}_{Ii}(t_{before}), \mathbf{P}_{Ii}(t_{before}))$. Asymptotically they are branched into pieces as

$$\Phi_I(\mathbf{r};\mathbf{R}(t_{before}))\delta(\mathbf{R}-\mathbf{R}_{Ii}(t_{before}))$$
$$\rightarrow \sum_{K}^{\text{AES on } K} \sum_{k} C_{Kk}(t_{after})\,\Phi_K(\mathbf{r};\mathbf{R}_{Kk}(t_{after}))\delta(\mathbf{R}-\mathbf{R}_{Kk}(t_{after}))\,, \quad (6.44)$$

where AES stands for adiabatic electronic state. (The dependence of C_{Kk} on the momentum $\mathbf{P}_{Kk}(t_{after})$ is not explicitly expressed.) The function of Eq. (6.44) fulfils the desired condition of Eq. (6.31) except that the nuclear wavefunction is not yet quantized.

6.2.3 *Branching conditions*

6.2.3.1 *High energy paths*

In our former studies in which the energy of incident energy is well higher than that of the avoided crossings, the path-branching condition was set to

$$|\dot{R}X_{IJ}| < A \quad (6.45)$$

with A being a parameter that should be predetermined appropriately. This method has been proved to be very useful for high energy cases compared to the barrier height to obtain relevant paths and transition probability. Moreover, Eq. (6.45) has been applied mainly to the path branching at the exit of nonadiabatic region. This is because for high energy passage of nonadiabatic region an averaged path gives a good approximation as long as the nonadiabatic electronic state mixing is strong. Therefore path branching that terminates the averaging was usually applied once at an exit point from the region of strong coupling. However, in case where the energy is comparable with or lower than the crossing energy, the branching paths in the strong coupling region should be able to well represent precise phenomena induced by the bifurcating quantum wavepackets. For instance, when the total energy is comparable to the barrier height, one path may surmount the barrier while the other comes back. We therefore need to figure out conditions that are theoretically acceptable and numerically tractable.

6.2.3.2 *Low energy paths*

It is quite natural to judge that path-averaging is no longer good when the averaged path deviates appreciably from the unaveraged paths that are born at the same time and propagated according to their own individual eigenforces. Suppose that an averaged path lies at a phase space point $(\mathbf{R}(t), \mathbf{P}(t))$. From this point on, we can keep generating an averaged path. But also unaveraged paths can emanate that are driven by the individual eigenforces $f_K(\mathbf{R})$. With $(\mathbf{R}(t), \mathbf{P}(t))$ being the initial point a path given by an eigenforce, say, $f_K(\mathbf{R})$ is propagated as

$$\mathbf{R}_K(t+\Delta t) = \mathbf{R}_K(t) + \Delta\mathbf{R}_K, \tag{6.46}$$

$$\mathbf{P}_K(t+\Delta t) = \mathbf{P}_K(t) + \Delta\mathbf{P}_K, \tag{6.47}$$

where K is fixed for each path. We call these trajectories side paths. The side paths are as many as that of eigenforces. The side paths are made restart at each starting point of the average paths. They gradually deviate from each other, reflecting the diversity of the relevant paths due to the nonadiabatic interactions. Some of the side paths are expected to run on the edge of the cascade of branching paths. Therefore we monitor a pair of side paths that are most separate in phase space, and if the distance exceeds a predetermined value Θ such that

$$\sum_{K<L} |\mathbf{R}_K - \mathbf{R}_L|^2 + |\mathbf{P}_K - \mathbf{P}_L|^2 > \Theta, \tag{6.48}$$

we judge that the real paths surrounding the average path are too widely distributed and that it is a time to terminate the path-averaging. The choice of Θ is not determined by the theory itself, and we need a systematic exercise in advance of extensive calculations.

6.2.4 *Energy-conserving path-branching with the force averaging*

6.2.4.1 *Initialization of the electron wavepacket right after branching*

We here consider only a two-state case for simple illustration, and extension to more states is rather straightforward. Suppose that at a phase space point $(\mathbf{R}(t), \mathbf{P}(t))$ we have a SET path, and also the force diagonalization gives a set of $\{D_1(t), D_2(t)\}$ for the force eigenstates $\{\lambda_1(\mathbf{r}, \mathbf{R}(t)), \lambda_2(\mathbf{r}, \mathbf{R}(t))\}$, giving rise to an electron wavepacket

$$\Phi(t) = D_1(t)\lambda_1(\mathbf{r}, \mathbf{R}(t)) + D_2(t)\lambda_2(\mathbf{r}, \mathbf{R}(t)). \tag{6.49}$$

We here consider the following two propagation of the state: (1) Continue the Ehrenfest dynamics for the next short time interval Δt (denoted as path 0), and (2) let the path branch into two pieces (path 1 and 2). For the path-branching, along with the associated electronic state mixing, we prepare two vectors as the initial electronic states (at this time t)

$$(D_1(t), 0) \quad \text{and} \quad (0, D_2(t)) . \tag{6.50}$$

The wavefunctions for them are, respectively,

$$\Phi_1(t) = D_1(t)\lambda_1(\mathbf{r}, \mathbf{R}(t)) \quad \text{and} \quad \Phi_2(t) = D_2(t)\lambda_2(\mathbf{r}, \mathbf{R}(t)). \tag{6.51}$$

With these initial conditions, two additional SET paths are generated that are different from the SET path 0. Then at time $t + \Delta t$ we compare the energies thus branched.

6.2.4.2 *Energy conservation of a path after branchings*

In branching from an averaged path as above, those paths can have different (higher and lower) energies than that of the averaged (parent) path. Therefore we should survey how the populations and energies should be allocated to each path, and in so doing it is confirmed that the total energy summed up over all the branched paths are conserved.

First we consider the path 0 (the parent SET path). The total Hamiltonian for the SET dynamics is

$$\begin{aligned} H_0(\mathbf{R}, \mathbf{P}) &= \frac{\mathbf{P}^2}{2} + \frac{\langle \Phi(t)) \left| H^{(el)} \right| \Phi(t) \rangle}{\langle \Phi(t) | \Phi(t) \rangle} \\ &= \frac{\mathbf{P}^2}{2} + \frac{\langle (D_1(t)\lambda_1(\mathbf{R}) + D_2(t)\lambda_2(\mathbf{R})) \left| H^{(el)} \right| (D_1(t)\lambda_1(\mathbf{R}) + D_2(t)\lambda_2(\mathbf{R})) \rangle}{\langle \Phi(t) | \Phi(t) \rangle}, \end{aligned} \tag{6.52}$$

where again the bra-ket inner product is to be performed over the electronic coordinates alone. Likewise the Hamiltonians to drive the path 1 and path 2 are, respectively,

$$H_K(\mathbf{R}, \mathbf{P}) = \frac{\mathbf{P}^2}{2} + \frac{\langle \Phi_K(t)) \left| H^{(el)} \right| \Phi_K(t) \rangle}{\langle \Phi_K(t) | \Phi_K(t) \rangle} = \frac{\mathbf{P}^2}{2} + \left\langle \lambda_K(\mathbf{R}) \left| H^{(el)} \right| \lambda_K(\mathbf{R}) \right\rangle \tag{6.53}$$

for $K = 1, 2$. It is well established that the individual energy is conserved along each SET path.

6.2.4.3 *Potential energy*

In the force-state representation, we usually do not directly refer to the value of adiabatic potential energy surfaces. This is a practice common to any other electron wavepacket dynamics that does not rest on the electronic adiabatic representation. Therefore the potential energy that each path bears is to be estimated with integration of the force over a shifted distance. With the above Hamiltonian, Eq. (6.52) for path 0, let us write down the time-derivative of the momenta, that is, the forces as

$$\dot{\mathbf{P}}_0^k = -\frac{\left\langle (D_1(t)\lambda_1(\mathbf{R}) + D_2(t)\lambda_2(\mathbf{R})) \left| \frac{\partial H^{(el)}}{\partial \mathbf{R}^k} \right| (D_1(t)\lambda_1(\mathbf{R}) + D_2(t)\lambda_2(\mathbf{R})) \right\rangle}{\langle \Phi(t) | \Phi(t) \rangle}. \tag{6.54}$$

For the force eigenstates, the off-diagonal terms are close to zero, that is,

$$\left\langle \lambda_1(\mathbf{R}) \left| \partial H^{(el)} / \partial \mathbf{R}^k \right| \lambda_2(\mathbf{R}) \right\rangle \simeq 0, \tag{6.55}$$

and only the diagonal parts remaining in the expression as

$$\dot{\mathbf{P}}_0^k = -\frac{\left\langle D_1(t)\lambda_1(\mathbf{R}) \left| \frac{\partial H^{(el)}}{\partial \mathbf{R}^k} \right| D_1(t)\lambda_1(\mathbf{R}) \right\rangle + \left\langle D_2(t)\lambda_2(\mathbf{R}) \left| \frac{\partial H^{(el)}}{\partial \mathbf{R}^k} \right| D_2(t)\lambda_2(\mathbf{R}) \right\rangle}{\langle \Phi(t) | \Phi(t) \rangle}. \tag{6.56}$$

Similarly we have the forces for path 1 and 2, respectively, as

$$\dot{\mathbf{P}}_K^k = -\left\langle \lambda_K(\mathbf{R}) \left| \partial H^{(el)} / \partial \mathbf{R}^k \right| \lambda_K(\mathbf{R}) \right\rangle \tag{6.57}$$

for $K = 1, 2$.

The variation of the potential energy can be estimated rather directly with an elementary expression as

$$\Delta V_0 = \langle \Phi(t) | \Phi(t) \rangle \sum_k \int_{\mathbf{R}^k}^{\mathbf{R}^k + \Delta \mathbf{R}^k} \dot{\mathbf{P}}_0^k d\mathbf{R}^k. \tag{6.58}$$

Note that the population of the state to be carried by the path 0 ($\langle \Phi(t) | \Phi(t) \rangle$) is multiplied to the classical expression of the potential energy. Likewise, we have

$$\Delta V_K = |D_K(t)|^2 \sum_k \int_{\mathbf{R}^k}^{\mathbf{R}^k + \Delta \mathbf{R}^k} \dot{\mathbf{P}}_K^k d\mathbf{R}^k \tag{6.59}$$

for path K ($K = 1, 2$). For a time interval as short as Δt, $\Delta \mathbf{R}^k$ can be approximated to be

$$\Delta \mathbf{R}^k = \mathbf{P}^k(\mathbf{t})\Delta t, \tag{6.60}$$

irrespective of the kind of these paths. Collecting Eqs. (6.56) to (6.57) and Eqs. (6.58) to (6.59) altogether, we have

$$\Delta V_1 + \Delta V_2 = \Delta V_0, \tag{6.61}$$

which shows that the variation in the sum of the potential energies of path 1 and path 2 is the same as that for path 0 in the first order.

6.2.4.4 *Kinetic and total energies*

Next we survey the variation of the kinetic energy. In the first order with respect to Δt, one can approximate as

$$\mathbf{P}(t + \Delta t) = \mathbf{P}(t) + \dot{\mathbf{P}}\Delta t, \tag{6.62}$$

which in turns give the first order expression of the kinetic energy as

$$\frac{1}{2}\mathbf{P}(t + \Delta t)^2 \simeq \frac{1}{2}\mathbf{P}(t)^2 + \mathbf{P}(t) \cdot \dot{\mathbf{P}}\Delta t. \tag{6.63}$$

Recalling here that at the branching time t all the momenta are taken equal, that is, $\mathbf{P}(t) = \mathbf{P}_0(t) = \mathbf{P}_1(t) = \mathbf{P}_{2.}(t)$, we compare only $\dot{\mathbf{P}}\Delta t$ terms. For path 0, it is

$$\begin{aligned}
&\dot{\mathbf{P}}_0\Delta t \times \langle \Phi(t) | \Phi(t) \rangle \\
&\quad = -\left(\left\langle D_1(t)\lambda_1(\mathbf{R}) \left| \frac{\partial H^{(el)}}{\partial \mathbf{R}^k} \right| D_1(t)\lambda_1(\mathbf{R}) \right\rangle\right. \\
&\quad\quad \left. + \left\langle D_2(t)\lambda_2(\mathbf{R}) \left| \frac{\partial H^{(el)}}{\partial \mathbf{R}^k} \right| D_2(t)\lambda_2(\mathbf{R}) \right\rangle\right) \Delta t,
\end{aligned} \tag{6.64}$$

while

$$\dot{\mathbf{P}}_K^k \Delta t = -\left\langle \lambda_K(\mathbf{R}) \left| \partial H^{(el)} / \partial \mathbf{R}^k \right| \lambda_K(\mathbf{R}) \right\rangle \Delta t \tag{6.65}$$

for path 1 and path 2 with $K = 1$ and 2, respectively. The same arithmetic in the potential energy can be applied to the present case, and the result is

$$\mathbf{P}(t)\cdot\dot{\mathbf{P}}_0\Delta t \langle \Phi(t) | \Phi(t) \rangle = \mathbf{P}(t)\cdot\dot{\mathbf{P}}_1\Delta t \left| D_1(t) \right|^2 + \mathbf{P}(t)\cdot\dot{\mathbf{P}}_2\Delta t \left| D_2(t) \right|^2. \tag{6.66}$$

Therefore, in the first order, the kinetic energy is conserved as

$$\mathbf{P}_0(t+\Delta t)^2 = \mathbf{P}_1(t+\Delta t)^2 + \mathbf{P}_2(t+\Delta t)^2. \tag{6.67}$$

Thus, the total energy E_0 at the point $(\mathbf{R}(t), \mathbf{P}(t))$ with the population $\langle \Phi(t)|\, \Phi(t)\rangle$ is conserved as

$$E_0 = E_1 + E_2, \tag{6.68}$$

where E_K $(K = 1, 2)$ is the energy carried by path K, whose population is $\langle \Phi_K(t)|\, \Phi_K(t)\rangle = |D_K(t)|^2$.

In summary the total energy at a branching point is to be partitioned in proportional to the populations of the to-be-branched paths, and each is carried in a manner to conserve their own energy. Note that these branching paths should be guided by the eigenforces. Under the conservation rule, the individual non-Born–Oppenheimer paths can thus penetrate into classically forbidden regions through nonadiabatic interactions.

6.2.4.5 *Schematic illustration*

Now that we know how the electronic population should be allocated to each path, we can consider qualitatively important cases within the framework of the path-branching theory. An illustrative schematic example is presented in Fig. 6.1: A path running on the lower potential surface comes in the interaction region and at some point below the top of the adiabatic potential (left white small circle) branches into two pieces: One with a smaller component has a higher energy. The energy of the other path of larger population is lowered a little to compensate the energy taken away by the high energy counterpart. Note that these branched paths are continuously connected with each other at the branching point in phase space, although the higher energy path graphically seems to jump abruptly in the potential curve diagram. Therefore this continuous branching should not be confused with the surface-hopping motion. (Note also that although the adiabatic potentials are drawn in the figure in order to help imagination the dynamics are performed with use of the force diabatization but not in the adiabatic representation.) The higher energy path can proceed to totally different pathway from that of the lower energy counterpart, and can bifurcate again at somewhere else (right white circle). So does the lower path. Note again that even after many-time branchings, all the cascade paths can be continuously traced back to the original path with the total energy being conserved in the above sense.

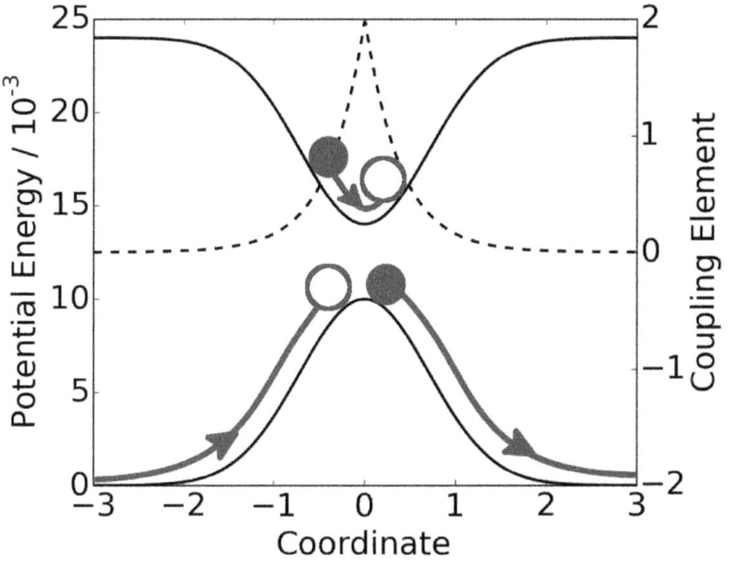

Fig. 6.1 Schematic representation of energy conserving path-branching in configuration space. The branching paths are continuously connected in phase space. Although the adiabatic potential energy curves are drawn to facilitate the graphical imagination, the practice for path-branching dynamics is performed in terms of the force matrix and the diabatic represention based on it (see the text). (Reprinted with permission from K. Yamamoto *et al.*, J. Chem. Phys. 140, 124111 (2014)).

6.2.5 *What is the decoherence in nonadiabatic transitions after all?*

The main results of the studies described above concerning the kinematically coupled mixed quantum–classical dynamics can be summarized as follows. A quantum subsystem undergoes a pure quantum mixing among the electronic states when a slow change of the classical subsystem (electronic-state coherence is involved). On the other hand, the dynamics of the classical subsystem results in an infinite number of cascades of path branching as a manifestation of the bifurcation of the nuclear wavepackets. Such path branching never arises in the well-posed initial value problem such as the Newtonian equation. This can happen because the "classical" dynamics couples with the quantum subsystem. This is known as "coherence" and it penetrates into the classical subsystem. Electronic-state coherence should always be accompanied by path branching. Meanwhile as the nonadiabatic couplings diminish, the branching terminates automatically (or, in other words, decoheres) and converges so that each path eventually runs on one of the adiabatic potential energy surfaces. This is a picture of spontaneous decoherence we have obtained in the theory of path branching.

6.3 Numerical examples of branching paths and transition probability

This section follows the above basic theory and method of path-branching representation of nonadiabatic dynamics with numerical examples. We also show the performance of the semiclassical Ehrenfest theory to compare with. What we present below is illustrative numerical realization rather than optimized practice for actual applications.

6.3.1 *Systems of two and three electronic states nonadiabatically coupled*

To study a realistic system as much as possible, we apply the method to LiH molecule within the singlet spin state manifold. For the sake of computational flexibility, we first fit the energy of the individual CSF (configuration state function) and the relevant coupling elements (electronic and kinematic off-diagonal elements) with the zeroth order Shepard interpolated functions [232] as a function of the inter-nuclear distance, and then re-sum them into the form of adiabatic representation. The basis set used is of CIS/6-31G(d,p) level in GAMESS package, [357] and 200 interpolation points are sampled with equal distances over the region from 1.2 to 14.2 Bohrs for LiH length. We have picked three excited states of Σ^+ symmetry, namely, the 3, 4 and $5\Sigma^+$ states for this study.

Using these three electronic states, we set up two systems; one is a three-state model using all the $3\Sigma^+$, $4\Sigma^+$, and $5\Sigma^+$ states and a two state model made by picking the $3\Sigma^+$and $5\Sigma^+$ states out of the three. Figure 6.2 show the adiabatic potential energy surfaces for these three states along with the spatial distribution of the nonadiabatic coupling elements defined in Eq. (6.7). In both models, the derivative coupling between the adiabatic states are broad in space, and is not localized in the narrow vicinity of the seemingly avoided crossings. More precisely, the derivative coupling elements between $3\Sigma^+$ and $5\Sigma^+$ give a rather widely coherent region with the peak at 3.1 Bohrs and has a long tail to the right hand side (longer bond-distance). Notice that the peak of the coupling term is located at approximately 1.5 Bohrs left to the apparent avoided crossing point between these two adiabatic states. Figure 6.2 also indicates that the couplings between the states of $3\Sigma^+/4\Sigma^+$ and $4\Sigma^+/5\Sigma^+$ are approximately of the half magnitude to that of $3\Sigma^+/5\Sigma^+$. With these broad features of the coupling elements and the regions of avoided crossing, complicated electronic transitions are expected to emerge.

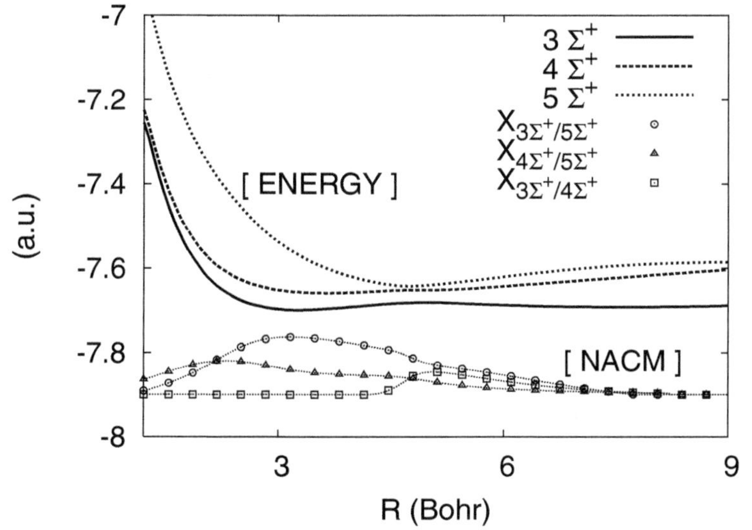

Fig. 6.2 The adiabatic potential energy curves and derivative coupling versus the internuclear distance for the three state model. The curves have been obtained by fitting the data for the 3, 4 and $5\Sigma^+$ states of LiH molecule calculated with CIS/6-31G(d,p) level. The solid, dashed and dotted lines are for 3, 4 and $5\Sigma^+$ states, respectively. The three lower curves represent the derivative couplings among $3/4\Sigma^+$, $4/5\Sigma^+$ and $3/5\Sigma^+$ states, whose values are shifted by as much as -7.9. (Reprinted with permission from T. Yonehara *et al.*, J. Chem. Phys. 129, 134109 (2008)).

6.3.2 *Practices in the semiclassical Ehrenfest theory and the full quantum dynamics of nuclear wavepacket dynamics*

6.3.2.1 *Semiclassical Ehrenfest dynamics*

The electronic equations of motion, Eq. (6.14) (setting the second-order derivatives are all zero, $Y^k_{IJ} = 0$), and the mean-field nuclear dynamics,

Eq. (6.23), should be solved in a coupled manner. In solving these, the time step to integrate the equations of motion for nuclei is set to $dt_{nuclei} = 2.4 \times 10^{-4}$ fs. The integrations have been carried out up to 12 fs in total, which is sufficiently long for the lowest energy trajectory point to reach the dissociation region. The propagation of the electronic wavefunction was performed with the Chebyshev polynomial expansion method [430]. At each time to integrate the nuclear dynamics with the interval dt_{nuclei}, we reset the electronic Hamiltonian matrix at the renewed nuclear positions. The sixth order symplectic integrator [497] was employed for the time propagation for nuclear part.

6.3.2.2 *Quantum mechanical transition probability based on nuclear wavepacket dynamics in adiabatic representation*

The coupled nuclear Schrödinger equations of motion for the nuclear wavefunction vector $(|\chi_1\rangle\ |\chi_2\rangle\ \cdots\ |\chi_M\rangle)^T$, which is equivalent to Eq. (2.6), has the form of

$$i\hbar\frac{\partial}{\partial t}\begin{pmatrix}|\chi_1\rangle\\|\chi_2\rangle\\\cdot\\\cdot\\|\chi_M\rangle\end{pmatrix}=\begin{pmatrix}\hat{H}_1 & \hat{D}_{12} & \cdot & \cdot & \hat{D}_{1M}\\\hat{D}_{21} & \hat{H}_2 & \cdot & \cdot & \hat{D}_{2M}\\\cdot & \cdot & \cdot & \cdot & \cdot\\\cdot & \cdot & \cdot & \cdot & \cdot\\\hat{D}_{M1} & \hat{D}_{M2} & \cdot & \cdot & \hat{H}_M\end{pmatrix}\begin{pmatrix}|\chi_1\rangle\\|\chi_2\rangle\\\cdot\\\cdot\\|\chi_M\rangle\end{pmatrix}, \tag{6.69}$$

where

$$\hat{H}_\alpha = \hat{T} + E_\alpha - \frac{\hbar^2}{2\mu}Y_{\alpha\alpha} \tag{6.70}$$

and

$$\hat{D}_{\alpha\beta} = -\frac{\hbar^2}{2\mu}\left(Y_{\alpha\beta} + 2X_{\alpha\beta}\partial\right). \tag{6.71}$$

Here $\hat{T}$ is a kinetic energy operator and E_α is the adiabatic potential energy for the α-th state. μ is the reduced mass of LiH molecule. M is the number of adiabatic electronic states to be considered. Note that E_α, $X_{\alpha\beta}$ and $Y_{\alpha\beta}$ depend on the internuclear distance of LiH. The coupling elements $Y_{\alpha\beta}$, as defined in Eq. (2.7), and $\langle\partial\chi_\alpha|\,\partial\chi_\beta\rangle$ are consistently neglected in the present calculations. Also, we hermitize the coupled Hamiltonian operators in Eq. (6.69) for numerical stability. [189, 315] Thus we reexpress $\hat{H}_\alpha$ and $\hat{D}_{\alpha\beta}$, respectively, as

$$\hat{H}_\alpha \approx \hat{T} + E_\alpha \tag{6.72}$$

and

$$\hat{D}_{\alpha\beta} \approx -\frac{\hbar^2}{2\mu}\left\{(\overleftarrow{\partial})X_{\alpha\beta} + X_{\beta\alpha}(\overrightarrow{\partial})\right\}, \tag{6.73}$$

where the derivative operators, $\overrightarrow{\partial}$ and $\overleftarrow{\partial}$, are supposed to act on a function to the right and left hand sides, respectively. The hermitization implies that the wavepacket under study is supposed to remain in a finite size range.

The Chebyshev expansion method [430] is applied to the propagation of the nuclear wavepacket subject to the above Hamiltonian for both the two- and three-state models. Nuclear wavefunctions, the potential functions, the matrix to represent the first and second nuclear derivatives arising from $\hat{T}$ of Eq. (6.70) are all expressed in the sinc discrete variable representation (sinc-DVR) [92]. The time length for one-step integration of nuclear wavepacket is set to 0.02 fs. The 1200 DVR grid points are employed within a range from -3 to 14 Bohrs. For a practical reason, the potential function is cut off in the range shorter than 1.2 Bohrs.

All the initial electronic state populations are set to be localized at the energetically highest adiabatic state to be compatible with the calculation for PSANB and the semiclassical Ehrenfest theory. The momentum of the wavepacket is varied by changing the initial wave number as k =16.05, 32.11, 48.16, 64.22, 80.28, 96.33, 112.39, 128.44, 144.50 and 160.56 a.u., which specifies the initial Gaussian wavepacket of minimum uncertainty

$$\chi = \left(\frac{1}{2\pi\delta^2}\right)^{1/4} \exp\left(-\left(\frac{R-R_c}{2\delta}\right)^2\right) \exp\left(i\,kR\right), \tag{6.74}$$

where $R_c = 1.3$ and $\delta = 10/k$ for all initial conditions.

6.3.3 *Nonadiabatic transition probability*

With the scheme of PSANB, one can calculate the nonadiabatic transition probability as follows. For the sake of simpler explanation, we describe the procedure for the systems of two electronic states. Suppose we have an asymptotic wavefunction irrespective of their history of branching such that

$$\begin{aligned}\Psi(\mathbf{r},\mathbf{R},t) = &\sum_k c_k^{(1)}\psi_1(\mathbf{r};\mathbf{R}_k^{(1)}(t))\chi_k\left(\mathbf{R}-\mathbf{R}_k^{(1)}(t)\,,t\right)\\ &+\sum_l c_l^{(2)}\psi_2(\mathbf{r};\mathbf{R}_l^{(2)}(t))\chi_l\left(\mathbf{R}-\mathbf{R}_l^{(2)}(t)\,,t\right),\end{aligned} \tag{6.75}$$

where $\psi_1(\mathbf{r};\mathbf{R}_k^{(1)}(t))$ denotes the first adiabatic electronic wavefunction (normalized) at a nuclear position $\mathbf{R}_k^{(1)}(t)$, with $\chi_k\left(\mathbf{R}-\mathbf{R}_k^{(1)}(t),t\right)$ being a (normalized) nuclear wavepacket localizing nearby $\mathbf{R}_k^{(1)}(t)$. The similar notations are employed for the second potential energy surface. A probability to find all the wavepackets lying on the first potential surface is

$$P^{(1)} = \int d\mathbf{r}d\mathbf{R}\left|\sum_k c_k^{(1)}\psi_1(\mathbf{r};\mathbf{R}_k^{(1)}(t))\chi_k\left(\mathbf{R}-\mathbf{R}_k^{(1)}(t),t\right)\right|^2. \tag{6.76}$$

The theoretical scheme proposed so far has not yet given the explicit functional form of $\chi_k\left(\mathbf{R}-\mathbf{R}_k^{(1)}(t),t\right)$, and hence we have no rigorous way to proceed further within the present PSANB scheme. However, as seen above in the numerical studies, the trajectories tend to asymptotically come close to one another on each potential energy surface. Therefore we may use an approximation

$$\int d\mathbf{R}\chi_k\left(\mathbf{R}-\mathbf{R}_k^{(1)}(t),t\right)\chi_l\left(\mathbf{R}-\mathbf{R}_l^{(1)}(t),t\right) = 1 \tag{6.77}$$

on each potential surface for any choice of k and l. This approximation retains the coherent sum in Eq. (6.76), giving rise to

$$P^{(1)} = \left|\sum_k c_k^{(1)}\right|^2. \tag{6.78}$$

This form of transition probability will be used throughout the present quantum-classical scheme. The explicit incorporation of the nuclear wavepacket $\chi_k\left(\mathbf{R}-\mathbf{R}_k^{(1)}(t),t\right)$ will be studied later in Sec. 6.7.

6.3.4 *Force diabatization*

As described in the preceding section, the eigenforces and their associated eigenfunctions should be calculated simultaneously with the electron wavepacket dynamics as the counterpart of the entire coupled scheme. Nevertheless, by intentionally ignoring the procedure of the electronic-state mixing, one can perform the force matrix dynamics alone in such a way that all the reference paths are to be self-determined only by the eigenforces. Furthermore, since the force matrix does not depend on $\dot{R}$ in one-dimensional system (recall the second term in the right hand side of Eq. (6.19) vanishes), the eigenforces and associated eigenfunctions can be globally graphed as a function of the internuclear distance R. Therefore, before examining the dynamics of the scheme of phase-space averaging and subsequent natural branching, we demonstrate how the eigenforces and the potential functions represented in terms of the eigenforce-states look.

For the α- and β-th adiabatic states having the energy E_α and E_β, respectively, the element of the force matrix is given as

$$\mathcal{F}_{\alpha\beta} = -\left\{(E_\beta - E_\alpha)X_{\alpha\beta} + \frac{\partial E_\alpha}{\partial R}\delta_{\alpha\beta}\right\}. \tag{6.79}$$

With this matrix and its eigenforces, denoted as $\{f_K | K = 1, \cdots, M\}$, we consider a diabatization. (By force diabatization, we do not mean the ordinary diabatization, in which the derivative coupling matrix elements X_{IJ} are all zero.) To examine the global feature of them, we calculate the spatial distribution of the eigenforces f_K and $\langle \lambda_K | H^{(\text{el})} | \lambda_K \rangle$, with λ_K being the eigenfunction belonging to f_K.

In Fig. 6.3(a) and (b), f_K and $\langle \lambda_K | H^{(\text{el})} | \lambda_K \rangle$ for $K = 1, 2$ are plotted, respectively, against the LiH distance. For comparison, the adiabatic forces, $-\partial E_\alpha / \partial R$ $(\alpha, \beta = 1, 2)$ are also plotted in panel (a). Likewise the adiabatic potential energy curves E_α and E_β are shown in panel (b) along with $X_{\alpha\beta}$. From panel (a) it turns out that the global feature of the eigenforce is very similar to the adiabatic forces, with a very important exception that the eigenforces avoid crossing. On the other hand, panel (b) indicates that the curves of $\langle \lambda_K | H^{(\text{el})} | \lambda_K \rangle$ have slight deviation even in the region of $R \sim 3$ Bohrs, which is pretty far from the avoided crossing region. The asymmetric spatial distribution of the nonadiabatic coupling element $X_{\alpha\beta}$ having the peak around $R \sim 3$ Bohrs accounts for this deviation. However, the truly remarkable feature of $\langle \lambda_K | H^{(\text{el})} | \lambda_K \rangle$ is that the crossing takes place almost all of sudden in a very narrow range, highlighting the essential feature of force diabatization.

As seen in Eq. (6.79), the diagonalization of the force matrix seems to be determined in the balance of $X_{\alpha\beta}$ and $-\partial E_\alpha / \partial R$ (the adiabatic force). However, simultaneous inspection of panels (a) and (b) of Fig. 6.3 strongly suggests that the crossing of the forces $-\partial E_\alpha / \partial R$ $(\alpha = 1, 2)$ in the adiabatic representation (panel (b)) should be the main origin of the level repulsion of the eigenforces (panel (a)), which in turn gives rise to the crossing between $\langle \lambda_1 | H^{(\text{el})} | \lambda_1 \rangle$ and $\langle \lambda_2 | H^{(\text{el})} | \lambda_2 \rangle$.

The above observation can be confirmed more clearly in the similar results arising from the three-state model, which are depicted in panels (c) and (d) in Fig. 6.3. Corresponding to the three avoided crossing points at $R = 2.3$, 4.8 and 7.3 Bohrs for the curves of the eigenforces in panel (a), $\langle \lambda_K | H^{(\text{el})} | \lambda_K \rangle$ undergo crossing at three places (panel (b)). Again, correspondingly at these places, crossing of the forces in the adiabatic representation $-\partial E_\alpha/\partial R$ takes place.

The diabatic energy curves in the three-state model, $\langle \lambda_K | H^{(\mathrm{el})} | \lambda_K \rangle$ $(K = 1, 2, 3)$, feature a very interesting characteristic: In the crossing area at around $R = 4.8$ the three states seem to couple with each other simultaneously. A closer look shows, however, that only two states, namely, $3\Sigma^+$ and $5\Sigma^+$ exchange their characters with respect to the eigenforce. In other word, only $3\Sigma^+$ and $5\Sigma^+$ are re-connected smoothly by the corresponding eigenstates of the force matrix, but the state of $4\Sigma^+$ is left alone from them, although it fluctuates significantly due to the presence of the other two states. Thus, the diabatization using the force eigenstates may offer an interesting characterization to more complicated nonadiabatic systems.

We emphasis that the results presented in this subsection is totally independent of PSANB, which will be numerically realized below.

6.3.5 *Geometry of branching paths*

We launch many electron wavepackets starting from the $5\Sigma^+$ state (the highest state) both in the two- and three-state models. While the initial position to start from is $R_{init} = 1.3$ Bohrs for all cases, the initial momenta are set to $16.05, 32.11, 48.16, 64.22, 80.28, 96.33, 112.39, 128.44, 144.50$, and 160.56 a.u., which are exactly the same as those chosen for the quantum wavepacket calculations.

6.3.5.1 *Geometry of branching paths of higher generations in phase space*

After finishing the phase-space averaging, we let the path naturally branch so that they eventually run on the relevant adiabatic potential surfaces. We first survey the two-state model and show how the successive branching looks. Just for the sake of presentation, we let the trajectories branch when $|\dot{R}X_{3\Sigma^+/5\Sigma^+}|$ takes the following threshold values, 0.0025, 0.020, 0.015, and 0.010, giving birth to the fourth generations of paths. We stop running these branching paths at points where the threshold takes the value of 0.005.

In Fig. 6.4, we show the phase-space paths starting with the ten different initial momenta as before. Figure 6.4(a) exhibits the global feature of the individual paths in phase space (R, P). Panel (b) shows a magnification of the lowest curve (the lowest energy path) in panel (a). As expected, 16

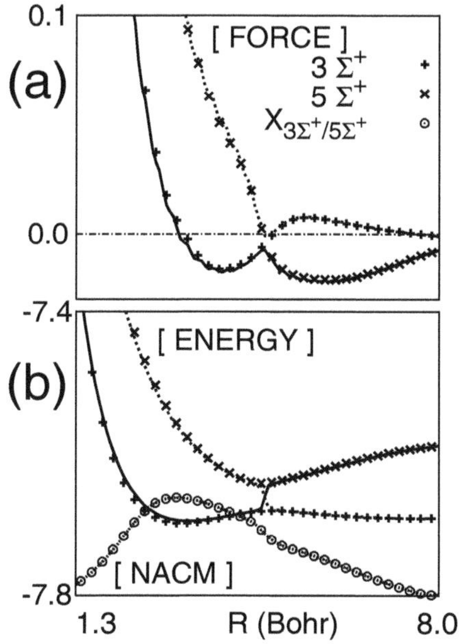

Fig. 6.3 Global feature of the eigenforces and associated electronic energies in the force diabatization: (a) The eigenforces in the two-state model, which avoid crossing. The adiabatic forces (based on the energy gradient in the adiabatic representation) of the states of $3\Sigma^+$ and $5\Sigma^+$ are also depicted in dotted curves. (b) The crossing electronic energies (diagonal elements of the electronic Hamiltonian) of the eigenforce states are presented for the two state model. The adiabatic energies of $3\Sigma^+$ and $5\Sigma^+$ are also shown for comparison in dotted curves, which avoid crossing. The derivative coupling is also shown in panel (b). (c) The same as (a) for the three state model ($3\Sigma^+$, $4\Sigma^+$, and $5\Sigma^+$) (d) The same as (b) for the three state model. (Reprinted with permission from T. Yonehara *et al.*, J. Chem. Phys. 129, 134109 (2008)).

paths have been generated from each initial trajectory. The paths are all driven by the eigenforces.

We next study the branching in the three-state model. The successive bifurcation of the paths has been made with the following automatic procedure. We first define a threshold value of nonadiabatic coupling element D,

$$D = 0.3 \times \left(\left| \dot{R} \right| \left(\left| X_{4\Sigma^+/5\Sigma^+} \right| + \left| X_{3\Sigma^+/5\Sigma^+} \right| \right) \right)_{\max}. \tag{6.80}$$

A path is forced to branch successively when the coupling element becomes as low as D, $4D/5$, $3D/5$, $2D/5$ and $D/5$, and so on, which, respectively, gives rise to the zeroth (g0), first (g1), second (g2), and third (g3) generation of the branching paths. Figure 6.5, panels (a) and (b) show the global feature and an extended view of the lowest energy curve, respectively.

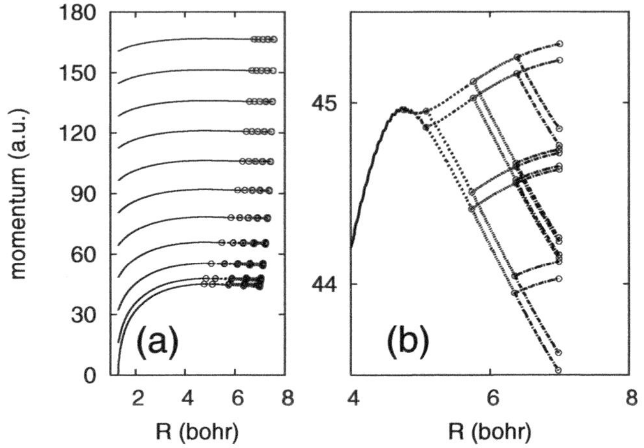

Fig. 6.4 Phase space view of branching paths given by PSANB scheme for the two-state model. (a) The global feature of the eleven paths with different initial conditions as explicitly shown. The marks on each path locate the positions where the branchings are made; actually they are the points where $|\dot{R}X_{12}|$ becomes 0.025,0.020, 0.015, 0.010 and 0.005 at the region of $R > 4$ Bohrs. The branching pieces are not visibly distinguished due to the large scale chosen. (b) A magnification of the lowest curve of panel (a). The succesive generation of branching pieces is clearly observed. (Reprinted with permission from T. Yonehara *et al.*, J. Chem. Phys. 129, 134109 (2008)).

6.3.5.2 *Nonadiabatic transition probability compared with the full quantum values*

The convergence property of the nonadiabatic transition amplitudes with respect to the number of branching generation is next studied. We here show the results only for the three-state model, since the two state model offers only an easier test. The transition probability of PSANB has been calculated with Eq. (6.78) with use of the branching paths of Fig. 6.5(a). This approximation is justified because the endpoints of the paths are generally very close to each other on the individual potential curves. To estimate the accuracy, we compare the probabilities with those obtained in the full quantum calculations described above.

In Fig. 6.5, panels (c) and (d) depict the nonadiabatic transition probabilities from $5\Sigma^+$ state to the $3\Sigma^+$ and $4\Sigma^+$ states asymptotically observed as a function of the initial kinetic energy. As observed in panel (d), the transition probability to $4\Sigma^+$ does not increase monotonically as the initial kinetic energy. Again, the semiclassical Ehrenfest theory gives excellent estimate of the transition probabilities. However, its deviation from the full quantum method is noticeable. The PSANB shows a good convergence to the exact values with respect to the number of generation, although the PSANB probabilities in $3\Sigma^+$ approaches the quantum values from above, while that of $4\Sigma^+$ from below.

We notice that the transition probability (and its convergence as well) to the $4\Sigma^+$ state given by the path of the lowest energy under study is not as good as those of the higher energies. This is presumably because the dispersion of the final location of the branching paths on the $4\Sigma^+$-state potential energy curve is so significant that the approximation of Eq. (6.77) is deteriorated. However, we may conclude that the global feature of the convergence of the nonadiabatic transition probabilities given by the PSANB procedure is quite good.

6.3.6 *An example in which the semiclassical Ehrenfest fails*

As observed above, both the semiclassical Ehrenfest and PSANB give highly accurate electronic transition probabilities, which are quite close to the full quantum counterparts. In particular, the semiclassical Ehrenfest offers a simpler and convenient computational scheme as far as the electronic transition probability is concerned. Indeed, we have been studying electron current dynamics in chemical reactions with this method. [307] On the other hand, there are situations in which the semiclassical Ehrenfest theory totally fails to reproduce correct dynamics. Even in such a case, PSANB provides not only accurate transition probability but also physically correct view of nuclear motion. We next show such an illustrative example.

6.3.6.1 *A coupled potential energy curves*

We study nonadiabatic dynamics on a pair of nonadiabatically coupled potential curves as shown in the upper panel of Fig. 6.17, which has been derived by schematically deforming the potential curves of LiH, $3\Sigma^+$ and $5\Sigma^+$. The coupling element has been simplified so as to localize at the avoided crossing region, which is also shown in the figure. The characteristic feature of them is that the lower adiabatic potential (denoted as E_1) gets flat right to the avoided crossing, while the upper one (E_2) rises in both sides of

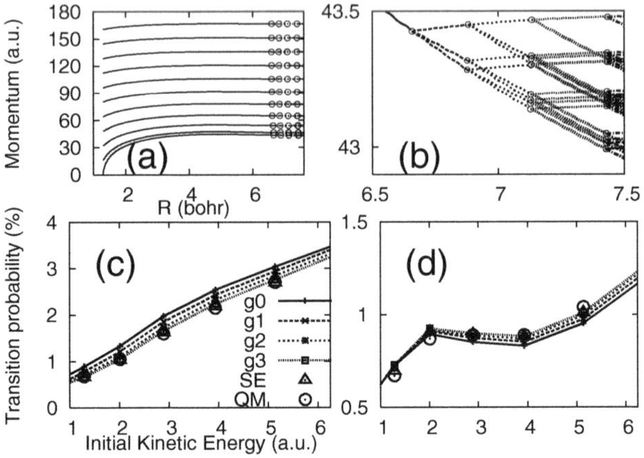

Fig. 6.5 Phase space view of branching paths given by PSANB scheme for the three-state model. (a) The global feature of the eleven paths with different initial conditions. The marks on each path indicate where the branchings are made. (b) A magnification of the lowest curve of panel (a). (c) and (d) The nonadiabatic transition probabilities from $5\Sigma^+$ to $3\Sigma^+$ and $4\Sigma^+$, respectively, calculated with phase-space averaging and succesive branchings up to the fourth generation. As the generation becomes higher starting from the zeroth generation (g0) to the fourth (g4), the increasing accuracy of the transition probability is observed. As a reference, the transition probabilities given by the full quantum wavepacket method along with the semiclassical Ehrenfest theory (SE) are also shown. (Reprinted with permission from T. Yonehara *et al.*, J. Chem. Phys. 129, 134109 (2008)).

it. The center of the initial wavepacket and the position of the corresponding classical trajectory are both prepared at $R = 2.0$ on E_1. By changing the initial momentum $k = 16.1 - 32.1$ (see Table I for the precise values), we calculate the first escape probability of the wavepacket, a probability for a wavepacket to be found in the asymptotic region ($R \to \infty$), after the first passage through the avoided crossing.

6.3.6.2 *Nuclear paths and transition probability*

Even after the passage over the avoided crossing, the nuclei treated in the semiclassical Ehrenfest theory are supposed to feel the average force arising from E_1 and E_2. By construction, the force from E_1 is zero while that from E_2 is negative large in the right region to the crossing. Therefore, unless the transition probability is perfectly zero (the entire packet remains lying on E_1), the Ehrenfest path should feel a force so as to be bounced back toward the left hand side. The dotted curves in the left panel of Fig. 6.6

represent such Ehrenfest paths for different initial k's. Indeed all of them are returning toward the basin area. Since they do never penetrate into the asymptotic region ($R \to \infty$), the resultant escape probability is simply zero.

We next run PSANB paths in two ways. (1) PSANB_0; PSANB of single branching at a point where $\left|\dot{R}X_{12}\right|$ becomes smaller than a predetermined threshold value $0.1 \times \left|\dot{R}X_{12}\right|_{\max}$, which generates two paths at each k. (2) PSANB_3; Further branching up to the third branching paths, which amount two groups consisting of eight paths each. The solid curves in the left panel of Fig. 6.6 are the trajectories of the paths of PSANB_3 for all the k's studied. It is clearly observed that as time passes they are bifurcated into two groups, each of which contains 8 paths. The turning points of the bouncing-back paths of PSANB are all located at left to those of the Ehrenfest paths. This reflects the fact that the Ehrenfest paths actually feel the weaker reflecting forces than the returning paths of PSANB, simply because the former feels an average between the forces arising from E_1 and E_2.

To examine the quality of the PSANB paths and transition probabilities, we have carried out the corresponding quantum wavepacket calculations. The numerical method is essentially the same as already described above. The right panel of Fig. (6.6) shows the time-propagation of the square modulus of the nuclear wavepackets, that is, a simple superposition of $|\chi_1(R,t)|^2$ and $|\chi_2(R,t)|^2$ for $k = 32.1$. At each time t, we have plotted points the height of $|\chi_1(R,t)|^2$ and $|\chi_2(R,t)|^2$ of predetermined values and connected them to make contour curves in (R,t)-space. The right and straight contour map represents $|\chi_1(R,t)|^2$ on the ground energy surface, while the bouncing back packet represents the contour map of $|\chi_2(R,t)|^2$ running on E_2. It is interesting to compare the quantum contour map for the nuclear wavepackets with the trajectories of PSANB. Although they should not be compared directly, it is nice to observe their high geometrical similarity.

The first escape probabilities thus calculated are tabulated in the Table. The semiclassical Ehrenfest fails completely in this tough situation. On the other hand, PSANB reproduces the probability remarkably well. No further comment is necessary for the quality of the transition probability given by PSANB, since it has been already described in a great detail. Thus, it turns out that PSANB scheme may be used as a promising alternative to the semiclassical Ehrenfest theory.

Table 6.1 The first escape probability for the wavepacket to be observed at the asymptotic region.

k	16.1	19.3	22.5	25.7	28.9	32.1
Ehrenfest	0.0	0.0	0.0	0.0	0.0	0.0
QM	73.3	67.7	61.4	56.1	51.0	46.5
$PSANB_0$	73.4	66.7	60.3	54.3	49.0	44.5
$PSANB_3$	73.3	67.0	61.1	55.7	50.6	46.1

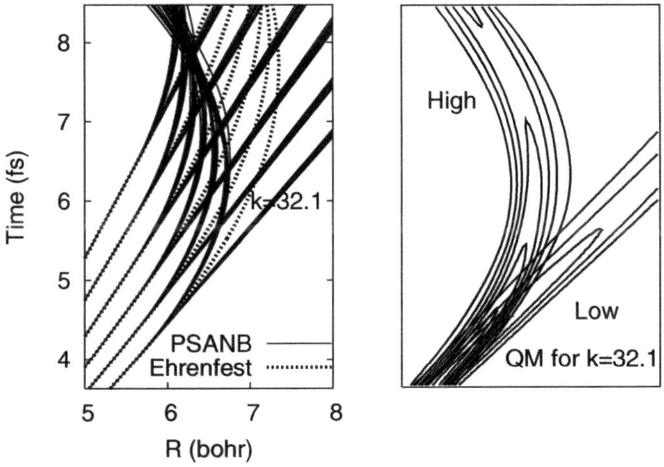

Fig. 6.6 (Left) Bunch of branching paths given by the PSANB method (solid curves) and those of the semiclassical Ehrenfest dynamics (dotted curves) for initial momentums $k = 16.06$, 19.3, 22.5, 25.7, 28.9 and 32.1, starting from the initial position $R = 2.0$ on the ground state potential. (Right) The branching of the quantum wavepackets by the non-adiabatic transition with the initial central momentum of $k = 32.1$. A simple super position of $|\chi_1(R,t)|^2$ (denoted as Low) and $|\chi_2(R,t)|^2$ (denoted as High) are drawn. (Reprinted with permission from T. Yonehara *et al.*, J. Chem. Phys. 129, 134109 (2008)).

6.4 Highly degenerate coupled electronic states

To the best of our knowledge, most of the general theories of nonadiabatic transition has been formulated to treat one-dimensional nonadiabatic systems of an isolated avoided crossing, the Landau–Zener theory being an example. On the other hand, the theory of PSANB (and SET as well as its approximation) is not limited to those specific cases. Nevertheless, we need to examine how practically PSANB works by comparing its results with the full quantum calculations in a couple of one-dimensional dense systems. The method of comparison is similar to that we did before [493] In this section, we adopt the simplest practice of path branching, which was introduced in our previous paper [495]. Not only PSANB but the FSSH (the fewest surface switches hopping) method in the original version [445] is examined, since FSSH is widely recognized as one of the best practical methods to provide the nonadiabatic transition probabilities. Besides, FSSH is designed to make use of the information of electron dynamics to access the hopping probability, and is therefore self-contained as PSANB is.

6.4.1 *System functions and computational details*

We build a one-dimensional model system in a similar manner to the work of SHEEP [133], in which one diabatic state couples with the other densely packed states, while the coupling among the latter are neglected. The model is designed as follows.

$$\begin{aligned} V_{11}(R) &= \mathrm{Dg}\left[\exp(-\alpha(R-\mathrm{Re})-1\right]^2 \\ V_{nn}(R) &= \exp(-\alpha R)+\eta n+\mathrm{De} \\ V_{1n}(R) &= \mathrm{Dc}\ \exp(-\lambda(\mathrm{R}-\mathrm{R}_{1n})^2)=V_{n1}(R) \quad \text{for} \quad n=2\sim M, \end{aligned} \tag{6.81}$$

with $V_{mn}(R) = 0$ (if $m \neq n$ and $m, n = 2 \sim M$). R_{1n} denote the crossing points between diabatic V_{11} and V_{nn}. The parameters are set to $\mathrm{Dg} = 0.2$, $\mathrm{De} = 0.05$, $\mathrm{Dc} = 0.05$, $\mathrm{Re}= 3$, $\alpha = 0.4$, $\lambda = 1$ and $\eta = 0.004$. The mass is set to 2000 a.u. The geometric dependence of the system functions is depicted in Fig. 6.7. The present dynamical calculations are carried out in the diabatic representation for the sake of simplicity.

The full quantum dynamics for the system was performed with use of the Chebyshev expansion method. [430] 800 grid points in the discrete variable representation were equally distributed from 0 to 15 Bohr. The details of the matrix structure can be found in our previous article. [493] Time step (Δt) for quantum dynamics was set to 5 a.u. The convergence against time step (Δt) and the grid interval (ΔR) were confirmed by comparing the results with shorter quantities, $0.5\Delta t$ and $0.75\Delta R$, respectively. In both PSANB and FSSH, we commonly use the nuclear time step of 0.5 a.u. and an electronic Hamiltonian was updated at every nuclear step. The electron dynamics were carried out by using the Chebyshev expansions method. The quantum time propagation operator is expanded in 50 Chebyshev polynomials, which assures the numerical accuracy.

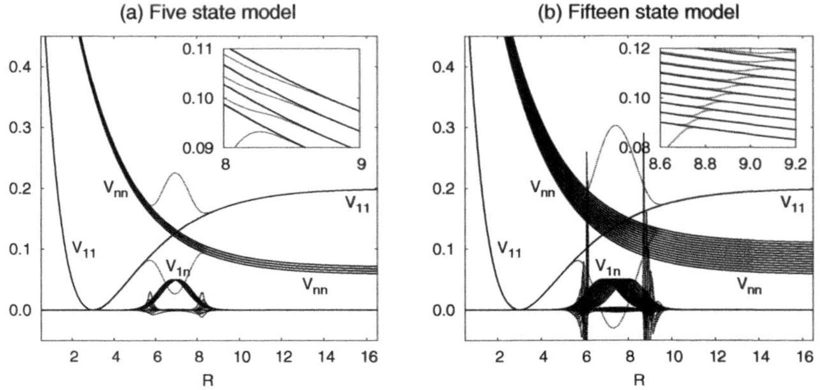

Fig. 6.7 Global feature of the potenial energy curves (in Hartree) and coupling matrix elemts among them are plotted against the bond distance R (in Bohr) in the diabatic representations (solid line). The adiabatic potential curves arising from them are also diplayed for comparion in dotted curves, along with the nonadiabatic coupling elements in dotted curves with a strong localization at two small areas. The insets show the magnified view of the crossing region. Panel (a) and (b) are for the 5 and 15 state model, respectively. (Reprinted with permission from T. Yonehara *et al.*, J. Chem. Phys. 137, 22A520 (2012)).

To be self-contained as much as possible we here make a very brief review of our practice in the application of the FSSH scheme: A diabatic representation is used for the dynamics of electron wavefunction. The electron wavefunction is propagated coherently on a reference nuclear trajectory throughout the dynamics. The part of density matrix created by the wavefunction is utilized to determine the probability for an otherwise classically continuous path to hop from one PES to another within every time interval $[t, t+\Delta t]$. To be more precise, the switching probability is given as

$$P_{K\to J}[t:t+\Delta t] \equiv \frac{\max\{B_{JK}\Delta t, 0\}}{\rho_{KK}} \tag{6.82}$$

with

$$B_{KL} \equiv \frac{2}{\hbar}\mathrm{Im}(\rho_{KL}^* V_{KL}) - 2\mathrm{Re}(\rho_{KL}^* \dot{\mathbf{R}}(t)\cdot \mathbf{d}_{KL}) \tag{6.83}$$

and the density matrix

$$\rho_{KJ} \equiv C_K C_J^*, \tag{6.84}$$

where C_K are the coefficients of the selected electronic basis functions. The dot over a symbol denotes the time derivative. The electronic Hamiltonian and nuclear derivative coupling matrices are correspondingly provided in the forms of $V_{KL} \equiv \langle \Phi_K | \hat{H}^{(el)} | \Phi_L \rangle$ and $\mathbf{d}_{KL} \equiv \langle \Phi_K | \nabla_{\mathbf{R}} | \Phi_L \rangle$, respectively. Note that the latter is assumed to vanish within a diabatic representation. Within this method, the state population is evaluated as the ratio of the number of the trajectory staying in the states of interest against the total number of trajectories initially sampled. A velocity scaling to conserve the total energy is performed at every switching. Such scaling is made in the direction of coupling vector. However, since the present example is just one-dimensional, there is no ambiguity in this aspect anyway.

In both PSANB and FSSH, we use the so-called importance sampling technique to prepare the initial trajectory points to mimic the Wigner phase-space distribution function of an initial Gaussian function, whose width is taken from the ground state eigenfunction of the vibrational motion in the harmonic approximation to the first diabatic state at its bottom point. We use the six-order symplectic integrator [497] to perform the nuclear path dynamics both for PSANB and FSSH. The total simulation time was 800 a.u. both for exact quantum mechanics and two types of semiclassical calculations. The gross features of dynamics are characterized by a single passage of wavepacket through over a state coupling region.

6.4.2 *Dynamics in five state model*

We start from the five state model. The central position of the initial Gaussian wavepacket is placed at $R = 13$ with the initial momentum having -50 (all in atomic units) on the second diabatic state (V_{22}). Figure 6.8 shows the comparison of population dynamics by PSANB and FSSH with the full quantum mechanical counterparts. Numbers of initial sampling points are 20 and 100 for PSANB and FSSH, respectively. The method of sampling schemes is common to these two methods. Each initial PSANB path ends up with branching paths as many as the total number of coupled states (five in this case). Each FSSH path has undergone about 5.8 hops in average before the state-coupling is over. The first branching time is about 320 a.u. Excellent agreement is observed between the results by PSANB and QM. As seen in panel (a), only 20 trajectories of PSANB reproduce the QM result very nicely. Panel (b) indicates that FSSH with 100 initial sampling points does a very good job.

Next, we examine the geometrical property of a PSANB path. Figure 6.9 shows a comparison between the trace (space-time history) of the time-propagation of the quantum wavepacket mechanical (QM) and the corresponding PSANB branching paths in the dynamics of Fig. 6.8. The

PSANB path drawn here is taken from the case study of single initial point. The path and the central position of the wavepackets show excellent geometrical agreement. Such a perfect agreement is not attained easily when a quantum wavepacket undergoes bifurcation on a single potential surface right after the passage of the nonadiabatic region.

The initial sample points does not have the same total energy at t = 0 because they are picked up randomly so as to reproduce the pseudo phase space (PS) structure of initial quantum wavepacket. In order to quantify the effect of distribution in PS on dynamics, we calculated the standard deviation of state populations in the PSANB scheme at t = 300 and 600. The former and latter roughly correspond to the times in the state coupling and off-coupling regions, respectively. The result is summarized in Tab.6.2.

Table 6.2 The standard deviation of populations in PSANB scheme with phase space sampling(PSANB-PSS).

state	$(\eta_k^{20}(300), \eta_k^{100}(300))$	$(\eta_k^{20}(600), \eta_k^{100}(600))$
1	(0.0272, 0.0316)	(0.0164, 0.0213)
2	(0.0459, 0.0543)	(0.0371, 0.0467)
3	(0.0074, 0.0108)	(0.0087, 0.0107)
4	(0.0073, 0.0096)	(0.0074, 0.0090)
5	(0.0072, 0.0087)	(0.0050, 0.0060)

$\eta_k^{Ms}(t) = \sqrt{\sum_{s=1}^{Ms}(p_k^{(s)}(t) - p_k(t))^2/Ms}$, for the case in Fig. 6.8. Here, s denotes a label of sampled trajectory and Ms is a total number of initial sampling points. $p_k^{(s)}(t)$ stand for the kth state population at time t evaluated along bifurcating trajectories starting from the initial sampling point labeled with s. Note that $p_k(t) = \sum_k p_k^{(s)}(t)/Ms$ as the average over all sampling points is meaningful population values in PSANB-PSS.

6.4.3 *Nonadiabatic dynamics in fifteen state model*

We proceed to more stringent examination of the methods, in which 15 diabatic states are involved, as in Fig. 6.7(b). The initial wavepacket is placed on V_{11} at $R = 13$ and the momentum is -30 (all in atomic units). The initial packet starts falling down the slope on the V_{11}. This illustration begins with the case of a single initial path for PSANB and 100 for FSSH. Figure 6.10, panel (a) and (c) make a comparison of population dynamics obtained by PSANB and FSSH, respectively, with that of the exact quantum mechanics calculation. The states shown for illustration are 1, 2, 3, 6, 9 and 15-th diabatic states.

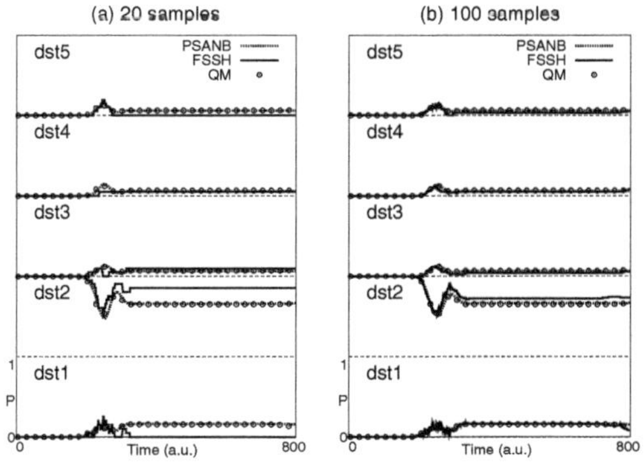

Fig. 6.8 Population dynamics in the five-state model resulting from PSANB (dotted curve) and FSSH (solid curve) procedures in comparison with those provided by the full quantum calculations (circle). The initial wavepacket is prepared at $R = 13$ with the momentum -50 on the second diabatic state (all in atomic units). The horizontal axis indicates the passage of time (in a.u.) for a prepared packet to run on the potential curves. The individual box labeled with dst$i$ ($i = 1, \cdots, 5$) display the population found in the ith diabatic state. The number of initial sampling points is (a) 20 and (b) 100 for both PSANB and FSSH. (Reprinted with permission from T. Yonehara *et al.*, J. Chem. Phys. 137, 22A520 (2012)).

PSANB already gives moderate agreements with QM result for the 6, 9, 15-th diabatic states, except for the oscillatory behavior during the passage through the interaction region due to the use of only a single reference trajectory. Incidentally FSSH does not make a meaningful data, by definition, with a single initial trajectory. Incidentally, that even a single initial sampling in PSANB gives a reasonably accurate transition probability suggests that SET can be made good use of within its validity range, that is, in a time domain in which the path branching is not required to track.

To examine how the results are refined by increasing the number of sampling points, we show the results in Fig. 6.10; panel (b) for 20 sampling points in PSANB and panel (d) for 2000 sampling points in FSSH. Each FSSH hopped about 12.2 times in average. It turns out the PSANB gives essentially the same results with the full quantum calculations after the passage of the interaction region. FSSH also works nicely but not always. Thus we have observed that the appropriate use of PSANB reduces the number of sampling paths, yet bringing about very accurate results.

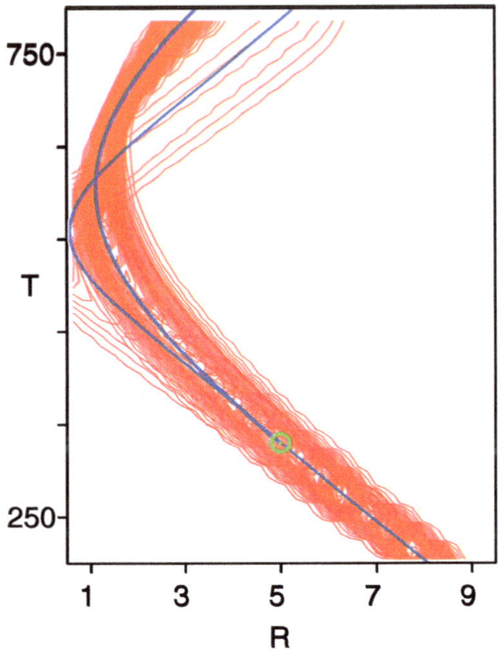

Fig. 6.9 Space-time history of a branching path in the space of (atomic distance (R), and time (T)) in the five state model. The path represents the central poistion of the initial Gaussian function as in Fig. 6.8. The point marked with a circle denotes the branching position, which is determined in terms of the condition of Eq. (6.45). The contour plot in curves represents the same space-time history of the corresponding quantum wavepacket. (Reprinted with permission from T. Yonehara *et al.*, J. Chem. Phys. 137, 22A520 (2012)).

We also calculated the standard deviation of state populations in the PSANB scheme, $\eta_k^{Ms}(t) = \sqrt{\sum_{s=1}^{Ms}(p_k^{(s)}(t) - p_k(t))^2/Ms}$, with Ms=20 in order to quantify the effect of distribution in PS on dynamics. The result is summarized in Table 6.3.

6.5 Electronic phase interference between different branching paths: Dynamics around conical intersections

To study the essential features of electronic-phase interference due to a conical intersection, we employ the same model functions as used recently

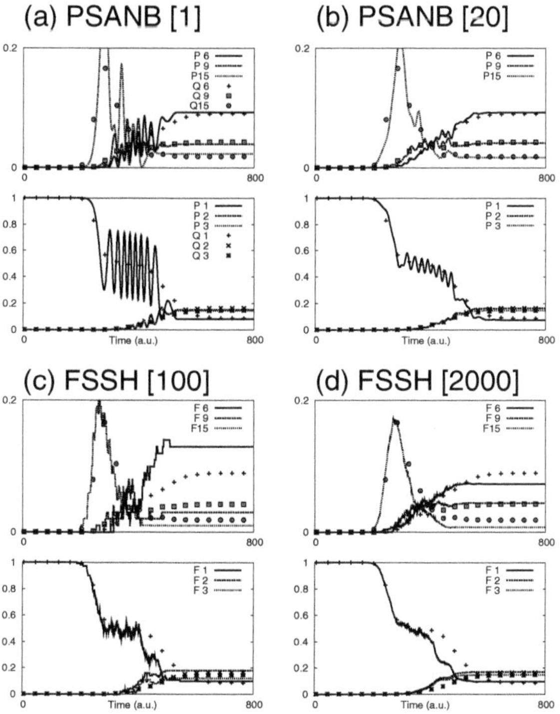

Fig. 6.10 Population dynamics in the fifteen state model resulting from PSANB and FSSH procedures in comparison with those provided by the full quantum calculations. The curve Pi of $i = 6, 9, 15$, for instance, indicate the population on the 6th, 9th and 15th diabatic state, respectively, given by PSANB. Likewise, Fi indicates those given by FSSH, and Qi does the values by the full quantum mechanics. The initial wavepacket is prepared at $R = 13$ with the momentum -30 on the first diabatic state (all in atomic units). The horizontal axis represents time in a.u. Panel (a) PSANB with only one sampling point. (b) PSANB with 20 sampling points. (c) FSSH with 100 sampling points, and (d) FSSH with 2000 sampling points. (Reprinted with permission from T. Yonehara *et al.*, J. Chem. Phys. 137, 22A520 (2012)).

in the work on the optimal spawning scheme by Martínez's group [489]. This model system is a modified version of that proposed in the paper of Ferretti et al. [131], which is based on a linear triatomic molecule of ABA type with two symmetric and an antisymmetric stretching mode. A similar model has also been used by Kapral et al. in a study of the environmental effect on the electron dynamics around a conical intersection [206]. The model is different from the simplest Teller model in that the potential function

Table 6.3 The standard deviation of populations in PSANB scheme with phase space sampling (PSANB-PSS) in Fig. 6.10.

state	$\eta_k^{20}(300)$	$\eta_k^{20}(600)$
1	0.1817	0.0301
2	0.0011	0.0130
3	0.0017	0.0095
4	0.0059	0.0095
5	0.0142	0.0086
6	0.1041	0.0082

The definition of symbols are the same as Table 6.2

is quadratically dependent on both the reaction and bath coordinates. It allows multiple recurrence of the wavepacket around the conical intersection point to take account of dynamics in bound states.

In the present system the potential functions of the diabatic states 1 and 2 are defined as

$$V_{11}(\mathbf{R}) = \frac{1}{2}k_x(x_1 - X_{c1})^2 + \frac{1}{2}k_y x_2^2, \tag{6.85}$$

$$V_{22}(\mathbf{R}) = \frac{1}{2}k_x(x_1 - X_{c2})^2 + \frac{1}{2}k_y x_2^2 + \Delta, \tag{6.86}$$

respectively. Here, $\mathbf{R} \equiv (x_1, x_2)$. The state coupling matrix element is given by

$$V_{12}(\mathbf{R}) = \gamma\, x_2 \exp(-\alpha(x_1 - X_{c3})^2 - \beta x_2^2), \tag{6.87}$$

where the parameters are set to $X_{c1} = 4$, $X_{c2} = 3$, $X_{c3} = 3$, $k_x = 0.01$, $k_y = 0.1$, $\Delta = 0.01$, $\alpha = 3$, $\beta = 1.5$, and $\gamma = 0.01$. The diabatic potential energies, V_{11}, V_{22}, and the coupling matrix elements V_{12} are depicted in Fig. 6.11. The masses assigned to the x_1 and x_2 coordinates are set to m_1=20000 and m_2=6667 in atomic units, respectively. Note that only the point $(x_1, x_2) = (2.5, 0)$ is degenerate in the adiabatic energy surfaces, which are obtained by diagonalizing the diabatic electronic Hamiltonian. The key feature of this system is that the diabatic coupling matrix has two maxima with opposite signs at symmetric positions with respect to a mirror axis ($x_2 = 0$) and therefore contains a node along this axis. This system imposes more stringent tests for numerical calculations than is seemingly expected.

The profiles of potential functions are graphically presented in Fig. 6.11, showing how the conical intersection looks geometrically in the diabatic representation (panel (a) and (b)). Notice the symmetry in panel (b), which arises from the simple relation

$$V_{KK}(x_1, x_2) = +V_{KK}(x_1, -x_2),$$
$$V_{KL}(x_1, x_2) = -V_{KL}(x_1, -x_2) \qquad (K \neq L), \qquad (6.88)$$

where K and L take 1 or 2. Imposing the above symmetry does not lose generality of the present work. Note that the diagonal potential elements are symmetric with respect to $x_2 = 0$ as expressed by the former equation, while the off-diagonal coupling potential elements are antisymmetric.

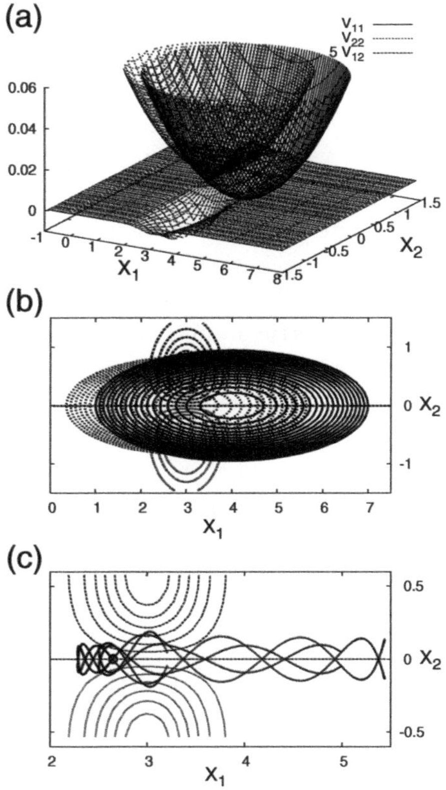

Fig. 6.11 Diabatic potential energies, V_{11}, V_{22}, and coupling matrix elements $V_{12} \times 5$. (a) Bird's eye view of V_{11}, V_{22}, and V_{12}. (b) Projection of them onto the plane of $V = 0$. Increment of contour lines are 0.0025. (c) Close-up of V_{12} and typical branching paths on the (x_1, x_2)-plane. Dotted ($X_2 < 0$) and dashed ($X_2 > 0$) contour lines distinguish the (positive or negative) sign of V_{12}. See text for the details of branching paths. (Reprinted with permission from T. Yonehara *et al.*, J. Phys. Chem. A 117, 8599 (2013)).

To characterize the essential property of parity arising from the electronic phase described in terms of PSANB dynamics, we run a symmetric pair of trajectories starting from two symmetric initial conditions $\mathbf{R}^{\pm}(t = 0) = (x_1(0), \pm x_2(0))$ accompanied with corresponding symmetric velocity vectors, $\dot{\mathbf{R}}^{\pm}(t = 0) = (\dot{x}_1(0), \pm \dot{x}_2(0))$. The first diabatic state was chosen as the initial state with the corresponding electronic coefficient being expressed as $C_K^{\pm}(t = 0) = \delta_{K1}$ (K is 1 or 2). For simplicity, trajectory branch is performed once at the time t^* according to the prescription described in Eq. (6.45). After branching we have four symmetric trajectories running such that

$$\mathbf{R}_J^{\pm}(\tau) = (x_{1J}(\tau), \pm x_{2J}(\tau)), \tag{6.89}$$

on the Jth state ($J = 1, 2$). After the branching ($t > t^*$), the *electronic state*, is represented as

$$|\Psi_K^{\pm}(t)\rangle = \sum_{J=1}^{2} A_K^{J\pm} |\Phi_K\left(\mathbf{R}_J^{\pm}(t)\right)\rangle,$$
$$\text{with} \quad A_K^{J\pm} \equiv C_K^{J\pm}(t - t^*) C_J^{\pm}(t^*) = \langle \Phi_K\left(\mathbf{R}_J^{\pm}(t)\right) |\Psi_K^{\pm}(t)\rangle, \tag{6.90}$$

where $\{C_K^{J\pm}(\tau)\}_{K=1,2}$ denote the state coefficients of electron wavepackets evolving on the path $\mathbf{R}_J^{\pm}$. The initial conditions to restart the electron dynamics are set anew to $C_K^{J\pm}(0) = \delta_{JK}$ at t^*.

First, let us note the symmetric property of the potentials. The potential functions in the diabatic representation is composed of the diagonal part $V_{JJ}(\mathbf{R})$ ($J = 1, 2$) and the off-diagonal coupling term $V_{12}(\mathbf{R}) = V_{21}(\mathbf{R})$. As mentioned above, $V_{12}(\mathbf{R})$ and $V_{KK}(\mathbf{R})$ are odd and even, respectively, with respect to the $x_2 = 0$ axis, and therefore observe

$$V_{12}(\mathbf{R}_J^{+}(\tau)) = -V_{12}(\mathbf{R}_J^{-}(\tau)) \quad \text{and} \quad V_{KK}(\mathbf{R}_J^{+}(\tau)) = V_{KK}(\mathbf{R}_J^{-}(\tau)). \tag{6.91}$$

Next, to track the time dependent relation among the symmetrically paired trajectories, we define a mean force (MF) for the trajectories labeled with $L\pm$ as

$$f_{kL}^{MF\pm} \equiv \sum_{I,J}^{2} \left(C_I^{L\pm}\right)^* F_{IJ}^k(\mathbf{R}_L^{\pm})\, C_J^{L\pm}, \tag{6.92}$$

where $k = 1, 2$ label the nuclear degrees of freedom. The definition of the force matrix given in Eq. (6.18) along with the symmetry property of this diabatic system with $X_{IJ}^k = 0$ directly lead to the following parity relations about force matrix elements

$$\begin{aligned} F_{II}^1(\mathbf{R}_L^{+}) &= +F_{II}^1(\mathbf{R}_L^{-}), & F_{II}^2(\mathbf{R}_L^{+}) &= -F_{II}^2(\mathbf{R}_L^{-}), \\ F_{IJ}^1(\mathbf{R}_L^{+}) &= -F_{IJ}^1(\mathbf{R}_L^{-}), & F_{IJ}^2(\mathbf{R}_L^{+}) &= +F_{IJ}^2(\mathbf{R}_L^{-}) \end{aligned} \tag{6.93}$$

where I and J stand for the electronic states with $I \neq J$.

The iterative use of symmetry properties of the system provides the relations on the electron wavepackets and effective force on trajectories $\mathbf{R}_J^{\pm}$ as [491],

$$C_K^{J+}(t-t^*)C_J^+(t^*) = (-1)^{K-1}C_K^{J-}(t-t^*)C_J^-(t^*), \tag{6.94}$$

and

$$f_{1J}^{MF+} = +f_{1J}^{MF-}, \qquad f_{2J}^{MF+} = -f_{2J}^{MF-}. \tag{6.95}$$

It is obvious that all these properties should be shared by any theoretical methods in which symmetric paths are generated at a conical intersection in a systematic manner as in the present PSANB scheme.

Finally, we note an important feature of the dynamics of electronic phase in the PSANB scheme, namely, the proper use of electronic amplitude $C_K(t^*)$ at branching points. Recalling $C_K^{\pm}(t=0) = \delta_{K1}$, we immediately find the simple rule on the even-odd property at the time of *continuous* transition, which is reflected in $A_K^{J\pm}$: $A_1^{1\pm}, A_1^{2\pm}, A_2^{1\pm}$, and $A_2^{2\pm}$ are correspondingly zero, two, one, and one. This means that the parity relations are the same for $A_1^{1\pm}$ and $A_1^{2\pm}$. So is the relation between $A_2^{1\pm}$ and $A_2^{2\pm}$. Note again the state transition inverts the parity of the electronic wavepacket part with respect to x_2 due to the odd parity of W for x_2 inversion.

As reference data for comparison, the full quantum mechanical wavefunctions of the coherent-type Gaussian form as in Eq. (6.106) are numerically propagated. The wavefunction is chosen to start at a point $(x_1, x_2) = (5.5, 0.0)$ with zero initial momentum on the first diabatic state. The point of conical intersection is located at $(x_1, x_2) = (2.5, 0.0)$, the potential energy of which is exactly the same as the value at $(x_1, x_2) = (5.5, 0.0)$. This symmetric dynamics is performed to purely illustrate how the quantum wavefunctions are branched behind the conical intersection. The contour plots of the density of nuclear wavefunctions at selected times are given in Fig. 6.12 (solid curves).

On the other hand, to mimic the full Gaussian quantum wavefunction, we employed the so-called frozen Gaussian approximation to generate nuclear (basis) functions, that is,

$$\begin{aligned}
\langle \mathbf{R}|\, g(\Omega(t))\rangle &= g(\mathbf{R};\Omega_{I_m}^m(t)) \\
&= \left(\frac{\pi^N}{\det|2\gamma|}\right)^{-1/4} \\
&\times \exp\left(-(\mathbf{R}-\mathbf{R}_c(t))^T\cdot\gamma\cdot(\mathbf{R}-\mathbf{R}_c(t)) + \frac{i}{\hbar}\mathbf{P}_c(t)^T(\mathbf{R}-\mathbf{R}_c(t)) + i\frac{W(t)}{\hbar}\right),
\end{aligned} \tag{6.96}$$

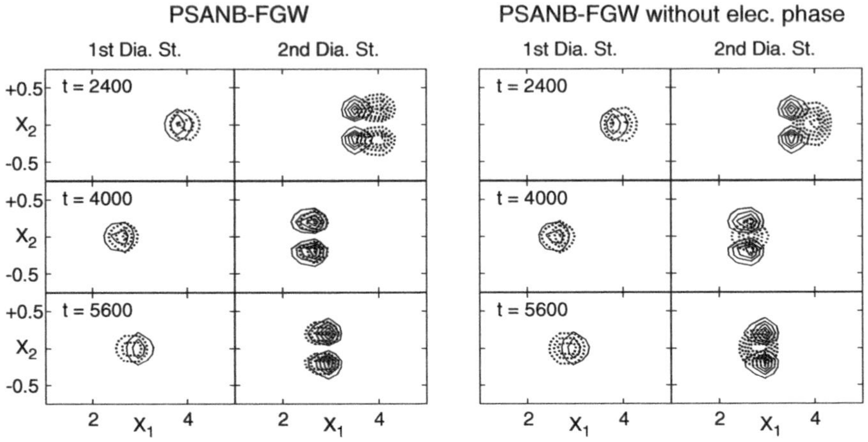

Fig. 6.12 Probability densities of the frozen Gaussian wavepackets (dotted lines) carried by PSANB paths and the full quantum counterpart (solid lines). PSANB electronic wavefunctions presented in the left column take the proper account of the HLH phase, while those in the right column are given after deliberately removing the phase. Snapshots correspond to the times before ($t = 2400$), on-the-way ($t = 4000$), and just after ($t = 5600$) the transition. See the text for the initial condition. The sampling number of initial paths in this presentation are $Ms = 2$. The contour plot of the second diabatic state is relatively magnified to give a better view. The left and right columns in each panel correspond to the first and second diabatic state, respectively. The PSANB with electronic phase nicely reproduces the nodal pattern in the electronic wavefunction, except for a slight retardeation due to the so-called late-kick-off procedure (see the text). (Reprinted with permission from T. Yonehara *et al.*, J. Phys. Chem. A 117, 8599 (2013)).

where $\Omega(t) = (\gamma, \mathbf{R}_c(t), \mathbf{P}_c(t), W(t))$ and $W(t)$ stands for a nuclear classical action defined as a time integral of the effective Lagrangian, which is the kinetic energy of PSANB trajectory minus the electronic mean-field potential. A constant diagonal matrix related to the width γ, γ_{11} and γ_{22} are set to those for the harmonic oscillations in the corresponding coordinates. The semiclassical nuclear wavepackets on state $\Psi_i^{sc}(\mathbf{R}, t)$ are constructed with the frozen Gaussians on the PSANB trajectories in such a way that

$$|\Psi_i^{sc}(t)\rangle = \mathcal{N}_{sc} \sum_{m}^{Ms} \sum_{I_m}^{N_b(t;m)} C_{iI_m}^{m} |g(\Omega_{I_m}^{m}(t))\rangle\langle g(\Omega^m(t=0))|\Psi_0\rangle, \tag{6.97}$$

with i being a label of electronic state. $\langle \mathbf{R}|\Psi_0\rangle$ denotes an initial wavefunction. The normalization constant $\mathcal{N}_{sc}$ at t satisfies

$$\sum_{i=1}^{2} \int d\mathbf{R}\, |\langle \mathbf{R}|\, \Psi_i^{sc}(t)\rangle|^2 = 1. \tag{6.98}$$

The Gaussian parameters attached to trajectory points are $\Omega^m_{I_m}(t) = \{\gamma, \mathbf{R}^m_{I_m}(t), \mathbf{P}^m_{I_m}(t), W^m_{I_m}(t)\}$.

Before comparing the wavefunctions, we return to panel (c) of Fig. 6.11 to see the behavior of path branching (see also Fig. 13(b) of Ref. [491]), along which $\Psi^{sc}_i(t)$ of Eq. (6.97) are carried. Two paths begin to run on the first diabatic state from two mutually symmetric points, $(x_1, x_2) = (5.42826, \pm 0.10816)$ with $(k_1, k_2) = (2.12978, \mp 2.02927)$, which were sampled randomly within the Wigner phase-space distribution function. They undergo smooth bifurcation at a point marked by the black circle, each branching into two pieces (therefore four in total); one remaining mostly on the first diabatic state and the other proceeding mainly into the second diabatic state. Note that in contrast to paths generated by a hopping procedure, for example, each of these paths on a certain diabatic state have a small component contributed from the other diabatic state, but they smoothly converge to the pure diabatic (adiabatic) state after the nonadiabatic coupling is reduced to zero. Note also that the branching point in this study is located behind the point of conical intersection. Therefore, before branching, each path is propagated as a phase-space averaging path in the nonadiabatically coupled region [491], in which electronic-state mixing between the first and second diabatic states is fully taken into account up to the electronic phase.

On the basis of the above pair of phase-space averaging and natural branching paths, we illustrate in Fig. 6.12 the electronic-phase interference that manifests itself in the nuclear wavepackets. As seen in the panels for the wavefunctions on the second diabatic state (the left six panels in Fig. 6.12), the present method well reproduces the nodal lines as a consequence of destructive interference, although the so-called late kick-off phenomenon, which was first noticed by Martínes [489, 495], is apparent as a price of our approximation. (The relevant transition begins only when the center of the wavepacket represented by PSANB comes across the nonadiabatic region, whereas the full quantum wavepacket begins to undergo transition as soon as its tail reaches the region.)

To further confirm the role of the HLH phase, we have computed the same quantities after intentionally removing the electronic phases by simply replacing $C^m_{iI_m}$ with $|C^m_{iI_m}|$ in Eq. (6.97). The results are displayed in the right column of Fig. 6.12. As clearly demonstrated, the interference pattern due to the HLH phase has disappeared completely. Thus the PSANB method with the frozen Gaussian gives qualitatively correct description of the geometrical symmetry of quantum mechanical phase associated with the electronic transition. It is also seen that the PSANB wavepacket left in the first diabatic state retains the even symmetry. These features well illustrate the correct evolution of electron quantum amplitude within the present scheme.

6.6 Quantum effects manifesting in the nuclear branching paths

6.6.1 *Interactions and initial conditions*

We apply the above algorithm to a model chemical reaction on one-dimensional, two-state interaction potential, which is constructed as follows: We first determine a pair of adiabatic potential energy curves (APEC) to be represented as

$$\epsilon_1(R) = V_{TS} \exp\left(-\left(\frac{R}{a}\right)^2\right)$$
$$\epsilon_2(R) = -\epsilon_1(R) + b, \qquad (6.99)$$

where the parameters V_{TS}, a, b are set to $V_{TS} = 0.01$, $a = 1$ and $b = 0.024$ in atomic units, and thus the top of $\epsilon_1(R)$ is 0.01 while the bottom V_B of $\epsilon_2(R)$ is 0.014. As usual the matrix U

$$\mathrm{U} = \begin{pmatrix} \cos\theta(R) & \sin\theta(R) \\ \sin\theta(R) & -\cos\theta(R) \end{pmatrix} \qquad (6.100)$$

transforms the APEC to diabatic potential energy curves (DPEC) with the coupling element X_{12}, which is given as

$$X_{12} = \frac{d\theta}{dR} = -X_{21}, \qquad (6.101)$$

where the rotation angle θ as a function of R is defined as

$$\theta(R) = \frac{\pi}{4}\left(2 - \exp\left(-\frac{4dR}{\pi}\right)\right) \quad \text{for } x \geq 0 \qquad (6.102)$$

and

$$\theta(R) = \frac{\pi}{4}\exp\left(-\frac{4dR}{\pi}\right) \quad \text{for } x < 0, \qquad (6.103)$$

with a parameter d being set to $d = 2$. The resulting APEC and DPEC are shown in Fig. 6.13. In this scheme therefore we can scan the intensity of the nonadiabatic coupling elements keeping the adiabatic potentials unchanged.

We can thus uniquely define the total energy of the non-Born–Oppenheimer paths with respect to the height of the potential barrier V_{TS}.

The initial conditions are chosen so that the total energy of the path is comparable to the barrier height V_{TS}. The initial paths are generated on the ground state. The nuclear mass is set to $M = 1836$, mimicking a proton. The initial position to launch the paths is set at $R_0 = -9$, and let them start with a momentum $P_0 = \sqrt{2ME_0}$ and the total energy E_0 to the positive direction (from left to right in the figures). We here demonstrate four typical paths:

(1) $E_0 = 0.0080 = 0.8V_{TS}$: below the barrier top (V_{TS})

(2) $E_0 = 0.0099 = 0.99V_{TS}$: close to the barrier top

(3) $E_0 = 0.0125 = 1.25V_{TS}$: higher than the top (V_{TS}) and lower than the bottom of the excited state (V_B)

(4) $E_0 = 0.0145 = 1.45V_{TS}$: a little higher than V_B.

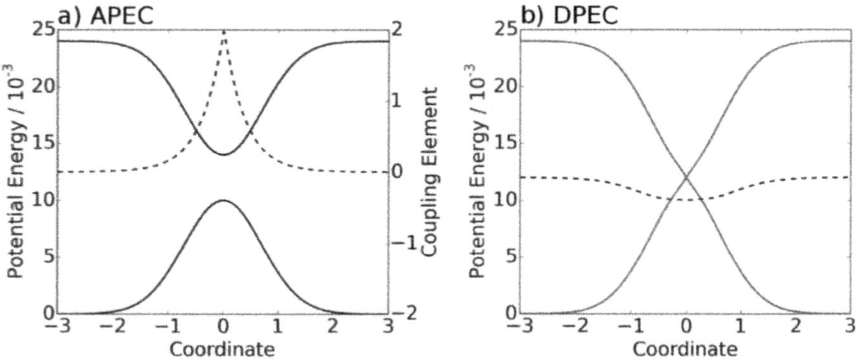

Fig. 6.13 Potential energy curves versus the coordinate R. (a) Adiabatic representation. The solid curves denote APEC ϵ_i, while the dashed curve does nonadiabatic coupling element X_{12}. (b) Diabatic representation. The solid curves represent the DPEC (ascending to the right: V_{11}, descending: V_{22}) and the dashed curve shows the electronic coupling element V_{12}. (Reprinted with permission from K. Yamamoto *et al.*, J. Chem. Phys. 140, 124111 (2014)).

6.6.2 *Practices of the dynamic calculations*

To avoid confusion about the computational procedures, we should describe the distinction between the common practices in full electron wavepacket dynamics [491] and the actual procedure we adopt in the present section for a demonstration of nonclassical dynamics due to nonadiabatic path-branching.

Given many-body electronic basis functions, for which we usually adopt the so-called configuration state functions (CSF, an appropriate linear combination of the Slater determinants), one can calculate the electronic matrix elements $H_{IJ}^{(el)}$ and the nuclear kinematic coupling elements X_{IJ}. The adiabatic electronic wavefunctions are not necessary, although they can be readily obtained by diagonalizing $H_{IJ}^{(el)}$. Once both $H_{IJ}^{(el)}$ and X_{IJ} become available on-the-fly, all the dynamical equations for the electronic state mixing, Eq. (6.14) (neglecting Y_{IJ} though), and the force matrix, Eq. (6.23) or Eq. (6.24), the general procedure in Sec. 6.2 gives a set of approximate solutions of the electron wavepackets along branching paths.

To perform the following illustrative calculations, on the other hand, we first prepared the two adiabatic states as above in Eq. (6.99) and assumed the presence of a set of diabatic matrix elements through the rotation angle, Eq. (6.102), which automatically gives both $H_{12}^{(el)}$ and X_{12}. We thus can perform all the necessary procedures for nonadiabatic dynamics including the electronic state components $\{C_1, C_2\}$ of Eq. (6.13) without necessity of referring to the explicit functional form of the electronic bases $\Phi_I(\mathbf{r}; \mathbf{R})$.

In this section therefore we demonstrate only the nuclear branching paths and the full nuclear wavepackets. However once again we stress that the direct investigation of the electron wavepackets are indispensable in the full study of chemical reactions.

6.6.3 *Surmounting a potential barrier by lower energy paths; a behavior looking like quantum tunneling*

We study first the branching dynamics of the two lower energy paths of $E_0 = 0.8V_{TS}$ and $E_0 = 0.99V_{TS}$. As suggested above, quantum paths can branch into even classically inaccessible domains, the dynamics of which is caused by the electronic state mixing due to the derivative coupling X_{IJ}^k. The surface hopping model and the related theories based on it would miss this phenomenon. The branch parameter Θ (see Eq. (6.48)) is set to 8 for the path of $E_0 = 0.8V_{TS}$ and to 7 for that of $E_0 = 0.99V_{TS}$. These parameters have been chosen to make the demonstrations clear.

As observed in Fig. 6.14, the paths come in and return to the reactant site with a large electronic-state population (the thickest red curves). However, in spite of the fact that the energy of these paths are lower than the barrier height, they begin to branch and some of the paths eventually reach the other side of the barrier. Inspection over this figure also shows that some other branched paths go back to the reactant site tracking different routes from the parent one. Therefore these dynamics represent a situation in which a part of the initial wavepacket is bifurcated and surmounts the barrier.

The paths that reach the other side of the barrier are assisted by the forces from the excited states. Let us focus on the path ending at around $t = 5 \times 10^3$ and $R = 1.5$ of Fig. 6.14(a). The precise history of this path is as follows. First, the path entered the interaction area with the first diagonal force (dominated by that of the ground state), and then at a branching point the path resumes to run driven by the second diagonal force (close to that of the excited state), and at the second branching point the driving force comes back to the first diagonal component (essentially coming from the ground state). In this way the path could go over the barrier although it was classically forbidden.

The overall phenomenon effectively looks like the so-called deep tunneling. [46, 140, 149, 151, 269] However, of course, they are independent phenomena. In the quantum deep tunneling, the paths go over the barrier through imaginary space and/or time, as in the instanton theory. [46] On the other hand, all the non-Born–Oppenheimer paths in this study are supposed to run in the real space, although the present theory of path branching can be generalized so that the paths can penetrate into complex spaces.

Another important difference between "surmounting" and "deep tunneling" is their dependences on the potential curves. Deep tunneling tends to be more effective as the potential barrier is thinner, while the nonadiabatic surmounting does not have such a simple characteristic. For instance, the path ending at $t = 5 \times 10^3$ and $R = 0$ in Fig. 6.14(b) is about to go back to the reactant side. This path getting back to the ground state does not contribute to the increasing of reaction probability. Another difference is the time-delay of paths, which has nothing to do with the so-called tunneling time. We will return the problem of time-delay in the next section.

6.6.4 *Trapping above the potential barrier: Time-delay in reaction dynamics*

We next examine the case where the total energy is slightly higher than the barrier height, namely, $E_0 = 1.2V_{TS}$ (lower than the bottom of the excited state though) and $1.45V_{TS}$ (higher than the bottom too). Only to the last case, the surface hopping model can access). The branch parameter Θ (see Eq. (6.48)) is set to 5 and 4, respectively. As shown in Fig. 6.15, some paths

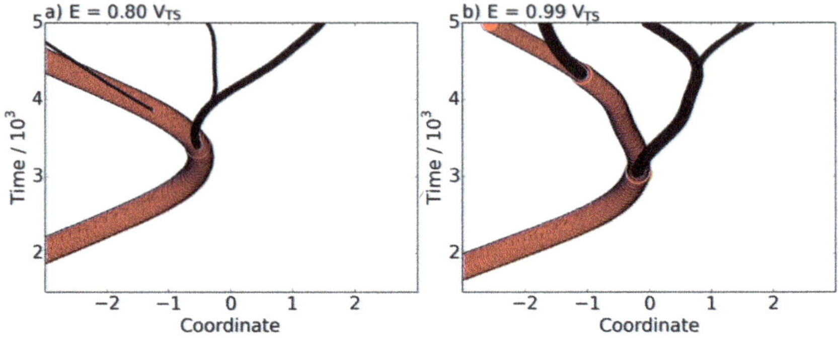

Fig. 6.14 Branching paths of the total energy lower than the barrier height. The width of the paths is depicted so as to be proportional to the square root of the electron population on each path. The paths in red (black) are driven by the forces mainly coming from the ground (excited) state. (Reprinted with permission from K. Yamamoto *et al.*, J. Chem. Phys. 140, 124111 (2014)).

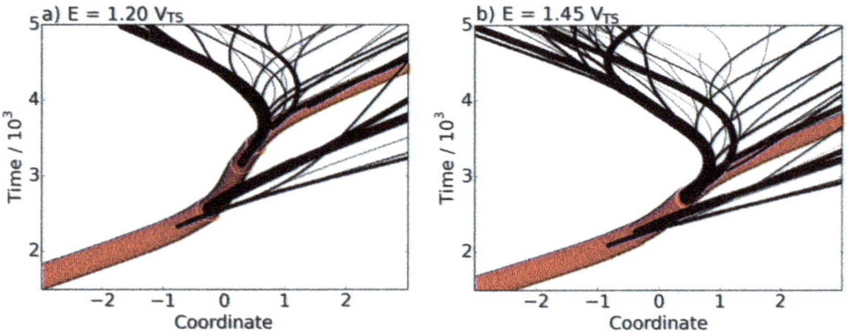

Fig. 6.15 The same as Fig. 6.14, except that the branching paths have the total energy higher than the barrier height. (Reprinted with permission from K. Yamamoto *et al.*, J. Chem. Phys. 140, 124111 (2014)).

are pushed back to the reactant side. These returning paths obviously work to reduce the reaction probability, and can be regarded as a friction to the reactive motions.

The precise mechanism is as follows. The paths come about the strong interaction area, where the electronic-state mixing is induced. The forces originated from the excited state trap some paths as though they undergo vibrational motion on the excited state adiabatic potential curve. However, again due to the nonadiabatic interactions, the trapped paths gradually spill out to the both ends of the potential barrier of the ground state. From the view point of chemical reaction therefore this delayed dynamics looks like a result of friction.

6.6.5 *Full-quantum dynamics to verify the branching phenomena*

We now proceed to confirm the physical picture drawn by the path-branching representation is indeed consistent with the full-quantum results. The nuclear quantum wavepackets can be readily calculated, provided that both $H_{12}^{(el)}$ and X_{12} are available globally, in terms of the Born–Huang expansion

$$\Psi_{\text{total}}(\mathbf{r}, R, t) = \sum_I \chi_I(R, t)\Phi_I(\mathbf{r}; R)|_{R=R(t)} \tag{6.104}$$

and the resultant coupled equations

$$i\hbar\frac{\partial}{\partial t}\chi_I(R,t) = -\frac{\hbar^2}{2M}\frac{d^2}{dR^2}\chi_I(R,t) + \sum_J \left(H_{IJ}^{(el)} - \frac{\hbar^2}{M}X_{IJ}\frac{d}{dR}\right)\chi_J(R,t), \tag{6.105}$$

where the second-order derivative terms Y_{IJ} have been neglected consistently. Note again that these coupled equations of motion are valid irrespective of the choice of the orthonormal set $\{\Phi_I(\mathbf{r}; R)\}$. We compare only the quantitative aspects of the phenomena observed above.

6.6.5.1 *Method and initial conditions*

The initial conditions for the quantum dynamics corresponding to those in the preceding section are as follows. Suppose that a total wavefunction $\Psi_{\text{total}}(\mathbf{r}, R, t)$ is expanded as in Eq. (6.104) and we are interested only in the nuclear wavepackets $\chi_1(R, t)$ and $\chi_2(R, t)$. We propagate them with the extended split operator method. [77] The initial nuclear wavepackets $\{\chi_I(R, t)\}$ are chosen to be a coherent-type Gaussian function only on the adiabatic ground state as

$$\begin{aligned}\chi_1(R,0) &= (\pi\Delta R^2)^{-1/4}\exp\left[-\frac{(R-R_0)^2}{2\Delta R^2} + i\frac{P_0 R}{\hbar}\right],\\ \chi_2(R,0) &= 0.\end{aligned} \tag{6.106}$$

R_0 and P_0 are the same quantities as those defined in Sec. 6.6.1 for the path-branching representation. The wavepacket width in coordinate space is set to $\Delta R = 1$. The number of grid points for FFT is 512 on the space ranging from -32 to 32. Two absorbing imaginary potentials V_a are placed at the both ends, which are defined as

$$V_a(R) = -30i \exp\left(-\left(\frac{R \pm 20}{2}\right)^2\right). \tag{6.107}$$

V_a works not only to avoid the artificial reflection of the wavepackets from the boundaries but in the calculation of the reaction probability by integrating the absorbed components.

6.6.5.2 *Time-delay induced by the trapping motion above the transition state*

The trails of the quantum wavepackets in (t, R)-coordinates are drawn in Fig. 6.16 left column, corresponding to the above four energies: (a) $E_0 = 0.008V_{TS}$, (b) $E_0 = 0.99V_{TS}$, (c) $E_0 = 0.0125 = 1.25V_{TS}$, and (d)$E_0 = 1.45V_{TS}$. The color chart in each diagram clearly indicates the wavepacket bifurcations, whose patterns are dependent on the energy.

In the right column, corresponding to each left panel, are depicted the square root of the population of the state ($|\Psi|$) that passes across the line $R = 9$ (red and solid curve, measured from right to left along the horizontal axis) and $R = -9$ (green and dashed curve, measured from the left to the right) as functions of time. The former is proportional to the magnitude of $|\Psi|$ reaching the product area, while the latter represents that of the original reactant area. (We do not resort to the population $|\Psi|^2$ itself to prevent the small components from being invisible.)

In the lowest energy case, we can readily identify the packet components that are flowing out to the product site (of large positive R values). It is the usual practice that we attribute this seemingly leaking dynamics to deep tunneling. That this is not the case can be readily confirmed by running the full quantum dynamics without both the excited state and nonadiabatic coupling elements (simple dynamics on the ground state alone with the wavepackets whose average energies are lower than the barrier). It turns out that with this width of the potential barrier, the pure (deep) tunneling turned out to be virtually invisible, and instead the so-called shallow tunneling dominates (passing over the barrier by wavepacket components of energy higher than the barrier height). However, the presence of the nonadiabatic surmounting dynamics is clearly confirmed in terms of the time-delay components in the wavepacket dynamics, that is, the shoulders of $|\Psi(9, t)|$ after the passage of the potential barrier (see Fig. 6.16, red curves after $t = 6$ in the right boxes of panels (a) and (b)). Such shoulders are more

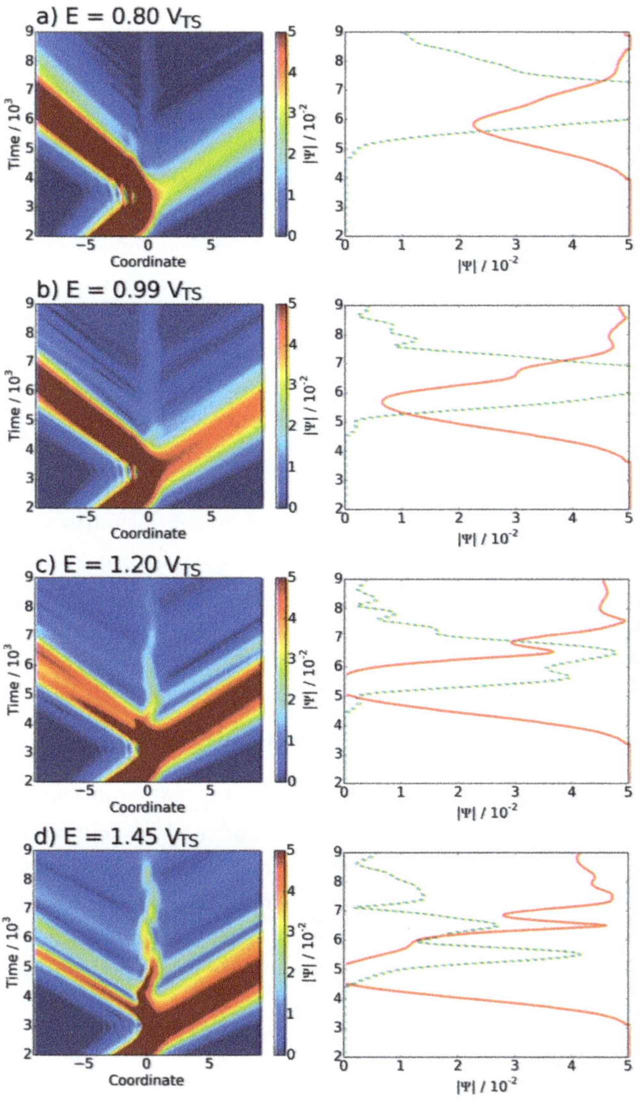

Fig. 6.16 The space-time pathways of the full-quantum wavepackets for various total energies. The color maps on the left hand side are the plot of $|\Psi(R,t)|$. The plots on the right hand side are $|\Psi(R=9,t)|$ (solid line, measure to the left) and $|\Psi(R=-9,t)|$ (dashed line, measure to the right). The solid line is scaled as $|\Psi| = 2.5\times10^{-2}$. (Reprinted with permission from K. Yamamoto *et al.*, J. Chem. Phys. 140, 124111 (2014)).

prominent in case of $E_0 = 0.99V_{TS}$ (panel (b)). The shoulders are seen not only in the graph of $|\Psi(9,t)|$ but in that of the reflecting waves $|\Psi(-9,t)|$. Without the nonadiabatic coupling, the nuclear wavepackets would split only into two pieces and no those shoulders are found.

In the higher energy regime, the evidence of the tentatively vibrational motion in the excited adiabatic curve is clearly observed as the oscillatory structure around $R = 0$ in panel (d) left. The major component in this energy is found in the product site. A remarkable consequence of the tentative vibrational motion in the excited state is sequential appearance of pulse-train wavepackets with "time-delay", and the small shoulders arise from the delayed pulses. The time interval of the appearance of the pulses is approximately equivalent to the vibrational period of the excited state, although a complicated quantum interference gives rise to deviation from such a simple rule. This phenomenon of pulse-train is essentially equivalent to those observed by Zewail et al. in pump-probe experiments for the photodynamics of NaI. [340] However, the present dynamics arises from collision events starting from the ground state and therefore may give additional pieces of information about transition state and associated excited state if observed experimentally.

6.7 Quantization of non-Born–Oppenheimer paths

Our next task is to quantize the non-Born–Oppenheimer paths by associating quantum wavepackets with individual nuclear paths. The difficulty here lies in the fact that the paths in non-Born–Oppenheimer theories are generally nonclassical and does not satisfy the Newtonian equation. To do so, we apply the theory of action decomposed function. [415, 427]

6.7.1 *Action Decomposed Function (ADF)*

6.7.1.1 *Basic theory*

We adopt the following representation of a nuclear wavefunction on its general Cartesian coordinate $\mathbf{q}$ as

$$\Psi(\mathbf{q},t) = F(\mathbf{q},t)\exp\left(\frac{i}{\hbar}S(\mathbf{q},t)\right), \tag{6.108}$$

where S is assumed to satisfy the Hamilton-Jacobi equation, and thereby $\exp\left(\frac{i}{\hbar}S\left(\mathbf{q},t\right)\right)$ is regarded as a transformation function to determine $F(\mathbf{q},t)$, giving an equation of motion for it as

$$\frac{\partial F(\mathbf{q},t)}{\partial t} = \left(-\mathbf{v}\cdot\nabla - \frac{1}{2}(\nabla\cdot\mathbf{v})\right) F(q,t) + \frac{i\hbar}{2}\nabla^2 F(\mathbf{q},t), \tag{6.109}$$

where

$$\mathbf{v} = \nabla S\left(\mathbf{q},t\right) \tag{6.110}$$

which is the velocity [424]. Equation (6.109) is readily rewritten in a manner of the Lagrangian view of fluid mechanics

$$\begin{aligned}\frac{dF(\mathbf{q},t)}{dt} &= \left(\frac{\partial}{\partial t} + \mathbf{v}\cdot\nabla\right) F(\mathbf{q},t)\\ &= -\frac{1}{2}(\nabla\cdot\mathbf{v})F(\mathbf{q},t) + \frac{i\hbar}{2}\nabla^2 F(\mathbf{q},t),\end{aligned} \tag{6.111}$$

which suggests that this equation of motion can be integrated along a path in the velocity field $(\mathbf{q}\left(t\right),\mathbf{v}\left(t\right))$. More explicitly, with the help of the Trotter decomposition [358], the short time propagation of $F(\mathbf{q},t)$ can be written as

$$\begin{aligned}&F(\mathbf{q}-\mathbf{q}\left(t+\Delta t\right),t+\Delta t)\\ &\simeq \exp\left[\frac{i\hbar}{2}\Delta t\nabla^2\right]\exp\left[-\frac{1}{2}(\nabla\cdot\mathbf{v})\Delta t\right]F(\mathbf{q}-\mathbf{q}\left(t\right),t)\end{aligned} \tag{6.112}$$

for a very small time step Δt.

6.7.1.2 *ADF equation of motion in laser field*

For an extension of ADF to a case including a laser field described by vector potential, [495] we study the following non-relativistic Schrödinger equation for nuclear motion on the coordinates $\mathbf{R}$

$$i\hbar\frac{\partial}{\partial t}\chi\left(\mathbf{R},t\right) = \left[\frac{1}{2}\sum_k\frac{1}{m_k}\left(\hat{P}_k - \frac{Z_k e}{c}A_k\left(\mathbf{R}\right)\right)^2 + V\left(\mathbf{R}\right)\right]\chi\left(\mathbf{R},t\right), \tag{6.113}$$

where $\hat{P}_k$ is the quantum momentum operator

$$\hat{P}_k = \frac{\hbar}{i}\frac{\partial}{\partial R_k} \tag{6.114}$$

and should be distinguished from its classical canonical momentum P_k. Here again, we resort to the mass weighted coordinates, rescaling all the masses to

$$m_k = 1. \tag{6.115}$$

The velocity of a particle is correlated with its momentum in such a way that

$$v_k = P_k - \frac{Z_k e}{c} A_k. \tag{6.116}$$

The classical counterpart of this dynamics is represented in the following Hamilton-Jacobi equation

$$\frac{\partial}{\partial t} S(\mathbf{R}, t) + \frac{1}{2} \sum_k \left(\frac{\partial S(\mathbf{R}, t)}{\partial R_k} - \frac{Z_k e}{c} A_k \right)^2 + V(\mathbf{R}) = 0, \tag{6.117}$$

where the momentum is generated through the action integral as

$$P_k = \frac{\partial S(\mathbf{R}, t)}{\partial R_k}. \tag{6.118}$$

It is convenient to summarize the classical relations as

$$\left(\frac{\partial S(\mathbf{R}, t)}{\partial R_k} - \frac{Z_k e}{c} A_k \right) = v_k = P_k - \frac{Z_k e}{c} A_k \tag{6.119}$$

subject to the condition Eq. (6.115).

We set the total wavefunction in the form

$$\chi(\mathbf{R}, t) = \exp\left(\frac{i}{\hbar} S(\mathbf{R}, \mathbf{t}) \right) F(\mathbf{R}, t). \tag{6.120}$$

The left hand side of the Schrödinger equation is

$$i\hbar \frac{\partial \chi(\mathbf{R}, t)}{\partial t} = \exp\left(\frac{i}{\hbar} S(\mathbf{R}, t) \right) \left(-\frac{\partial S}{\partial t} F + i\hbar \frac{\partial F(\mathbf{R}, t)}{\partial t} \right). \tag{6.121}$$

Next, we evaluate the right hand side in Eq. (6.113). Since the kinetic energy parts are the origin of confusion, we should treat them with some care. First, we examine the operations on $\exp\left(\frac{i}{\hbar} S\right)$ as

$$\left(\hat{P}_k - \frac{Z_k e}{c} A_k(\mathbf{R}) \right) \exp\left(\frac{i}{\hbar} S \right) = \exp\left(\frac{i}{\hbar} S \right) \left(P_k - \frac{Z_k e}{c} A_k(\mathbf{R}) \right), \tag{6.122}$$

and

$$\begin{aligned} &\frac{1}{2} \left(\hat{P}_k - \frac{Z_k e}{c} A_k(\mathbf{R}) \right)^2 \exp\left(\frac{i}{\hbar} S \right) \\ &= \frac{1}{2} \left(\hat{P}_k - \frac{Z_k e}{c} A_k(\mathbf{R}) \right) \left(P_k - \frac{Z_k e}{c} A_k(\mathbf{R}) \right) \exp\left(\frac{i}{\hbar} S \right) \\ &= \frac{1}{2} \exp\left(\frac{i}{\hbar} S \right) \left[\left(P_k - \frac{Z_k e}{c} A_k(\mathbf{R}) \right)^2 + \hat{P}_k \left(P_k - \frac{Z_k e}{c} A_k(\mathbf{R}) \right) \right]. \end{aligned} \tag{6.123}$$

Then the following straightforward formulation goes as

$$\left[\frac{1}{2}\sum_k\left(\hat{P}_k-\frac{Z_k e}{c}A_k\right)^2+V(\mathbf{R})\right]\exp\left(\frac{i}{\hbar}S(\mathbf{R},\mathbf{t})\right)F(\mathbf{R},\mathbf{t})$$
$$=F\exp\left(\frac{i}{\hbar}S\right)\left\{\sum_k\left[\frac{1}{2}\left(P_k-\frac{Z_k e}{c}A_k\right)^2+\frac{1}{2}\hat{P}_k\left(P_k-\frac{Z_k e}{c}A_k\right)\right]+V(\mathbf{R})\right\}$$
$$+\exp\left(\frac{i}{\hbar}S\right)\frac{1}{2}\sum_k^2\hat{P}_k^2F+\exp\left(\frac{i}{\hbar}S\right)\sum_k\left(P_k-\frac{Z_k e}{c}A_k\right)\left[\hat{P}_kF\right].$$
(6.124)

Combining Eqs. (6.121) and (6.124), we see

$$-\frac{\partial S}{\partial t}F+i\hbar\frac{\partial F}{\partial t}$$
$$=F\left\{\sum_k\left[\frac{1}{2}\left(P_k-\frac{Z_k e}{c}A_k\right)^2+\frac{1}{2}\hat{P}_k\left(P_k-\frac{Z_k e}{c}A_k\right)\right]+V(\mathbf{R})\right\}$$
$$+\frac{1}{2}\sum_k\hat{P}_k^2F+\sum_k\left(P_k-\frac{Z_k e}{c}A_k\right)\left[\hat{P}_kF\right].$$

With the help of the Hamilton-Jacobi equation (6.117), we have (using Eq. (6.115))

$$\frac{\partial F(\mathbf{R},t)}{\partial t}=\left[-\mathbf{v}\cdot\nabla-\frac{1}{2}[\nabla\cdot\mathbf{v}]+\frac{i}{2}\hbar\nabla^2\right]F. \tag{6.125}$$

Note that this ADF equation is of exactly the same form as that for field-free case. This is not surprising, since the ADF equation essentially represents kinematics. Therefore, the technique developed before for approximately solving the ADF equation in nonadiabatic problem can be readily extended to nonadiabatic dynamics in laser field.

6.7.2 *Normalized Variable Gaussians (NVG) as a simple approximation*

Let us return to Eq. (6.112) to consider a tractable approximation. We first consider the propagation by the momentum-gradient term, that is,

$$\exp\left[-\frac{1}{2}(\nabla\cdot\mathbf{v})\Delta t\right]F(q-q(t),t), \tag{6.126}$$

which gives a semiclassical theory. The formal solution of Eq. (6.126) is given only in terms of the information of velocity field such that

$$F(q-q(t),t)=F(q-q(0),0)\exp\left[-\int_0^t\frac{1}{2}(\nabla\cdot\mathbf{v})ds\right], \tag{6.127}$$

assuming that our treated problem does not contain the explicit dependence of time as in the molecular dynamics in a laser field.

For a one-dimensional system, one can proceed further with the following approximation.

$$\int_0^t \frac{1}{2}(\nabla v)ds = \frac{1}{2}\int_0^t \frac{\partial v}{\partial q}ds = \frac{1}{2}\int_0^t \frac{ds}{dq}dv = \frac{1}{2}\int_{v(0)}^{v(t)} \frac{dv}{v}. \tag{6.128}$$

This is just

$$\int_0^t \frac{1}{2}(\nabla v)ds = \frac{1}{2}\log v\left(t\right) - \frac{1}{2}\log v\left(0\right). \tag{6.129}$$

So, for a given time interval Δt, we have

$$F(q - q\left(t + \Delta t\right), t + \Delta t) = F(q - q\left(t\right), t)\left|\frac{v\left(t + \Delta t\right)}{v\left(t\right)}\right|^{-1/2} \exp\left[-i\pi\frac{M}{2}\right], \tag{6.130}$$

where M is the Maslov index in this sense. This approximation can be applied to a path solution by regarding it as a one-dimensional system along the curve in configuration space. However for multidimensional systems more elaboration is needed. One should consult with Refs. [415, 427].

Next we should consider the term of quantum diffusion $\exp\left[\frac{i\hbar}{2}\Delta t\nabla^2\right]$ of Eq. (6.112), which essentially gives rise to the Feynman kernel for a free particle. The Feynman path summation is known to generally suffer from violent oscillations and lack of the integral measure. However, it is also well known that it can be rigorously and compactly treated for a Gaussian function. The price to utilize a Gaussian functions is, of course, that it is no longer an exact solution of Eq. (6.111) in general. Nevertheless, our numerical calculations shows that the Gaussian approximation to ADF or the linear combinations of them give accurate results. So, we resume the dynamics of ADF with a Gaussian function reconsidering the momentum-gradient term before the quantum diffusion is taken into account.

We first note that the norm of any ADF should be conserved, since

$$\begin{aligned}\int dq\left|\Psi\left(q\right)\right|^2 &= \int dq\left|\exp\left[\frac{i}{\hbar}S\left(q,t\right)\right]F(q - q_k\left(t\right), t)\right|^2 \\ &= \int dq\left|F(q - q\left(t\right), t)\right|^2 = 1.\end{aligned} \tag{6.131}$$

We then assume the form of the Gaussian, but here its exponent is allowed to vary in such a way that

$$F(q - q\left(t\right), t) = \left(\frac{2}{\pi}\right)^{1/4} f\left(q_k\left(t\right)\right)\exp\left[-\gamma\left(t\right)\left(q - q\left(t\right)\right)^2\right], \tag{6.132}$$

where $\gamma(t)$ is a complex number. Suppose that the height of this function is multiplied by a positive factor R such that

$$F^{(R)}(q - q(t)) = RF^{(1)}(q - q(t)) \\ = R\left(\frac{2}{\pi}\right)^{1/4} f(q(t)) \exp\left[-\gamma(t)(q - q(t))^2\right] \quad (6.133)$$

Then, we want to renormalize this function by varying the exponent as

$$\int dq \left|F^{(R)}(q - q(t), t)\right|^2 = \int dq R^2 \left|F^{(1)}(q - q(t), t)\right|^2 \\ = \sqrt{\frac{2}{\pi}} |f(q(t))|^2 \sqrt{\frac{\pi R^4}{2\mathrm{Re}(\gamma(t))}} = 1. \quad (6.134)$$

Thus, if the exponent is rescaled as

$$\gamma(t) \to R^4 \gamma(t), \quad (6.135)$$

then this Gaussian is normalized again. We refer to this normalized Gaussian with variable exponent as ADF-NVG or simply NVG.

6.7.2.1 *Momentum-gradient term revisited*

In the propagation process of Eq. (6.112), we first perform the short time propagation with the momentum-gradient term, that is,

$$F^{vg}(q - q(t + \Delta t)) \equiv \exp\left[-\frac{1}{2}(\nabla v)\Delta t\right] F(q - q(t)) \\ = \left(\frac{2}{\pi}\right)^{1/4} \exp\left[-\frac{1}{2}(\nabla v)\Delta t\right] f(q_k(t)) \exp\left[-\gamma(t)(q - q(t))^2\right]. \quad (6.136)$$

Then, using the rescaling discussed above, we have

$$F^{vg}(q - q(t + \Delta t)) = \left(\frac{2}{\pi}\right)^{1/4} f^{vg}(q(t + \Delta t)) \exp\left[-\gamma^{vg}(q - q(t + \Delta t))^2\right], \quad (6.137)$$

where

$$f^{vg}(q(t + \Delta t)) = f(q(t)) \left(\frac{v(t)}{v(t + \Delta t)}\right)^{1/2}. \quad (6.138)$$

and

$$\gamma^{vg}(t + \Delta t) = \gamma(t) \left(\frac{v(t)}{v(t + \Delta t)}\right)^2. \quad (6.139)$$

6.7.2.2 *Quantum Diffusion*

We now apply the Feynman kernel to the above obtained new Gaussian such that

$$F(q-q(t+\Delta t),t+\Delta t)=\int dyK(q,y,\Delta t)\,F^{vg}(y-q(t+\Delta t)),\quad (6.140)$$

which simply gives

$$\begin{aligned}&F(q-q(t+\Delta t),t)\\&=\left(\frac{2}{\pi}\right)^{1/4}f(q(t))\left(\frac{v(t)}{v(t+\Delta t)}\frac{1/\gamma^{vg}}{1/\gamma^{vg}+i/a}\right)^{1/2}\\&\quad\times\exp\left[-\frac{1}{1/\gamma^{vg}+i/a}(q-q(t+\Delta t))^2\right],\end{aligned}\quad (6.141)$$

where

$$a=\frac{m}{2\hbar\Delta t},\quad (6.142)$$

which is regarded as a parameter characterizing the quantum diffusion for a short time interval Δt.

Thus ADF running at a point $(q(t),v(t))$ in the velocity field is explicitly approximated as above in terms of an NVG. Furthermore, ADF-NVG can be cast into more compact form allowing for a clear physical interpretation. [416] Write the NVG in the following form

$$F(q-q(t),t)=\left(\frac{2}{\pi}\right)^{1/4}f(q(t))\exp\left[-\frac{1}{c(t)+id(t)}(q-q(t))^2\right],\quad (6.143)$$

and we have

$$c(t)=\left(\frac{v(t)}{v(0)}\right)^2c(0),\quad (6.144)$$

$$d(t)=\frac{d(0)}{v(0)^2}v(t)^2+\frac{2\hbar}{m}v(t)^2\int_0^t\frac{1}{v(s)^2}ds,\quad (6.145)$$

and

$$f(q(t))=f(q(0))\left(\frac{v(t)}{v(0)}\right)^{1/2}\left(\frac{\gamma(t)}{\gamma(0)}\right)^{1/2}.\quad (6.146)$$

6.7.3 *Illustrative application to a two-state model*

We now unify the theory of electron wavepackets on the non-Born–Oppenheimer branching paths (PSANB) and the nuclear wavepacket dynamics running along these paths (ADF). This is readily done as

$$\Psi(\mathbf{r},\mathbf{R},t) = \sum_{K}^{\text{AES}} \sum_{k}^{\text{path}} C_{Kk}(t)\, \Phi_K(\mathbf{r};\mathbf{R}_{Kk}(t)) \times F(\mathbf{R}-\mathbf{R}_{Kk}(t),t) \exp\left[\frac{i}{\hbar} S(\mathbf{R}-\mathbf{R}_{Kk}(t))\right]. \tag{6.147}$$

In what follows, we use the ADF-NVG described above as $F(\mathbf{R}-\mathbf{R}_{Kk}(t),t)$ in Eq. (6.147).

To avoid unnecessary complexity and confusion, we here study an illustrative two-state nonadiabatic quantum wavepacket dynamics, the adiabatic potential functions of which are drawn in Fig. 6.17.

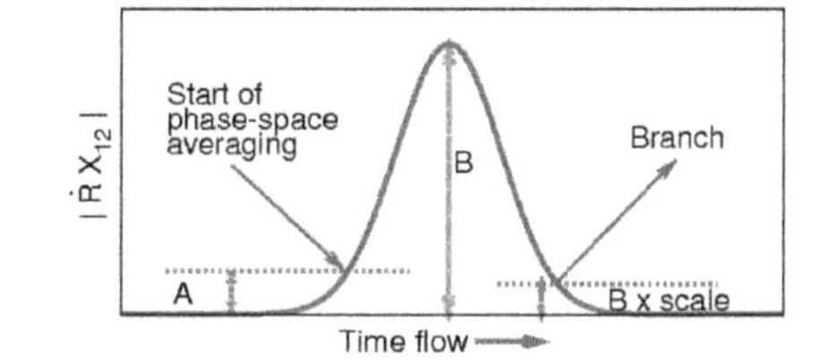

Fig. 6.17 The upper panel: Two adiabatic potential curves that couple with each other through the coupling element (the Gaussian-like curve). The lower panel magnifies the coupling function, which illustrates the starting point of the phase-space averaging and the exit from which for two branching paths to emerge. (Reprinted with permission from T. Yonehara *et al.*, J. Chem. Phys. 130, 214113 (2009)).

This system was artificially modeled based on the ab initio potential curves of LiH molecule. [493] The spatial distribution of the nonadiabatic coupling element is superposed on the two potential curves. In this system, the semiclassical Ehrenfest theory (the mean-field approximation) completely fails, because any Ehrenfest path starting from the basin area can never proceed to the asymptotic region of the dissociation channel. [493]

We explore two physical events in this system: One is a vibrational decay through the nonadiabatic coupling, the initial wavepacket of which is prepared at the left cliff of the lower curve ($R = 2$, with the initial momentum $\hbar k = 25$). The other one is a collision event, with an initial packet coming in from the dissociation channel (from $R = 7$, with $\hbar k = -45$). For these events, we construct the total wavepacket states with PSANB-ADF-NVG according to Eq. (6.147). These total wavefunctions are projected onto the nuclear configuration space and thereby compared with the full quantum nuclear wavepacket attained through the standard method, Eq. (6.5).

The full quantum nuclear wavepacket to be compared as a reference is a Gaussian function initially prepared at a position R_0, with the initial wave number k_0. The width is set to $\delta_0 = 10/k_0$ so as to be in the same order of its corresponding de Broglie wave length. The explicit form is

$$\chi_0^{\mathrm{QM}} = \left(\frac{1}{\pi}\right)^{1/4} \exp\left[-\left(\frac{R-R_0}{2\delta_0}\right)^2\right] \exp\left(\frac{i}{\hbar} S_0\right), \tag{6.148}$$

where

$$S_0 = mv_0(R-R_0) + \frac{1}{2}\left(-\frac{V_0'}{v_0}(R-R_0)^2\right). \tag{6.149}$$

Here V_0' means a potential gradient at $R = R_0$ on the initial electronic state energy surface. S_0 appearing in the phase factor is chosen so as to make a consistent comparison with PSANB-ADF-NVG method.

As for the initial nuclear wavepacket for ADF-NVG, we prepare the following function

$$\chi_0^{\mathrm{NVG}} = \left(\frac{1}{\pi}\right)^{1/4} f_0 \exp\left(-\gamma_0(R-R_0)^2\right) \exp\left(\frac{i}{\hbar} S_0\right), \tag{6.150}$$

with $f_0 = 1$, $c_0 = 4\delta_0^2$, $\gamma_0 = \dfrac{1}{c_0}$, $d_0 = 0$ and $\gamma_0 = \dfrac{1}{c_0 + id_0}$. S_0 is chosen to be the same as that of Eq. (6.149).

The electron dynamics has been integrated with the Chebyshev method. [430] The time propagation of multi-state coupled nuclear wavefunction has been performed with use of the Chebyshev expansion method. [430] Since the ground electronic state $|1\rangle$ involves molecular dissociation, we used an optical potential near the edge of the grid space and the damping operator $\exp\left(-\hat{\Gamma}(R)\frac{dt}{\hbar}\right)$ was multiplied to the vector in each time step. Here $\hat{\Gamma}(R) = 0.1(R-10)^2\theta(R-10)|2\rangle\langle 2|$ with $|2\rangle$ being the electronic excited state, and θ is the Heaviside step function. Time step dt was set to 1, and total propagation time was 1200. The derivative operator and position operator was treated in sinc discrete variable representation [92] with 600 grid points equally distributed from $R = -3$ to 14. We have checked the convergence of the results with respect to grid-density and time step.

6.7.3.1 *Branching paths*

We present first the basic behavior of branching paths generated by the PSANB scheme. Suppose that a classical path denoted as j is running on one of the adiabatic potential surfaces, say, the J-th one. Running on this way, the path acquires only the trivial phase

$$\exp\left[-\frac{i}{\hbar}\int^t ds\langle\Phi_J(R_j(s))|\hat{H}^{el}(s)|\Phi_J(R_j(s))\rangle\right], \tag{6.151}$$

but there is no mixing among the adiabatic electronic states. However, as soon as it penetrates into the coupling region, the judging condition for which is predetermined (see Fig. 6.17), we begin the phase-space averaging according to the procedure of Eqs. (6.35) and (6.36). Let this time be t_{in}, and suppose that we have the electronic wavepacket $C_{Jj}(t_{in})\Phi_J(R_j(t_{in}))$ at this time. Then the electronic state mixing also begins by Eq. (6.14) (with setting $\frac{\hbar^2}{4}\sum_k\left(Y_{IJ}^k + Y_{JI}^{k*}\right)$ to zero) with this packet $C_{Jj}(t_{in})\Phi_J(R_j(t_{in}))$ being the initial condition to integrate for it. The mixed state is represented as

$$\left[C_1(t)\Phi_1(\mathbf{r};\langle R(t)\rangle) + C_2(t)\Phi_2(\mathbf{r};\langle R(t)\rangle)\right]\delta(R - \langle R(t)\rangle) \tag{6.152}$$

for this two-state model. While this averaged path proceeds through the strong coupling region, it is allowed to bifurcate into many pieces using the forces of Eq. (6.43). However, for the sake of simple (actually the simplest) presentation, we do not let it branch but simply continue the phase-space averaging up to the *exit* point of the coupling region, at which the threshold magnitude of the nonadiabatic coupling $\left|\dot{R}X_{IJ}(R)\right|$ is also predetermined (see Fig. 6.17). At this time point, say t_{out}, we let them branch using the eigenforces such that two independent electronic wavepackets starting from

$$C_1(t_{out})\Phi_1(\mathbf{r};R(t_{out})) \tag{6.153}$$

and

$$C_2\left(t_{out}\right)\Phi_1(\mathbf{r};R\left(t_{out}\right)). \tag{6.154}$$

And they are carried by the individual paths running eventually on their own adiabatic potential surfaces, accumulating the phases just as in Eq. (6.151). After that time the electronic mixing does not occur until they return to the coupling region. In the present two-state model, paths return the coupling region many times to undergo bifurcation, except for those paths running out to the dissociation channel, which carry the population of the basin area to the outside gradually. Thus it is crucial to track the history of the individual branching (and therefore increasing) paths. These paths should be coherently summed up after the nuclear wavepackets of ADF-NVG are assigned.

Figure 6.18 shows such branching paths emanating from the above initial conditions: The left-most panel in the upper row (a) is for the vibrational decay, and the lower one (b) for the collision. The corresponding densities from the full quantum calculation, that is $|\chi_K(R,t)|^2$ $(K=1,2)$ are also drawn in this figure; the middle columns for the density on the lower potential curve, while the right-most one for the upper curve. As seen in the figure, the branching paths are already quite similar geometrically to the trajectory of the quantum densities $|\chi_K(R,t)|^2$. For more precise and general behavior of the branching paths of this system, and those arising from the ab initio potential curves of $3\Sigma^+$, $4\Sigma^+$, and $5\Sigma^+$ of LiH molecule as well, we refer to Ref. [493].

6.7.3.2 *Propagating ADF-NVG along the branching paths*

With the branching paths thus obtained the ADF-NVG is associated. Below we discuss some technical matters that is not automatically deduced from the theory presented above, since the paths we are treating are not purely classical.

First of all, we choose the action function for a PSANB path $(R(s),P(s))$ such that

$$\begin{aligned} &S(R,t)\\ &\equiv\int_0^t\left(\frac{P^2(s)}{2m}-\langle\Phi(R(s))|\hat{H}^{el}(s)|\Phi(R(s))\rangle\right)ds+P(t)(R-R(t))\\ &\quad+\frac{1}{2}(R-R(t))^2m\frac{\dot{P}(t)}{P(t)}, \end{aligned} \tag{6.155}$$

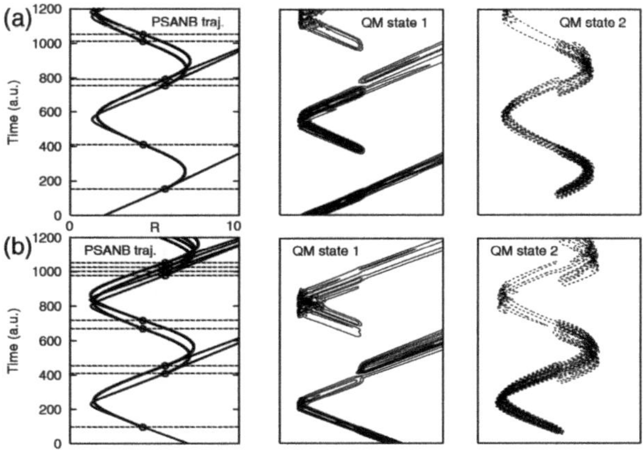

Fig. 6.18 PSANB paths (the left-most column), the space-time plot of $|\chi_1(R,t)|^2$ (the middle column), and that of $|\chi_2(R,t)|^2$ (the right-most column). These paths are tracked along (a) the vibrational decay event with $(x_0, k_0) = (2, 35)$ and (b) the collision event with $(x_0, k_0) = (7, -45)$. The green small circles on the horizontal straight lines denote the branching points for the PSANB paths. (Reprinted with permission from T. Yonehara *et al.*, J. Chem. Phys. 130, 214113 (2009)).

as a counterpart of the classical action expanded up to the second order with respect to position variable R. $\dot{P}(s)$ is numerically evaluated from the trajectory history. Note however that this action function actually does not affect the propagation of the Gaussian part, since the equation of motion for the variable Gaussian may be regarded formally as an independent one. Although the last term in the integrand of Eq. (6.155) takes an infinite value at a turning point, all of these singularities are supposed to be smoothed by the imaginary part of the complex exponent in ADF-NVG function. This leads to an apparently smooth change of the wavefunction at the turning points. We also have confirmed that this singularity of the phase from the pseudo action does not have significant role in the analysis of physical properties under consideration.

In determining the NVG, we encounter the divergence in the momentum-gradient ∇v at the turning points. It is smoothed by the quantum diffusion term (the last term of Eq. (6.109)). For this smoothing to be realized numerically, we have to achieve a good integration of Eq. (6.145). There are actually many ways to do so. Here is one of such methods: Since Eq. (6.145) can be analytically performed in the case of harmonic potential, we approximate the PSANB paths by regarding them as a trajectory on a harmonic potential only in the vicinity of the turning points. The approximated harmonic potential parameters $V_s^{'}$ and $V_s^{''}$ at a switching point are obtained in such a way of $V_s^{'} \sim -m\dot{v}(t_s)$ and $V_s^{''} \sim -\frac{m}{v(t_s)}\ddot{v}(t_s)$ where $\dot{v}$ and $\ddot{v}$ are determined numerically from the history of a PSANB trajectory. The dot denotes the time derivative and $V_s^{'}$ and $V_s^{''}$ mean the first and second order derivative with respect to R, respectively.

Theoretically ADF-NVG can be used as basis functions, and indeed one can place many different Gaussians in their linear combinations. Takahashi and Takatsuka have shown that such linear combination of ADF-NVG can reproduce quite well the time propagation of oscillatory wavefunctions. [416] Nevertheless, again for the simplest presentation, we place a single Gaussian on each branching path. This is the lowest level approximation that PSANB-ADF-NVG offers.

6.7.3.3 *Comparison with the full quantum results*

To make a comprehensive comparison between the two representations of a total wavefunction, namely, Eqs. (6.5) and (6.147), we need a high dimensional space composed of both nuclear and electronic degrees of freedom. However, it is practically impossible, and therefore we alternatively compare only the nuclear part of Eq. (6.3) associated with the electronic eigenfunction $\Phi_K(\mathbf{r}; R)$, that is,

$$\chi_K(R, t) \tag{6.156}$$

and the nuclear part of Eq. (6.147) accompanied by $\Phi_K(\mathbf{r}; R_{Kk}(t))$. On the other hand, it is expected in the asymptotic region that the electronic structure does not vary much within the extending range of the NVG (recall that it is a Gaussian function), that is,

$$\Phi_K(\mathbf{r}; R_{Kk}(t)) F_{NVG}(R - R_{Kk}(t), t) \simeq \Phi_K(\mathbf{r}; R) F_{NVG}(R - R_{Kk}(t), t), \tag{6.157}$$

and Eq. (6.147) may be approximated as

$$\begin{aligned} &\Psi(\mathbf{r}, R, t) \\ &\sim \sum_{K}^{\text{AES}} \Phi_K(\mathbf{r}; R) \\ &\times \left(\sum_{k}^{\text{path on } K} C_{Kk}(t) F_{NVG}(R - R_{Kk}(t), t) \exp\left[\frac{i}{\hbar} S(R - R_{Kk}(t))\right]\right). \end{aligned} \tag{6.158}$$

Then, the following function $\phi_K(R,t)$ defined as

$$\phi_K(R,t) = \sum_{k}^{\text{path on } K} C_{Kk}(t) F_{NVG}(R - R_{Kk}(t), t) \exp\left[\frac{i}{\hbar} S(R - R_{Kk}(t))\right] \tag{6.159}$$

is a quantity to be compared with $\chi_K(R,t)$ of Eq. (6.156). With these nuclear wavepackets, we calculate the following quantities to examine the quality of our approximation.

A stringent test for PSANB-ADF-NVG is the direct comparison of the wavepackets up to the quantum phase. Figure 6.19 shows the comparison of the real parts of $\chi_K(R,t)$ and $\phi_K(R,t)$ in the case of vibrational decay. For the collision dynamics, the similar comparison is displayed in Fig. 6.20. The agreement up to the fine oscillatory structures is quite good. We may rather conclude that the agreement is excellent for this lowest level approximation in construction of the PSANB-ADF-NVG wavefunctions.

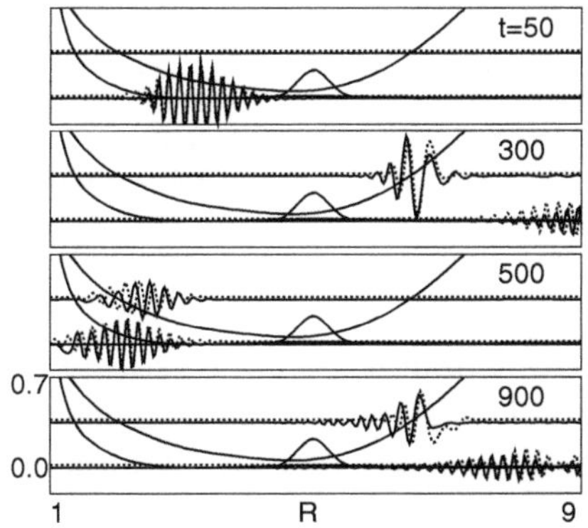

Fig. 6.19 Comparison of the real parts of $\phi_K(\mathbf{R},t)$ [broken curves] and $\chi_K(\mathbf{R},t)$ [solid curves] ($K = 1, 2$) for the vibratinal decay case with the initial condition $(x_0, k_0) = (2, 35)$. The wavepackets on the lower (upper) straight line is running on the lower (upper) potential surface. $\chi_K(\mathbf{R},t)$ are slightly shifted below $\phi_K(\mathbf{R},t)$ for clearer presentation. Time at which each snapshot is taken is indicated in each box. (Reprinted with permission from T. Yonehara *et al.*, J. Chem. Phys. 130, 214113 (2009)).

Figure 6.21 presents the time dependent electronic population for the excited state. It is defined as

$$P_2(t) = \frac{\int |\phi_2(R,t)|^2 dR}{\sum_K \int |\phi_K(R,t)|^2 dR} \tag{6.160}$$

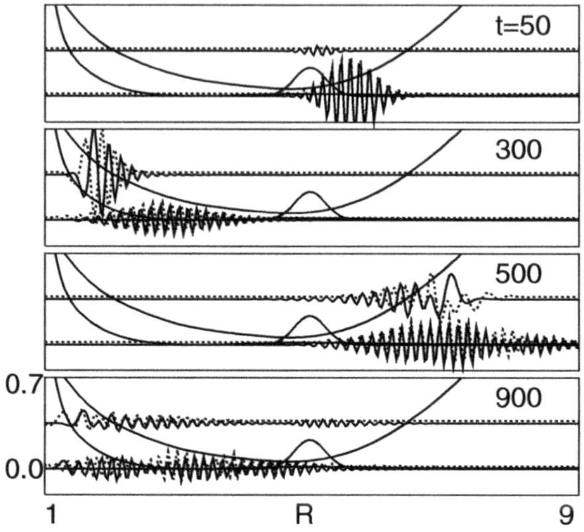

Fig. 6.20 Comparison of the real parts of $\phi_K(\mathbf{R},t)$ [broken curves] and $\chi_K(\mathbf{R},t)$ [solid curves] $(K=1,2)$ for the collision case with the initial condition $(x_0,k_0)=(x_0,k_0)=(7,-45)$. The wavepackets on the lower (upper) straight line is running on the lower (upper) potential surface. $\chi_K(\mathbf{R},t)$ are slightly shifted below $\phi_K(\mathbf{R},t)$ for clearer presentation. Time at which each snapshot is taken is indicated in each box. (Reprinted with permission from T. Yonehara *et al.*, J. Chem. Phys. 130, 214113 (2009)).

for PSANB-ADF-NVG (further abbreviated as PAN), and for the full quantum counterpart $\phi_K(R,t)$ in Eq. (6.160) should be replaced with $\chi_K(R,t)$. One can estimate $P_2(t)$ only in terms of PSANB without ADF-NVG by replacing $\phi_K(R,t)$ as

$$\phi_K(R,t) \to \sum_k^{\text{path on } K} C_{Kk}(t)\,\delta(R-R_{Kk}(t),t)\,. \tag{6.161}$$

These three different calculations show good agreement among them. This indicates that the present semiclassical dynamical methods can well reproduce the complicate electronic transition behavior. On the other hand, the deviation of the result of PSANB-ADF-NVG in the last stage of panel (b) (the collision event) should have been brought about by the severe restriction in this particular work, in which we have placed only a single NVG on each branching path. This would be remedied by using more NVG on each branching path.

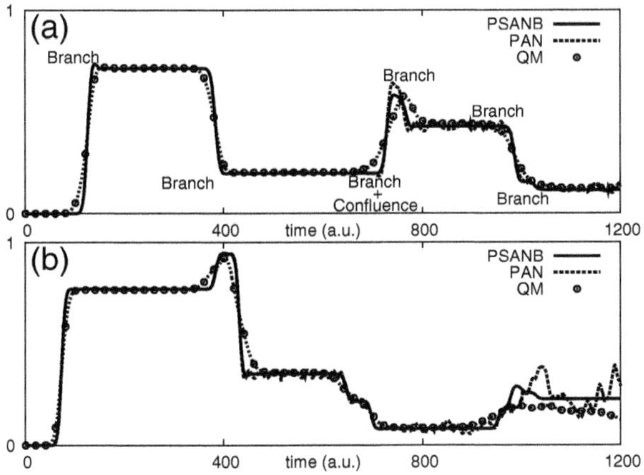

Fig. 6.21 The variation of the population on the upper elctronic state, $P_2(t)$ of Eq. (6.160) estimated with the PSANB without nuclear wavepacket (solid curve), PAN (PSANB-ADF-NVG, dash curve), and the full quantum method (circle). (a) Vibrational decay case with $(x_0, k_0) = (2, 35)$, (b) Collision case with $(x_0, k_0) = (7, -45)$. (Reprinted with permission from T. Yonehara *et al.*, J. Chem. Phys. 130, 214113 (2009)).

Finally, taking the advantage of quantization of the non-Born–Oppenheimer paths in terms of the NVG, we show the following four physical quantities. For both the vibrational decay (panel (a) with $(x_0, k_0) = (2, 35)$) and collision event (panel (b) with $(x_0, k_0) = (7, -45)$), Fig. 6.22 displays the following four quantities. (1) A mean time for the wavepackets to spend in the upper electronic state, that is,

$$\tau(t) \equiv \int_0^t ds \int dR |\chi_2(R, s)|^2. \tag{6.162}$$

(2) The flux of the wavepackets flowing out to the dissociation channel on the ground state potential energy surface. A flux operator is defined as

$$\hat{F}(t) = \frac{1}{i\hbar}[\hat{T}, \theta(R - 8)], \tag{6.163}$$

where θ is again the Heaviside step function and $\hat{T}$ denotes a kinetic energy operator $(-\frac{\hbar^2}{2\mu}\frac{\partial^2}{\partial R^2})$. (3) The accumulated population that is found in the dissociation channel, which is defined as

$$\int_0^t \hat{F}(s) ds. \tag{6.164}$$

(4) The time dependent dispersion of the upper state nuclear packets, which is measured with

$$(\langle R^2\rangle - \langle R\rangle^2)^{1/2}, \tag{6.165}$$

which is evaluated during the time when the excited state population exceeds 0.01. The average $\langle\cdot\rangle$ means the normalized expectation value.

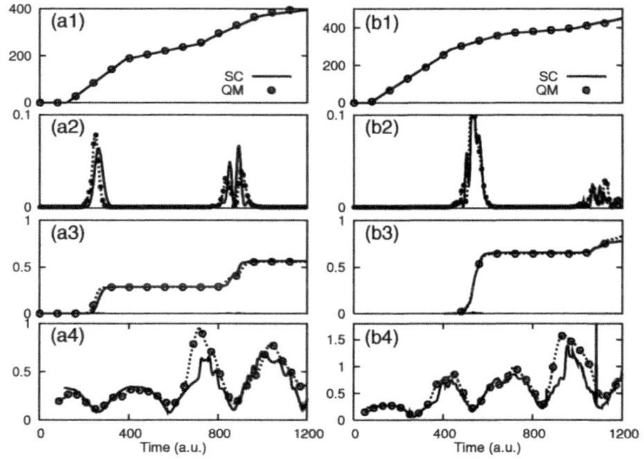

Fig. 6.22 Four miscellaneous quantities that are calculated in terms of the quantum nuclear wavepackets given in the present nonadiabatic dynamics. Panels (a1,b1); A time for a wavepacket to stay in the upper electronic surface. (a2,b2); Flux to the dissociation channel on the ground state (the values are multiplied by 10). (a3,b3); Acumulated population to the dissociation channel, which is given by the time integration of the above flux. (a4,b4); The dispersion of the wavepacket around its center, which is monitored on the upper surface. The left and right panels are for the vibrational decay $((x_0, k_0) = (2, 35))$ and collision $((x_0, k_0) = (7, -45))$, respectively. (Reprinted with permission from T. Yonehara *et al.*, J. Chem. Phys. 130, 214113 (2009)).

Most of these quantities assure that the present quantization of the nuclear wavefunction in the context of the repeated nonadiabatic transitions work quite accurately, reproducing the full quantum values well. At the same time, the figures of the dispersion of the NVG suggest that it is not an easy task to keep the wavepacket in an appropriate shape. In particular the very sharp peak of the NVG in panel (b4) at about $t = 1100$ reflects the divergent behavior of the momentum-gradient at the turning points. Nevertheless, we may conclude that the present scheme to generate non-Born–Oppenheimer electronic and nuclear wavepackets works quite well already in this lowest level of approximation.

6.8 Appendix A: Reduction of muti-dimensional PSANB to one-dimensional (1D) approximation

Evaluation of the multidimensional force matrix is often time consuming. We suggest a useful one dimensional reduction of the PSANB scheme without loss of dynamical information. The essential point is the projection of the force matrix vector onto the semiclassical Ehrenfest force, not onto the nuclear velocity or the momentum vector. The validity of the scheme is verified numerically. This one-dimensional project-out version of PSANB in multidimensional space is referred to as 1D-PSANB.

In Fig. 6.23, the equivalence of branching trajectories among the two methods with respect to a bifurcation from the second to the second or the third state in the LiH and H^+ collision dynamics accompanied with bond rearrangement is shown. Along with Fig. 6.24 showing the adiabatic energies on the branching trajectories, the properties agree well with each other, especially the time and energy profiles around the avoid crossing point at $t = 22.5$ fs.

The 1D-PSANB and PSANB schemes provide quantitatively the same level of results as shown in Figs. 6.23 and 6.24; the former method is superior from a practical view point if one uses an on-the-fly ab initio quantum chemical calculation.

From a practical view point of using on-the-fly quantum chemistry calculation, the timing of entering and ending of state mixing can be safely measured by monitoring the inverse of the state energy difference instead of direct calculation of the state coupling magnitude. The criteria for branch employed is included in the caption of Fig. 6.23.

The second highest natural orbital of a branching electron wavepacket are depicted in Fig. 6.25. The time dependent behavior of the spatial distribution of the electronic density of natural orbital depicted here depends on whether the branching paths are from the second to the second or the third state. The latter shows a clear oscillation among the H_2 moiety in the product region after significant nonadiabatic transition around the time $t = 22.5$ fs while the former is rather stationary in the H_2 moiety. This indicates that PSANB supports a real time and real space description of not only the nuclear branch trajectories but also the electronic branch dynamics.

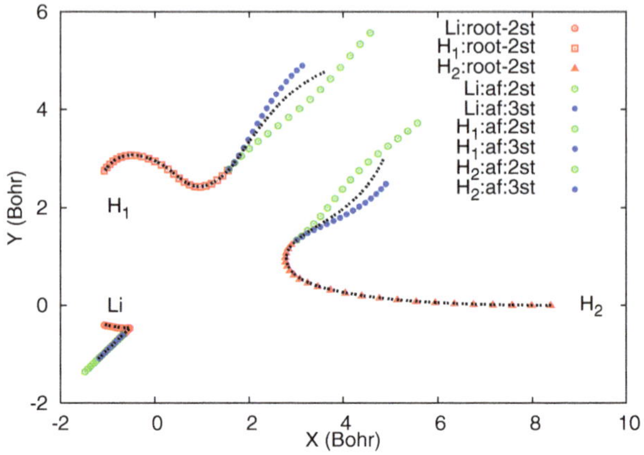

PSANB

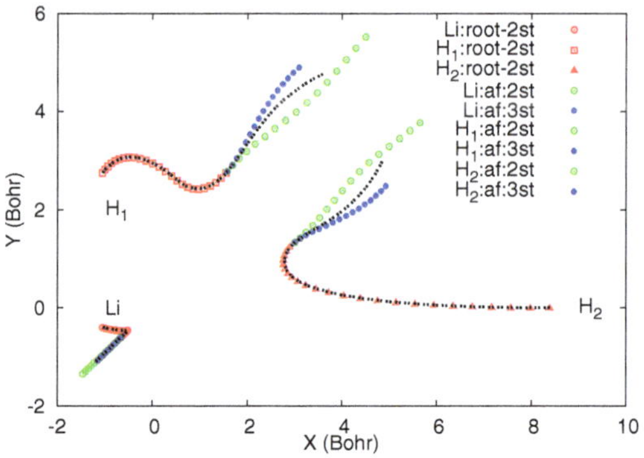

od-PSANB

Fig. 6.23 PSANB against 1D-PSANB in the trajectory branching in LiH+H^+ collision dynamics. CISD/6-31G. Initial condition is the second state and the same as the example for θ=90° used in the Advances in Chemical Physics review article. Dotted lines correspond to semiclassical Ehrenfest theory. Branching paths construction in the multi-dimensional PSANB is totally the same as in previous articles involved with the conical intersection. Employed inverse energy thresholds for coupling entrance and branching were 1/0.010 and 1/0.015 Hartree^{-1}, respectively.

6.9 Appendix B: Quantum chemical calculations of the matrix elements of nonadiabatic interactions

We will now briefly review some of the methods used to calculate the nonadiabatic coupling elements. As can be easily surmised, the computational scheme of nonadiabatic coupling elements depends heavily on the methods of electronic state calculation such as configuration interaction, multi configurational self consistent field (MCSCF) method, and so on.

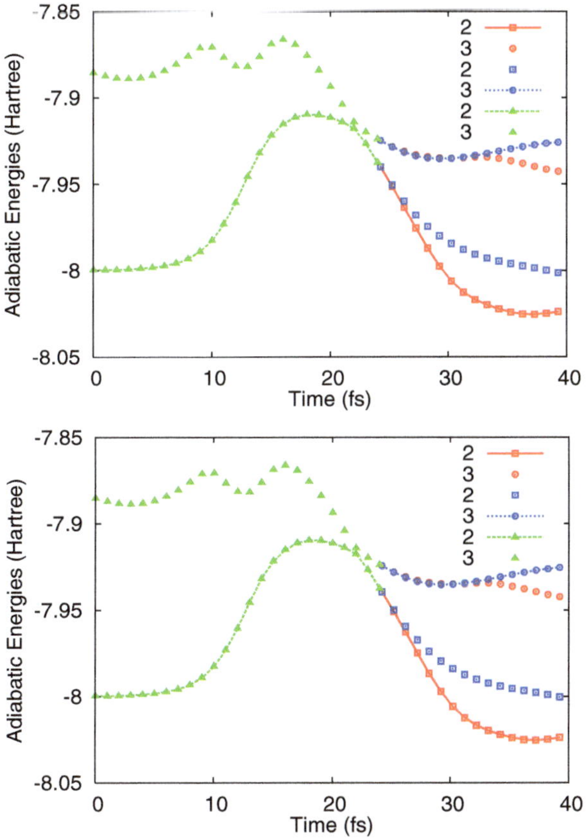

PSANB

od-PSANB

Fig. 6.24 PSANB against 1D-PSANB in $LiH+H^+$ collision dynamics: time evolution of adiabatic potential energy along branching trajectories of Fig. 6.23. The green dashed curve shows the initial trajectory on the second electronic state while the solid red and dotted blue curves after 22.5 fs show the two bifurcated trajectories on the second and third electronic states, respectively. The point sequences without joining lines denote the part on the other adiabatic state along the reference branching trajectories.

Computational efficiency and/or limitations are often determined by the choice of methods, and hence the relevant methodological study is important from a practical point of view and requires much effort to develop it. Since the relevant calculations are somewhat technical, we describe them here in an Appendix. Nevertheless, we do want to emphasize that it is possible to catch a glimpse of the deep aspects of quantum mechanical nonadiabatic dynamics in such technical details.

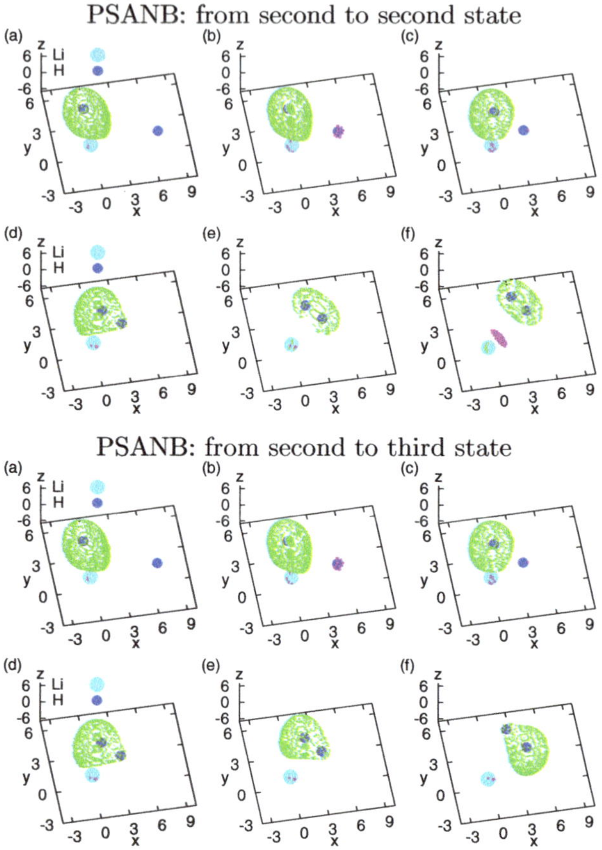

Fig. 6.25 The second highest natural orbital snapshots from LiH+H^+ collision dynamics. PSANB, CISD/6-31G. Initial condition (trajectory starting on the second state) is the same as the example for θ=90° used in the Advances in Chemical Physics review article. (a,b,c,d,e,f) correspond to t = 5, 10, 15, 20, 25 and 30 fs, respectively. Reasonable descriptions are attained between SENB and PSANB approaches.

6.9.1 *On diabatic representation*

The nuclear derivative couplings emerge from a connection matrix for the orthogonal electronic basis states $|\Phi_I\rangle$, which are generated along a trajectory that is characterized in terms of an adiabatic parameter vector $\mathbf{R}$. Suppose that $\mathbf{R}$ depends smoothly on time t and the electron basis set is determined at each point in $\mathbf{R}$-space, namely, $\mathbf{R}(t)$. The wavefunction

$|\Phi(t;\mathbf{R}(t))\rangle$ can be expanded in terms of this basis set in the form of

$$|\Phi(t;\mathbf{R}(t))\rangle = \sum_I C_I(t)|\Phi_I(\mathbf{R}(t))\rangle. \tag{6.166}$$

It should obey the Schrödinger equation with a system Hamiltonian operator, $\hat{H}(\mathbf{R}(t))$, which also depends on $\mathbf{R}(t)$, as shown below

$$i\hbar\partial_t|\Phi(t;\mathbf{R}(t))\rangle = \left(\hat{H} - i\hbar\dot{\mathbf{R}}(t)\nabla_{\mathbf{R}}\right)|\Phi(t;\mathbf{R}(t))\rangle. \tag{6.167}$$

Multiplication of $\langle\Phi_I(\mathbf{R}(t))|$ in this equation of motion from the left side yields

$$i\hbar\dot{C}_I(t) = \sum_J \left(H_{IJ} - i\hbar\dot{\mathbf{R}}(t)\mathbf{X}_{IJ}(\mathbf{R}(t))\right)C_J(t), \tag{6.168}$$

where

$$X_{IJ}(\mathbf{R}(t)) \equiv \langle\Phi_I(\mathbf{R})|\nabla_{\mathbf{R}}|\Phi_J(\mathbf{R})\rangle, \tag{6.169}$$

is the adiabatic parameter derivative coupling. A dot denotes a time derivative. This term originated from

$$\langle\Phi_I(\mathbf{R}(t))|\partial_t|\Phi_J(\mathbf{R}(t))\rangle. \tag{6.170}$$

Furthermore, the time derivative appears as an element in

$$\begin{aligned}\langle\Phi_I(\mathbf{R}(t))&|\Phi_J(\mathbf{R}(t+\Delta t))\rangle \\ &= \langle\Phi_I(\mathbf{R}(t))|\Phi_J(\mathbf{R}(t))\rangle + \langle\Phi_I(\mathbf{R}(t))|\partial_t|\Phi_J(\mathbf{R}(t))\rangle\Delta t.\end{aligned} \tag{6.171}$$

Indeed the overlap integral on the left-hand-side of this equation appears as an origin of the nonadiabatic coupling in the path integral/indexpath integral formalism [164]. Therefore a comparison of these expressions from Eqs. (6.169) to (6.171) gives a clear geometrical meaning in the form of a connection matrix between neighboring electron basis sets in the adiabatic parameter space. Concomitantly, there are two basic ways to evaluate the nonadiabatic coupling elements, either through Eqs. (6.169) or (6.170), and both of them are outlined below. If we set $\mathbf{R}(t)$ to a referential nuclear classical trajectory point at t, $\mathbf{R}(t)$, $\mathbf{X}_{IJ}$ is referred to as the nuclear derivative coupling matrix elements. In what follows, we discuss two representations of the electronic states, namely, the diabatic and adiabatic.

We first consider a formal equation for such a transformation. (See Ref. [27, 28] for an extensive review of the transformation by Baer.) We start with an adiabatic basis set $\{\Phi_I(\mathbf{r};\mathbf{R})\}$, each term of which is associated with the eigen-energy E_I. If we expand the total wavefunction in the form $\Psi(\mathbf{r},\mathbf{R},t) = \sum_I \Phi_I(\mathbf{r};\mathbf{R})\chi_I(\mathbf{R},t)$, then the Schrödinger equation becomes

$$i\hbar\frac{\partial}{\partial t}\chi_I = \left[\sum_j \frac{-\hbar^2}{2M_j}(\nabla^{(j)} + \mathbf{X}^{(j)})^2 + H^{el}(\mathbf{R})\right]_{IJ} \chi_J. \tag{6.172}$$

The diabatic basis $\{F_a(\mathbf{r};\mathbf{R})\}$ is such a basis set that eliminates the derivative coupling (only locally). If we write the transformation matrix U, such that $\sum_J \Phi_J U_{Ja} = F_a$, then the associated nuclear wavefunctions ζ_a are related by $\chi_J = \sum_a U_{Ja}\zeta_a$. For an $\mathbf{X}$ in the adiabatic representation, a transformation matrix that satisfies the equation

$$\nabla U + \mathbf{X}U = 0 \tag{6.173}$$

makes the nuclear derivative coupling vanish, which is to be integrated with an appropriate boundary conditions. If the dimension of the nuclear derivative ∇ is greater than one, then there is an integrability condition [26]

$$\frac{\partial X^\mu}{\partial R^\nu} - \frac{\partial X^\nu}{\partial R^\mu} + [X^\mu, X^\nu] = 0 \text{ for all } \mu < \nu. \tag{6.174}$$

The condition Eq. (6.174) actually poses a certain limitation on the existence of the diabatic representation. One of its outcome is the inexistence of perfect diabatic basis in the finite dimensional expansion. To see this, we denote the projection on the finite basis set explicitly by $\mathcal{P} \equiv \sum_{J\in\Omega} |\Phi_J\rangle\langle\Phi_J|$, where Ω represents the label set of the basis. The requirement for diabatization becomes

$$\mathcal{P}\{\nabla U + \mathbf{X}U\}\mathcal{P} = 0, \tag{6.175}$$

and the integrability condition then becomes

$$\mathcal{P}\{\frac{\partial X^\mu}{\partial R^\nu} - \frac{\partial X^\nu}{\partial R^\mu} + [X^\mu, X^\nu]\}\mathcal{P} = 0 \text{ for all } j < k. \tag{6.176}$$

Here, the last term on the right hand side of this equation inevitably involves such terms that cannot be calculated within the projected space, that is,

$$\mathcal{P}X^\mu(1-\mathcal{P})X^\nu\mathcal{P} - \mathcal{P}X^\nu(1-\mathcal{P})X^\mu\mathcal{P}. \tag{6.177}$$

Thus the perfect diabatic representation does not in fact generally exist unless the basis set is complete [198]. Practical theories of diabatic representation, including those for multi-dimensional systems, have been studied extensively by Smith [376] and Baer [27, 28]. Another scheme was proposed using a different perspective [23, 292]. In this approximate but practical treatment, a diabatization is pursued by requiring the basis states to retain their individual characters smoothly, rather than minimizing the magnitude of derivative coupling.

6.9.2 *Evaluation of the nuclear derivative coupling matrix elements with canonical molecular orbitals*

There exist both similarity and difference between the computational schemes for the energy gradient with respect to the nuclear coordinates [486] and the nuclear derivative coupling elements (NDC) [57, 233]. The common difficulty lies in the treatment of the implicit geometry-dependence of the variational coefficients, such as those of molecular orbitals, CSFs, and so on. These coefficients depend only indirectly on the nuclear positions through the dynamical equations used to determine them. Therefore, in order to make such dependence explicit, the following methods are usually taken: (i) The first order expansion of the one- or multi-electron basis set with respect to the slight displacement of the nuclear positions (giving rise to NDC) is represented within the same basis functions of the original position. (ii) Or, the similar first order expansion is determined by solving dynamical equations. These in turn are obtained by the derivative of orthogonal relations which must be satisfied by the properties of the basis functions. These equations are generally known as coupled-perturbed equations.

First, we consider the case of the configuration state function (CSF) to be used in configuration interactions as the basis functions. These are an antisymmetrized product of one-electron orbitals or a symmetry-adapted linear combination of them. The linear equations, with respect to the infinitesimal nuclear displacement, are derived from the nuclear differentiation of orthogonality relation between the occupied and virtual molecular orbitals of the Fock operator. This suffices to obtain the nuclear derivative coupling matrix element in the CSF representation. The computation of the nonadiabatic coupling elements for general state-averaged multi-configuration self-consistent wavefunctions is more involved. In this scheme, the total energy is defined by the weighted sum of state energies and it is optimized simultaneously with respect to molecular orbitals and the CSF coefficients. The nuclear derivatives of the stationary conditions of this averaged energy with respect to these two sets of coefficients yield equations to determine the perturbation coefficients. Although complicated, these methodologies are indeed necessary for the compact and accurate representation of nonadiabatic chemical dynamics [119, 188].

In the evaluation of the nuclear derivative coupling elements around a crossing seam, the electronic phase should be appropriately taken into account. In addition, it should be noted that NDCs are singular at crossing seams. If we treat the dynamics of passing across a crossing seam, then the state coupled dynamics within an adiabatic representation does not work. Hence, some appropriate diabatic representation is required. Our experience [10, 451] suggests that the CSF representation is a good and practical alternative to a diabatic representation, if the molecular orbitals to be utilized are smoothly connected along a nuclear path (see below). Nonetheless, the nuclear-derivative coupling elements are not exactly zero and the electronic Hamiltonian matrix has large off-diagonal elements. It is recommended, therefore, that the nuclear derivative coupling elements $\mathbf{X}_{IJ}$ are always taken into account explicitly. A CSF representation has already been employed in study of electron dynamics within the semiclassical Ehrenfest theory [10, 286, 287, 307, 308, 425, 451, 492, 494, 496].

As mentioned above briefly, there is a method that does not resort to the direct application of the nuclear derivative of electronic wavefunctions. Instead, an overlap integral between two electronic states at slightly different times (or nuclear positions) as in Eq. (6.171), which is free of some of the difficulties such as the singularity at a crossing seam can be used. The physical meaning of theses transfer-like integrals between the states is clear. In addition, this overlap integral does not strongly depend on how the state function is constructed, either simple CI or MCSCF. In order to perform such an overlap integral for two electronic state functions, it is a rigorous requirement that all the basis functions be tied to each other in a one-to-one manner along a nuclear path. However, in the context of practical calculations, it is not always easy to make such a unique correspondence among them.

6.9.3 *Nuclear derivative coupling elements in CSF representation*

We here concentrate on the CSF representation in somewhat greater detail. The first order nuclear derivative CSF coupling elements are expressed in terms of the following equation

$$X_{IJ}^{k} \equiv \langle \Phi_I \left| \partial_k \right| \Phi_J \rangle = \sum_{i,j} a_{ij}^{IJ} \langle \phi_i \left| \partial_k \right| \phi_j \rangle , \tag{6.178}$$

where $a_{ij}^{IJ} = \langle \Phi_I | \hat{E}_i^{\dagger} \hat{E}_j | \Phi_J \rangle$ denotes a one-electron coupling constant with $\hat{E}_i^{\dagger}$ and $\hat{E}_i$ being the creation and annihilation operator for the ith molecular orbital ϕ_i. Orthonormality is imposed on them as $\langle \phi_i | \phi_j \rangle = \delta_{ij}$. The electronic spin is omitted for the sake of simplicity. ∂_k denotes a nuclear derivative with respect to the kth nuclear degree of freedom. It is sufficient to formulate the first order nuclear derivative coupling matrix element of the molecular orbital functions as

$$d_{ij}^k \equiv \langle \phi_i | \partial_k | \phi_j \rangle . \tag{6.179}$$

Since molecular orbitals $\{\phi_i\}$ are usually prepared as a linear combination of the atomic orbitals (AO) $\{\chi_\mu\}$ with the coefficients c_μ^i as

$$\phi_i = \sum_\mu c_\mu^i \chi_\mu, \tag{6.180}$$

the nuclear derivative couplings are written in terms of the AOs and their associated coefficients as

$$d_{ij}^k = \sum_\mu \sum_\nu c_\mu^i c_\nu^j S_{\mu\nu}^k + \sum_\mu \sum_\nu c_\mu^i \frac{\partial c_\nu^j}{\partial R_k} S_{\mu\nu}, \tag{6.181}$$

where

$$S_{\mu\nu} = \langle \chi_\mu | \chi_\nu \rangle \quad \text{and} \quad S_{\mu\nu}^k = \langle \chi_\mu | \partial_k | \chi_\nu \rangle. \tag{6.182}$$

It is convenient for later purposes to antisymmetrize d_{ij}^k as

$$d_{ij}^k = \frac{1}{2}(\langle \phi_i | \partial_k | \phi_j \rangle - \langle \phi_j | \partial_k | \phi_i \rangle), \tag{6.183}$$

which naturally arises from the derivative of orthonormalized molecular orbitals.

6.9.3.1 *The first order derivative of the molecular orbital coefficients*

Since the case for the AOs is rather straightforward, here we are concerned only with the first order derivative over molecular orbital coefficients with respect to the nuclear coordinates. There are four points to be considered: (i) Implicitly the term $\{c_\mu^i\}$ depends only on $\{R_k\}$. (ii) We should formulate their response to a slight nuclear displacement $\mathbf{R}$ using a perturbation theory. (iii) Consequently the spaces thus spanned by the perturbed molecular orbitals are determined by the unperturbed (original) Fock matrix as the linear combination of the original AOs. (iv) It follows then that the derivative of the molecular orbital coefficients $\{c_\mu^i\}$ should be determined by the quantities related to a "derivative" of the Fock matrix. These are the basic ideas behind the application of the coupled perturbed Hartree–Fock method to this matter. Their generalization is included in the literature [233, 234, 243].

We represent the nuclear derivative of molecular orbital coefficients in the form of the following equation

$$\partial_k c_\mu^i = \sum_m^{MO} c_\mu^m U_{mi}^k, \tag{6.184}$$

where U_{mi}^k are coefficients to be determined. In particular, we need to determine U_{mi}^k between c_μ^i for occupied molecular orbitals and c_μ^m for unoccupied molecular orbitals. The orthonormality condition gives the conditions

$$U_{ij}^k + U_{ji}^k + S_{ij}^k = 0, \tag{6.185}$$

where

$$S_{ij}^k \equiv \sum_{\mu\nu}^{AO} c_\mu^i c_\nu^j \frac{\partial S_{\mu\nu}}{\partial k} = \sum_{\mu\nu}^{AO} c_\mu^i c_\nu^j \left(S_{\mu\nu}^{kR} + S_{\mu\nu}^{kL}\right), \tag{6.186}$$

with $S_{\mu\nu}^{kR} \equiv \langle\chi_\nu|\partial_k\chi_\nu\rangle$ and $S_{\mu\nu}^{kL} \equiv \langle\partial_k\chi_\nu|\chi_\nu\rangle$. By using them, the derivative couplings are reduced to the form

$$d_{ij}^k = \sum_\mu\sum_\nu c_\mu^i c_\nu^j S_{\mu\nu}^k + U_{ij}^k. \tag{6.187}$$

6.9.3.2 *Using the CPHF equation to determine* U_{ij}^k

Next the terms U_{ij}^k between the molecular orbitals in both occupied and unoccupied space are obtained using the coupled perturbed Hartree–Fock method (CPHF). First, we recall that the Fock matrix for the closed-shell system is written as

$$F_{ij} = h_{ij} + \sum_k^{d.o.} \left\{2(ij|kk) - (ik|jk)\right\}, \tag{6.188}$$

where

$$h_{ij} = \langle\phi_i|\left(-\frac{1}{2}\Delta_r + \sum_A \frac{Z_A}{r_A}\right)|\phi_j\rangle \tag{6.189}$$

with Z_A being the nuclear charge of atom A and

$$(ij|kl) = \int dr_1 dr_2 \phi_i^*(r_1)\phi_j(r_1) r_{12}^{-1} \phi_k^*(r_2)\phi_l(r_2). \tag{6.190}$$

For the sake of simplicity, we introduce the following notation

$$h_{ij}^{a} \equiv \sum_{\mu\nu} c_{\mu}^{i} c_{\nu}^{j} (\partial_a h_{\mu\nu}), \tag{6.191}$$

$$(ij|kl)^{a} \equiv \int dr_1 dr_2 \phi_i^*(r_1)\phi_j(r_1)(\partial_a r_{12}^{-1})\phi_k^*(r_2)\phi_l(r_2), \tag{6.192}$$

$$F_{ij}^{a} \equiv h_{ij}^{a} + \sum_{k}^{d.o.} \{2(ij|kk)^{a} - (ik|jk)^{a}\}, \tag{6.193}$$

and

$$A_{ij,kl} \equiv 4(ij|kl) - (ik|jl) - (il|jk) = A_{kl,ij} = A_{ij,lk}, \tag{6.194}$$

where d.o. in Eq. (6.193) means a summation over the doubly-occupied molecular orbital space. Finally, with the help of the stationary condition F_{ij} with respect to the perturbation a, namely, $\partial F_{ij}/\partial a = 0$ $(\neq F_{ij}^{a})$, we arrive at the following compact form of the equation for the virtual (virt.) and doubly occupied (d.o.) orbital pairs

$$\sum_{k}^{virt.} \sum_{l}^{d.o.} \tilde{A}_{ij,kl} U_{kl}^{a} = B_{0,ij}^{a}, \tag{6.195}$$

where ($i \in$ virt., $j \in$ d.o.) and

$$\tilde{A}_{ij,kl} \equiv (\epsilon_j - \epsilon_i)\delta_{ik}\delta_{jl} - A_{ij,kl}, \tag{6.196}$$

$$B_{0,ij}^{a} \equiv F_{ij}^{a} - S_{ij}^{a}\epsilon_j - \frac{1}{2}\sum_{k}^{d.o}\sum_{l}^{d.o.} S_{kl}^{a} A_{ij,kl}. \tag{6.197}$$

Equation (6.195) gives $\{U_{kl}^{a}\}$ by inversion, unless $\epsilon_j - \epsilon_i = 0$.

6.9.4 *Practical calculation of* X_{IJ}^{k}

Using the properties introduced above, the nuclear derivative coupling elements for molecular orbital space now become

$$d_{ij}^{k} = S_{ij}^{kR} + U_{ij}^{k}, \tag{6.198}$$

with

$$S_{ij}^{kR} \equiv \sum_{\mu\nu}^{AO} c_{\mu}^{i} c_{\nu}^{j} S_{\mu\nu}^{kR}, \tag{6.199}$$

which directly yields the CSF derivative coupling elements,

$$X_{IJ}^{k} = \sum_{ij}^{MO} a_{ij}^{IJ} \left(U_{ij}^{k} + S_{ij}^{kR}\right). \tag{6.200}$$

However, one problem now arises. The U_{ij}^k terms for the same occupied or unoccupied molecular orbital space are not determined uniquely from the CPHF equation based on the orthogonality of occupied and virtual space. However, this problem can be avoided by using the following approximation. Namely that the symmetric part of $\{U_{ij}^k\}$ is a physically meaningful part of a CSF derivative coupling matrix element, $\langle\phi_i|\partial_k|\phi\rangle$, for the $i-j$ pairs that belong to the same molecular orbital space. This approximation allows the replacement of $-\frac{1}{2}S_{ij}^k$ with U_{ij}^k and for any inner space rotation pair, $(i,j) \in$ occ. or virt., and we thus obtain a derivative coupling matrix element for molecular orbitals in such an approximate manner as

$$\begin{aligned}\langle\phi_i|\partial_k\phi_j\rangle &= U_{ij}^k + S_{ij}^{kR} \\ &\approx \frac{1}{2}(U_{ij}^k + U_{ji}^k) + S_{ij}^{kR} \\ &= -\frac{1}{2}S_{ij}^k + S_{ij}^{kR} = -\frac{1}{2}(S_{ij}^{kR} + S_{ij}^{kL}) + S_{ij}^{kR} = \frac{1}{2}(S_{ij}^{kR} - S_{ij}^{kL}),\end{aligned} \tag{6.201}$$

in the antisymmetric form. Finally, the nuclear derivative coupling matrix elements for CSF can be written down as

$$\begin{aligned}\langle\Phi_I|\partial_k|\Phi_J\rangle &= \left(\sum_{ij}^{occ} + \sum_{ij}^{vir}\right) a_{ij}^{IJ} d_{ij}^k + \sum_i^{occ}\sum_j^{vir} a_{ij}^{IJ} d_{ij}^k + \sum_i^{vir}\sum_j^{occ} a_{ij}^{IJ} d_{ij}^k \\ &= \left(\sum_{ij}^{occ} + \sum_{ij}^{vir}\right) a_{ij}^{IJ} \frac{1}{2}\left(S_{ij}^{kR} - S_{ij}^{kL}\right) \\ &\quad + \sum_i^{occ}\sum_j^{vir} \left(a_{ij}^{IJ} - a_{ij}^{JI}\right)\left(U_{ij}^k + S_{ij}^{kR}\right).\end{aligned} \tag{6.202}$$

6.9.5 *Nonadiabatic coupling without use of a nuclear derivative*

With increasing system size, the implementation of ab initio electron wavepacket dynamics, such as the semiclassical Ehrenfest theory, using nuclear derivative coupling tends to be computationally more demanding because of the necessity of solving coupled perturbed equations. We therefore propose a useful treatment of nonadiabatic coupling, in which one can avoid the tedious coupled perturbed equations for the nuclear derivative of molecular orbitals and CSFs.

Let us resume with the Schrödinger equation for an electronic wavefunction with nonadiabatic state coupling written as follows

$$i\hbar\frac{d}{dt}|\Phi\rangle = \hat{H}|\Phi\rangle \quad \Leftrightarrow \quad i\hbar\dot{C}_I = \sum_J (H_{IJ} - i\hbar D_{IJ})C_J. \tag{6.203}$$

The electronic wavepacket is expanded in terms of CSFs as $|\Phi\rangle = \sum_I C_I|\Phi_I\rangle$. A dot denotes the time derivative. The time-derivative overlap matrix, which causes state mixing, has the form of

$$D_{IJ} \equiv \langle\Phi_I|\dot{\Phi}_J\rangle = \sum_{ij}^{MO} a_{ij}^{IJ}\langle\phi_i|\dot{\phi}_j\rangle \equiv \sum_{ij}^{MO} a_{ij}^{IJ} d_{ij}, \tag{6.204}$$

where molecular orbital nuclear derivative coupling matrix element is defined by

$$d_{ij} \equiv \langle\phi_i|\frac{d}{dt}|\phi_j\rangle = \frac{A_{ij} - A_{ji}}{2}, \tag{6.205}$$

with

$$A_{ij}(t) \equiv \langle\phi_i(t)|\dot{\phi}_j(t)\rangle = \sum_{\mu\nu}^{AO}\left[c_\mu^i(t)\dot{c}_\nu^j(t)S_{\mu\nu}(t) + c_\mu^i(t)c_\nu^j(t)\dot{R}S_{\mu\nu}^R(t)\right]. \tag{6.206}$$

Here the orthogonality of real molecular orbital functions is exploited. $S_{\mu\nu}(t) \equiv \langle\chi_\mu(t)|\chi_\nu(t)\rangle$ and $S_{\mu\nu}^R(t) \equiv \langle\chi_\mu(t)|\nabla_R|\chi_\nu(t)\rangle$ are atomic orbital (AO) overlap matrix elements and nuclear derivative overlap matrix elements, respectively. Thus, we can avoid the coupled perturbed Hartree–Fock equations or their analog to obtain the nuclear derivative over molecular orbital coefficients. Also, the computation of the time-derivative is much less time consuming. However, the price for gaining these advantages is an efficient and accurate computation of $\{\dot{c}_\mu^i\}$. They are numerically available with the aid of orbital continuity along the time coordinate in the propagation of the electronic wavepackets. In this scheme, the derivative coupling matrix for CSF state vectors does not appear. In a further simplified implementation of it, one can neglect the force component originating from the nuclear derivative coupling; that is assuming that they are generally small.

6.10 Appendix C: Tracking the continuity of molecular orbitals along a nuclear path

In propagating an electronic wavepacket along a path, it is generally expanded with molecular orbitals as one-electron basis functions. molecular orbitals are, however, generated at each nuclear configuration (geometry) as eigenfunctions of the static one-electron Hamiltonian, for example, with the Fock operator. Therefore, no information about time is included at this level. Hence, any phase factor can be attached keeping the constituent total density matrices invariant. This is because individual molecular orbitals are determined independently irrespective of neighboring counterparts. Nevertheless, one needs to identify how the molecular orbitals are actually connected with each other in terms of their orbital character. Without correct correspondence of the successive molecular orbitals, that is, $\phi_i(\mathbf{R}(t)) \leftrightarrow \phi_i(\mathbf{R}(t+\Delta t))$, one may lose correct identification of the CSF $\Phi_I(\mathbf{R}(t))$ or the total electronic state, which could result in wrong time propagation of Eq. 6.166. A most serious situation arises when two or more molecular orbitals are energetically degenerate. This is exactly the situation where nonadiabatic transition is expected to be most significant. Therefore, the tracking of the unique continuity of molecular orbitals is extremely vital for the successful time propagation of the electronic wavepacket.

One typical example, that may appeal to readers and that emphasizes the seriousness of the matter, is as follows: Let us consider two degenerate π molecular orbitals in a diatomic molecule, whose bond direction is in the z-coordinate. At each internuclear distance, a computer program can generate a pair of π orbitals with an arbitrary orientation in the xy-plane. Arbitrary, of course, unless one forces them to align along, say, the x and y directions. Without this special care, the resultant CSFs including them will become useless as expansion bases in the practical propagation of an electronic wavepacket.

A method for obtaining a set of molecular orbitals appropriate for dynamical simulation was proposed by Truhlar's group. [292, 293] It uses a variational functional to obtain a set of diabatic molecular orbitals as a function of the nuclear geometry. An advantage of this method is that such diabatization is possible without prior knowledge of the nuclear paths.

6.10.1 *Concept of unique-continuity of molecular orbitals*

Let us now concentrate on this problem with respect to the more technical aspects. It should be noted that this kind of issue arises in any quantum dynamics problem, in which a time-dependent non-orthogonal basis set is used to track the wavepacket dynamics. Suppose that we have two sets of molecular orbitals, $\{\phi_i(t-\delta t)\}$ and $\{\phi_i(t)\}$ at slightly different times. Here, the suffix i runs from 1 to M, which is equal to the number of atomic orbitals employed. t and $t + \delta t$ are a particular time and its next neighbor, respectively. The usual practice is to sort the molecular orbitals in order of their energy. However, here we need to track their identity smoothly in terms of their physical character (usually referred to as the orbital character), instead of their energy. The smooth connection can be achieved by ordering the new molecular orbitals in such a way as to maximize the overlap between $\{\phi_i(t-\delta t)\}$ and $\{\phi_i(t)\}$. Let the resulting set of new molecular orbitals be $\{\widetilde{\phi_j}(t)\}$, which are represented by appropriate linear combinations of $\{\phi_i(t)\}$. If the time step δt is sufficiently small, we can thus obtain a natural connection between the two orbital sets. By repeating this method precisely, a sign-matching between the corresponding molecular orbitals as well as labeling of molecular ordering can be achieved.

The new molecular orbitals may be obtained as follows. As trial functions, $\{\widetilde{\phi_j}(t)\}$ are given by the linear combination of $\{\phi_i(t)\}$ so that the overlap error estimation function may be minimized. The cost function for the overlap matrix is defined as the square of the difference from the unit matrix

$$f(\underline{\underline{\mathbf{c}}}) = ||\underline{\underline{\mathbf{I}}} - \underline{\underline{\widetilde{\mathbf{s}}}}||^2 \equiv \sum_{i,j=1}^{M} (\delta_{ij} - \widetilde{s}_{ij})^2, \tag{6.207}$$

where δ_{ij} is the Kronecker delta and the overlap concerned is defined by

$$\widetilde{s}_{ij} \equiv \langle \phi_i(t-\delta t)|\widetilde{\phi_j}(t)\rangle, \tag{6.208}$$

with

$$|\widetilde{\phi_j}(t)\rangle = \sum_{k=1}^{N} |\phi_k(t)\rangle c_{kj}. \tag{6.209}$$

$[\underline{\underline{\mathbf{c}}}]$ is the molecular orbital mixing matrix to be optimized and $[\underline{\underline{\mathbf{c}}}]_{ij} = c_{ij}$ is the contribution of the ith primitive molecular orbital to the jth trial orbital at time t. Here a bold symbol with double underbars denotes an $M \times M$ square matrix. One can elaborate further on the overlap matrix, but we here employ the present form to focus on the concept of orbital uniformity (unique-continuity) along a nuclear path.

Next, we introduce the primitive molecular orbital overlap matrix as

$$o_{ik} \equiv \langle \phi_i(t-\delta t)|\phi_k(t)\rangle. \tag{6.210}$$

Substituting this into the cost function above yields

$$f(\underline{\underline{\mathbf{c}}}) \equiv \sum_{i,j=1}^{N} (\delta_{ij} - \sum_{k=1}^{N} o_{ik} c_{kj})^2. \tag{6.211}$$

The requirement of the overlap matrix being closest to the unit matrix is written as

$$\frac{\partial f}{\partial c_{ij}} = 0, \tag{6.212}$$

for all i and j. Its formal solution is given by

$$\underline{\underline{\mathbf{c}}} = \underline{\underline{\mathbf{o}}}^{-1}. \tag{6.213}$$

Conceptually, this provides unique continuity of the neighboring molecular orbitals.

6.10.2 *Practical implementation of unique-continuity of molecular orbitals*

Suppose we use linear combinations of atomic orbitals (LCAO) in constructing molecular orbitals. We also make the rather optimistic assumption that the time interval δt is so small that atomic orbital overlaps between $t - \delta t$ and t can be approximated safely as $\langle \chi_\mu(t - \delta t)|\chi_\nu(t)\rangle \approx \langle \chi_\mu(t)|\chi_\nu(t)\rangle$. Here, $\{\chi_\mu\}$ is a set of atomic orbitals. This can be employed in evaluating an approximation of the overlap $\underline{\underline{\mathbf{o}}}$ among the molecular orbitals at different times. Since the constructed overlap matrix is not exactly unitary, it is preferable to introduce a natural modification of the molecular orbital overlap $\underline{\underline{\mathbf{o}}}$ to achieve orthonormality of resultant transformed orbitals. In fact, this orthonormality plays a key role, not only in the relevant mathematical properties, but also as working equations for constructing the gradient properties of electrons. One way to achieve orthogonality is to apply the Gramm-Schmidt technique for the columns of vectors of $\underline{\underline{\mathbf{o}}}$. This is equivalent to the unitarization of $\underline{\underline{\mathbf{o}}}$. We refer to the unitary matrix obtained this way as $\widetilde{\underline{\underline{\mathbf{o}}}}$. Because of the unitarity of $\widetilde{\underline{\underline{\mathbf{o}}}}$, the mixing matrix $\underline{\underline{\mathbf{c}}}$ is effectively set to

$$\underline{\underline{\mathbf{c}}} \approx \underline{\underline{\mathbf{o}}}^{T}, \tag{6.214}$$

where T on the shoulder of a matrix means taking its transpose.

The proposed scheme is then reduced into a form taking linear combinations within a restricted molecular orbital sub-space in which the molecular orbitals strongly correlate with each other during an adiabatic change. This is aimed at removing negligible mixing to attain numerical stability. In doing so, we merely need to construct the matrix $\widetilde{\underline{\underline{\mathbf{o}}}}$ for strongly correlating molecular orbitals. Practically, we can select them by checking the diagonal part of $\widetilde{s}_{ij}$. Our practical procedure is summarized below:

(1) For extracting molecular orbital sub-space with a view to strong mixing among them, we pick up the set of orbital indices $\Omega = \{k\}$, which satisfy

$$1 - |\langle \phi_k(t - \delta t)|\phi_k(t)\rangle| > p, \tag{6.215}$$

where p is a preset threshold value. Note that the mixing space can be readily divided into sub-blocks among which they do not correlate by examining the off-diagonal elements, $|\langle \phi_k(t - \delta t)|\phi_k(t)\rangle|$.

(2) Next calculate the matrix

$$o_{ij} = \langle \phi_i(t - \delta t)|\phi_j(t)\rangle, \tag{6.216}$$

with i and j being elements of Ω.

(3) Then prepare $\underline{\underline{\widetilde{o}}}$ by making $\underline{\underline{o}}$ unitary as explained above.

(4) Next obtain a reconstructed canonical molecular orbital set at t, which is uniformly connected to the set at $t - \delta t$,

$$|\widetilde{\phi_j(t)}\rangle = \sum_{k=1}^{N} |\phi_k(t)\rangle [\widetilde{o}]^T_{kj} = \sum_{k=1}^{N} |\phi_k(t)\rangle \widetilde{o}_{jk}. \tag{6.217}$$

The orbital energy, $\widetilde{\epsilon}_i$, corresponding to the uniformed orbital, $\widetilde{\phi}_i(t)$, is set by

$$\widetilde{\epsilon}_i = \left\langle \widetilde{\phi}_i(t)|\hat{F}|\widetilde{\phi}_i(t)\right\rangle = \sum_k \epsilon_k |\widetilde{o}_{ik}|^2, \tag{6.218}$$

with ϵ_k being the canonical orbital energy of ϕ_k. Here we have used the unitarity of $\underline{\underline{\widetilde{o}}}$.

Chapter 7

Dynamical Electron Theory for Chemical Reactions

In the preceding section we have studied the rather formal theory for adiabatic electron wavepacket dynamics and its technical aspects. Electronic theory of organic chemistry had been initiated even before wave (quantum) mechanics and was brought to be the final form by Robinson and Ingold. It still constitutes the very basic heart of the theoretical understanding of chemical reactivity. The valence bond theory, resonance theory, the notion of electron negativity by Pauling and the so-called π-electron theories by Coulson and many other authors are the first monumental works of quantum mechanical reaction theory. Then, after the progress of molecular orbital theory based on the Hartree–Fock method by Mulliken et al., the frontier orbital theory by Fukui and the Woodward–Hoffmann rule have established the idea on how critical the role of quantum phases is to comprehend and even predict chemical reactions. Yamaguchi and Fueno have developed a theory of non-concerted chemical reactions (the Woodward–Hoffmann forbidden reaction), in which biradicals (even in totally spin singlet case) are generated in the course of reactions mainly by nonadiabatic interactions. Recent progress in chemical reaction theory relies more on accurate energetics such as the relative relations of the transition states with respect to the initial molecular arrangement, and so on. All these theories are based on the Born–Oppenheimer approximation, and therefore no chemical or physical situation beyond it is considered, such as time-fluctuation of electronic states due to the breakdown of the BO approximation, which becomes prominent for dynamics in the manifold of highly quasi-degenerate electronic states.

Here in this chapter we show examples of applications of electron dynamics to some typical chemical reactions for deeper understanding of the basic mechanisms. We hope the readers to find that even a classic problem like proton transfer dynamics reveals a new aspect that was not known within the existing stationary-state quantum chemistry.

7.1 Electron flux in chemical reactions

7.1.1 *Definitions*

A particle density $\rho(\mathbf{q}, t)$ arising from a quantum wavefunction

$$\rho(\mathbf{q}, t) = \psi^*(\mathbf{q}, t)\psi(\mathbf{q}, t) \tag{7.1}$$

forms a probability-conserving (incompressive) flow in general configuration space $\mathbf{q}$. This flow defines an flux vector as usual as

$$-\frac{\partial}{\partial t}\int_V \rho(\mathbf{q}, t)\, d\mathbf{q} = \int_{\partial V} \vec{j}(\mathbf{q}, t) \cdot d\vec{S}, \tag{7.2}$$

where V is a volume in space and ∂V is its surface with $d\vec{S}$ being an infinitesimal surface element on ∂V. With the help of the Schrödinger equation, the probability density current (or simply flux) $\vec{j}(\mathbf{q}, t)$ can be given as

$$\vec{j}(\mathbf{q}, t) = \frac{\hbar}{2im}[\psi^*(\mathbf{q}, t)\nabla\psi(\mathbf{q}, t) - \psi(\mathbf{q}, t)\nabla\psi^*(\mathbf{q}, t)]. \tag{7.3}$$

Extension to many-particle systems is rather straight forward, and will be described later. It is physically obvious that $\vec{j}(\mathbf{q}, t)$ should be zero (no current) for the stationary state wavefunction, which is a superposition

$$\begin{aligned}\psi_{stationary}(\mathbf{q}) &= \psi(\mathbf{q}, t) + \psi(\mathbf{q}, -t)\\ &= \psi(\mathbf{q}, t) + \psi^*(\mathbf{q}, t),\end{aligned} \tag{7.4}$$

which is a real-valued function. Therefore no direct information is given by stationary-state wavefunctions such as the adiabatic electronic wavefunctions based on the Born–Oppenheimer approximation. Quantum chemistry has been thus quite a poor means to provide electron dynamics. In contrast, a time-dependent wavefunction, which is complex-valued as in our nonadiabatic electron wavepacket approaches, can give a very vivid view on how electron distributions are deformed and flow in and in between molecules in real-time scale of chemical events.

In many cases there are dynamics involving the coupling of states, examples including nuclear kinematic (nonadiabatic), optical, and spin-orbit couplings. Due to the superposition of electronic states, the dynamics of electron affect deeply the chemical reaction dynamics on the time scale of electronic motion. We call these electrons dynamical electrons. The longer the lifetime of electronic state mixture persists, the larger the chemical dynamics is affected by the quantum dynamics of electrons. Thus we require a chemical theory which can appropriately treat electron dynamics coupled to the reaction dynamics. The time evolution of the electronic phase is often significant.

Electron flux is expressed through a mixture of electronic state wavefunctions: An electron flux principally arises from nonadiabatic electronic state-mixing and smooth change along molecular deformation. To see it more precisely, Nafie noted [283] that a complete adiabatic (CA) component parametrically depends on the nuclear phase space point, namely, not only position but also momentum. In his theory, the CA electronic wavefunction on the electronic coordinates $\mathbf{r}$ should have the following form

$$\Psi_e^{CA}(\mathbf{r};R,P) = \Psi_e^0 + \sum_{s\neq e} \Psi_s^0 c_{se}^0 \left(R + i\hbar P/E_{se}^0\right), \tag{7.5}$$

with R and P the nuclear position and momentum variables. Here Ψ_e^0 denotes the electronic wavefunction of the eth BO adiabatic state at the reference geometry $R = 0$. The nonadiabatic component is given by

$$c_{se}^0 = \langle \Psi_s^0 | \frac{\partial \Psi_e^0}{\partial R} \rangle. \tag{7.6}$$

Let the energy gap between the eth and sth BO electronic states be defined as $E_{se}^0 = E_s^0 - E_e^0$. Using this extended CA wavefunction including the molecular motion, the electronic charge density $\rho_e^{CA}(\mathbf{r};R,P)$ and the electronic charge current density $\mathbf{j}^{CA}(\mathbf{r};R,P)$ can be constructed in a natural way as,

$$\begin{aligned}\rho_e^{CA}(\mathbf{r};R,P) &\equiv -e\left[\Psi_e^{CA}(\mathbf{r};R,P)\right]^* \left[\Psi_e^{CA}(\mathbf{r};R,P)\right] \\ &= -e\left[(\Psi_e^0)^2 + 2\Psi_e^0 \sum_{s\neq e} \Psi_s^0 c_{se}^0 R + \sum_{s\neq e}\sum_{s'\neq e} \Psi_s^0 \Psi_{s'}^0 c_{se}^0 c_{s'e}^0 (R^2 + \frac{\hbar^2 P^2}{E_{se}^0 E_{s'e}^0})\right],\end{aligned} \tag{7.7}$$

and

$$\mathbf{j}_e^{CA}(\mathbf{r}; R, P) \equiv \mathrm{Re}\left[\frac{ie\hbar}{2m}\left[\Psi_e^{CA}(\mathbf{r}; R, P)\right]^* \nabla_r \Psi_e^{CA}(\mathbf{r}; R, P)\right]$$
$$= \frac{e\hbar^2}{m}\left[\sum_{s\neq e}(\Psi_e^0 \nabla_r \Psi_s^0 - \Psi_s^0 \nabla_r \psi_e^0)\frac{c_{se}^0}{E_{se}^0}P\right.$$
$$\left. - \sum_{s\neq e}\sum_{s'\neq e}(\Psi_s^0 \nabla_{r\Psi s'}^0) c_{se}^0 c_{s'e}^0 RP\left(\frac{1}{E_{se}^0} - \frac{1}{E_{s'e}^0}\right)\right]. \qquad (7.8)$$

Especially seen in the latter, the momentum adiabaticity is essential for the description of adiabatic electron flux. This was originally introduced to resolve the puzzle of the vibronic coupling in infrared intensity missing in the usual BO picture [283, 284]. Extension to the general case including significant nonadiabatic transition current density was done by the same author [285]. Using this extended adiabatic wavefunction, the CA concept allows a natural form of conservation law for the charge density distribution with use of the electron current density. We can consider that this type of complete adiabatic electronic flow is also involved in the general case accompanied by significant nonadiabatic dynamics.

Diestler and his coworkers also revisit this problem sophisticating the formalism of adiabatic electron flux [116]. They have provided a rather practical adiabatic formalism for describing intramolecular electron flux in a diatomic molecule [115],

$$\mathbf{j}_e(\mathbf{r}, t) \equiv \frac{1}{2}\int d\mathbf{R}[\Delta_b(\mathbf{r}; \mathbf{R}) - \Delta_a(\mathbf{r}; \mathbf{R})]\mathbf{j}_{b,a}(\mathbf{R}, t), \qquad (7.9)$$

where the electron flux is formulated as an electron density difference integrated over the nuclear intramolecular coordinate weighted by a nuclear flux density

$$\mathbf{j}_{b,a}(\mathbf{R}, t) \equiv \frac{\hbar}{2i}\int d\mathbf{R}_L \delta(\mathbf{R} - (\mathbf{R}_b - \mathbf{R}_a))$$
$$\times \left\{\langle \Psi(\mathbf{R}_L, t)|(1/M_b)\nabla_b - (1/M_a)\nabla_a|\Psi(\mathbf{R}_L, t)\rangle - c.c.\right\}. \qquad (7.10)$$

Here, the composite coordinates of the total molecule in the laboratory frame is described by $\mathbf{R}_L \equiv (\mathbf{R}_a, \mathbf{R}_b, \mathbf{r})$. a and b denote atoms in the molecule. In this formulation, the electron density partitioning for atoms

$$\Delta_\alpha(\mathbf{r}; \mathbf{R}) \equiv \Psi(\mathbf{r}; \mathbf{R})\Phi_\alpha(\mathbf{r}; \mathbf{R}) \qquad (7.11)$$

is in terms of the decomposition of the real adiabatic electronic wavefunction

$$\Psi(\mathbf{r};\mathbf{R}) = \sum_{\alpha=a,b} \Phi_\alpha(\mathbf{r};\mathbf{R}). \tag{7.12}$$

Another impressing study on electron current during a reaction process in a realistic molecular system was performed by Hege et al. [12, 172] They investigated the adiabatic electron flux during a pericyclic chemical bond rearrangement reaction of cyclooctatetrane in the tunneling energy limit of nuclear wavepacket motion. The relevant process accompanied by single and double bond alternations is chemically important as a fundamental examination of the nature of quantum flux. The wavepacket for describing electron flux was prepared virtually as a direct product of electron and nuclear subspace with parametric dependence on molecular geometry. Its real electronic component was statically obtained at each molecular geometry and constructed within a pericyclic valence orbital space, namely, one electron Hilbert space which can capture the Lewis structures. The time dependent nuclear amplitudes and quantum phase was integrated into the electron quantum amplitude, and the electron flux was unfolded by the time derivative of these densities divided among the bonding areas. The analysis yielded a counter-intuitive result of the electron current from single to double bond regions during the bond alternation process along this pericyclic reaction. They found through a further analysis of the electronic component that its origin exists in the electron supplement from outer C–H σ bonds. These types of predictions are highly expected to become observable by future imaging techniques of femto- to attosecond time-resolved spectroscopy.

The time shift flux introduced by Okuyama and Takatsuka [307] is another approach has been designed to extract information about flux from a pure adiabatic state that is carried on an ab-initio molecular dynamics. The definition of this operator is expressed by

$$\begin{aligned}&\vec{j}_{ts}(\mathbf{r},t,-\Delta t)\\ =&\frac{\hbar}{2im}\left[\psi^*\left(\mathbf{r},t+\frac{\Delta t}{2}\right)\nabla\psi\left(\mathbf{r},t-\frac{\Delta t}{2}\right)-\psi\left(\mathbf{r},t-\frac{\Delta t}{2}\right)\nabla\psi^*\left(\mathbf{r},t+\frac{\Delta t}{2}\right)\right]\\ =&\vec{j}_{ts}(t,\Delta t)^*,\end{aligned} \tag{7.13}$$

which can be considered as a numerical extension of the usual flux. Note that $\psi^*\left(\mathbf{r},t+\frac{\Delta t}{2}\right)$ for instance can be an adiabatic electronic wavefunction determined at each time, and therefore they may be real-valued. In such cases, we expect

$$\vec{j}_{ts}(\mathbf{r}, t, \Delta t) \to 0 \quad \text{as} \quad \Delta t \to 0. \tag{7.14}$$

But on the other hand, if ψ is indeed a time-dependent wavepacket,

$$\vec{j}_{ts}(\mathbf{r}, t, \Delta t) \to \vec{j}_{ts}(\mathbf{r}, t) \quad \text{as} \quad \Delta t \to 0, \tag{7.15}$$

the limit is therefore gives a real value flux. This extension allows to treat general cases with or without nonadiabatic transition. This operator, Eq. (7.13), represents a loss of the overlap of a wavefunction specified at time $t - \dfrac{\Delta t}{2}$ and that at time $t + \dfrac{\Delta t}{2}$ through the boundary $\partial\Omega$. This itself takes a complex value. Its numerical implementation performs well with quantum chemistry ab initio calculations for fundamental chemical reaction systems involved with bond formation and deformation. Tracking the electronic and nuclear simultaneous dynamics coherently is critical for the realistic understanding of highly quantum mechanical phenomena.

7.1.2 *An example: Collision of Na and Cl*

The semiclassical Ehrenfest theory coupled with this representation was applied in an electron flux analysis in chemical reactions where large charge transfer occurs caused by significant nonadiabatic transition. The chemical systems treated in the summaries below are Na + Cl and formic acid dimer (FAD). The time shift flux operator stated in the previous subsection was utilized in an analysis of the microscopic electron dynamics in this chemically representative case.

The first example treats the electronic current density map in a diatomic reaction in the NaCl molecular system. This system includes a typical chemical process of electron transfer dynamics due to the potential curve crossing between the ionic $Na^+ + Cl^-$ state and the neutral $Na + Cl$ state. The avoided crossing is located at an internuclear distance of 7.4 Å. The case considered was the neutral Na and Cl hit with a small kinetic energy (about 0.095 eV) but with zero impact parameter. SET wavepacket has been propagated in time in terms of 703 CSFs of CISD with molecular orbitals of 6-31G basis set. Figure 7.1 (a) shows the electron current at $R = 8.31$ Å, which is significantly before the crossing. As seen, the electrons are already flowing from the Na to the Cl side through such a distant space. It is noteworthy that electrons around the chlorine atom is receding from the nucleus, and there is a confluent point between the two flux lines from Na and Cl, which is naturally located close to the Cl nucleus. In the internuclear distance of $R = 7.39$ Å (panel (b)), which is very close to the avoided crossing point, the confluent point is found almost at the center of the bond distance. After the nuclear path has passed over the crossing region, at $R = 6.02$ Å (panel (c)), the qualitative feature of the flux does not change much from that of panel (b).

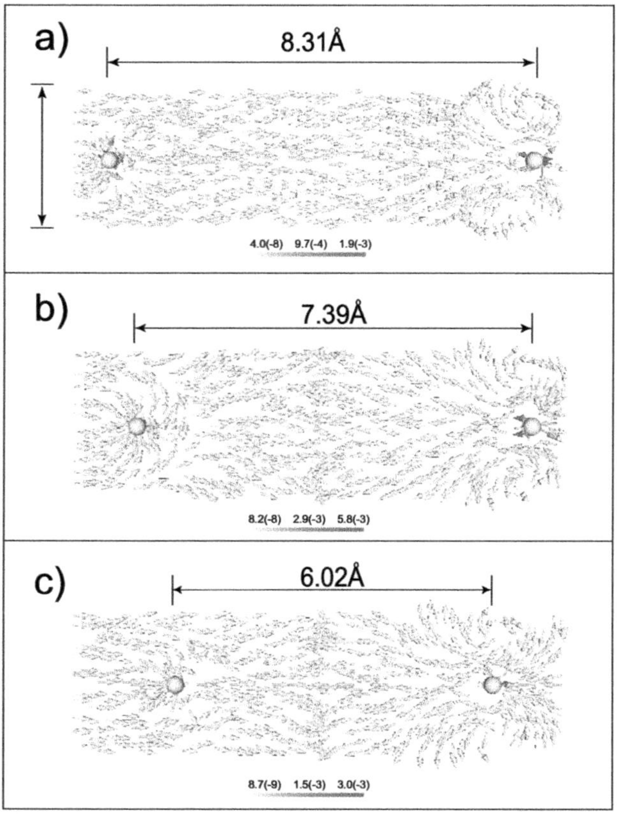

Fig. 7.1 The Schiff flux estimated with SET in the NaCl molecule (left and right spheres graphically represent Na and Cl nuclei, respectively. The flux vectors are drawn at randomly chosen points.): Snapshots taken at geometry (a) before crossing, (b) at the crossing point, and (c) after crossing. The vertical line-segment in panel (a) indicates 2.65 Å. (Reprinted with permission from M. Okuyama *et al.*, Chem. Phys. Lett. 476, 109 (2009)).

The usual flux operator and the time-shift flux was examined. In Fig. 7.2, the first, the second, and the third row, represent the cases of $R = 8.31$ Å, $R = 7.39$ Å, and $R = 6.02$ Å, respectively. The panels in the left column (a1), (b1), and (c1) display the induced flux ($\Delta t = 50$ as), which can be directly compared with Schiff fluxes of Fig. 7.1. The right column panels exhibit the adiabatic flux. Again the topological similarity between the adiabatic and induced fluxes is apparent.

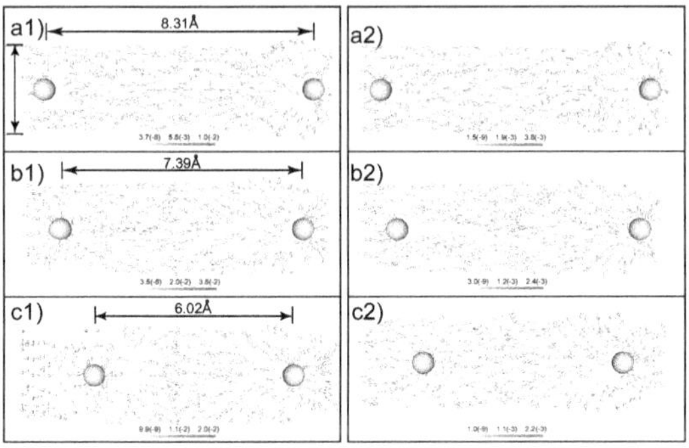

Fig. 7.2 The time shift flux with SET in NaCl molecule with $\Delta t = 50$ as, the real part (induced flux, left column) and the imaginary part (adiabatic flux, right column). The flux vectors are drawn at randomly chosen points. The nuclei of Na and Cl are located towards the left and right, respectively. Snapshots taken at geometry (a1, a2) before crossing, (b1, b2) at the crossing point, and (c1, c2) after crossing. The vertical line-segment in panel (a1) indicates 2.65 Å. (Reprinted with permission from M. Okuyama *et al.*, Chem. Phys. Lett. 476, 109 (2009)).

7.2 Real-time dynamics of electron migration in a model water cluster anion system

Electron hydration is quite an interesting and important issue and many experimental and theoretical studies have already been devoted to the issue. [93, 123, 154, 166, 167, 199, 209, 341, 374, 382] It is a prototype of solution-phase chemistry in polar solvent. Also, the dynamics are expected to provide an important insight into charge-transfer chemistry.

Static structural calculations for the water cluster anion have revealed the quantum mechanical origin of the binding force for the excess electron and various equilibrium structures. [87, 158, 195, 211, 339] It is now established that the excess electron is bound principally by the dipole field formed by the water molecules. In their elaborate studies on the potential landscape of anion water clusters, Choi and Jordan [87] explored a large number of local minima as well as the transition states on the potential energy surface of $(H_2O)_6^-$. They identified the minimal energy pathway from the neutral isomer to low-energy anion isomers.

With respect to the dynamical properties of the hydrated electron in cluster systems, the first principle dynamics using ab initio molecular dynamics and so on have been extensively applied. [135, 180, 371, 408, 446] They revealed information about the structure and relative stabilities of the isomer clusters. Nonadiabatic dynamics of a solvated electron in various photochemical processes has also been studied experimentally. [62, 123, 294, 329] Rossky and co-workers [327, 468] also studied the relaxation dynamics of excess electrons using quantum molecular dynamics simulation techniques. Here the nonadiabatic interactions were taken into account basically within the scheme of surface hopping technique. [444]

On the other hand, there are very few studies available on nonadiabatic dynamics in cluster deformation dynamics. We expect that, in the course of cluster deformation and the accompanying change in electronic orbitals, nonadiabatic transition may in fact play a critical role. The dynamics of the hydrated electron in a water cluster anion consisting of two or three molecules. The electronic structure of such a small cluster, however, is known to have a very small energy scale (the vertical detachment energy (VDE) is of order 10^1 to 10^2 meV [211, 458]). Furthermore, it is known that the accuracy of the theoretically calculated energy levels sensitively depends on the choice of the size and quality of the basis functions. On the other hand, the SET calculations that will be shown in this section are so much limited by the computational capacity available to us that the results are not necessarily accurate enough to be predictive at a quantitative level. In spite of these limitations, SET calculations are performed because it is expected that they can give a qualitative view of the nonadiabatic migration of the excess electrons within the model water clusters.

7.2.1 *Nonadiabatic dynamics of hydrated electron*

7.2.1.1 *Electronic states*

Our dynamical simulation with SET is implemented into the GAMESS package. [357] The basis functions used were those of the TZV basis set [120], to which an s-type diffusion base and two sets of p-type polarization functions were added on hydrogen atoms while s- and p-type diffusion functions were augmented on oxygen atoms. (The exponent is 0.03600 for the s-type diffusion base, 0.08450 for the s- and p-type diffusion bases, respectively, 2.00 and 0.50 for the p-type polarization bases.) The total number of (contracted) basis functions is 38 for each water molecule. We admit that the size of basis set is considerably smaller than that adopted in previous studies on static electronic structure calculations, [211] and therefore the computational results will not necessarily be accurate at the quantitative level.

The electronic states and the relevant matrix elements have been determined using the Restricted Open-shell Hartree–Fock (ROHF) method, [337] followed by a configuration interaction (CI) calculation with double excitations. The active space is limited to 10 molecular orbitals (MO), consisting of 2 occupied, 1 singly-occupied, and 7 unoccupied MOs. Excitations to the higher MOs are neglected. The total number of configuration state functions (CSFs) in the active space amounts to 479.

7.2.1.2 *Static properties of dimer anion $(H_2O)_2^-$*

We here study $(H_2O)_n$ with monomer number $n = 2$ (dimer) and $n = 3$ (trimer). These clusters are known to have a couple of locally stable structures. [211] In this study, we will focus on two distinct equilibrium structures of water anion dimers as shown in Fig. 7.3. The structures shown here are stable within the accuracy of our calculation, whereas previous calculations with larger basis set indicate different structures. The planar structure is close to "C2h" in Ref. [211]. According to the calculations using a larger basis set, this was not in fact an equilibrium structure. Nevertheless this planar structure is still of interest because the binding mechanism of the excess electron is clearly different from that of the nonplanar structure, In addition, other water molecules surrounding the excess electron with their electronic dipole momenta pointing inward may stabilize the planer structure in larger systems. Indeed it is also argued that related types of binding structures may exist in larger clusters and/or in the presence of other ions. [195] Therefore the small cluster under investigation may be regarded as a model subsystem being sampled out of a larger cluster.

7.2.1.3 *Electronic propagation*

The initial conditions for SET dynamics were chosen as follows. Let us consider a system with three water monomers with an excess electron; one dimer anion plus one neutral molecule at the beginning. The first two monomers (m1 and m2) are placed in the configuration of the planar anion dimer. The other monomer (m3) is set to approach the dimer from a

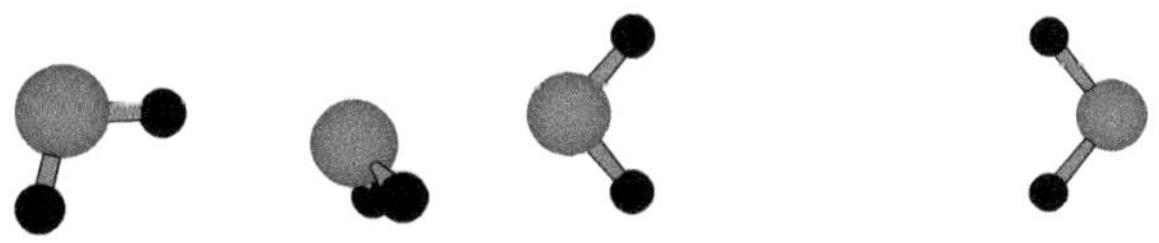

Fig. 7.3 Equilibrium structures of water-anion dimers. Here and throughout this section, small black circles indicate hydrogen atoms while large gray circles indicate oxygen atoms. Left dimer; nonplanar configuration. Atoms in the left monomer and the oxygen atom in the right monomer are on a single plane while the two hydrogen atoms in the right monomer are out of the plane. Right; planar configuration. All the atoms reside in a single plane. (Reprinted with permission from T. Yonehara *et al.*, Chem. Rev. 112, 499 (2012)).

distance in such a manner that m3-m2 eventually forms the nonplanar anion configuration (see Fig. 7.4). This arrangement was chosen with the expectation that electron migration is possibly more favorable than other random orientations. We also examined other collision geometries that are not explicitly shown here. The relative translational velocity given to the colliding monomer (m3 in Fig. 7.4) is set 0.17 Bohr/fs. Although this translational velocity is large, in the sense that the corresponding energy is much larger than the VDE predicted by a more accurate calculation such as Ref. [211], it does not directly affect the binding structure since it is a translational energy. The vibrational modes are then determined using the standard mode analysis on the equilibrium structure of dimer and monomer. Each vibrational mode is treated approximately as an independent classical harmonic oscillator. Each oscillator is given an energy equal to the zero-point energy with a randomly chosen phase. We prepared a set of initial conditions scanning only the phases, which gives rise to about 30 collision paths in total, in order to confirm that the qualitative conclusions drawn are not pathological. No statistical sampling is taken with a view to calculating the average rate. Finally, the time steps for integrating the coupled equations for SET are set to 0.125 fs and 0.005 fs for nuclear and electronic dynamics, respectively. It was confirmed that finer time steps did not change the results qualitatively.

7.2.2 *Mechanisms of migration of the hydrated electron*

As a result of our SET simulation, basically two types of migration mechanisms for the excess electron was identified. One is what can be called the adiabatic migration, which is driven simply by the adiabatic change of the total electronic state. It has a rather long migration time scale reflecting

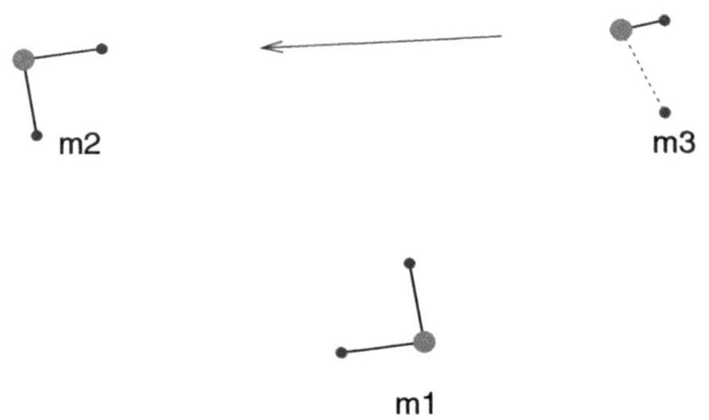

Fig. 7.4 Configuration of collision between the planar dimer anion consisting of m1 and m2 and a water monomer m3. (Reprinted with permission from T. Yonehara *et al.*, Chem. Rev. 112, 499 (2012)).

the time scale of cluster deformation. Adiabatic electron migration has already been thoroughly studied by Tachikawa [408] and Head-Gordon. [180] Hence here we will consider the nonadiabatic mechanism in depth from the analysis on the collision shown in Fig. 7.4.

To demonstrate that this collision really induces nonadiabatic electron transfer, we expand the electronic wavepacket of SET in the adiabatic electronic states at each nuclear configuration and thereby attain the state population for the relevant adiabatic states. Fig. 7.5 shows such population dynamics for the ground, the first excited, and the second excited state configurations. It is clearly observed that the interchange between the ground and the first excited states took place two times in between 10 and 15 fs. A large fluctuation in the state populations follows after that. In this graph, two characteristic time scales are observed; one with period as short as about 8 fs, whereas the large amplitude component of period can be as long as about 20 fs. The shorter time period is almost the same as that of the antisymmetric stretching mode of a water monomer. The longer time scale will be addressed below.

7.2.3 *Complex-valued natural orbitals in electron wavepacket dynamics*

The natural orbitals, which are eigenfunctions of the first order density matrix, are among the well-known quantities in standard quantum chemistry. However, the natural orbitals resulting from a time-dependent electronic wavefunction such as SET and PSANB are complex-valued in general, and

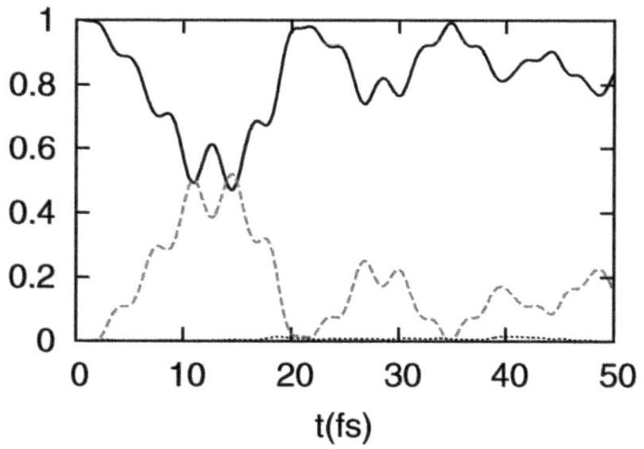

Fig. 7.5 Time-dependent behavior of the adiabatic state populations during the collision of a water dimer anion and a water monomer as shown in Fig. 7.4. Black solid line, gray dashed line, and thin gray dotted line represent that of the ground, first excited, and second excited states, respectively. Changes in the populations of these states indicate nonadiabatic transitions. (Reprinted with permission from T. Yonehara *et al.*, Chem. Rev. 112, 499 (2012)).

through this the state-mixing is well realized, as actually shown later in this subsection. The first-order spin-free density matrix for an (instantaneous) electronic wavefunction $\Phi(t)$ is defined as

$$\begin{aligned}\rho(\mathbf{r},\mathbf{r}',t) &\equiv N\int d\omega_1 dq_2\cdots dq_N \\ &\quad\times \Phi^*\left(\mathbf{r}',\omega_1,\mathbf{r}_2,\omega_2,\cdots,\mathbf{r}_N,\omega_N\right)\Phi\left(\mathbf{r},\omega_1,\mathbf{r}_2,\omega_2,\cdots,\mathbf{r}_N,\omega_N\right) \\ &= \sum_{\mu,\nu}\rho_{\mu\nu}(t)\chi_\mu(\mathbf{r}')\chi_\nu(\mathbf{r}), \qquad (7.16)\end{aligned}$$

where $\mathbf{r}_i$ and ω_i are the spatial and spin coordinates, respectively, of the ith electron. Here, $\int dq_i$ is a short-hand notation for spatial integration and spin summation. χ_μ are basis functions such as atomic orbitals, which are chosen to be real functions. The natural orbitals in electron dynamics can become complex and are given as eigenfunctions of the matrix $\rho_{\mu\nu}$ such that

$$\sum_\nu \rho_{\mu\nu}\xi_\nu^{(\lambda)} = n^{(\lambda)}\xi_\nu^{(\lambda)}, \qquad (7.17)$$

where $\xi_\mu^{(\lambda)}$ is the λth eigenvector of $\rho_{\mu\nu}$ with a real eigenvalue (occupation number) $n^{(\lambda)}$, giving rise to a natural orbital

$$\phi_\lambda(\mathbf{r}) = \sum_\nu \xi_\nu^{(\lambda)}\chi_\nu(\mathbf{r}). \qquad (7.18)$$

Equation (7.17) does not determine the phase of the natural orbital. It is necessary, therefore, to fix the phase factor so that the real part of ξ has the largest norm.

To see how the electronic states change nonadiabatically, a comparison of SOMO, LUMO, and the singly occupied natural orbital (SONO) are useful, as shown in Fig. 7.6. Here, SONO is defined as the natural orbital whose occupation number is closest to unity. In these figures we observe a characteristic change in the phase of SONO. Let us track the shape of SOMO, LUMO, and the real part of SONO as shown in Fig. 7.6 and see the deformation in the electron migration. It is clear that the topology of either the SOMO or LUMO does not appreciably change during the course of the collision, whereas that of the SONO changes significantly. This is obviously due to a mixing of the SOMO and LUMO. It should also be noted that the occupation number of SONO remains very close to unity. Thus it is in the smooth change in the shape of SONO that the electron-dynamical effect of the nonadiabatic transition manifests itself most clearly.

7.2.4 *Nuclear motion inducing nonadiabatic transitions*

We next study the geometrical aspects of the observed nonadiabatic transition. The observed nonadiabatic transitions originates from the matrix element of the nonadiabatic coupling operator $\hbar\dot{\mathbf{R}} \cdot \mathbf{X}$ between the two adiabatic states; $|\Psi_g(\mathbf{R}(t))\rangle$ and $|\Psi_1(\mathbf{R}(t))\rangle$. It is therefore pertinent, in the first instance, to study a spatial vector of $\hbar\mathbf{X}_{g1} \equiv \hbar\langle\Psi_g(\mathbf{R}(t))|\hat{\mathbf{X}}|\Psi_1(\mathbf{R}(t))\rangle$. Since the nonadiabatic coupling element $\hbar\dot{\mathbf{R}}\cdot\mathbf{X}$ is an inner product between the nuclear velocity vector $\dot{\mathbf{R}}$ and the coupling element vector $\hbar\mathbf{X}_{g1}$, nuclear modes that are parallel to the latter vector should strongly enhance the nonadiabatic transition.

For the adiabatic functions expanded in the CSF basis set as $|\Psi_\alpha\rangle = \sum_I C_{\alpha,I}|\Phi_I\rangle$, the nonadiabatic coupling is decomposed into

$$\langle\Psi_\alpha(\mathbf{R})|\frac{\partial}{\partial R^\mu}|\Psi_\beta(\mathbf{R})\rangle = \sum_{I,J} C^*_{\alpha,J}X^\mu_{JI}C_{\beta,I} + \sum_I C^*_{\alpha,I}\frac{\partial}{\partial R^\mu}C_{\beta,I}. \quad (7.19)$$

Therefore, even if the transition through the above elements of X^μ_{IJ} are set to zero, the left hand side is not necessarily reduced to zero, because of the presence of the last term. In Fig. 7.7, we plot the first term on the right

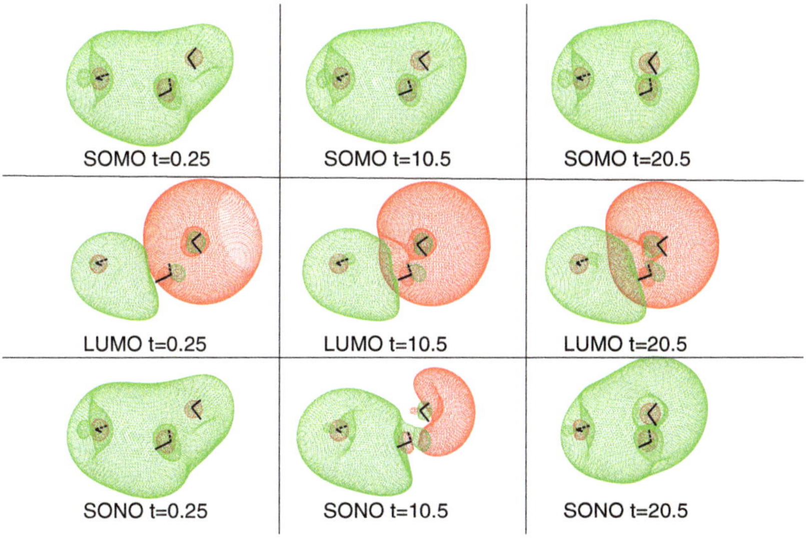

Fig. 7.6 Selected snapshots of the spatial distribution of SOMO (upper row), LUMO (middle row), and the real part of SONO (bottom row) during the collision of a water dimer anion and a water monomer. Green and red surfaces indicate the isocontour surfaces of the electronic orbitals at the amplitude +0.01 $\mathrm{Bohr}^{-\frac{3}{2}}$ and −0.01 $\mathrm{Bohr}^{-\frac{3}{2}}$, respectively. Water molecules are indicated as a pair of black thin lines, each representing an OH bond. From left to right, times are $t = 0.25$, $t = 10.50$ and $t = 20.50$ (in units of fs). Although either SOMO or LUMO do not change their characters during the reaction, SONO changes significantly. (Reprinted with permission from T. Yonehara *et al.*, Chem. Rev. 112, 499 (2012)).

hand side of Eq. 7.19 times $\hbar$, redefined below more explicitly

$$\hbar\tilde{\mathbf{X}}_{g1} \equiv \hbar \sum_{I,J} C^{*}_{gJ} X^{\mu}_{JI} C_{1I}, \tag{7.20}$$

where $C_{gJ} \equiv \langle \Psi_g(\mathbf{R}) | \Phi_J(\mathbf{R}) \rangle$, and $C_{1I} \equiv \langle \Psi_1(\mathbf{R}) | \Phi_I(\mathbf{R}) \rangle$. It is immediately clear from this figure that $\hbar\tilde{\mathbf{X}}_{g1}$ strongly couples to the OH anti-symmetric stretching mode of the water dimer.

The nonadiabatic coupling term $\hbar\dot{\mathbf{R}} \cdot \mathbf{X}_{g1}$ undergoes a rapid oscillation mainly due to the fast oscillatory motion of the nuclei. Its angular frequency has a value around 0.0017 in atomic units, which clearly reflects the frequency of the OH anti-symmetric stretching (as) mode, ω_{as} in dimer anions; $\omega_{as}/2\pi \sim 3700\,\mathrm{cm}^{-1}$. As a result, both the spatial distribution of $\hbar\tilde{\mathbf{X}}_{g1}$ and the time-series of $\hbar\dot{\mathbf{R}} \cdot \langle \Psi_g(\mathbf{R}) | \hat{\mathbf{X}} | \Psi_1(\mathbf{R}) \rangle$ strongly suggest that the OH antisymmetric stretching motion should trigger the nonadiabatic migration of the hydrated electron.

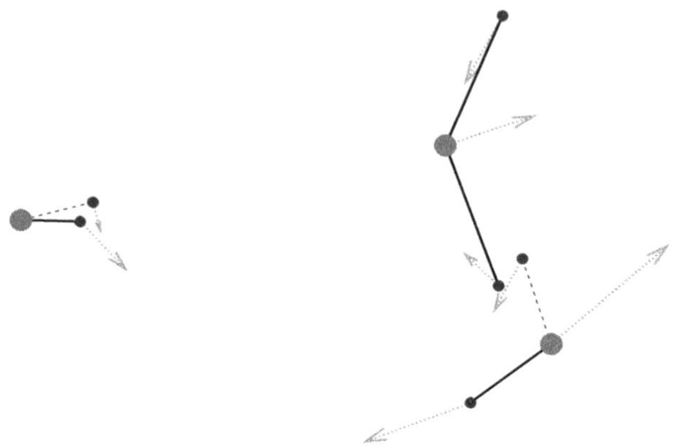

Fig. 7.7 Approximate nonadiabatic coupling vector $\tilde{\mathbf{X}}_{g1}$ (see the text for the definition) at $t = 18.625$ fs, at which the nonadiabatic coupling term reaches its maximum value in the collision of a water dimer anion and a water monomer. Although the vectors are time-dependent, their spatial orientations change only minimally during the periods of the vibration motions. (Reprinted with permission from T. Yonehara *et al.*, Chem. Rev. 112, 499 (2012)).

7.2.5 *Isotope effects*

The present electron migration is heavily dependent on the nonadiabatic coupling elements, which is in turn a function of the nuclear (reduced) masses. Therefore, it is readily anticipated that the migration should be significantly affected by exchanging the protons by deuterons. Indeed it is the case as we will exemplify below. We track the dynamics of such isotope effects by choosing the same initial configuration, orientation and kinetic energies as those in Figs. 7.4 and 7.5. The results in Fig. 7.8 clearly show a significant suppression in nonadiabatic transitions. To see it more clearly, in Fig. 7.9 we show the snapshots of SOMO, LUMO, and SONO for this collision at selected instances. It is obvious that neither the SOMO or LUMO is affected by deuterium substitution, whereas the SONO most definitely is (compare with Fig. 7.6). This highlights one of the crucial differences between the SET calculations and ab initio MD type (or CI type) static calculations, in which virtually no change in electronic properties is expected to be induced by deuterium substitution. As seen in the figure, the SONO in the current case is quite similar to SOMO. This clearly illustrates that the SOMO and LUMO do not mix much into the SONO (compare again with SONO in Fig. 7.6).

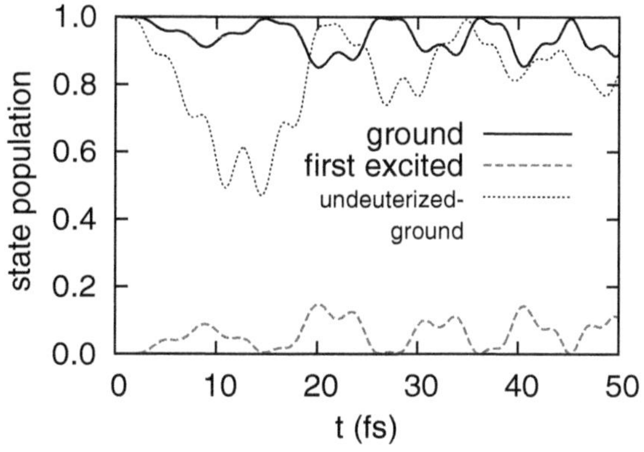

Fig. 7.8 Time-dependent behavior of the adiabatic state populations during the collision of a fully deuterized molecules (water dimer anion and water monomer). The electronic state populations in the adiabatic basis are shown. Black solid line, gray dashed line represent that of the ground, first excited states, respectively. The time evolution of the ground state population obtained from the corresponding calculation with undeuterized molecules is also plotted with thin gray dotted lines for comparison (see Fig. 7.5). (Reprinted with permission from T. Yonehara *et al.*, Chem. Rev. 112, 499 (2012)).

As illustrated above, nonadiabatic dynamics exhibits vividly how electrons move in and between molecules. Complex natural orbitals, in particular SONO in the present case, clearly illustrate how the electronic wavefunction evolves in time. In addition to the time scale, the driving mechanism for the electron migration has also been illustrated. By clarifying such complex electron behavior, not available using stationary-state quantum chemistry, our understanding of realistic chemical reactions is greatly enhanced. As a result, it has clearly been shown that nonadiabatic electron wavepacket theory is invaluable in the analysis of non-rigid and mobile electronic states of molecular systems.

7.3 Single and relayed proton transfer in peptide

We next study water-assisted (relayed) proton transfer in formamide, comparing with a forced proton transfer within formamide in gas phase (Fig. 7.10), which are also associated with the tautomerization. We start from the enol (imidic acid) form HO-CH=NH and track the transfer process to the keto (amide) form O=CH-NH_2. We first survey the above very basic puzzles with this system, identifying that this is indeed a "proton

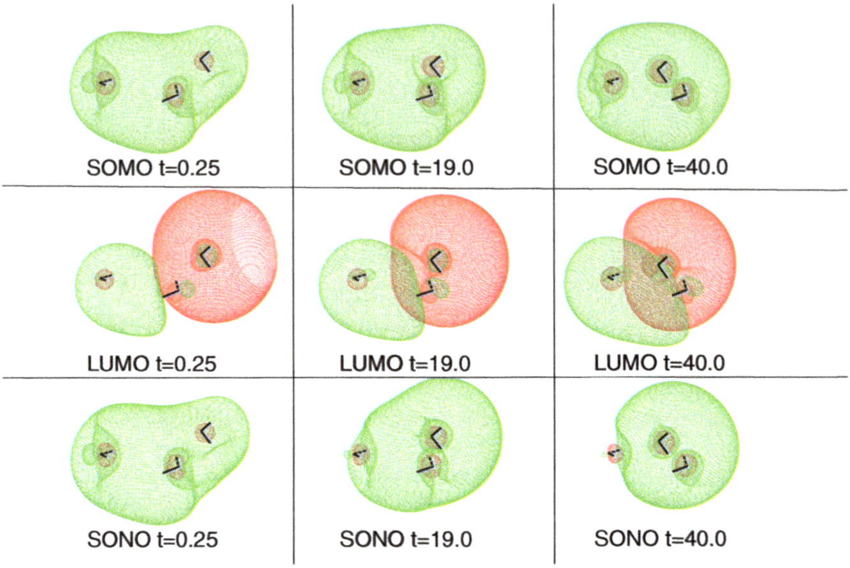

Fig. 7.9 Selected snapshots of the spatial distribution of SOMO (upper row), LUMO (middle row), and the real part of SONO (bottom row) for the collision of fully deuterized molecules (water dimer anion and water monomer). Green and red surfaces indicate isocontour surfaces of the electronic orbitals at the amplitude +0.01 Bohr$^{-\frac{3}{2}}$ and −0.01 Bohr$^{-\frac{3}{2}}$, respectively. Water molecules are indicated as a pair of black thin lines, each representing an OH bond. From left to right, times are $t = 0.25$, $t = 19.00$ and $t = 40.00$ (in units of fs). This figure should be compared to the undeuterized counterpart, Fig. 7.6. (Reprinted with permission from T. Yonehara *et al.*, Chem. Rev. 112, 499 (2012)).

transfer process" in terms of the spatial distribution of radicals (unpaired electron density) [417] on the transferring proton. Then, we show a collective electron quantum flux of electrons, which flows backward against the proton motion, thereby suggesting the compensation of the tightly covering the proton as represented in terms of the Mulliken charge. Besides these basic issues, this molecular system is of a particular relevance to protein dynamics. The keto form formamide is one of the simplest species in the group O=CR$_1$-NHR$_2$, which is a unit of polypeptide. In gas phase, the nitrogen atom can have a pyramidal structure as in ammonium molecule, and therefore the C-N bond may allow free (low barrier) rotation along it. On the other hand, this rotation is obviously prohibited by the formation of the double bond C=N induced by the proton transfer. Not only the proton transfer itself, which we confirm that this process is indeed the case, but also the electronic structures left behind are greatly affected by the presence of water molecule(s) and polar solvents. In discussing the relative stability of the formamide after the proton transfer, the resonance structures O^--$CH{=}N^+H_2 \longleftrightarrow O{=}CH\text{-}NH_2$ are usually considered. Although there is no need to emphasize that the Pauling resonance theory is extremely useful and constitutes a basic language of chemistry, we here revisit it to reveal the dynamical manifestation of the present resonance structures a little deeper in terms of our dynamical electron theory.

7.3.1 *Quantities characterizing the electron dynamics*

7.3.1.1 *Unpaired electron density*

We begin with the spin-free reduced density matrix $\rho(\mathbf{r}',\mathbf{r};t)$ is just the electron density. In chemical applications, it is often expanded in atomic orbitals $\chi_{aA}(\mathbf{r})$ as

$$\rho(\mathbf{r},t) = \sum_{\mathrm{A}} \sum_{a}^{\text{on A}} \sum_{\mathrm{B}} \sum_{b}^{\text{on B}} \rho_{aA,bB}(t)\, \chi_{aA}(\mathbf{r})\, \chi_{bB}(\mathbf{r}), \qquad (7.21)$$

where A and a represent an atom and the ath function on A, respectively, with $\rho_{aA,bB}(t)$ being the associated coefficient. The Mulliken population on an atom A is

$$M_A(t) = \sum_{a,a'}^{\text{on A}} \rho_{aA,a'A}(t)\, S_{aA,a'A} + \frac{1}{2} \sum_{a}^{\text{on A}} \sum_{\mathrm{B}\neq\mathrm{A}} \sum_{b}^{\text{on B}} \rho_{aA,bB}(t)\, S_{aA,bB} \qquad (7.22)$$

where $S_{aA,bB}$ is the overlap integral between $\chi_{aA}(\mathbf{r})$ and $\chi_{bB}(\mathbf{r})$.

By extracting from $\gamma(\mathbf{r};t)$ only the product terms between the atomic orbitals belonging two different atoms, say A and B, one can define bond-order density in such a way

$$B_{AB}(\mathbf{r},t) = \sum_{a}^{\text{on A}} \sum_{b}^{\text{on B}} \rho_{aA,bB}(t)\, \chi_{aA}(\mathbf{r})\, \chi_{bB}(\mathbf{r}). \qquad (7.23)$$

The positiveness of its spatial distribution represents a formation of chemical bonding. The familiar bond-order itself is recovered by integrating $B_{AB}(\mathbf{r},t)$ over $\mathbf{r}$.

It is well known that even the spin singlet ground-state can have a biradical character due to electron correlation. To monitor the distribution of such radicals, or unpaired electrons, at a given site $\mathbf{r}$, we use the so-called unpaired electron density [417], which is defined as

$$D(\mathbf{r}) = 2\rho(\mathbf{r},\mathbf{r}) - \int d\mathbf{r}'\rho(\mathbf{r},\mathbf{r}')\,\rho(\mathbf{r}',\mathbf{r})\,, \tag{7.24}$$

where $\rho(\mathbf{r},\mathbf{r}')$ is an off-diagonal element of the first order spin-free density matrix. $D(\mathbf{r})$ is represented in an invariant manner by means of the natural orbitals $\{\lambda_i(\mathbf{r})\}$, which diagonalize $\rho(\mathbf{r},\mathbf{r}')$, as

$$D(\mathbf{r}) = \sum_i (2-n_i)\,n_i\,|\lambda_i(\mathbf{r})|^2\,, \tag{7.25}$$

where n_i is the occupation number of $\lambda_i(\mathbf{r})$. It is obvious from this expansion, that $n_i = 1$ maximizes $(2-n_i)\,n_i$, and $n_i = 2$ (double occupancy) and $n_i = 0$ (vacancy) makes null contribution to $D(\mathbf{r})$. Therefore it is natural to define the spatial distribution of "radicals" using $D(\mathbf{r})$. When applied to chemical reaction in a singlet state, $D(\mathbf{r})$ can also distinguish its reaction mechanism; concerted reaction (with low unpaired electrons) or radical reactions (with high values) in a quantitative manner. It is straightforward to define the unpaired electron density on an atom A, say D_A by expanding $D(\mathbf{r})$ in atomic basis functions as

$$D(\mathbf{r}) = \sum_{A,B}^{\text{atoms}}\sum_{a}^{\text{on A}}\sum_{b}^{\text{on B}} D_{aA,bB}\,\chi_{aA}(\mathbf{r})\,\chi_{bB}(\mathbf{r})\,. \tag{7.26}$$

The number of unpaired electrons on atom A, say D_A, is defined as

$$D_A = \sum_{a,a'}^{\text{on A}} D_{aA,a'A}(t)\,S_{aA,a'A} + \frac{1}{2}\sum_{a}^{\text{on A}}\sum_{\text{B}\neq\text{A}}\sum_{b}^{\text{on B}} D_{aA,bB}(t)\,S_{aA,bB} \tag{7.27}$$

(We will report a detailed account of the theory along with several examples in our future publication [287].)

7.3.1.2 *Time fluctuation of the electron density*

In an analysis using the electron flux, we sometime face a situation in which the driving force causing the flux is not clear. In such a case, it is helpful to study the time-dependent local change of the electron density. We quantify such increasing or decreasing of the electron density at $\mathbf{r}$ as follows. First, we expand the instantaneous density matrix of the electronic state $\gamma(\mathbf{r},t;\mathbf{R})$ at each nuclear position in "molecular orbitals" $\phi_i(\mathbf{r};\mathbf{R})$ (not the atomic orbitals) such that

$$\rho(\mathbf{r},t;\mathbf{R}(t)) = \sum_{i,j} P_{ij}(t)\phi_i(\mathbf{r};\mathbf{R})\phi_j(\mathbf{r};\mathbf{R})\Big|_{\mathbf{R}=\mathbf{R}(t)}\,, \tag{7.28}$$

where the coefficients $P_{ij}(t)$ are the function of the coefficient of CSF. It is

$$P_{ij}(t) = \left\langle \Phi(t) \left| \sum_{\sigma} \hat{X}^{\dagger}_{\sigma i} \hat{X}_{\sigma j} \right| \Phi(t) \right\rangle = \sum_{I,J} C^{*}_{I}(t) A^{ij}_{IJ} C_{J}(t), \tag{7.29}$$

where

$$A^{ij}_{IJ} = \left\langle \Phi_I \left| \sum_{\sigma} \hat{X}^{\dagger}_{\sigma i} \hat{X}_{\sigma j} \right| \Phi_I \right\rangle \tag{7.30}$$

with $\hat{X}^{\dagger}_{\sigma i}$ and $\hat{X}_{\sigma i}$ being the creation and annihilation operators, respectively, of a molecular orbital i of spin σ. We then consider the time derivative of γ, that is

$$\frac{\partial}{\partial t}\rho(\mathbf{r}, t; \mathbf{R}) \equiv \sum_{i,j} \dot{P}_{ij}(t)\phi_i(\mathbf{r}; \mathbf{R})\phi_j(\mathbf{r}; \mathbf{R}). \tag{7.31}$$

It should be noted that we take account of only the time derivative of $P_{ij}(t)$, while the derivative arising from $\partial \mathbf{R}(t)/\partial t$ is discarded. This is because we want to see only the electronic change due to the variation of chemical bonding, which is directly represented in $P_{ij}(t)$. [Note that the change of the electron density emerging from a simple shift of atomic position is not necessarily of direct relevance. For instance, suppose a translational motion of a free hydrogen atom, which necessarily causes the (trivial) spatiotemporal change of the electron density.] $\dot{P}_{ij}(t)$ can be estimated as follows: With the help of Eq. (6.14), we have

$$\begin{aligned} \dot{P}_{ij} &= \sum_{I,J} \dot{C}^{*}_{I} A^{ij}_{IJ} C_J + C^{*}_{I} A^{ij}_{IJ} \dot{C}_J \\ &= \sum_{I,J,K} \frac{i}{\hbar} C^{*}_{I} \left[(H^{el}_{IK} + G^{cp}_{IK}) A^{ij}_{KJ} - A^{ij}_{IK} (H^{el}_{KJ} + G^{cp}_{KJ}) \right] C_J, \end{aligned} \tag{7.32}$$

where

$$G^{cp}_{IJ} = -i\hbar \sum_{k} \dot{R}_k X^{k}_{IJ} - \frac{\hbar^2}{4} \sum_{k} \left(Y^{k}_{IJ} + Y^{k*}_{JI} \right). \tag{7.33}$$

Neglecting the term proportional to $\hbar^2$ in G^{cp}_{IJ} as usual as

$$G^{cp}_{IJ} \simeq -i\hbar \sum_{k} \dot{R}_k X^{k}_{IJ}, \tag{7.34}$$

we obtain

$$\begin{aligned} \dot{P}_{ij} =& \frac{i}{\hbar} \sum_{I,J,K} C^{*}_{I} \left[H^{el}_{IK} A^{ij}_{KJ} - A^{ij}_{IK} H^{el}_{KJ} \right] C_J \\ &+ \sum_{I,J,K} C^{*}_{I} \left[\sum_{k} \dot{R}_k (X^{k}_{IK} A^{ij}_{KJ} - A^{ij}_{IK} X^{k}_{KJ}) \right] C_J. \end{aligned} \tag{7.35}$$

Bringing this quantity back into Eq. (7.31), $\frac{\partial}{\partial t}\gamma(\mathbf{r}, t; \mathbf{R})$ is finally evaluated, whose spatial distribution will be used along with the flux.

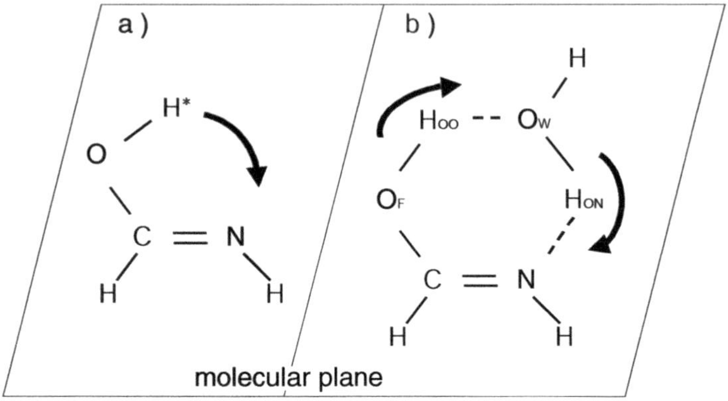

Fig. 7.10 (a) Forced proton transfer of formamide in gas phase. (b) Relayed proton transfer mediated by a water molecule (water-assisted proton transfer). The proton transfer is associated with tautomerization from enol form HO-CH=NH to keto form O=CH-NH_2. (Reprinted with permission from K. Nagashima *et al.*, J. Phys. Chem. A 113, 15240 (2009)).

7.3.2 *Computational details*

7.3.2.1 *System studied*

We study electron wavepacket dynamics associated with "proton transfer" from enol HO-CR_1=NR_2 to keto O=CR_1-NHR_2 (R_1 =H and R_2 =H) with and without mediation by a water molecule as depicted in Fig. 7.10. (By "proton transfer" we tentatively mean only the shift of the position of nucleus proton, before we decide whether it is really proton transfer or hydrogen-atom migration in the next subsection.) Although the keto form is particularly interesting because it is a unit of the polypeptide, we start from the enol form, and then investigate the relations between them through the dynamical analysis of the Pauling resonance structure.

7.3.2.2 *Basis set, molecular orbitals, configuration state functions*

The electronic wavepackets are determined in an expansion with the configuration state functions (CSF) of single and double excitations (CISD) with the STO-6G basis set. This basis set is obviously small, but the main concern in this work is not the accuracy but qualitative insights about the dynamical electrons in the course of chemical reactions. The program codes for the semiclassical Ehrenfest theory (SET) have been implemented in the GAMESS package [357].

The 1s-like inner shell orbitals of C, O and N atoms have been frozen and totally neglected in the process of electronic mixing. Furthermore, in water-bridging system, the lowest 10 molecular orbitals (MO) have been also neglected in the SET dynamics, and 1653 CSFs arising from the resultant 25 MOs are actually used. Likewise, for the gas-phase reaction, we have used 1540 CSFs based on 18 MOs. Although we have carried out extensive numerical calculations including several excited states, we here report only the dynamics of the low energy state, starting from the ground electronic state as an initial condition.

7.3.2.3 *Initial conditions for the nuclear motion*

The initial geometry for formamide was set to the enol form as in Fig. 7.10. The atoms O, C, and N make a molecular plane, and the bridging water molecule is also placed initially so as to lie in that plane. In what follows, we refer to the molecular orbitals approximately lying on the plane and to those approximately perpendicular to the plane as σ and π orbitals, respectively. Likewise, using only π orbitals in $\gamma(\mathbf{r},t)$ and $B_{AB}(\mathbf{r},t)$, we estimate the π electron density and π bond-order, respectively. Similarly, the σ electron density and σ bond-order are made available. This distinction between the σ and π subspaces is just a matter of convenience, and of course they are not physical observables individually, since C_S symmetry is not imposed on the molecular system. On the contrary, all the vibrational modes are active in the present SET calculations. Since the aim of this study is not to estimate the reaction probability but the mechanism of the electron dynamics associated with proton transfer, we chose somewhat artificial initial conditions of nuclear motion to sample as many paths achieving proton transfer as possible.

Water-assisted dynamics: We first prepared formamide and water molecule, which are geometrically optimized in each species (not optimized as a super-molecule), and configure them so that the distances between $O_W - H_{OO}$ and $N - H_{ON}$ are both as close as about 1.4 Å. We sampled about 70 paths by scanning the initial velocity of H_{OO}, varying the energy in the antisymmetric vibrational mode of the water molecule, and giving or not giving the zero-point energies to the normal modes. Fifty out of these seventy paths have actually reached the proton transfer structure. The potential barrier is of course far lower than that of the gas-phase reaction [259].

7.3.3 *Proton transfer in the ground state*

Among the sampled paths chosen as above for the semiclassical Ehrenfest theory, we pick a couple of examples as a generic case-study of "proton transfer".

7.3.3.1 *Over-all feature of the reactions of Water assisted transfer*

We first present the overall feature of an example of the dynamics in water-assisted proton transfer in Fig. 7.11. Panel (a) indicates that the distance of the proton H_{OO} from O_F becomes longer than that from O_W at time about 2 fs, while panel (b) shows that the proton H_{ON} leaves from the site of O_W and arrives at the vicinity of N at about 5 fs. Thus the first transfer took place in the site of $O_F - O_W$ and that of $O_F - N$ followed in this particular example of proton relay. This is not always the case, though. That is, proton transfer in the site of $O_W - N$ can precede that of $O_F - O_W$. The π bond-order displayed in panel (c) claims the double bond has shifted from $C - N$ to $O_F - C$ at time about 7 fs. Since the tautomerization is completed in this way, we may judge that the present proton-relay has been achieved successfully along this path.

We next survey the basic behavior of the electronic population on the relevant atoms in terms of the Mulliken (electronic) population in panel (d) through (f). In panel (d), we notice that the Mulliken population on both H_{OO} and H_{ON} is kept almost constant throughout the course of transfer process. This clearly indicates that these protons are covered with electrons as much as about 0.7 each. Panel (e) suggests that the electron density on O_W in the molecular plane remains almost constant, while that on O_F (N) increases (decreases) to some extent. On the other hand, panel (f) indicates that π-electron density is shifted a little from O_F to N, while that on C does not change in the course of π-bond alternation. We note that this motion of the π-electron density is in the reverse direction to the shift of the position of π-bond itself.

7.3.3.2 *Proton transfer versus hydrogen-atom migration*

In the study of proton transfer dynamics in a collision between $H_3O^+ + H_2O$, Ushiyama and Takatsuka [451] have observed that the transferring proton is always covered with electrons as much as about 0.6 throughout the collision with the Mulliken population analysis. The similar phenomenon is confirmed for other proton transfer systems in our laboratory. Since the

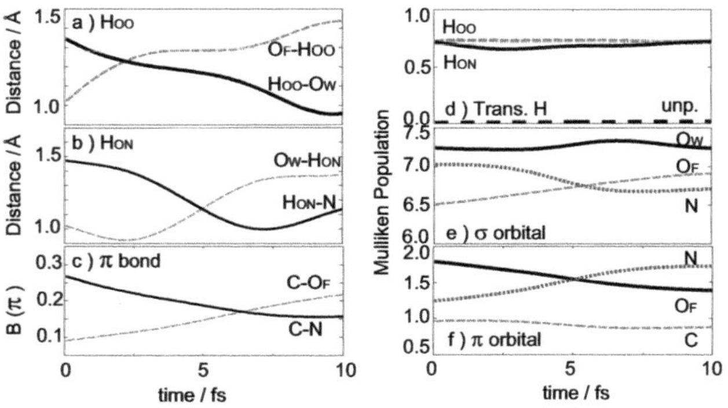

Fig. 7.11 Some selected quantities profiling the time dependence of the water-assisted proton transfer along a reaction pathway. a) Distance of the transferring proton H_{OO} from the neighboring atoms, O_F and O_W. b) Distance of the transferring proton H_{ON} from the neighboring atoms, O_W and N. c) π-bond order, $B(\pi)$, in two bonds, $O_F - C$ and $C - N$. d) Mulliken population on H_{OO} and H_{ON} (solid line and dashed lines represent the number of total electrons and that of the unpaired electrons, respectively.) e) Mulliken population on the skeletal atoms arising from the σ orbitals on the molecular plane. f) Mulliken population on the skeletal atoms arising from the π orbitals perpendicular to the molecular plane. (Reprinted with permission from K. Nagashima *et al.*, J. Phys. Chem. A 113, 15240 (2009)).

Mulliken population is considerably dependent on the basis set chosen, the amount of electrons carried by the nucleus proton is not necessarily determined uniquely. Instead of the Mulliken analysis, we have carried out numerical integration to track the amount of electrons around the nucleus proton and found the similar behavior to that found with the Mulliken analysis, although a little more fluctuation of the electron density has been observed [286]. This conclusion is consistent with the earlier population analysis based on Bader's theory of atoms in molecules [25] reported by Wong et al. [478]. In any analysis, it is commonly concluded that the proton is never transferred as a bare proton in a naked fashion, which is not surprising in view of the high binding energy of 1s electron in hydrogen atom, that is 13.6 eV. As described above, the similar phenomenon has been reconfirmed in the present system, where electrons as much as about 0.7 has been found to consistently follow the nucleus proton tightly. Then, the following naive questions arise: Is this "proton transfer" a really proton transfer? If so, in what manner and through which route are such covering electrons rendered back to the proton-launching site. Or, is this process a hydrogen-atom migration in an electron deficient state? If yes, how and through which route is the additional electron pushed to a destination site? As far as the present system is concerned, chemical intuition suggests that it is unlikely for the hydrogen-atom migration to take place. However, it is certainly critical to confirm whether it is indeed the case, since the electronic mechanism of "proton transfer" is closely linked to the electron dynamics of tautomerization.

Therefore we here propose a rather quantitative condition to distinguish proton transfer and hydrogen-atom migration not by the number of electrons involved but the qualitative difference in their electronic states. For hydrogen-atom migration, we literally take it as a radical transfer in the hydrogen-atom-like state. As an illustrative example, let us consider a schematic "proton transfer" reaction H-A- B $\rightarrow$ A-B-H. In the course of this reaction, H radical (H·) should be formed to some large extent in the hydrogen-atom migration. On the other hand, we regard proton transfer as a concerted reaction. That is, it undergoes a transfer through a smooth bond exchange, in which an electronic closed-shell structure (α- and β-spin pairing state, or simply doubly occupied state irrespective of the number of involved electrons) is realized around the nucleus proton without formation of radicals. To summarize, proton transfer is like a concerted reaction maintaining a closed shell electronic structure, while the hydrogen-atom migration can be termed as a radical or singlet biradical reaction. (Therefore the closed-shell Hartree–Fock calculation cannot reproduce the state of hydrogen-atom migration in this definition.) We will report in a greater detail the account of the theory along with several examples in our future publication [287].

To monitor the distribution of radicals, or unpaired electrons, at a given atomic site $\mathbf{r}$ in a molecule, we use the unpaired electron density [417] defined as in Eq. (7.24) through Eq. (7.27). This density was figured out to characterize the non-concerted (Woodward–Hoffmann forbidden) reactions. Therefore, any indicator to monitor such spatial distribution of unpaired electrons (biradicals) will be equally acceptable. Indeed, several other studies on the unpaired electron have been reported in the theory of chemical reactions proceeding via biradical states [180, 227, 245, 328, 387–389, 442]. In particular, the extensive studies by Davidson and his coworkers [387–389] have set a landmark in this field.

In Fig. 7.11, panel (d) for the water bridging system, we have plotted D_A on the two transferring protons. It turns out that both of them remain to be almost zero during the transfer. (Recall that the present SET calculation contains single and double electronic excitations.) Therefore, there is virtually no radical character around the proton, and we conclude on this quantitative basis that this double "proton transfer" is indeed proton transfer. Incidentally, hydrogen-atom migration is frequently observed in reactions of electronically excited states.

7.3.4 *Reversal electron current against the proton motion*

As for the electron flux on the molecular plane (σ-flux) for the water-bridging system, it is clearly observed in Fig. 7.12 that the electron flux induced by the nuclear dynamics is flowing in the direction opposite to the proton motion as though they are compensating the electrons carried by the nucleus proton. In particular, panel (b) of Fig. 7.12 seems to have a collective electron flow passing through the mediating water molecule. This is the first observation of the electronic backward flux associated with proton transfer.

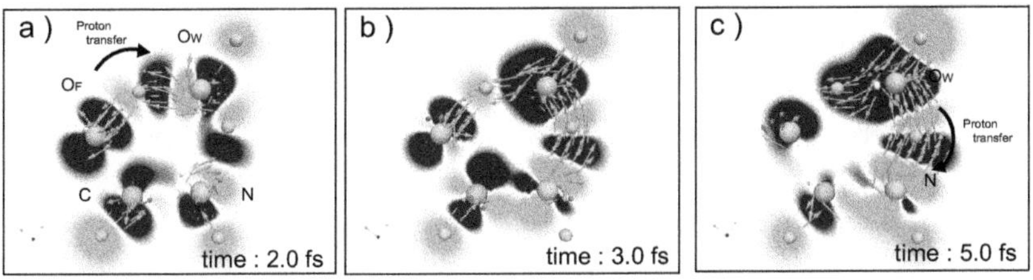

Fig. 7.12 Snapshots of electron flux on the molecular plane arising from σ orbitals in the course of water-assisted proton transfer dynamics. They are induced mainly in the area connecting the atoms O_F, H_{OO}, O_W, H_{ON}, and N. Time at which snapshot are taken is indicated in the individual panels. Note that the flow direction is opposite to that of the motion of the protons, and a collective flow covering the entire area of proton transfer is observed. The time dependent fluctuation of the σ-electron density, estimated with $\frac{\partial}{\partial t}\rho(\mathbf{r}, t; \mathbf{R})$ in Eq. (7.31), which indicates the increasing (in black region) and decreasing (in light gray region) of σ-electron density. (Reprinted with permission from K. Nagashima *et al.*, J. Phys. Chem. A 113, 15240 (2009)).

7.3.5 *Rearrangement of π-bonds: Dynamical manifestation of the Pauling resonance structures*

An electronic state of the zwitter ionic form left behind right after the proton transfer, which should be represented in terms of the left-hand-side structure of Fig. 7.13 in the Pauling resonance (valence bond) theory. On the other hand, the corresponding hydrogen-atom migration should leave a biradical state $\dot{O}$-$\dot{C}$H-NH_2 after the migration and then turns to the keto form. (In fact, however, no such significant unpaired density expected from the above biradical structure has been observed at any instance.) This keto structure O=CH-NH_2 is more stable than the zwitter ionic one [259]. Nevertheless, proton transfer rather than hydrogen-atom migration actually took place as shown above. This fact suggests that the proton transfer is a process of kinetic control, and that the bond-breaking to create hydrogen-atom-like radical state needs a higher energy to surmount.

In the Pauling valence bond theory, however, these two extreme structures are expected to resonate with each other as shown in Fig. 7.13. The resonance theory is such a useful theory in organic chemistry as it can give the essential feature of the electronic state without quantum mechanical calculation. However, the information provided is essentially static. We therefore investigate the dynamical implication of the resonance structures. First we consider the bond order density $B_{AB}(\mathbf{r},t)$ defined in Eq. (7.23), which is merely a spatial distribution of the bond-order, representing the quality and strength of a chemical bond between atoms A and B. Furthermore, we here concentrate on the π-bond order to track the dynamics of double bonds lying in the region of the skeletal atoms O, C, and N. In Fig. 7.14, panels (A)-(C) gives a time series of the propagation of the π-bond order; starting from the purely double bond in $\mathrm{C}-\mathrm{N}$ site right after the proton transfer. Due to the resonance, the double bond penetrates into $\mathrm{O}-\mathrm{C}$ site, and then fluctuate of the double bonds continues in between these two sites (see panels (B) and (C)).

More dramatic time-propagation of the π-electron dynamics can be extracted using the flux as shown in Fig. 7.15. The left column panels provide the flux in a side view (seen in a direction parallel to the molecular plane $\mathrm{O}-\mathrm{C}-\mathrm{N}$), while the right column gives a top view (seen in the direction perpendicular to the molecular plane) of the flux of the same timing. At 2.5 fs (panel (a)), the electrons are flowing into the vicinity of C atom from both ends. But, at 5.0 fs (panel (b)), the uniform electron current takes place from left (O side) to right (N side), and at 7.5 fs (panel (c)) the reverse flow is observed. Further, at 10 fs (panel (d)), the electrons flow out from the region of C atom, which is reversal to the flux at 2.5 fs. In this manner, the electron flux continues to fluctuate. Thus, electron flow behind the resonance theory has been exposed by means of the electron wavepacket dynamics coupled with nuclear motion.

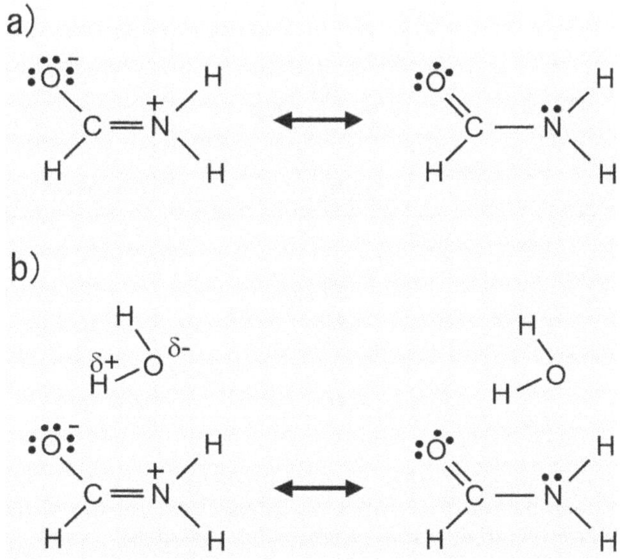

Fig. 7.13 Dominant resonance forms after the proton transfer. (Reprinted with permission from K. Takatsuka *et al.*, Phys. Chem. Chem. Phys. 13, 4987 (2011)).

7.3.6 *Stabilization of the zwitter ionic resonance structure by the mediating water*

Returning to the resonance structures in Fig. 7.13, we immediately notice that the zwitter ionic structure may be stabilized far better than the keto structure by polar solvents. In our water-bridging system, a water molecule is already there to mediate proton transfer. After the proton transfer is completed, this water molecule can now serve as a stabilizing polar solvent as schematically drawn in Fig. 7.16. Such an effect should be much weaker for the keto form. Without the water molecule, the zwitter ionic structure is energetically higher than the keto form, but due to the stabilization, their difference turns out to be smaller. This is why the double bonds remain in the both sites of O – C and C – N, and the π-electron flux continues to fluctuate among these atoms.

To see how crucial the stabilization effect is, we compare it with the forced proton transfer in gas phase. Figure 7.14, panel (a)-(c) display the time propagation of the π-bond order. As in the water-assisted proton transfer, the π-bond is first prepared in C – N site upon proton transfer. It begins to penetrate into the site of O – C and after some fluctuation time, say 30

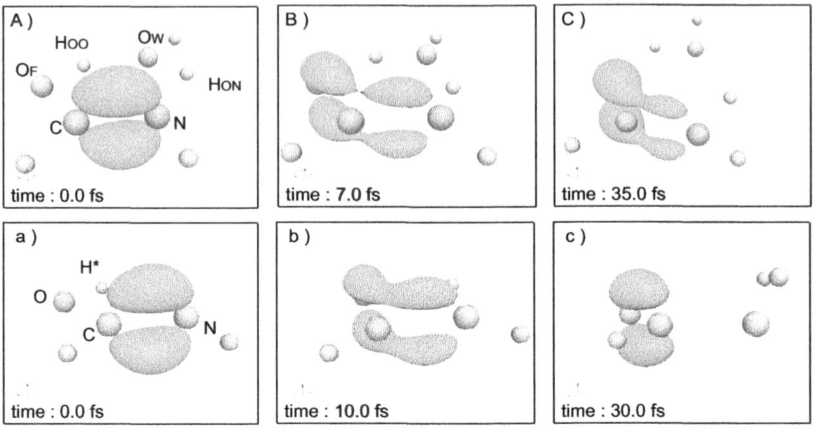

Fig. 7.14 Time propagation of the π bond order densities in O – C bond and C – N bond at each time indicated in the panels. (Upper row, panels (A)-(C)); the water-assisted proton transfer. (Lower row, panels (a)-(c)); forced proton transfer in gas phase. At time 30 fs in case of the gas phase reaction, the character of π bond between C and N atoms has virtually disappeared and the pyramidal structure as in the ammonium molecule has shown up around the N atom. The iso-value surfaces on which the bond order density takes 0.01 are drawn. (Reprinted with permission from K. Nagashima *et al.*, J. Phys. Chem. A 113, 15240 (2009)).

fs, it is localized almost completely to O – C as confirmed in panel (c). Furthermore, this panel shows explicitly that the bonding state around the nitrogen atom turns out to be triangular-pyramidal ammonium molecule. Therefore, it is virtually not reversible for the double bond at O – C site to be brought back to C – N site. The electron flux of the gas phase system also supports the one-way propagation of the state. Only the flux directed from O atom to N atom through C is observed (up to 11 fs) in clear contrast to the fluctuating behavior of Fig. 7.15.

An immediate suggestion from the above study is that the rotation along the axis of C – N in polypeptide becomes far more difficult by the presence of water molecules for two reasons: They may stabilize the zwitter ionic structure as a solvent effect. Besides, water molecules may cause proton relay transfer in the reverse manner of the present state, that is, from keto form (peptide) to enol form, which results in creation of a strong double bond in C – N. Thus the presence of water molecules around polypeptide can make the dramatic difference from gas phase reaction. Although the extent of possibility of free rotation around the C – N bond in polypeptide is well-known as one of the classic issues in protein science [244], one should

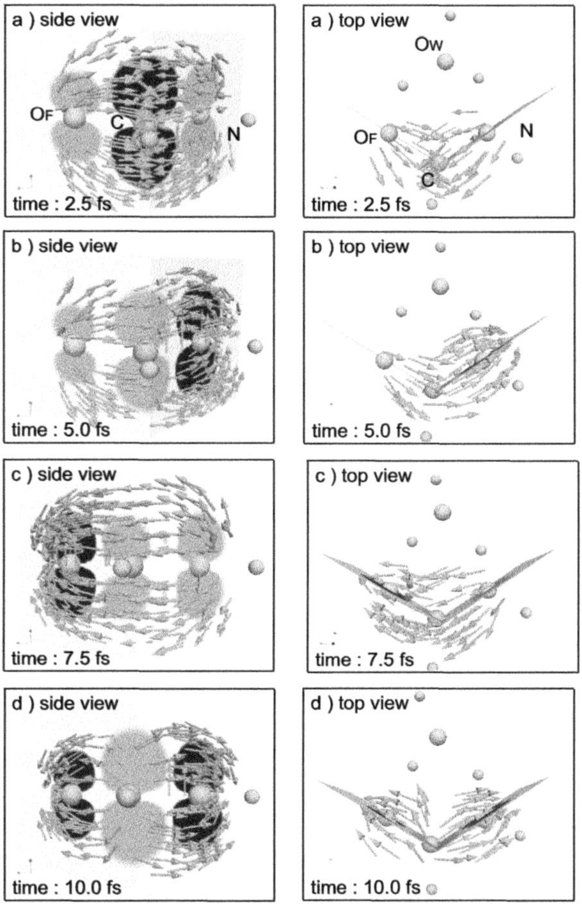

Fig. 7.15 Electron flux arising from π orbitals in water assisted proton transfer dynamics at each time indicated in the panels. Left column; a side view parallel to the $O - C - N$ plane. Right column; the top view. In the left column, the time dependent fluctuation of the π-electron density, which indicates the increasing (in black region) and decreasing (in light gray region) of π-electron density. In this reaction, a frequent change of the direction of the flux takes place with a period of about several hundred attoseconds. (Reprinted with permission from K. Nagashima *et al.*, J. Phys. Chem. A 113, 15240 (2009)).

consider once again these chemical facts in the numerical study of protein folding dynamics.

We track the nonadiabatic electron dynamics from HO-CH=NH to O=CH-NH_2. Main findings through study are as follows. (i) This dynamics

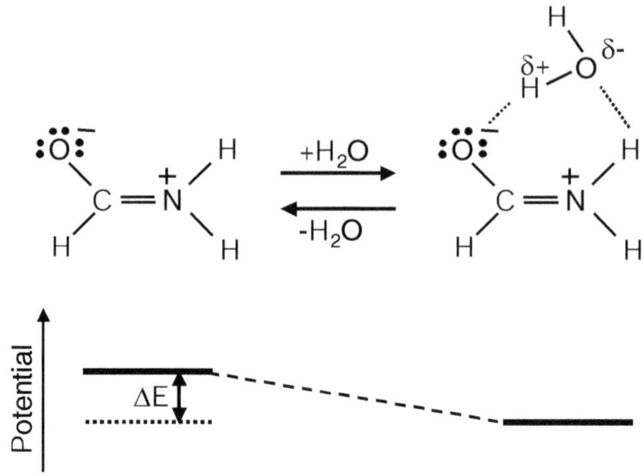

Fig. 7.16 Schematic representation of stabilization of the resonance structure of the zwitter ionic form by the bridging water molecule. (Reprinted with permission from K. Nagashima *et al.*, J. Phys. Chem. A 113, 15240 (2009)).

has been identified to be indeed a "proton transfer process" rather than hydrogen-atom migration in terms of the unpaired electron density. (ii) We have shown a collective quantum flux of electrons, which flows backward against the proton motion, thereby compensating the electrons tightly covering the proton as represented in terms of the Mulliken charge. (iii) We have revealed the dynamical manifestation of the resonance structures O^--$CH{=}N^+H_2 \longleftrightarrow O{=}CH\text{-}NH_2$ by means of our dynamical electron theory. The dynamics and relaxation process behind the static view of the resonance theory has been uncovered. (iv) The keto form formamide is one of the simplest species in the group $O{=}CR_1\text{-}NHR_2$, which is a unit of polypeptide. In gas phase, the nitrogen atom may have a pyramidal structure as in ammonium molecule, and therefore the C-N bond may allow low barrier rotation along it. On the other hand, this rotation is strongly prohibited by the formation of the double bond C=N induced by the proton transfer, which is enhanced by the presence of bridging water molecule.

7.4 Double proton transfer in formic acid dimer

There are many molecules in nature that have multiple hydrogen bonding like DNA in a double helix structure. In those systems, multiple proton transfer may take place. In particular, double proton transfer in formic acid dimer (FAD, Fig. 7.17) has been studied extensively as a typical example of multiple proton transfer. In the early studies on this dimer, optimized geometry, transition states, barrier heights were main objects in quantum chemistry [75, 171, 407]. The tunneling splitting has also been calculated with experimental [51] and theoretical approaches [369]. Ushiyama and Takatsuka have studied the timing of double proton transfer and associated electronic structures with use of the *ab initio* molecular dynamics simulation [449], finding that double proton transfer in FAD takes place rather successively (see Ref. [2] for a latest discussion on this subject).

In this subsection, we study the dynamical electron mechanism of the elementary process (in gas phase) of double proton transfer in FAD with an electron wavepacket method. To begin with, we ask ourselves a question; what is the basic mechanism of this double proton transfer (Fig. 7.18(a)), hydrogen-atom migration (Fig. 7.18(b)), hydride transfer (Fig. 7.18(c)), or else? (We term the event of Fig. 7.18(a) "proton" transfer tentatively until we can determine the mechanism.) At a glance none of the above three mechanisms seems totally satisfactory. What is wrong with these classic mechanisms?

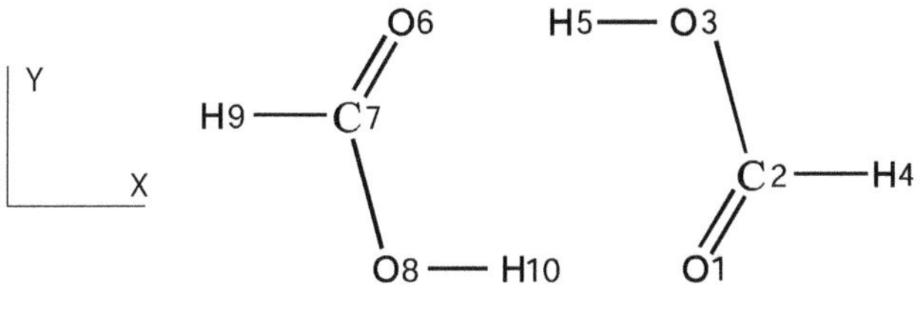

Fig. 7.17 Schematic structure of formic acid dimer placed on X-Y plane. (Reprinted with permission from M. Okuyama *et al.*, Bull. Chem. Soc. Jpn. 85, 217 (2012)).

7.4.1 *Electronic configurations and initial conditions*

In the dynamical calculations with SET, which has been implemented into GAMESS package [357], we use the electronic configurations of single and double excitations (CISD) to construct configuration state function (CSF), for which the method by graphical unitary group approach (GUGA) has been applied [310–312, 366, 367]. The basis set for molecular orbitals we used are basically STO-6G. In CISD calculation with STO-6G, the number of the canonical molecular orbital (CMO) is 34 with 24 and 10 occupied

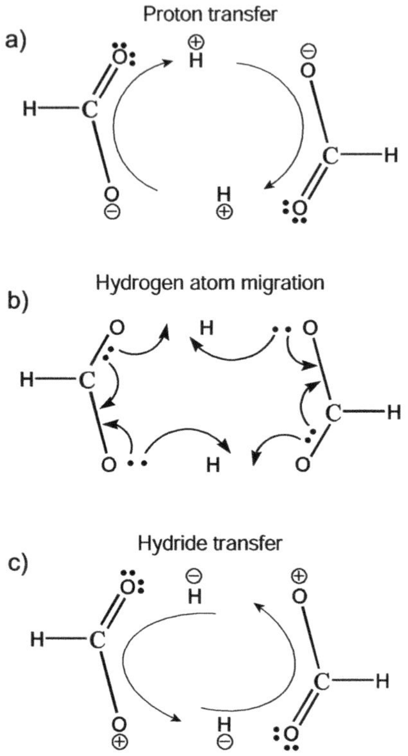

Fig. 7.18 Schematic picture for (a) double proton transfer, (b) hydrogen-atom migrations, and (c) double hydrid transfer. The arrows schematically suggest the directions of electron flow. (Reprinted with permission from M. Okuyama *et al.*, Bull. Chem. Soc. Jpn. 85, 217 (2012)).

and unoccupied orbitals, respectively. We selected 20 frozen core orbitals out of 34 CMOs, which implies only four MO's, HOMO, HOMO-1, HOMO-2, HOMO-3, are treated as active orbitals. This results in 864 CSF's for the time-dependent configuration mixing in SET. This choice of active space is obviously too small for the computation of molecular properties such as energy as in electronic correlation. However, it is good enough for the present system to track the electronic motion, since, as we see later, the correlation effect is actually small in the present system. Indeed, we have tested other choices of the set of core orbitals and found that the overall results did not differ much from each other, particularly in qualitative conclusions. Besides, we concentrate only on the dynamical change of the electronic (wavepacket) state in such a way that the discussions are made mostly on the basis of density matrix.

The sampling of the initial conditions and the computational procedures to solve numerically the equations of motion, Eqs. (6.14) and (6.24) for SET are as follows. First of all, we note that the purpose of the sampling is different from that of the usual chemical dynamics. In the latter, we need to choose sampling points to simulate the physical conditions such as collision in gas phase of canonical ensemble to compute the reaction probability or rate. On the other hand, the present study on the mechanism demands collection of an ensemble of SET trajectories that successfully undergo double proton transfer in order to analyze the qualitative pathways of electron flux in this event. To find such paths, we first prepare the electronic ground state at its optimized geometry of FAD using CISD. At this optimized geometry, the kinetic energy equivalent to the zero point energy of each normal mode is given, which totally amounts to about 50.05 kcal/mol. Then an additional energy about 10 kcal/mol is given to a randomly selected mode. The directions of all these momenta are sampled in a random manner. With these energies distributed, the dimer begins to break into two FAMs (formic acid monomers). After some time (about 180 fs), we terminate these separation dynamics, and invert all the momenta at the same nuclear configuration for them to approach each other. Thus the entire dynamics begins with collision (dimerization) and separation after possible proton transfer. Obviously, the SET paths thus sampled pass through the energetically-optimized symmetric geometry of FAD, which should favor simultaneous proton transfer rather than the successive transfer. 16 paths in total have been launched to evolve the electron wavepacket in SET calculations, and 11 paths out of these 16 successfully undergo double proton transfer. Although the detailed dynamics along these 11 individual paths are all different from each other, the electron dynamical mechanism of double proton transfer has been found to be essentially common among them. It is this qualitative mechanism that we want to analyze, rather than the accurate numerical quantities like the reaction rate and probability. The integrators used to solve Eq. (6.24) is the 5th order Gear method with the time-step of 30 attoseconds (as), which is accurate enough to conserve the total energy.

Among such 16 SET paths, we here pick up a single one that seems to serve as a typical example of a series of dynamics of dimerization and double proton transfer as in Fig. 7.19. To track the position of protons, H5 and H10 in Fig. 7.17, the relative coordinate, X_1 and X_2 are defined and used as follows [449].

$$X_1 = \frac{R_{\mathrm{O6H5}}}{R_{\mathrm{O3O6}}} \cos\left(\theta_{\mathrm{O3O6H5}}\right), \tag{7.36}$$

and

$$X_2 = \frac{R_{\mathrm{O8H10}}}{R_{\mathrm{O1O8}}} \cos\left(\theta_{\mathrm{O1O8H10}}\right), \tag{7.37}$$

where $R_{\mathrm{A_I B_J}}$ is the internuclear distance between the I-th atom A and J-th atom B, whereas $\theta_{\mathrm{A_I B_J C_K}}$ is an angle between lines $\mathrm{A_I B_J}$ and $\mathrm{B_J C_K}$. $X_1 > 0.5$ and $X_2 < 0.5$ imply that H5 and H10 stay at the O3 and O8 sides, respectively. Hence $X_1 = 0.5$ is regarded as a criterion to judge whether H5 exceeds the bisector of O3-O6. The path exemplified in Fig. 7.19 is actually composed of four stages as labeled on top of the upper panel: (A) two FAM are approaching to form a FAD (the time interval $0 < t <\sim 120$), (B) In their strong interaction region, the protons are moving back and forth, and this process ends up with a double proton transfer in $\sim 130 < t <\sim 170$. (C) The temporarily formed FAD spends some long time around the stable structure of FAD during $\sim 180 < t <\sim 250$. (D) Then the final double proton transfer takes place and two FAMs recede from each other in $\sim 260 < t <\sim 360$. In what follows we first study the dimerization process of (A), and then proton transfer process from (C) to (D).

Before proceeding, we make the following convention in showing the numerical results: As shown at the top of Fig. 7.19, the approaching FAMs to dimerize are placed in such a way that the upper proton belongs to the left monomer (in (A)). On the other hand, the proton transfer after dimerization in the process (C) and (D), the upper proton belongs to the right FAM (in (C)) and shifts to the left (in (D)). Since such difference in the nuclear configuration can make confusion in our visual comprehension, we make a convention, making use of the mirror image of the system (D_{2h} symmetry in the optimized geometry of FAD), that we view the FAM in such a way that the upper proton belongs to the right monomer in the beginning of each process of dimerization (A) and proton transfer (B→C) also, as in Fig. 7.17.

7.4.1.1 *Note on the basis set chosen*

Obviously, the STO-6G basis set is small. In addition, we have examined other small but a little larger basis sets like 3-21G at fewer sampling points. It turns out only numerically that the dynamics represented by the STO-6G tends to be well separated into three subprocesses; dimerization, proton transfer, and mutual separation. But for the 3-21G set, proton-transfer often takes place nearly simultaneously with dimerization. This difference arises because the energy of the potential basin of FAD is estimated deeper by STO-6G than by the 3-21G set, and therefore the lifetime of FAD in STO-6G is longer. Thus it turns out that the STO-6G dynamics is convenient for the sake of more precise analysis of electron dynamics, since the dimerization and proton-transfer tend to be better separated. Therefore we track the STO-6G dynamics in what follows, realizing that our studies in this stage cannot serve as an accurate and quantitative simulation of the energy, reaction rate, and so on.

7.4.2 *Electron dynamics in dimerization process*

7.4.2.1 *Electronic reorganization from monomers to dimer*

We first survey the relaxation of electronic structure in the dimerization process of two FAMs to the stable FAD. To do so, we study the difference density in the spatial distribution of one electron density [483] between FAD and two FAMs: Suppose that there is a FAD at a given geometry. Then we assume the presence of two FAMs at this geometry without any mutual interaction. Then $\Delta\rho(\mathbf{r})$ between FAD and FAM is simply defined as follows

$$\Delta\rho(\mathbf{r}) = \rho_{\mathrm{FAD}}(\mathbf{r}) - \rho_{\mathrm{FAMs}}(\mathbf{r}), \tag{7.38}$$

where $\mathbf{r}$ is the Cartesian coordinate of one electron, and $\rho_{\mathrm{FAD}}(\mathbf{r})$ and $\rho_{\mathrm{FAMs}}(\mathbf{r})$ are the diagonal spin-free first order density matrix for FAD and for the two noninteracting FAMs, respectively, at this geometry.

Figure 7.20 shows the difference density, where two FAMs are specified to be the left fragment consisting of O6C7O8H9H10 and the right fragment of O1C2O3H4H5 (the upper proton H5 is transferring from right to left, and H10 from left to right.) These FAM are regarded as two noninteracting FAMs that have been brought to the given positions and geometries. The spatial distribution of the difference densities at the stable structure, panel (a1)–(a2), and those for the transition state, panel (b1)–(b2) are shown. Panels (a1) and (b1) display the contour plot sliced at $Z = -1.0$ Bohr from the molecular plane, which represent the difference density mainly of π-bonding. On the other hand, panels (a2) and (b2) shows those at $Z = 0.0$ Bohr, which detect the difference density in the σ-bonding. The value of d in each panel denotes the contour width. Red and blue closed curves in all panels indicate positive and negative values, respectively. The difference

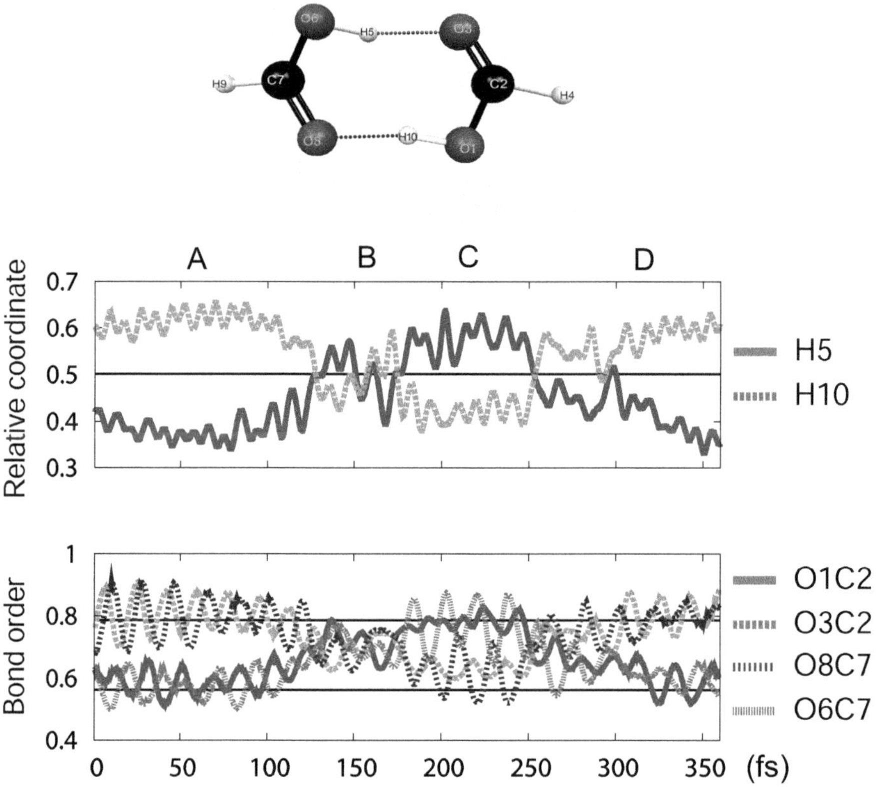

Fig. 7.19 Time series of an illustrative example of dimerization and proton transfer with an initial geometry on top of the figure. The middle panel indicates the relative position of the transferring protons (H5 and H10). The bottom panel represents the bond-order of selected bonds, synchronized with the position of the protons. This entire dynamics is separated into four stages: (A) Approaching of FAMs to dimerize to FAD, $0 < t <\sim 120$. (B) Protons are shifting forward and backward, $\sim 130 < t <\sim 170$. (C) Sitting around the stable dimer, $\sim 180 < t <\sim 250$. (D) Receding to FAMs after proton tranfer, $\sim 260 < t <\sim 360$. (Reprinted with permission from M. Okuyama *et al.*, Bull. Chem. Soc. Jpn. 85, 217 (2012)).

density thus obtained have provided the geometrically clear information of the rearrangement of the electronic state in an invariant manner with respect to the choice of molecular orbitals and the choice of electronic configurations. (Incidentally, CI calculation with single and double excitations from 12 active orbitals based on 3-21G basis set gives the same topology (not shown) as that of Fig. 7.20.)

Below we describe the electronic reorganization solely in terms of the chemical terminology like sp^3 hybridization, π-bonding, and so on without resort to the detailed numerical values of the density. (Note that once we refer to the magnitude of the coefficients of the atomic orbitals, the discussion may depend on the basis function chosen. By hybridization like sp^3 and sp^2 used below we simply mean the directionality or symmetry of the spatial distribution of an one-electron state as represented in the first-order density matrix.)

Several characteristic features are noticed in Fig. 7.20; (i) [panel (a1)] The electron density in the direction out of the plane on O1 (O6) increases despite the large shift of the π-bonding character from O1=C2 (O6=C7) to the C2-O3 (C7-O8). Likewise, the density out of the plane on C2 (C7) and O3 (O8) decrease. (ii) [panel a(2)] On the σ-plane, the density on O3 (O8) pointing to H5 (H10) decreases in spite of the fact that the bond of O3-H5 (O8-H10) should be weaker in FAD than in FAM. Conversely, σ-density on O1 (O6) decreases. (iii) The density on the transferring H5 (H10) decreases only slightly. (The difference electron density on H5 is always negative because H5 has been consistently treated as though it belongs to the site of the right FAM.) By comparing all the panels, no big difference is observed between the difference density in the stable state of FAD and that of the transition-state of "proton transfer".

The rather counter-intuitive results of the above (i) and (ii) findings may be summarized and partly interpreted as follows. There was a clear double bond (consisting of sp^2 and π orbitals) in between O1 and C2 (O6 and C7) in the isolated FAM, and this hybridization made two lone pairs (sp^2 orbitals) on the molecular plane. On the other hand, O3 (O8) was surrounded by three sp^3 orbitals, one being used to tie H5 (H10), and the other two pointing out of the molecular plane. By forming the stable FAD, sp^3 character penetrates in O1 (O6), while the components of sp^2 and π orbitals decrease. Thus in forming FAD the electronic structure has been significantly reorganized so that the release of H5 (H10) from O3 (O8) making a lone pair on O3 (O8) and the accepting of H10 (H5) by O1 (O6) making a new bond have become easy and smooth. Besides, it may be concluded from the numerical result that the out-of-the-molecular-plane component of the electron density from two sp^3 should be larger that that of a single π orbital. This explains why the density of O1 (O6) in panel (a1) of Fig. 7.20 increases. Likewise, on-the-molecular-plane component of the three sp^2 orbitals should be larger than that of the sum of three sp^3 orbitals, which explains the panel (a2).

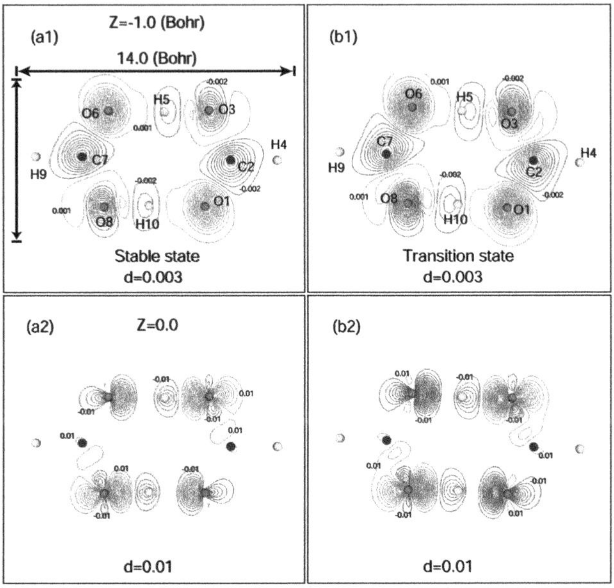

Fig. 7.20 Spatial distribution of the difference density based on one electron density of FAD and FAMs at the optimized geometry (a1)-(a2) and in the transition states (b1)-(b2). Panels (a1) and (b1) are the plot on X-Y plane at $Z = -1.0$ (Bohr) mostly represnting the π-electron density, while (a2) and (b2) are taken on X-Y plane at $Z = 0.0$ (Bohr), which is the σ-plane. The light and dark curves represent positve and negative values, respectively, and d is the width of the contour plot. The vertical line segment in panel (a1) indicates 10 (Bohr). (Reprinted with permission from M. Okuyama *et al.*, Bull. Chem. Soc. Jpn. 85, 217 (2012)).

7.4.3 *Net electron flow across a geometric cross-section*

We next accumulate the above probability current across a cross-section. Let $j_x(x_0, y, z, t)$ be a flux at time t passing across an x–y plane that is set at a point $x = x_0$ in the Cartesian coordinates. Define the following function

$$S_x(x_0, t) = \int_S dydzc_x(x_0, y, z, t). \tag{7.39}$$

This integral means the total flux that passes through the surface of section, S at $x = x_0$ at time t. Likewise, one can define a similar quantity like $j_y(x, y_0, z, t)$ and $S_y(y_0, t)$. The numerical integral of Eq. (7.39) is approximated by the sum

$$\int_a^b dx f(x) \cong \sum_{i=1}^{n} \Delta x_i f(x_i), \tag{7.40}$$

where x_i is coordinate, $f(x_i)$ is integrand at x_i and $\Delta x_i = x_{i+1} - x_i$. In the present calculations of Eq. (7.39), we select Δx_i, Δy_i and Δz_i of Eq. (7.40) with 0.0667 Bohr and select the integral range in each coordinate as -4.0 to 4.0 Bohr.

We further define a net flow accumulated during a given time-interval in such a way that

$$Q_x(y_0, t) = \int_{t_0}^{t} d\tau S_x(y_0, \tau), \tag{7.41}$$

where t_0 means an initial time and τ is time variable later than t_0. We call a point such as x_0 and $S_x(x_0, t)$ the section point and surface of section, respectively. In what follows, we set the section points at the midpoint on a segment connecting two neighboring atoms like C7 and O6. However, only in between O3 and O6 (and O1 and O8 as well) a section point is placed at the midpoint connecting these two points. This is to see how much electron flows between the two molecular sites irrespective of the position of the transferring protons.

In Fig. 7.21 panels (a)–(c) are plotted the resultant values of Eq. (7.41), which are the total amount of current that is accumulated up to time t (net electron flow). These graphs actually show the net flow of electrons in the process of dimerization during $0 < t < 90$ fs (no proton transfer takes place in this interval). We plot this quantity because the instantaneous flux at each position, $\vec{j}(x, y, z, t)$, fluctuates from time to time and place to place in general, reflecting the very fast quantum dynamics of electrons sensitively responding to the shift of nuclear positions. A derivative of each graph with respect to time gives the instantaneous flow direction at each surface of cross-section, and as seen in the graphs, the electron flow is never steady and seems unpredictable. For instance, the graph of $Q_y(t)$ for C2→O3 in panel (a) of Fig. 7.21 indicates that it is globally negative, except in the early stage up to about 30 fs. Therefore, electron flowed from O3 to C2 in the y-direction. Likewise, we see an electron flow from C7 to O6 also in the y-direction. Panel (b) indicates a slight flow from O8 to C7 and another flow from C2 to O1. In panel (c) are observed a slight flow from O3 to O6 and from O8 to O1. These are in the adverse direction of the motion of transferring protons, H5 and H10, respectively.

Thus, we have observed a clear and qualitative (topological) directionality of the net electron flow, which is schematically summarized in red arrows in the bottom panel of Fig. 7.21. Interestingly, electrons flow from the single-bond site to the double-bond counterpart in each monomer (O8→O6 and O3→O1) as two FAM are coming closer to each other. This seems rather counter intuitive *at first sight.* On the other hand, only slight electron flow is observed in the proton-bridging site in the same direction as the motion of double proton transfer.

Note that in the present calculations we do not make a rigorous distinction between the flows of σ and π electrons because of the presence of the out-of-plane vibrations. Therefore Fig. 7.21 is simply summarized that electrons flow to the carbonyl oxygen atoms (O6 and O1) from single bonded oxygen atoms (O8 and O3). This is consistent with the difference density shown in Fig. 7.20. We thus have a very simple yet systematic account for the electron flow: the carbonyl oxygen appears to require more electrons in the reorganization from sp^2 hybridization to sp^3.

7.4.4 *Electron dynamics in double proton transfer*

We next study the mechanism of "proton" transfer starting from the dimer molecular structure, which is the stage of C to D in Fig. 7.19.

7.4.4.1 *"Proton" transfer as a concerted reaction*

We first study the theoretical distinction between proton transfer and hydrogen atom migration from the view point of electronic structure. The idea is that if hydrogen-atom migration happens, a radical center should appear on the transferring proton nucleus, as though a hydrogen atom is isolated on the way of transfer. Since this implies that a strong chemical bond OH (say O3-H5) is broken before forming a new OH bond (say O6-H5), this process is likely to be unfavorable in energetics. Nevertheless, let us investigate this possibility.

In Fig. 7.22 (panel (a1) and (a2)), we show the dynamical change of unpaired electron populations on H5 and H10, estimated with Eq. (7.27). From this figure, the unpaired electron populations on H5 and H10 are as small as

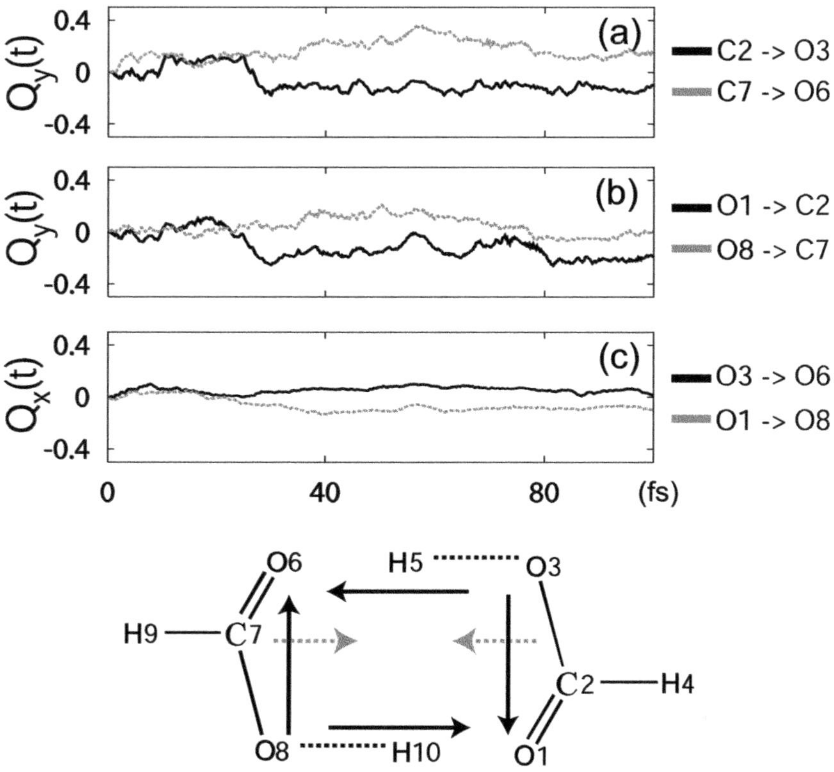

Fig. 7.21 Net electron flow within the molecular system in the process of **dimerization**. The upper three panels indicate the time dependence of net flow through selected chemical bonds. The bottom figure schematically indicates the directions of the over-all net flow. (Reprinted with permission from M. Okuyama *et al.*, Bull. Chem. Soc. Jpn. 85, 217 (2012)).

about 0.01. This fact means that the radical character of FAD is negligibly small in this reaction, which in turn indicates that much electronic-state mixing has not been induced either by the electronic Hamiltonian (or the configuration interaction) or by the nonadiabatic coupling. Therefore this reaction is essentially concerted, and virtually no radical center is formed on the hydrogen atoms. Thus it can be judged that any significant radical center does not appear on the transferring protons in the course of their transfer dynamics.

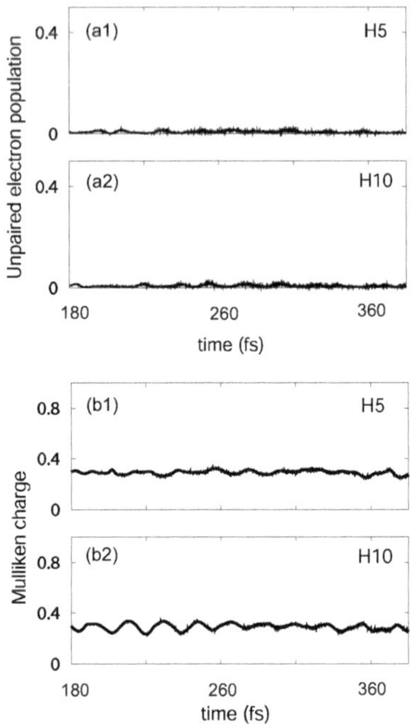

Fig. 7.22 Time series of the number of unpaired electrons on H5 (panel (a1)) and H10 (panel (a2)), and that of the Mulliken charge on H5 atom (panel (b1)) and H10 (panel (b2)). (Reprinted with permission from M. Okuyama *et al.*, Bull. Chem. Soc. Jpn. 85, 217 (2012)).

7.4.4.2 *Much electron density carried over by the proton nuclei*

We next track the time series of the Mulliken charge in Fig. 7.22 (panel (b1) and (b2)). A quite interesting fact is observed both in H5 and H10: They remain almost constant, as low as about 0.3 (not 1.0), throughout the proton transfer reaction. A straightforward interpretation of this observation should be that each proton nucleus is covered with electrons as much as about 0.7. Therefore proton in "proton" transfer is never a naked proton nucleus, but it is rather a positively charged hydrogen atom as far as only the electron population is concerned. Furthermore, the electronic population on the transferring protons is rather steady in the transition from one side to the other. This phenomenon was first reported by Ando and Hynes [11], and later by other authors in different systems [11, 286, 449]. The actual population of the covering electrons depends on a system under study and the basis set used. However, it is quite likely that the above view should be a phenomena quite common to the proton transfer dynamics, since detaching an electron from the hydrogen atom requires very high energy as much as about 13.6 eV.

7.4.4.3 *Electron current in the course of proton transfer*

We finally quantify the net electron flow within the molecular system during the double-proton transfer as in Fig. 7.23. Comparison between Fig. 7.21 and 7.23, panel by panel, shows that the direction of electron flows in each site is qualitatively the same between the dimerization process (Fig. 7.21) and proton transfer event Figure 7.23. The overall electron flow in the latter is summarized in the bottom panel of Fig. 7.23, which again highlights the same directionality.

As observed, the direction of electron flow from O3 to O6 (and from O8 to O1) is the same as that of the transferring proton H5 (and H10). This is qualitatively different from what we observed before in the proton transfer in Zundel cation [449] and water-molecule-mediated proton transfer in formamide [286], in which electrons flow in the reverse direction of proton motion. Therefore it appears that in the cyclic molecular field like FAD, much electron carried by proton nuclei does not have to be compensated by the backward intermolecular flow but mainly by reorganization within the individual monomers.

Thus we may conclude that the dimerization (Fig. 7.21) and concomitant proton transfer (Fig. 7.23) are smoothly continued in this regard. In other words, this smooth process is supported by the mechanism of concerted reaction, and there is not a conflicting factor in between the processes of dimerization and double proton transfer. As mentioned before, the STO-6G basis employed in this study tends to separate the dimerization and proton transfer as independent processes. However, these two processes can happen simultaneously in different basis sets. Even in such a case, the qualitative characteristics of the reaction mechanism must be essentially the same as we surveyed here, since the electron flow induced by these two processes are qualitatively equivalent to each other. (Of course, this does not justify the use of STO-6G basis for quantitative study to seek for accurate physical quantities.)

Finally, let us compare the quantity of the net electron flow in the two processes (Fig. 7.21 and 7.23). Since the present calculation is not good enough for the precise analysis of the electron flux, and since we have not taken a statistical average for the same reason, we just mention the balance of incoming and outgoing electron flows induced by the collision. Let us concentrate on the charge of O6 as an example. Figure 7.21(a) shows that only about 0.1 electrons have been carried from C7 to O6. On the other hand, far less electrons have flown in from the side of O3 (see panel (c) of Fig. 7.21). Therefore electrons as little as only about 0.1 have been pumped to O6 in the dimerization process. Likewise, in Fig. 7.23(a), we see about 0.2 electrons are accumulated onto O6 in the proton transfer process. Thus the intramolecular electron flow from O8 to O6 via C7 in the proton transfer is about two times larger than the counterpart in the dimerization. On the other hand, it turns out that the intermolecular electron flow (measured in between O3 and O6) is smaller in both processes than in the intramolecular flows. Thus the net electron-flow actually observed is very little, actually much less than anticipated in the proton transfer mechanism of Fig. 7.18. Besides, it is interesting to recall that the shifting proton, say H5, carries tightly electrons as much as about 0.7, indicating that the net charge on H5 is about +0.3. This is consistent with the total amount of electrons pumped onto O6 as described above. Thus they roughly compensate each other, and H5 should see the electronic state of O6 as though it is already that of O3 before collision of two FAMs.

7.4.5 *Summary of the mechanism*

Using the nonadiabatic electron wavepacket dynamics, we have identified the pathways of electrons and the dynamical electron mechanism of the "double proton transfer" in the colliding FAM in two stages; (i) electronic reorganization in dimerization and (ii) proton transfer and associated electron flow. The electron flow in both stages is induced in the same direction in the following manner: (1) Each of the shifting proton nuclei carries electron density as much as about 0.7 keeping the α- and β-spin pair (no singlet radical is generated). (2) However, only a very slight net electron flow takes place in between the two FAM's, which actually takes place in the same direction as the shifting proton nuclei. (3) The electronic-state reorganization in each FAM mainly on four oxygen atoms is associated with a weak electron flow in each of FAM from the site of singly-bonded oxygen to carbonyl oxygen. Besides, the total amount of electrons thus carried in each FAM is about 0.3, which compensates most of the positive charge ($\sim$0.3) carried by each shifting proton as above.

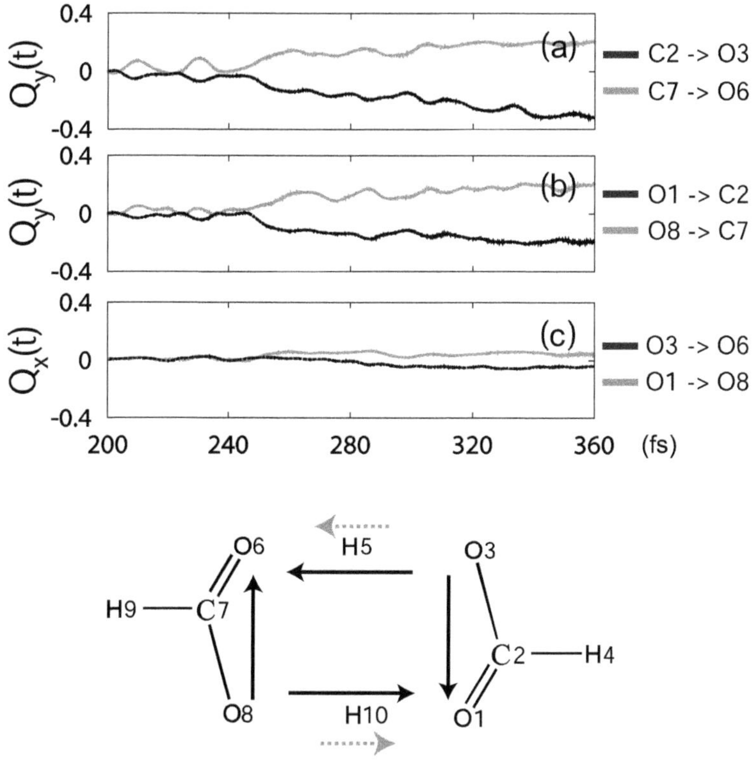

Fig. 7.23 Net electron flow within the molecular system in the process of **proton transfer**. The upper three panels indicate the time dependence of accumulated net flow through selected chemical bonds. The bottom figure schematically indicates the directions of the over-all net flow. (Reprinted with permission from M. Okuyama *et al.*, Bull. Chem. Soc. Jpn. 85, 217 (2012)).

In contrast to the view of panel (a) and (c) of Figure 7.18, a large negative or positive charge-center does not appear in the system, and it turns out that only a small amount of electron flow is induced by the electronic-state reorganization. In other words, the reaction seems to proceed as though it is guided by a "principle" that the net electronic flow is kept as low as possible so that it can nearly minimize charge separation within a molecular system.

None of the classic pictures of Fig. 7.18, which chemists commonly use, represents the physical reality in electron dynamics. However, if we regard a single dot in panel (b) of Fig. 7.17 as about a half of an α and β pair (about one electron in total), which is schematically defined as $\bullet = (\uparrow\downarrow) \times 0.5$, the reality is not far from that scheme. An advantage of this definition is that it qualitatively represents the bond-rearrangement with the minimum net flow of electrons within the concerted mechanism. On the other hand, this convention obviously fails to describe the true radical reactions and/or Woodward–Hoffmann forbidden reaction, in which radical centers indeed appear in the spin singlet state. Therefore, we need more conventional symbols to represent these states in a systematic manner.

7.5 Excited-state proton-electron simultaneous transfer

We here report a nonadiabatic electron wavepacket study on the excited state reaction in phenol that is hydrogen-bonded with ammonia clusters

$$\mathrm{Ph^*OH} \cdots (\mathrm{NH_3})_n \to \mathrm{PhO^\bullet} \cdots [\mathrm{H}(\mathrm{NH_3})_n]^{*\bullet} \qquad (7.42)$$

with $n \leq 5$. This reaction is now widely recognized as a typical reaction of excited-state hydrogen atom transfer. [71, 102, 228, 378–381, 443] It has been established that the initially promoted $\pi - \pi^*$ state conical-intersects with $\pi - \sigma^*$ state, which pumps the hydrogen atom from phenol to the ammonia cluster. An important and difficult aspect of this reaction is that the relevant low-lying electronic states are highly quasi-degenerate due to the Rydberg-like diffused states on the ammonia cluster. To study this system therefore we have carried out a large scale nonadiabatic electron wavepacket dynamics with many configuration state functions on the full-dimensional nuclear coordinates. Since the former studies were made within the static ab initio quantum chemistry working mostly on the lowest excited state alone (namely, nonadiabatically coupled two adiabatic states), the present work is complementary to them.

Phenol has been interesting many chemists as a model system to study how photo-excited bases in DNA can deactivate to the ground state without resulting in mutation. Not only in such biology-oriented studies, this molecule shows many other interesting features when put in surrounding molecules, clusters, and solvents. Small ammonia clusters are frequently used in place of water-molecule clusters, because ammonia is more proton-attractive than water, and very extensive studies have been performed for the type of reaction of Eq. (7.42). It has been found theoretically that the ground state of phenol with $(NH_3)_n$ of $n \geq 6$ is already zwitter ionic state, (phenol(-)·ammonia(+)), which is a proton transfer state from phenol to ammonia cluster. [265, 457] As for the excited state dynamics with $n \leq 5$ in Eq. (7.42), the series of pioneering works by Domcke and his coworkers are quite important. [71, 228, 378–381] In particular, they have ascribed the mechanism of the decay of the initial $\pi - \pi^*$ to excited state hydrogen atom transfer (ESHAT) through nonadiabatic transition between the first excited state ($\pi - \pi^*$) and $\pi - \sigma^*$ state mediated by the dissociation of OH bond. The H· thus produced should be rendered to the nearby ammonia cluster, thus explaining both the proton-nucleus shift and biradical formation. This mechanism is referred to as SDDJ model. [381] Before their papers were published, "proton transfer" was conceived to be the main mechanism of deactivation of phenol. In SDDJ model, charge separation is not expected since the shifting hydrogen atom is anticipated to carry both nuclear positive and electron negative charges at a time. Also, a radical (unpaired electron) should appear on H atom. Indeed, they have confirmed computationally that H· can be ejected from $\pi - \pi^*$ state of phenol molecule in the absence of the ammonia clusters. [102, 381] The $\pi - \sigma^*$ state further conically intersects with the ground state, thereby relaxing to the ground state (phenoxy radical and NH_4 radical cluster somewhere in the ammonia cluster) without radiation process.

The studies so far made for this ESHAT employed the ab initio configuration interaction (CI) and/or complete active space SCF (CAS-SCF) methods and examined only the two lowest coupled potential energy surfaces (and nuclear wavepacket dynamics simplified [228]). However, because of the fact that the relevant excited states are highly quasi-degenerate in reality, and in view of the complicated nature of electron dynamics associated with "proton" transfer as described above, it is worthwhile to re-examine the mechanism from the nonadiabatic electron wavepacket dynamics.

It turns out that the found mechanism is not as schematically simple as the SDDJ model, and it is more appropriate to be referred to as "coupled proton-electron transfer in excited state", rather than "hydrogen-atom transfer". This mechanism further requires the study of recombination dynamics between the separated charges in clusters and solvents.

7.5.1 *Electronic configurations and initial conditions*

To perform the dynamical calculations with SET, we use CSF of the electronic configurations of single and double excitations (CISD), for which the method by graphical unitary group approach (GUGA) has been applied. [310–312, 366, 367] Only the spin-singlet states are considered. Below we outline the practice of our computational scheme.

Basis set and electronic configurations: To generate molecular orbitals we have examined numerous basis sets, ranging from STO3G to the aug-cc-pDVZ, to find a compromise between tractability and accuracy. The severe constraint in choosing the basis comes from the fact that our electron wavepacket calculation is far more time-consuming than the stationary-state eigenvalue problem diagonalizing the electronic Hamiltonian. That is why we could use only single and double (CISD) excitation configurations. To determine the basis set and CSF, we carried out preliminary calculations for the energy profile along the O-H bond distance (denoted as R(OH)), as shown in panel (a) of Figs. 7.26 and 7.27. The potential barrier on the first excited state located near R(OH)= 1.4 Å comes from nonadiabatic interaction (conical intersection).

Finally selected as the practical basis set is the 6-31G, to which the Dunning-Hay type Rydberg basis (S and P) are added only on the N-atoms of ammonia molecules. [122] These diffused functions are crucial as we show later. No d-functions were used. On the other hand, the preliminary calculations have suggested that we do not have to add diffused functions onto the proton in this particular study [432]. The reasons are: (i) A diffused function added only on the shifting proton changes very slightly the energy profiles of all the excited states studied, presumably because we have already used the diffused functions on the N-atoms as above. (ii) Nevertheless, the diffused functions added to all the hydrogen atoms lower the energy-profile of the *second* excited state (denoted as S_2, the so called $\pi - \sigma^*$ state in the literature) alone to some extent but only *in the region before the nonadiabatic transition* (of short OH), while beyond this region the potential functions were not affected substantially. (iii) This is consistent with the reports by Domcke and his coworkers [380, 381] and by Daigoku et al. [102] that the diffused nature on the shifting proton in S_2 state converges quickly to the ordinary valence type 1s orbital, as the OH becomes longer coming closer to the crossing region. (iv) Besides, this S_2 state is optically dark from the ground state (S_0). Since we track the electron dynamics starting from the first excited (S_1, $\pi - \pi^*$) state, we thus do not anticipate to see the effect of the diffused function on the shifting proton after all. In addition, we have confirmed that the basis set thus selected are among those that give almost converging results with respect to addition of further basis functions. [The vertical excitation energy from S_0 to S_1 at the optimized geometry of S_0 along with the bond length of OH and NH are provided in Supporting Material just for a reference, although high accuracy required in static quantum chemistry calculations is not necessarily crucial in the present study.]

The single and double (CISD) excitation configurations have been generated using HOMO and HOMO-1 as the particle level (active space) and 40-72 virtual orbitals as hole levels. (HOMO-2, and HOMO-3 were also used in order to confirm that the results were virtually unaffected by further extension of the particle space.) The resultant numbers of CSFs in practice are about 10,000–30,000 for the wavepacket propagation. The 5th-order Gear method and Bulirsch-Stoer interpolation method have been applied to integrate the nuclear and electronic dynamical equations, respectively, with a time-step of 100 attoseconds (as). One entire step of propagation consists of about 4 electronic and one nuclear steps of propagation. The norm of the electronic wavefunctions is conserved within the accuracy of 10^{-8}.

Nuclear paths: The optimized structure for S_1 state (phenol plus ammonia clusters) is first prepared. Each path has been initiated by given a kinetic energy to the shifting proton nucleus of these structures, ranging from 0.5 to 1.8 eV, which is high enough to surmount the potential barrier (refer to the captions of Figs. 7.26 and 7.27). Therefore we did not study the mechanism of quantum mechanical tunneling. The zero point energy is not taken into account either in this particular case, since this study is devoted to understanding the qualitative mechanism but not to simulating the rate process quantitatively. The total energy given is scanned to survey the qualitative dependence of the event on the energy, but the results are essentially (qualitatively) common to each other. About 30 SET paths have been examined in the case of a phenol with three-membered ammonia cluster, while five paths are generated for five-membered one.

7.5.2 *Mechanism of transitions*

We now investigate the early-stage dynamics of what is so far called "hydrogen atom transfer". We first survey the static properties such as the potential energy surfaces and then proceed to the electron dynamics.

7.5.2.1 *Static properties*

Optimized geometry: We begin with the optimized geometry of the ground state of a phenol molecule with clusters of one, three and five ammonia molecules. These are close to the initial geometry of the SET dynamics in the $\pi-\pi^*$ state. There are several local minima for the five ammonia case, which are not shown graphically. In contrast to the system with ammonia molecules more than five [265, 457], the phenol is electronically neutral and no charge separation is observed. The proton nucleus, which is later to be shifted to the clusters, is located in the phenol area.

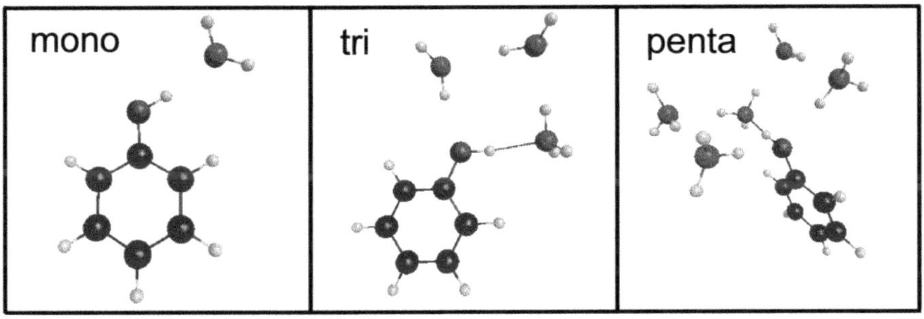

Fig. 7.24 Energetically optimized structures of phenol with one, three, and five ammonia molecules in the ground state. (Reprinted with permission from K. Nagashima *et al.*, J. Phys. Chem. A 116, 11167 (2012)).

To facilitate understanding the underlying mechanism, we show some relevant molecular orbitals in Fig. 7.25. The orbital that we call the valence LUMO is π^*, and the lowest energy LUMO, which we call the Rydberg LUMO, is literally a diffused Rydberg-like orbital mainly localized at the rightmost ammonia molecule, which is denoted as AM1 (see panel (b) in Fig. 7.26) Likewise, LUMO+1 (the next LUMO), LUMO+2, and LUMO+3 are all of the Rydberg character. In particular, LUMO+3 covers the entire region of three-ammonia cluster. It was actually hard to find a molecular orbital that clearly and solely represents the character of σ^* on the OH bond.

Potential energy surfaces and the characteristics of the low-lying states: A one dimensional cross-section of the global adiabatic potential energy hypersurfaces (PES) in the coordinate of the OH distance are shown in panel (a) of Figs. 7.26 and 7.27 under the presence of three and five ammonia molecules, respectively. In this diagram only,

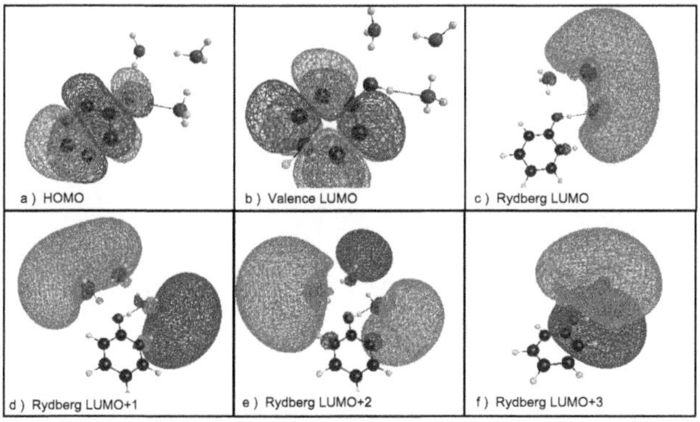

Fig. 7.25 Spatial distribution of some of the relevant molecular orbitals in phenol with three membered ammonia cluster. (Reprinted with permission from K. Nagashima *et al.*, J. Phys. Chem. A 116, 11167 (2012)).

the other degrees of freedom are all frozen at the geometry of the electronic ground state. These figures depict the first nine low-lying excited states. These states are highly quasi-degenerate, actually packed together in a narrow energy range. There are strong nonadiabatic couplings among those states, which will be surveyed with electron dynamics later in this section.

It is obvious that the lowest excited state in the initial geometry range (of the short OH distance) is $\pi - \pi^*$ state. On the other hand, the lowest excited state in the right-hand side potential basin (beyond the potential barrier and of a longer OH distance) is of the Rydberg-state nature. More precisely, the main electronic configuration of the four low-lying (adiabatic) excited states beyond the potential barrier are

The 1st excited state; $\quad (\text{HOMO})^1(\text{Rydberg LUMO})^1$

The 2nd excited state; $\quad \begin{cases} (\text{HOMO})^1(\text{RydbergLUMO}+2)^1 \\ +(\text{HOMO})^1(\text{RydbergLUMO}+1)^1 \end{cases}$

The 3rd excited state; $\quad \begin{cases} (\text{HOMO})^1(\text{RydbergLUMO}+1)^1 \\ +(\text{HOMO})^1(\text{RydbergLUMO}+2)^1 \end{cases}$

The 4th excited state; $\quad (\text{HOMO})^1(\text{Rydberg LUMO+3})^1$

Therefore, all the relevant excited states after the proton nucleus shifts to the cluster region are of the Rydberg nature. Again, we could not find in the precise track of these states a major contribution from configurations that mainly consist of OH σ^* level.

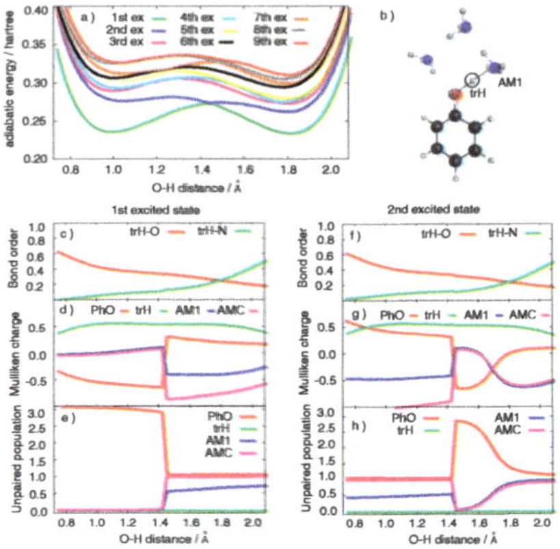

Fig. 7.26 Selected static properties of phenol·$(NH_3)_3$. (a) The potential energy surfaces for the nine low-lying excited states, one-dimenional projection in OH distance with the other degrees of freedom frozen at the ground state geometry shown in panel (b). The height of the potential barrier of the lowest excited state is about 0.0029 hartree (0.8 eV). (c) and (f); the bond order of OH and NH. (d) and (g); the Mulliken charge for the site of PhO, the transferring proton site (trH), the total ammonia cluster (AMC), and ammonia molecule AM1 that is hydrogen-bonded to phenol. (e) and (h); unpaired electron population at the same sites as in panel (d) and (g). Panels (c), (d), and (e) are for the first excited state, while (f), (g), and (h) exhibit for the second excited state. (Reprinted with permission from K. Nagashima *et al.*, J. Phys. Chem. A 116, 11167 (2012)).

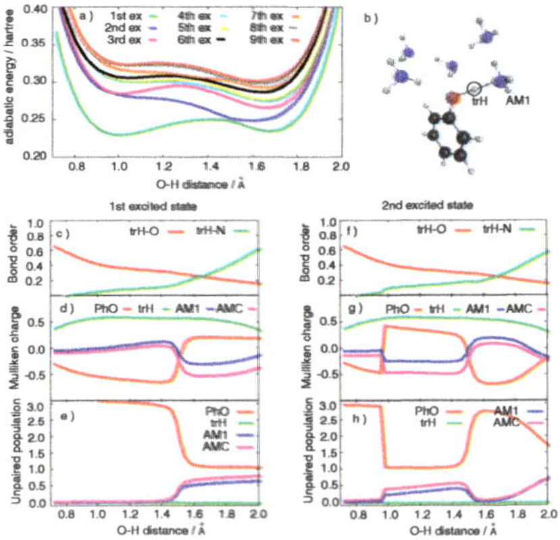

Fig. 7.27 The same quantities as those in Fig. 7.26, except that the ammonia cluster here is composed of five molecules. The height of the potential barrier of the lowest excited state is about 0.0021 hartree (0.56 eV). (Reprinted with permission from K. Nagashima *et al.*, J. Phys. Chem. A 116, 11167 (2012)).

7.5.3 Quantities related to electron density

To further characterize the electronic properties of the first and second excited states, let us examine the quantities related to electronic density. Panel (c), (d), and (e) of Fig. 7.26 exhibit the bond order at OH bond, the Mulliken charge of some selected parts, and the number of unpaired electrons of the first excited state of phenol and three membered ammonia cluster as a function of the bond length of OH. Likewise panel (f), (g), (h) show the similar quantities for the second lowest excited state, which strongly couples nonadiabatically with the first excited state. In panel (a) to (h) of Fig. 7.27 are displayed the similar graphs for phenol and five membered ammonia cluster.

Bond order: As clearly observed in panel (c) and (f) of Fig. 7.26, only smooth interchange between the bond orders of OH and NH takes place in positive values. This directly indicates that chemical bond is smoothly shifted from OH to NH site both in the first and second excited states. This kind of smooth interchange in positive value is often observed in the typical "proton transfer" in the ground states. Likewise, the similar interchange is observed in Fig. 7.27 for five membered ammonia cluster.

Mulliken charge: panel (d) and (g) of Fig. 7.26 show that the Mulliken charge on the shifting proton nucleus remains almost constant, as low as about 0.5 (not 1.0), throughout the proton transfer reaction. A naive interpretation of this observation should be that each proton nucleus is covered with electron density as much as about 0.5 in both the first and second excited states. (Recall that no diffused basis function is used on the proton.) Therefore the proton nucleus is not naked but is steadily dressed with a negative charge of this amount. This is again a typically observed phenomenon in the ground-state "proton" transfer. Indeed, the similar phenomenon has been observed quite ubiquitously in other proton transfer systems in the ground state [11, 286, 308, 449].

Panel (d) indicates that in the initial state ($\pi - \pi^*$ state) the charge is separated in such a way that about +0.5 on the shifting proton nucleus and about −0.5 in the site of C_6H_5-O (denoted as PhO in the figure). Beyond the crossing region, PhO becomes almost neutral and −0.5 charge goes to ammonia cluster (AMC), mainly on the right most ammonia molecule (denoted as AM1). Thus the charge separation remains such that roughly +0.5 on the shifting proton nucleus and −0.5 on the ammonia cluster.

Incidentally, the second excited state exhibits in panel (g) that right after the crossing point the charge separation gives rise to such that +0.5 on the shifting proton nucleus and −0.5 on the ammonia cluster, which is very similar to that of the first excited state right before the crossing point as described above. This suggests a smooth conversion of the first excited state to the second excited state at the crossing region: A clear manifestation of conical intersection.

Unpaired electron population: Panel (e) of Fig. 7.26 suggests that the initial $\pi - \pi^*$ state contains about three radicals. It is obvious that there should exist at least two radicals, and it is likely that the additional one radical center appears due to the strong electron correlation arising from the nearly degenerate orbitals in the benzene ring (near degeneracy in each manifold of π and π^* orbital). Beyond the barrier, one radical remains localized in PhO site and another is found in AMC (mainly on AM1), which is really the biradical state. [71]

Virtually no significant unpaired electron population is found on the shifting proton nucleus consistently, which strongly indicates that no radical center is ever created on the proton nucleus throughout the reaction.

7.5.3.1 *Critical role of the Rydberg states*

We next explore the electronic structure in the site of ammonia cluster. It is readily understood that such a high quasi-degeneracy of the potential energy surfaces in Figs. 7.26 and 7.27 comes from the Rydberg-state character of the ammonia molecules. [71, 228, 378–381] To see it more directly, we studied how the molecular orbitals of the composite system of phenol and AMC are made up from those of the individual subsystems. At the geometry where the proton nucleus is already shifted to the ammonia cluster site, actually attaching on AM1, Fig. 7.28 shows interaction between the MO's of PhO-H (the leftmost levels) and the three membered AMC (the rightmost one) to form those of the entire system. Those MO's of PhO-H and AMC are independently calculated without mutual counterparts. Note that LUMO of PhO-H and the valence LUMO of the entire system are both π^*orbital on the benzene ring, while the Rydberg LUMO is the energetically lowest unoccupied MO, which consists mostly of the Rydberg-type basis functions of AMC. As seen directly in the figure, the bunch of the low-lying unoccupied orbitals have originated from AMC.

A straightforward idea to emphasize the role of those Rydberg-type MO's is to construct the potential energy profile without them, which is

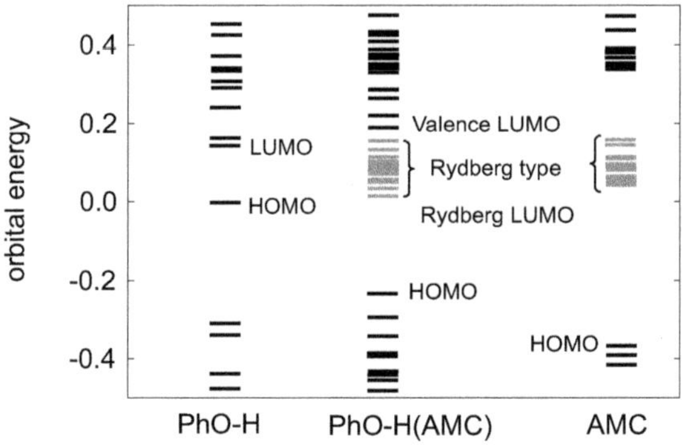

Fig. 7.28 (Center) The energy level of molecular orbitals formed in the interaction of phenol and three membered ammonia cluster at the geomtry after proton-nucleus is shifted to the ammonia site. (Left) The MO levels for PhO-H at the same geometry but without the ammonia cluster. (Right) Those of the ammonia cluster without phenol. (Reprinted with permission from K. Nagashima *et al.*, J. Phys. Chem. A 116, 11167 (2012)).

shown in Fig. 7.29. The effect is clearly identified in comparison of this figure with Fig. 7.26; the crossing due to conical intersection between the first and second excited states has disappeared. Furthermore, the level spacing is now much more sparse. There is of no doubt that the physics on PES without the Rydberg-type states should be far from reality.

7.5.4 *Electron dynamics*

Many things have been thus clarified in the above simple analysis of the one-dimensional projection of the potential energy surfaces. We then proceed to the study of nonadiabatic electron dynamics based on the full-dimensional nuclear paths that are nonadiabatically coupled with electrons. Although we have studied the cases of three- and five-membered ammonia clusters, we mainly report the tri-ammonia case in what follows.

Some time after excitation to $\pi - \pi^*$ state, the excess energy due to photon absorption along with that coming from molecular deformation will eventually transformed to the kinetic energy of the shifting proton nucleus, which will push it to the ammonia molecule (AM1). Instead of waiting for the natural energy transfer process, we push the proton nucleus by giving kinetic energy, which is high enough to surmount the potential barrier of the first excited state. The initial conditions to this end have already been described in Sec. 7.5.1.

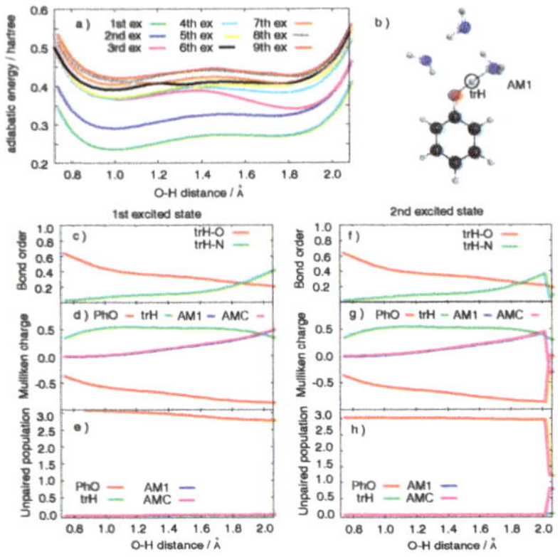

Fig. 7.29 The same quantities as in Fig. 7.26, except that the Rydberg-like diffused basis functions on the nitrogen atoms are deliberately excluded in the calculation of the electronic states. (Reprinted with permission from K. Nagashima *et al.*, J. Phys. Chem. A 116, 11167 (2012)).

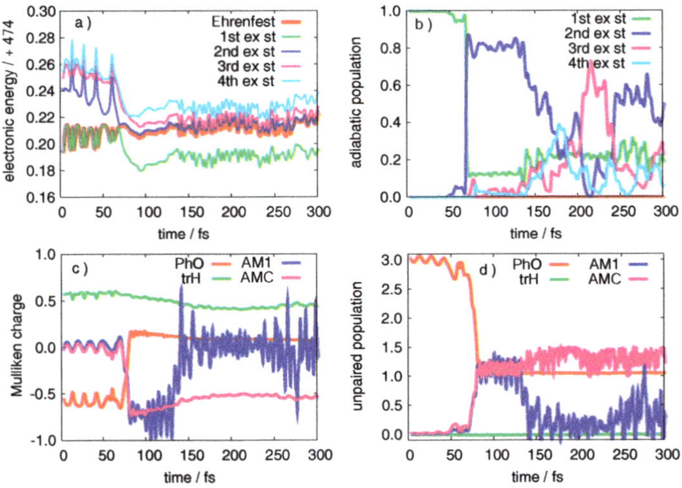

Fig. 7.30 Time series data for selected quantities in the semiclassical Ehrenfest electron wavepacket dynamics of phenol·$(NH_3)_3$. (a) The electronic energy of the total electron wavepacket along a path. The energy of the four low-lying excited states are also tracked along the same path. (b) The variation in the weight of the adiabatic states in the total electron wavepacket. (c) and (d) show the change of Mulliken charge and unpaired electron population on the sites of PhO, shifting proton (trH), ammonia molecule accepting the proton-nucleus from PhO (AM1), and the entire ammonia cluster (AMC). (Reprinted with permission from K. Nagashima *et al.*, J. Phys. Chem. A 116, 11167 (2012)).

Figure 7.30 displays a typical example of such electron dynamics; the time series data of (a) the total electronic energy of the SET wavepacket along a path and those of the four low-lying adiabatic states, (b) population of the adiabatic states involved in the propagating electron wavepacket, (c) the Mulliken charge of the selected sites, and (d) the unpaired electron population. We have recorded the graphs up to 300 fs. Correspondingly, the geometry has changed as in Fig. 7.31. Most of the paths we have obtained have been found to behave like this example.

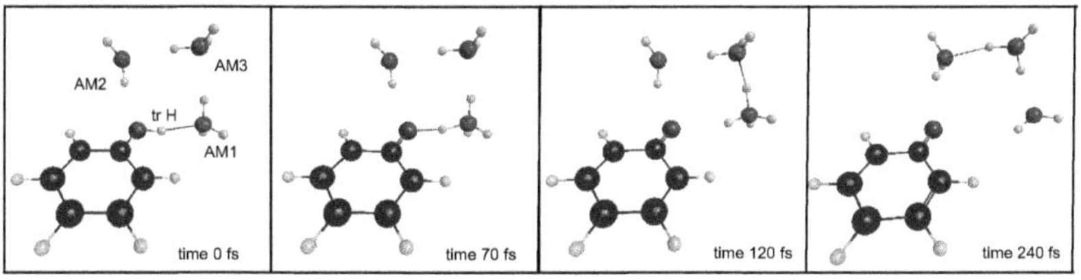

Fig. 7.31 Geometrical change along the path of Fig. 7.30. (Reprinted with permission from K. Nagashima *et al.*, J. Phys. Chem. A 116, 11167 (2012)).

7.5.5 *The successive nonadiabatic transitions*

The first excited state undergoes nonadiabatic transition with the second one at the distance of OH is about 1.4 Å and at time about 60 fs in case of Fig. 7.30. As observed in panel (b), the transition probability from the first to second state is as high as 80 %, which is consistent with the former identification that this nonadiabatic transition is due to conical intersection. Therefore, a simple minded method of dynamics like ab initio molecular dynamics tracking only on a single adiabatic PES should follow the minor part. Incidentally, both Figs. 7.26 and 7.27 suggest that the second adiabatic state beyond the first nonadiabatic transition soon nonadiabatically interact with the third and higher excited states.

The Mulliken charge shown in panel (c) of Fig. 7.30 again clarifies that the shifting proton nucleus is almost always covered with electron density as much as about 0.5e, and negative charge in PhO (ca −0.5e) moves to AMC after the conical intersection. This suggests that if one could directly pump S_0 to the second electronic states (S_2), only electron transfer can occur independently without proton-nucleus shift. However, the relevant oscillator strength is very weak. Proton-nucleus shifting and electron jumping are thus basically independent events.

Figure 7.30 suggests that after 130 fs in the electron dynamics simulation, the second nonadiabatic transition carries the electronic charge residing already on the local Rydberg state to the other side of a different Rydberg state that spreads globally over the ammonium cluster. Both panel (c) and (d) indicate that the charge and radical center have transferred from AM1 to the other place of AMC. An inspection on Fig. 7.31 shows directly that this nonadiabatic interaction has been induced by "proton transfer" from AM1 to AM2; relayed proton transfer within excited state ammonia cluster. On the other hand, panel (d) of Fig. 7.30 indicates that an unpaired electron keeps remaining in the ammonia cluster even after the second nonadiabatic transition, which suggests that the charge recombination is not yet finished.

7.5.6 *Analysis of electron dynamics*

Evolution of the unpaired electron distribution: Panel (d) of Fig. 7.30 dynamically tracks the positions of the radical center in this spin singlet state. The total number of the unpaired electrons is initially about 3.0 residing on PhO site before the first excited state passes across the conical intersection. After the passage, we see one on PhO and another on AMC mainly on AM1 (before $t = 130$ fs). Again no radical center is created on the shifting proton nucleus.

To further convince that this situation does not represent "hydrogen-atom" migration, we below show a typical example of such migration in a singlet excited state dynamics, in the first excited state dynamics of

$$[H_2O\text{-}H\text{-}NH_3]^+ , \tag{7.43}$$

which is depicted in Fig. 7.32. Panel (a) shows the unpaired electron population on the three relevant sites as a function of the OH distance. For a short OH, a large radical component is created on the ammonia site, while the smaller but not very small ones (about 0.5 each) are found on trH (transferring H) and water molecule. Near the middle point, the radical seems to suddenly jump from ammonia to water, and the unpaired population on trH becomes large reaching as high as about 1.0. Therefore the radical center remains on trH throughout the transfer. Panel (b) graphically exposes how large unpaired electron density is actually created. This is how hydrogen atom migration can be observed and identified, which is totally different from the case of phenol$\cdot(NH_3)_n$. Thus, we may conclude that hydrogen-atom migration does not actually take place in the present phenol-ammonia system.

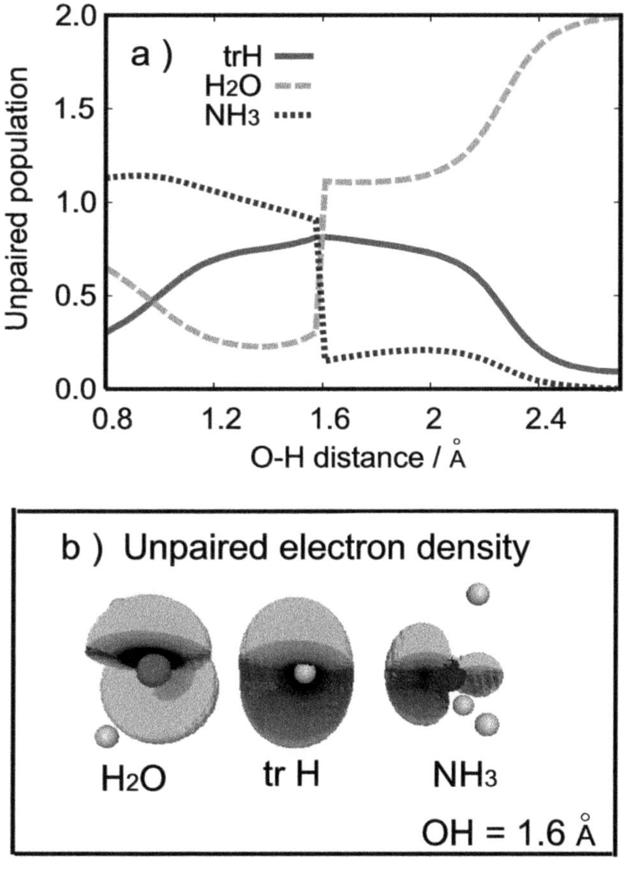

Fig. 7.32 Example of hydrogen atom transfer taken from the example of the spin-singlet excited state dynamics of $[H_2O\text{-}H\text{-}NH_3]^+$. (a) unpaired electron population, (b) and its spatial distribution. Cross-sections are drawn for the density plot in panel (b). (Reprinted with permission from K. Nagashima *et al.*, J. Phys. Chem. A 116, 11167 (2012)).

Returning to the phenol ammonia system, we next show in Fig. 7.33 how the spatial distribution of the unpaired electrons evolves in time. Before the first transition (see panel (a)) the unpaired electrons are distributed on the benzene ring and the oxygen atom. After the first transition (panel (b)), some large part of the unpaired electron density moves to AM1. At time 211 fs (after the second nonadiabatic transition, panel (d)), a large component of the unpaired electron is found on AM2, through a transient state at 148 fs (panel (c)).

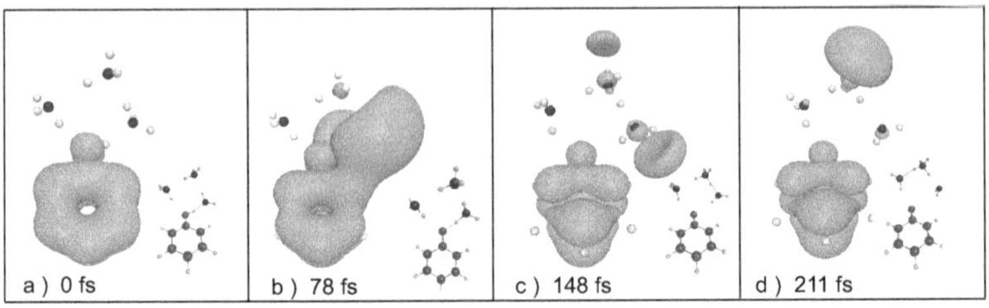

Fig. 7.33 Snapshots of the dynamics of the unpaired electron density arising from the semicalssical Ehrenfest electron wavepacket theory along the reaction path of Fig. 7.30. (a) the initial stage of $\pi - \pi^*$ optical transition, (b) at the moment of electron transfer from phenol to ammonia cluster. (c) and (d) show further fluctuation within the ammonia cluster. (Reprinted with permission from K. Nagashima *et al.*, J. Phys. Chem. A 116, 11167 (2012)).

Electron flux suggesting an independent path of electron: We have observed above that the proton nucleus shifts its position from phenol to ammonia, carrying the electron population to the amount of about 0.5e. Therefore we need to track how the remaining electron population (as much as about 0.5e) flows in the system. To do so, we have computed the Schiff probability current (shortly flux) [356] with the electron wavepackets. The flux analysis of electron and/or nuclear dynamics for chemical problems have been performed before by the present authors [286, 307, 425, 426, 494] and by Manz and his group [2, 38, 39, 172, 207, 208].

Figure 7.34 shows the electron flux at selected times, actually the same timing as that of Fig. 7.33. Note that the electrons carried tightly by the shifting proton nucleus gives rise to only a very small flux, since the net velocity of those electrons is very small (as slow as the nuclear velocity) and again the nature of these tightly bonding electrons is represented mostly in the 1s function that is real valued. At time 78 fs (panel (a)), right after the first nonadiabatic transition, an intensive flux is observed at the site of AM1 and in between AM1 and AM3. Although we do not show graphically, the electron flux generally fluctuates to some large extent both in space and time, the behavior of which is very different from a simple laminar flow of liquid in fluid mechanics. Nevertheless, the vector field of the flux suggests very firmly where and when the dynamics of electrons are in progress.

Panel (b), on the way of the second nonadiabatic transition, gives rise to a large flux wide-spread in the area of AM1 and AM2. After this transition, the flux at 211 fs, (panel (c)), suggests that a flow is then somewhat weaker.

Therefore it may be concluded that electron population as much as about 0.5 has been carried tightly by the proton nucleus and the other 0.5 has jumped from the phenol to ammonia site in a similar fashion to the well-known nonadiabatic long-distance electron jump as in $Na+Cl \rightarrow Na^{+}+Cl^{-}$. Thus the pathways of electron and the proton-nucleus have been identified mutually different, which is again not consistent with the view of hydrogen-atom migration. It should be therefore more appropriate to characterize this photoreaction as coupled proton and electron transfer in an excited state.

Finally, let us confirm an important difference between the present coupled proton-electron transfer and the ordinary proton transfer in the ground state. As shown in our study on the proton transfer in the ground state of formamide and its water-mediated relayed proton transfer, [286] the proton nucleus in these systems carries electron population as much as about 0.5e as a tightly covering component. In these systems, however, the same amount of electron is flowed back through an electron flux generated nearby. Thus it looks effectively as though one positive charge is carried by the proton nucleus. In the present system, on the other hand, such an back-flow does not occur, but electron population as much as about 0.5e jumps forward from phenol to ammonia, thereby the total charge is neutralized as though a single hydrogen atom is shifted without extra electronic charge.

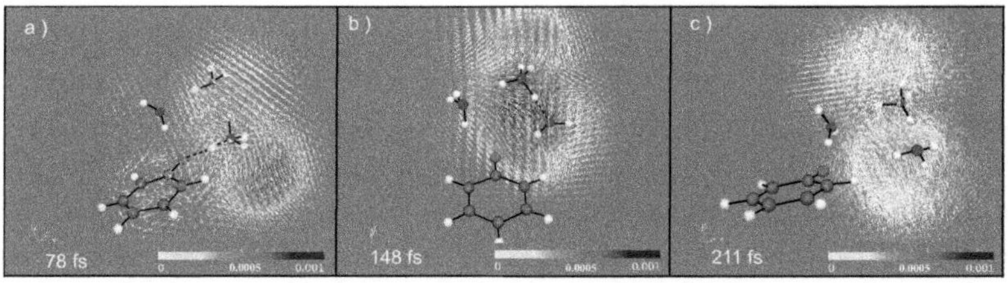

Fig. 7.34 Snapshots of the Schiff current of probability (electron flux) in the nonadiabatic electron dynamics, (a) at the moment of transfer from phenol to ammonia cluster, (b) and (c) further fluctuation of the flux within the ammonia cluster. (Reprinted with permission from K. Nagashima *et al.*, J. Phys. Chem. A 116, 11167 (2012)).

7.5.7 *Summary*

Through the above analysis based on nonadiabatic electron dynamics, it has been clarified:

1) Domcke and his coworkers' interpretation based on the conical intersection is basically correct.

2) However, the mechanism of the transfer is not the excited state hydrogen atom transfer (ESHAT) but should be more appropriately characterized as a coupled proton-electron transfer in excited state. (See Ref. [159, 192] for reviews of proton-coupled electron transfer.)
3) The Rydberg states of ammonia clusters are responsible for accepting the π^* electron of phenol. Indeed, our model calculation shows that the conical intersection is not formed without those Rydberg-like states.
4) Qualitative difference between the number of ammonia molecules 3 and 5 is not observed in the early process we have studied.
5) The successive proton-electron transfer takes place within the ammonia clusters, which can make difference among the different sizes of the ammonia cluster.

From the methodological view point, we would like to emphasize that the nonadiabatic electron dynamics in the branching path representation or its approximation SET is indeed necessary and useful to comprehend the dynamics of such highly quasi-degenerate electronic systems.

One of the most important aspects that are missing in the present study is the longer time-scale dynamics. As suggested from Fig. 7.31, the ammonium cluster can recede from phenol C_6H_6O radical. Then, the globally distributed Rydberg electron on the ammonia cluster can be eventually deactivated through a series of non-radiative transitions to the ground state by recombination with the "proton", which should make the molecular site of NH_4 somewhere in the cluster. Before coming to the end of the ground state, further relayed proton transfer may continue from one ammonium molecule to another. By the recombination, the cluster may end up with break-apart or ejection of one of the hydrogen atoms. The rate process of this deactivation may depend on the energy and size of the cluster. Therefore it should be quite interesting to explore how such charge recombination is realized. This study should be relevant to a basic aspect of the lifetime of excitons generated by photo-energy conversion and is under way in our laboratory.

7.6 Chemical dynamics for systems where notion of potential energy surfaces loses sense

As a case in which electron dynamics plays a critical role, let us consider molecular systems with adiabatic potential energy hyper-surfaces that are densely quasi-degenerate in a wide range of molecular geometry. In such a system, adiabatic electronic states tend to strongly interact nonadiabatically through frequent avoided crossings and crossing seams. Technically, therefore, it is often not clever to treat accurately the individual potential energy surfaces in a single by single manner. The more physically appropriate picture will be given by using the branched bunch of non-Born–Oppenheimer trajectories, part of which follows the field created by averaged electronic states in a time semi-local region. In fact, the mutual nonadiabatic couplings may spread wide in space due to the energy-time uncertainty relation. From a chemical view point, such dense set of electronic states should offer a very interesting molecular field in which chemical reactions proceed in characteristic manners.

One possible way to treat such a case is to use an approximated approach of the nonadiabatic electron wavepacket theory, the phase-space averaging and natural branching (PSANB) method [493], or the branching-path representation, in which the wavepackets propagate along non-Born–Oppenheimer branching paths.

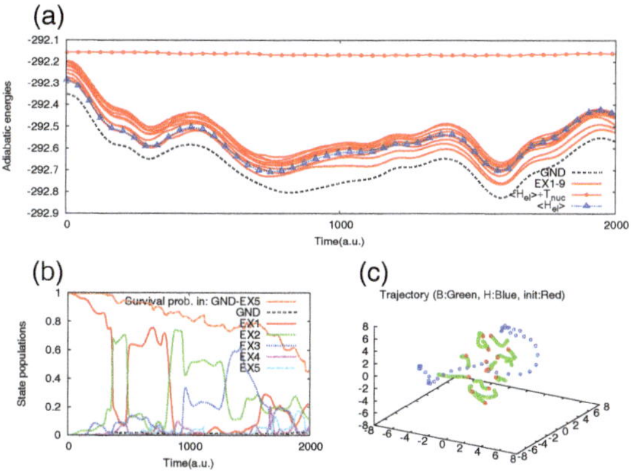

Fig. 7.35 SET dynamics of nonadiabatic rearrangement of hydrogen molecule embedded B_{12} cluster. (a) The time dependent mean potential $\langle H_{el} \rangle$ and its relation to the adiabatic potential energies of the ground state (GND: dashed black line) and excited states (EX1–9: red solid lines). The total energy $\langle H_{el} \rangle + T_{nuc}$ is also shown as an almost horizontal line. (b) Time dependent population of 6 adiabatic states; the ground (GND), first (EX1), second (EX2), third (EX3), fourth (EX4) and fifth (EX5) excited states. (c) Space-time history of the atoms in three-dimensional Cartesian coordinates (in Bohr units). The trajectories of 12 boron and 2 hydrogen atoms are expressed with green and blue points, respectively, the initial positions of which are marked with red circles embedded in the inner region. The hydrogen atoms are immediately pulled apart and each moves to the surface of the cluster. (Reprinted with permission from T. Yonehara *et al.*, J. Chem. Phys. 137, 22A520 (2012)).

How well the PSANB method works with such densely quasi-degenerate nonadiabatic systems was assessed. From the practical aspect, the PSANB scheme in fact reduces the number of initial points needed for qualitatively (but not quantitatively) converged result compared to the fewest switches surface hopping (FSSH) method, one of the most reliable and flexible methods to date. [496] This feature is preferable for tracing branches of electron wavepackets.

One chemically interesting system is B_{12}. The excited state manifold along its dynamics is expected to be densely quasi-degenerate. Though this feature depends on the symmetry of structure and the situation of dynamics under consideration, the picture is supported by the characteristics of the boron atom. Due to the vacancy in the $2p$ atomic orbitals of the boron atom $[1s^2 2s^2 2p^1]$, highly flexible bonding patterns emerge in the aggregate system. This is expected even more so in the excited states due to the same reason. In fact, it is known that boron atom doping in other molecular systems largely affects the electronic character related to orbital non-locality or aromaticity. It is considered that this expected high fluctuation of bonding pattern in the excited state manifold provides various types of chemical reaction fields.

With regard to this complicated nonadiabaticity, the electron wavepacket dynamics in the path-branching representation and the so-called semiclassical Ehrenfest theory was applied to a hydrogen molecule embedded in a twelve membered boron cluster (B_{12}) in the excited states [496]. The immediate bond breaking, subsequent deformation of the cluster, and B-H bond creation at the endhedral part observed through an on-the-fly SET approach indicates that the inside of this initial boron cluster has a highly active reaction field for the included molecule. This trend is well depicted in Fig. 7.35. The released nuclear kinetic energy along the structural relaxation without fission enhanced nonadiabatic transition, i.e., nuclear energy was effectively converted to energy for electronic activation. The complicated and lasting state couplings are shown in Fig. 7.36, indicating that the bond flexibility of the excited state cluster assisted the energy conversion from nuclei to electrons. It is highly expected that this kind of process has a potential for the creation of a highly fluctuated electron field which can serve as a unique chemical reaction field. As a final example, we see nonadiabatic branching of molecular entities. A nonadiabatic collision dynamics between H_2 and B_{12} shows an interesting phenomena indicating the possibility of different reactivity according to which branching the paths take. This is well illustrated in Fig. 7.37, which suggests an interesting functionality of boron clusters.

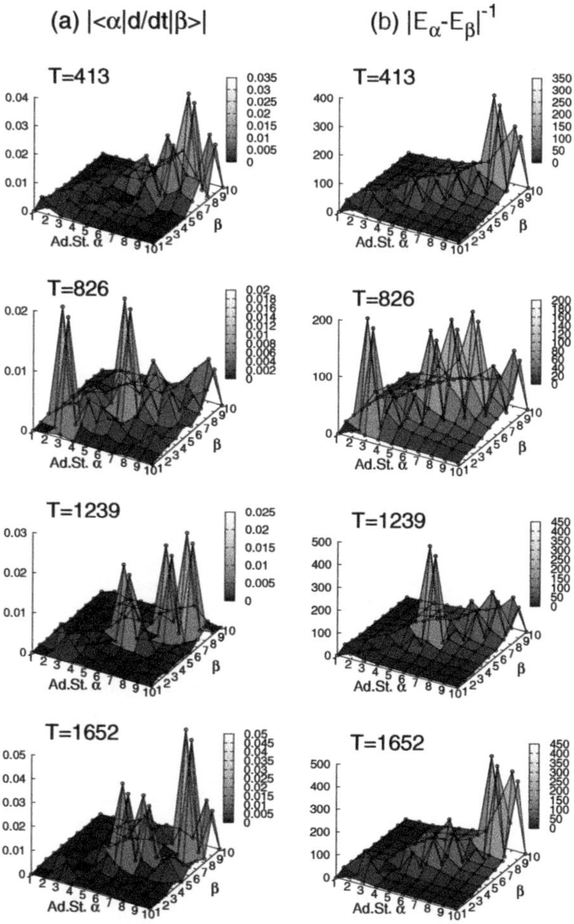

Fig. 7.36 (a) Snapshots of the absolute values of the coupling elements between the α and β adiabatic states ($\alpha, \beta \in [1, 10]$) at selected times. (b) The inverse of the energy difference, $\left|E_\beta - E_\alpha\right|^{-1}$. Values are set to zero for the same pair ($\alpha = \beta$). Energy in units of Hartree. (Reprinted with permission from T. Yonehara *et al.*, J. Chem. Phys. 137, 22A520 (2012)).

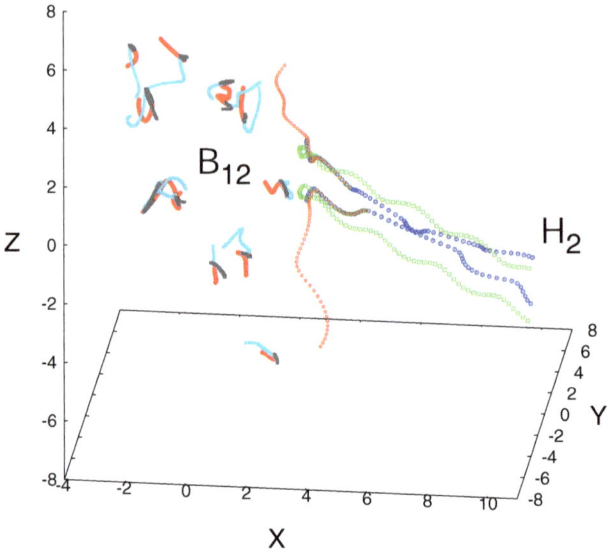

Fig. 7.37 Three dimensional view of a collision of H_2 and B_{12} cluster on an initially SET path in the excited state manifold. The coordinates X, Y, and Z are in units of Bohr. Blue circles represent in-coming hydrogen atoms (in a molecule) and red circles represent the trajectories of the boron atoms in the same timing. Branching paths emanate from a closer point, among which only two are drawn here: a path driven by the "ground state" (H atoms in green circle and B atoms in light-blue) shows that the hydrogen molecule is recoiled backward elastically; the other associated with the first excited state (H in orange and B in gray) indicates that as soon as the hydrogen molecule begin to penetrate into the cluster, it dissociates to two H atoms, subsequently running on the cluster surface. (Reprinted with permission from T. Yonehara *et al.*, J. Chem. Phys. 137, 22A520 (2012)).

Chapter 8

Molecular Electron Dynamics in Laser Fields

In Chapter 5, we have studied some of the effects of laser fields on chemical dynamics. In particular, we have investigated how time-resolved photoelectron spectroscopy can be used as a very good means to monitor the femtosecond-scale nuclear dynamics such as the passage across nonadiabatic regions. The modulation of nonadiabatic interactions (both avoided crossing and conical intersection) is also among the main subjects from the view point of control of chemical reaction. Chapter 7, on the other hand, has treated nonadiabatic electron wavepacket dynamics relevant to chemical reactions. Here in this chapter, we therefore rise to the theory of electron dynamics in laser fields mainly associated with chemical dynamics.

Recent advances in laser technology have made it possible to generate light pulse of the pulse-width as short as tens of attoseconds, which is short enough to track real time electron dynamics in chemical phenomena. Also it is well known that laser intensities higher than 10^{15} W cm^{-2}, which are not unusual now, can create electronic fields within molecules that are as strong as the native interactions among electrons and nuclei. These research fields, both experimental and theoretical, are quite active and each has been making a rapid progress. In particular, the very quantum nature of atoms and small molecules such as hydrogen molecule and its cation have been elucidated to a very precise level. These studies are even opening a paradigm of the next stage of laser science, in which the commonly used Schrödinger dynamics is not valid any more. Such profound knowledge about atoms and molecules should be made use of to comprehend of chemical reactivity in a higher level. It is therefore quite timely for us to unify the two different fields in physical chemistry: (1) quantum measurement and control processes of atto- and femto-second electron and nuclear wavepacket dynamics, and (2) theory of nonadiabatic transitions based on electron wavepacket dynamics. We thus have come up with the concept of unified developments between ultrafast molecular processes and theory of nonadiabatic chemical reactions in highly quasi-degenerate electronic states and/or laser fields. In so doing we should be able to establish a chemical reaction theory far beyond the seminal approach of Kenichi Fukui, by taking full account of the cutting-edge experimental technologies developed in physical chemistry and laser sciences.

We first take a brief look at the recent progress of experimental methods in a conceptual manner, and then consider what and how we can do theoretically to enhance the progress. Before starting though we note that the present chapter is not dedicated to an introduction to the vast field of photochemistry or excited-state chemistry, in which usually rather weak optical source is applied in a level of weak perturbation and create an excited state with a tuned wavelength resonant to the relevant energy gap. Also, the recent extensive progress in nonlinear optics and its applications to molecular science will be mentioned here.

8.1 Experimental progress and theoretical issues

8.1.1 *Attosecond laser tracking of molecular electronic states*

The time scale of valence electrons is typically of the order of 10^{-1} femtoseconds, while that of the inner shell is of course much faster and becomes close to the speed of light velocity for heavy atoms. Just as the femto-scale laser technology made a great contribution to the analyses of chemical dynamics of nuclear motions, the attosecond laser is anticipated to play the similar role for electron dynamics.

Attosecond dynamics is now one of the most active fields in science [222, 476] (see also the introductory section of Ref. [499], which shows a concise list of the studies covering many phenomena and relevant studies). In particular, tracking electronic motions in chemical dynamics is a fundamentally important process. To monitor those electron wavepacket dynamics in an attosecond intense laser field [476], Bandrauk and his coworkers have developed the theory of attosecond-scale time-resolved photoelectron spectroscopy [499, 500]. Photoionization dynamics of small molecules like H_2^+ has been studied so far, for which direct numerical integrations of the related (low-dimensional) time-dependent Schrödinger equations are possible. The photoelectron signals are extracted from those numerical solutions in terms of the electron flux (probability current density) that represents the flow rate of electrons leaving from the molecular site.

The electron dynamics along the chemical reaction process should be ideally traced in attosecond or subfemtosecond time scale. Experiments estimate or analyze electronic motions through the observation of dynamical properties connecting or roughly mapped to the integrated physical properties of these attosecond electrons in a chemical system. Though the experimental measurements are now limited to the small molecular system under a simple chemical process, the concepts are gradually to be applied for the moderate size of molecular systems under the optical coupling.

Visualization of electron rearrangement in space and time during chemical reaction process should fascilitate our intuitive understanding of chemical dynamics, and therefore vital to the progress of chemistry. For instance, Li, Becker and Murnane et al. observed experimentally the rearrangement dynamics of entire valence shell electrons during the dissociation reaction dynamics from molecular-like to atomic-like structure as a bond breaks [238]. They utilized the approach measuring total ionization signal and angular distributions of ionization fragments. They observed density changes over a long time and large distance in the photodissociation of Br_2, which disclosed that electrons do not localize onto the individual Br atoms until the fragments leave far apart, in a region where potential curves are nearly degenerate. This indicates a possibility of emergence of a quantum entanglement in a photodissociation reaction and its application to a new type of measurement. The correct understanding of the experimental findings has been supported by the theoretical estimation of the strong field ionization rate of this molecule. In fact, the phase diagrams for the pump-probe delay and kinetic release energies as well as alignment dependence of the ionization rate obtained by experiments have been qualitatively well explained by theoretical studies. The information about time dependent change of dominant orbitals as the sources of ionization electrons provide a charge distribution dynamics in the dissociation process. The peaks appearing in time dependent total ion counts were assigned to the characters of valence orbitals obtained through electronic structure calculations at each distance mapped to pump-probe delay times.

8.1.2 *Laser manipulation to create new states*

Weak electromagnetic fields created by stationary laser fields (continuous-wave laser) can efficiently excite or deexcite a state to another only when the photon energy is resonant with the difference of the two relevant states, provided that the optical transition is allowed by selection rule. However, with an increase of intensity of electronic resonant laser field, a nonlinear transition or multiphoton excitation can readily take place, yielding high energy states. Furthermore, for electronic state excitation among nearly degenerate states, nonadiabatic couplings among them can couple with such a high-intensity laser field. For instance, even an infrared (IR) laser, which usually induce only vibrational excitations, can induce electronic excitations efficiently in the presence of a strong nonadiabatic interaction by multiphoton absorption. This aspect should be always taken into account when we apply an intense laser to nonadiabatic molecular systems.

8.1.2.1 *Control of femtosecond nuclear wavepacket dynamics*

The theories of control of chemical reacrtions so far proposed make use of quantum mechanical manipulation of nuclear wavepackets by means of laser fields in femtosecond timescale [365, 372, 437]. For instance, Tannor and Rice developed a variational approach for making optimal form of laser field which maximize the a formation of a desired chemical species within a coherent two-photon process [433]. Similarly, Tannor, Kosloff and Rice proposed the reaction control scheme by using the resonant excited electronic state. The laser pulses are designed so that chemical reactions in ground state can be changed via designated intermediate states. The resultant chemical products can depend on the duration time between the resonant pulses [435]. Those devised fields characterized in terms of proper parameters work as a kind of a dynamical filter for desired chemical reaction dynamics. Brumer and Shapiro proposed a control scheme via modulation of the phase difference of two coherent laser fields of one and three photon resonance. They also provided a theoretical prediction for an entanglement control with use of chemical reaction system having degenerate electronic states [365].

A closed loop scheme using the iterative feedback of the experimental signal and the laser field parameters was introduced as a robust control scheme using the so-called genetic algorithm by Gerber's group [21]. The essential point of this scheme is that the optimal solutions can be obtained without a priori knowledge of molecular system and the experimental environments. In the same spirit, Rabitz's group also proposed the use of a genetic algorithm in the optimization of laser field and numerically demonstrated its use with a theoretical model [203]. They also applied for the control of selective bond dissociation and rearrangement in molecule with use of a strong field laser pulses and its optimally tailored form [236]. This scheme is also applied to the coherent X-ray generation via a robust creation of a HHG light field emitted from electrons [36].

Control of molecular orientation and/or alignement in space has been proposed and extensively studied by Seideman, in which she designs to creat a coherent standing wavepacket for rotational motion in terms of a non-resonant laser field [386].

8.1.2.2 *Control of electronic states with attosecond and/or intense laser fields*

A laser field having intensity higher than 10^{15} W cm^{-2} can apply as forces to nuclei and electrons as strong as their original interactions do. With these experimental techniques, researchers have been trying to modulate molecular electronic states or even create a new wavepacket state that does not exist in nature [487]. The excited electronic state manifold principally includes an infinite degrees of freedom of electrons. The mixture of them can produce any type of effective electronic field or the effective potential energy surface for molecular reaction. If one can manipulate its status or the tendency of electronic change in a robust manner, he/she may be able to discover unknown types of chemical reaction. Further, another point to be considered is that those created electronic states should immediately face additional nonadiabatic couplings through nuclear kinematic and optical couplings simultaneously. Again a very good theory of nonadiabatic electron wavepackets in laser fields and the ideas of chemical reactivity are demanded. Thus we need to theoretically and numerically seek for the principle of electronic state modulation.

The ultrafast laser control of electron dynamics in molecular systems were studied by Baumert group from the systems of atoms to solids [476]. By using laser fields precisely tailored in their phases, they performed the control of electronic wavefunction in a concept of relative phase modulation. They varied the interaction between the second laser field and time oscillating electron dipole due to a coherent mixture of states induced by the first laser pulse. In this scheme called the selective population of dressed state approach (SPOD) , external fields in-phase and out-of-phase guides the electronic wavepacket correspondingly into the lower and upper field dressed states. They categorize this into photon-locking controls as the analogy of quantum state control, or in the nuclear magnetic resonance

(NMR) research field, spin lock technique. The gradually diminishing field that follows this selection of population allows in turn the electronic state to launch an aimed electronic state and the chemical reaction characterized by each electronic state. This is one trial study of connecting between the electron dynamics control to the modification of chemical reaction by laser field. They also combine the information on the induced electron dynamics in molecular system with the imaging of photoelectron wavepacket [476].

As a fundamental study on field induced chemical reactions, Neidel and Vrakking et al. observed attosecond dynamics of electrons in a series of small- and medium sized neutral molecules by monitoring time-dependent variation of the parent molecular ion yields [296]. The information on electron dynamics was extracted from experimental data on the basis of the relation between the time dependent dipole and ionization. This was performed in the two-color femtosecond near infrared (NIR) pump–attosecond extreme ultraviolet (XUV) probe experiment. They claim that the time-dependent dipole induced by the moderately strong NIR pulse field is monitored with attosecond time resolution. The oscillations are interpreted in terms of a time dependent screening induced by the polarization of the molecule, which alters the photoionization yield of the neutral molecule. This scheme can be considered as the first example of molecular attosecond Stark spectroscopy.

To appropriately describe the quantum dynamics in such nonadiabatic chemistry in intense laser fields and highly degenerate states, a theory based on time-dependent nonadiabatic electron wavepackets is an essential requirement, which will be studied below in this chapter.

8.1.3 *Secondary effects of an induced electromagnetic field by external laser fields.*

The electron current within a molecule driven by an intense laser field should generate an induced electromagnetic field. This in turn will bring about the nontrivial secondary effects. The more intense is the external field, the more prominent these secondary effects should follow. However, they can be studied only by solving the coupled equations of the quantum electronic wavepacket dynamics and the (classical) Maxwell equations in a self-consistent manner. To calculate the generated electromagnetic field, one needs the induced electron currents. Bandrauk and his colleague have already started such a project [248].

8.2 Dressed electronic states and nonadiabatic nuclear dynamics on them driven by laser fields

Continuous-wave laser, or stationary laser, makes a periodic electromagnetic field in space-time. The external periodic field thus created in time-coordinate modulates the stationary electronic states to different characters and energy levels by having the electronic states dressed with optical fields. The periodically modulated dynamical states are well-known to be nicely treated in term of the so-called Floquet theory. Interestingly, the idea of dressed states can be generalized to a case of pulse lasers, provided that the pulse width is not too short. Here in this section, we first introduce the standard Floquet theory and then show the so-called two-time formalism to generalize the Floquet theory, which transform the chemical dynamics in a pulse laser field to nonadiabatic nuclear wavepacket dynamics among the quasi-potential energy surfaces made up with the dressed electronic states. Since the quasi-potential energy surfaces play roles to nuclear dynamics in a quite similar way to the ordinary potential energy surfaces in absence of laser fields, such generalization is of great significance.

8.2.1 *Introduction to the standard Floquet analysis*

We first consider an analysis of physical systems in periodic external fields using the Floquet theorem [134, 168]. As we shall see below, the theorem provides theoretical basis for the existence of "field-dressed" quasi-stationary state which expand the propagator. Earlier application of the Floquet theorem or related ideas to physical problems includes Refs. [224, 370], whereas the progress in this field is recently reviewed in Ref. [90]. Although it will later be extended to allow small non-periodic modulations, discussions in this section assumes perfect periodicity. The Schrödinger equation is given as

$$i\hbar\frac{\partial}{\partial t}\psi(t) = H(t)\psi(t), \tag{8.1}$$

where periodicity of the Hamiltonian, $H(t+T) = H(t)$, is assumed. For simplicity we consider a finite dimensional Hilbert space with dimension $\mathcal{N}_H$, although this restriction is not essential. The Floquet theorem asserts the existence of $\mathcal{N}_H \times \mathcal{N}_H$ dimensional (fundamental) matrix solution of Eq. (8.1) in the form

$$\underline{\Psi}(t) = \underline{F}(t)e^{-i\underline{\Lambda}t}, \tag{8.2}$$

where $\underline{F}(t)$ is a periodic (in the sense $\underline{F}(t+T) = \underline{F}(t)$) matrix and $\underline{\Lambda}$ is a constant diagonal matrix. It can be easily seen that the periodic part $\underline{F}(t)$ satisfies the eigenvalue equation

$$\mathcal{H}(t)\underline{F}(t) = \underline{F}(t)\underline{\Lambda}, \tag{8.3}$$

where $\mathcal{H}(t)$ is defined as $\mathcal{H}(t) \equiv H(t) - i\hbar\frac{\partial}{\partial t}$ and will be referred to as the Floquet operator. For the mathematical rigor and details of the Floquet theorem, we refer to references [168], whereas in the following discussions, we are more interested in gaining physical intuitions. For later convenience, we rewrite Eqs. (8.2), (8.3) for each "column vector" as

$$\psi_\alpha(t) = F_\alpha(t)e^{-\frac{i}{\hbar}\lambda_\alpha t}, \tag{8.4}$$

and

$$\mathcal{H}(t)F_\alpha(t) = \lambda_\alpha F_\alpha(t), \tag{8.5}$$

respectively.

Proof of the Floquet theorem Readers not familiar with the Floquet theorem are recommended to refer to textbooks on ordinary differential equations such as Ref. [168], where one would find a formal proof of the theorem different from the following discussions. Here we try to derive an intuitive but physically appealing *explanation* of the Floquet theorem based on Fourier analysis. We start with the Fourier transformation of the time dependent Schrödinger equation (8.1),

$$\hbar\xi\widetilde{\psi}_\xi - \int \frac{d\xi'}{2\pi}\widetilde{H}_{\xi-\xi'}\widetilde{\psi}_{\xi'} = 0, \tag{8.6}$$

where $\widetilde{\psi}_\xi$ and $\widetilde{H}_\xi$ are the Fourier transformation of $\psi(t)$ and $H(t)$, respectively. Due to the assumed periodicity, Fourier modes of $H(t)$ is restricted to integer multiplies of the fundamental frequency $\omega \equiv 2\pi/T$ hence

$$\widetilde{H}_\xi = \sum \overline{H}^n 2\pi\delta(\xi - n\omega), \tag{8.7}$$

where

$$\overline{H}^n \equiv \int_{-T/2}^{T/2} e^{in\omega} H(t)dt/T \tag{8.8}$$

is the nth Fourier mode of the periodic Hamiltonian. It follows that any pair of Fourier components of wavefunctions, $\widetilde{\psi}_\xi$ and $\widetilde{\psi}_{\xi'}$, are disconnected unless $\xi - \xi'$ is an integer multiple of ω, hence solutions of Eq. (8.6) can be constructed from Fourier modes of the form $\xi = \nu_\alpha + n\omega (n \in \mathbb{Z})$, where ν_α is some fixed frequency. The solution is then written in the form

$$\psi_\alpha(t) = \sum_n e^{-i(n\omega+\nu_\alpha)t} C_n, \tag{8.9}$$

which satisfies Eq. (8.2) with $\lambda_\alpha = \hbar\nu_\alpha$. Note that in order to complete the proof, one has further to prove that such solutions forms fundamental solution of the equation, which is not straightforward in this approach. The formal proof of the Floquet theorem in Ref. [168], which takes an approach very different from this, provides a more elegant proof of the theorem but the role of the fundamental frequency ω, which is crucial in physical applications, is less transparent.

Calculation of the Floquet states We now solve the eigenvalue equation (8.5) for Floquet states[1]. By virtue of the periodicity, each Floquet state $|u_\alpha(t)\rangle$ can be expanded in a Fourier series. We take an arbitrary time-independent orthonormal basis set that expands the Hilbert space of our interest, $\{|\phi_j\rangle\}$, then the Floquet state can be expanded using the basis states of the form $|\phi_j^m(t)\rangle \equiv |j\rangle e^{-im\omega t}$ with an integer m. We note that in this basis set, inner product (with arbitrary operator $\hat{O}$) is defined as

$$\langle\langle \phi_i^n|\hat{O}|\phi_j^m\rangle\rangle \equiv \int_{-T/2}^{T/2} e^{i(n-m)\omega t}\langle\phi_i|\hat{O}_t|\phi_j\rangle dt/T. \tag{8.10}$$

We expand the Floquet state as

$$|u_\alpha(t)\rangle = \sum_{j,m} |\phi_j^m(t)\rangle C_\alpha^{j,m}, \tag{8.11}$$

where $C_\alpha^{j,m} = \langle\langle \phi_j^m|u_\alpha\rangle\rangle$ are expansion coefficients. The eigenvalue equation (8.5) now becomes

$$\left[\overline{H}_{ij}^{n-m} - \hbar n\omega\delta_{ij}\delta^{nm}\right] C_\alpha^{jm} = \lambda_\alpha C_\alpha^{in}. \tag{8.12}$$

The Fourier representation of the Floquet operator thus obtained is an infinite dimensional matrix in general. Fortunately, however, it often takes a simple form when the time dependence of the Hamiltonian is monochromatic, or composed of a single mode with frequency ω; in such case $\overline{H}_{ij}^{n-m}$ have nonzero values only for $n - m = 0, \pm 1$.

For any Floquet state $|u_\alpha(t)\rangle$, its modulation, $e^{-in\omega t}|u_\alpha(t)\rangle$, with an arbitrary integer n, is also a Floquet state with quasi-energy $\lambda_\alpha - \hbar n\omega$.

[1] A perturbative approach to related problem is discussed in Ref. [24]

We refer to this modulated state as "$n\omega$-shift" and symbolically denote as $|u_{\alpha-n\omega}(t)\rangle$. It can be easily seen that

$$\langle\langle\phi_j^m|u_{\alpha-n\omega}\rangle\rangle = \langle\langle\phi_j^{m-n}|u_\alpha\rangle\rangle, \tag{8.13}$$

hence in this basis set expansion, any $n\omega$-shifted state is obtained by a mere shift of indices.

One of the most important consequence of this analysis is a propagator of the system. Using the complete set of Floquet states, where we exclude infinite series of $n\omega$ shifts, the propagator of the system is given as

$$K(\mathbf{r},t;\mathbf{r}',t') = \sum_\alpha \langle\mathbf{r}|u_\alpha(t)\rangle e^{-\frac{i}{\hbar}\lambda_\alpha(t-t')}\langle u_\alpha(t')|\mathbf{r}'\rangle. \tag{8.14}$$

Interpretation of Floquet States Equations (8.5) and (8.14) shows that the Floquet states are quasi-stationary states in the periodic field; if the physical system is $|u_\alpha(0)\rangle$ at time 0, then it is stationary up to periodic time evolution and phase gain, $|u_\alpha(t)\rangle e^{-\frac{i}{\hbar}\lambda_\alpha}$. In a shorter timescale than T, it may undergo rapid time evolution over many adiabatic states due to the interaction with the external field. Such behavior of the Floquet state is sometimes described as a field-dressed state. Its importance in analysis in the field-induced dynamics is analogous to that of the energy eigenstate under a time-independent Hamiltonian. We will further show that even a small non-periodic term can be incorporated in this framework in an analogous manner as the nonadiabatic transitions in the field-free dynamics.

One can obtain further physical insights by interpreting the Fourier mode number n of the basis state $|\phi_j^n(t)\rangle$ as the "photon number" [370]. We consider a larger Hamiltonian including the field degrees of freedom. It then follows that the time-dependence of the field is no longer an externally given function of time but an operator solution of quantum-mechanical Heisenberg equation. The total Hamiltonian is time-independent and conserves energy. We follow the standard procedure [174] of the electromagnetic field quantization using the Coulomb gauge. The total Hamiltonian is composed of material part H_{mat}, field part H_f and interaction part H_{int}. In the non-relativistic regime, material and interaction part becomes

$$H_{mat} + H_{int} = \frac{1}{2m}\sum_j \left(\mathbf{p}_j - \frac{q_e}{c}\mathbf{A}(\mathbf{r}_j)\right)^2 + V(\mathbf{r}), \tag{8.15}$$

whereas the field part is (assuming a large but finite $L \times L \times L$ cubic box to discretize photon modes)

$$H_f = \sum_{\boldsymbol{\lambda}} \frac{\hbar\omega_{\boldsymbol{\lambda}}}{2}\left[a^\dagger_{\boldsymbol{\lambda}} a_{\boldsymbol{\lambda}} + H.c.\right], \tag{8.16}$$

where $a_{\boldsymbol{\lambda}}(a_{\boldsymbol{\lambda}}^{\dagger}$ is the annihilation (creation) operator of (discretized) the transversal mode labeled with $\boldsymbol{\lambda}$ whose eigen frequency is $\omega_{\boldsymbol{\lambda}}$. We also note that $H.c.$ stands for Hermitian conjugate. The mode expansion is related to that of the vector field

$$\mathbf{A} = \sqrt{\frac{4\pi c^2}{L^3}} \sum_{\boldsymbol{\lambda}} \sqrt{\frac{\hbar^2}{2\omega}} \left(\boldsymbol{\varepsilon}_{\boldsymbol{\lambda}} a_{\boldsymbol{\lambda}} e^{i\mathbf{k}_{\boldsymbol{\lambda}}\cdot\mathbf{r}} + H.c. \right), \tag{8.17}$$

where $\boldsymbol{\varepsilon}_{\boldsymbol{\lambda}}$ is the polarization vector of the mode $\boldsymbol{\lambda}$. States of the system can be expanded by the basis states of the form $|\phi_i; \{n_{\boldsymbol{\lambda}}\}\rangle$, which is the product of material state ϕ_i and the field state $\{n_{\boldsymbol{\lambda}}\}$ (represented as the photon number of each mode). Moving to the interaction picture, the field operator $\mathbf{A}^{int}(t) \equiv e^{i(H_{mat}+H_f)t/\hbar}\mathbf{A}e^{-i(H_{mat}+H_f)t/\hbar}$ is solved as

$$\mathbf{A}^{int}(t) = \sqrt{\frac{4\pi c^2}{L^3}} \sum_{\boldsymbol{\lambda}} \sqrt{\frac{\hbar^2}{2\omega}} \left(\boldsymbol{\varepsilon}_{\boldsymbol{\lambda}} a_{\boldsymbol{\lambda}} e^{-i\omega_{\boldsymbol{\lambda}} t + i\mathbf{k}_{\boldsymbol{\lambda}}\cdot\mathbf{r}} + H.c. \right), \tag{8.18}$$

where $\boldsymbol{\varepsilon}_{\boldsymbol{\lambda}}$ is the polarization vector of the field. We now see that in the field-material interaction $\frac{q_e}{mc}\mathbf{p}^{int}(t) \cdot \mathbf{A}^{int}(t)$, the Fourier mode with frequency $\pm\omega_{\boldsymbol{\lambda}}$ are related to absorption and emission of one photon quantum with matrix elements given as

$$\begin{aligned} &\langle \phi_j; n_{\boldsymbol{\lambda}} - 1 | \frac{q_e}{mc}\mathbf{p}^{int}(t) \cdot \mathbf{A}^{int}(t) | \phi_i; n_{\boldsymbol{\lambda}} \rangle \\ &\quad = \langle \phi_j | \frac{q_e}{mc} \sqrt{\frac{4\pi c^2}{L^3}} \boldsymbol{\varepsilon}_{\boldsymbol{\lambda}} \sqrt{n_{\boldsymbol{\lambda}}} e^{i(E_j - E_i - \hbar\omega_{\boldsymbol{\lambda}})t/\hbar} | \phi_i \rangle, \end{aligned} \tag{8.19}$$

and

$$\begin{aligned} &\langle \phi_j; n_{\boldsymbol{\lambda}} + 1 | \frac{q_e}{mc}\mathbf{p}^{int}(t) \cdot \mathbf{A}^{int}(t) | \phi_i; n_{\boldsymbol{\lambda}} \rangle \\ &\quad = \langle \phi_j | \frac{q_e}{mc} \sqrt{\frac{4\pi c^2}{L^3}} \boldsymbol{\varepsilon}_{\boldsymbol{\lambda}} \sqrt{n_{\boldsymbol{\lambda}} + 1} e^{i(E_j - E_i + \hbar\omega_{\boldsymbol{\lambda}})t/\hbar} | \phi_i \rangle, \end{aligned} \tag{8.20}$$

respectively. The apparent photon number dependence $\sqrt{n_{\boldsymbol{\lambda}}}$ ($\sqrt{n_{\boldsymbol{\lambda}} + 1}$) can be replaced by a representative value $\sqrt{\bar{N}}$ when the photon numbers of states under consideration is centered at large value $\bar{N}$ and effects of one photon emission or absorption has negligible effect on the system behavior ($\bar{N} \gg 1$). The basis state $|\phi_i^n\rangle$ is identified as the material-photon product state $|\phi_i; \bar{N} - n\rangle$ in the limit of large $\bar{N}$, and the variation of n can be interpreted as photon emission or absorption.

8.2.2 *Generalization of the Floquet theory to treat dynamics in pulse lasers*

Laser-induced molecular dynamics we will examine below is driven by an oscillating electric field whose strength is comparable to the atomic Coulomb field. Solving for molecular dynamics in such strong optical field requires a non-perturbative approach.

8.2.2.1 *Expansion using Field-dressed periodic states*

Presumably the most straightforward approach to chemical dynamics in intense laser fields is to use the time-independent or time-dependent adiabatic states [352], which are the eigenstates of field-free or field-dependent Hamiltonian at given time points respectively, and solve the Schrödinger equation in a stepwise manner. However, when the laser field is approximately periodic, one can also use a set of field-dressed periodic states as an expansion basis. The set of quasi-static states in a periodic Hamiltonian is derived by a Floquet type analysis and is often referred to as the Floquet states [370]. Provided that the laser field is approximately periodic, advantages of using the latter basis set include (1) analysis and interpretation of the electron dynamics is clearer since the Floquet state population often vary slowly with the timescale of the pulse envelope and each Floquet state is characterized as a field-dressed quasi-stationary state, (2) under some moderate conditions, the nuclear dynamics can be approximated by mixed quantum–classical (MQC) nonadiabatic dynamics on the field-dressed PES. The latter point not only provides a powerful clue for interpretation of nuclear dynamics but also implies possible MQC formulation of intense field molecular dynamics.

The Floquet theorem, when applied to the quantum mechanics [370], implies the stationarity of Floquet states under a perfectly periodic Hamiltonian. We define the electronic Floquet operator as $\mathcal{H}_t \equiv H_t^{\mathrm{el}} - i\hbar\partial_t$ and the Floquet states as its periodic eigenstates which satisfy $\mathcal{H}_t|\Phi_\lambda(t)\rangle = \lambda|\Phi_\lambda(t)\rangle$. The above mentioned stationarity states that the solution of time-dependent Schrödinger equation $|\Psi_t\rangle$ can be expanded as

$$|\Psi_t\rangle = \sum_\lambda |\Phi_\lambda(t)\rangle C_\lambda e^{-\frac{i}{\hbar}\lambda t}, \tag{8.21}$$

with a set of time-independent coefficients $\{C_\lambda\}$. Note that $|\Phi_\lambda(t)\rangle$ is a periodic function and its time-dependence is calculated when one solves the eigenvalue problem. Each Floquet state evolves as $|\Phi_\lambda(t)\rangle e^{-\frac{i}{\hbar}\lambda t}$ and regarded as quasi-stationary field-dressed state. Here one can also see close analogy of adiabatic state in the field-free dynamics and the Floquet state in the dynamics in the periodic field. The point will be further discussed in the next subsection.

Unfortunately, however, the original form of Floquet theorem does not apply to the cases where H^{el} is not perfectly periodic. This includes the dynamics under pulsed laser field and/or dynamics coupled with nuclear motion, which are the cases we are most interested in. In order to apply the Floquet based analyses to such cases, one has to generalize the original formulation. Interestingly, as we will show below, the effects of nonperiodicities can be incorporated in terms of nonadiabatic transitions among Floquet states.

Generalization of the Floquet Hamiltonian method in this direction was first formulated by Ho and Chu [90, 186]. Later, Hanasaki and Takatsuka [165] reformulated the method using (t, t') formalism [319] to show wider generality and application range of this method. Here we introduce the latter formulation.

8.2.2.2 *Formulation of generalized Floquet Hamiltonian method*

We consider the time-dependent Schrödinger equation

$$i\hbar\frac{\partial}{\partial t}|\psi_t\rangle = H_t|\psi_t\rangle, \tag{8.22}$$

where $|\psi_t\rangle$ is the state vector of the system. H_t includes the interaction term with time-dependent external field which is not necessarily periodic.

We now introduce two time-like variables. Such formalism was originally developed in Ref. [319] by the name of (t, t') formalism. It provides a larger functional space and makes later discussion clearer. We extend the state vector $|\psi_t\rangle$ into a function with two time variables $|\widetilde{\psi}_{t,s}\rangle$. The central requirement is the equality to the original, or the physical state, at the point $t = s$;

$$|\widetilde{\psi}_{t,s}\rangle|_{t=s} = |\psi_t\rangle. \tag{8.23}$$

The Schrödinger equation for such extended state is given by

$$\left(i\hbar\frac{\partial}{\partial t} + i\hbar\frac{\partial}{\partial s}\right)|\widetilde{\psi}_{t,s}\rangle = H_t|\widetilde{\psi}_{t,s}\rangle, \tag{8.24}$$

or, equivalently,

$$i\hbar\frac{\partial}{\partial t}|\widetilde{\psi}_{t,s}\rangle = \mathcal{H}_{t,s}|\widetilde{\psi}_{t,s}\rangle, \tag{8.25}$$

with the Floquet-type operator $\mathcal{H}_{t,s} \equiv H_t - i\hbar\frac{\partial}{\partial s}$.

We next introduce the fundamental period T and assume the quasi-periodicity of the Hamiltonian in the sense $H_{t+T} = H_t + O(T/T_{\rm np})$, where $T_{\rm np}$ characterizes the time scale of nonperiodic variation and is assumed to be much larger than T. We then extend the Hamiltonian H_t to two time variables $\widetilde{H}_{t,s}$. Along with the central requirement $\widetilde{H}_{t,t} = H_t$, here we also require formal periodicity in the second variable s in the sense $\widetilde{H}_{t,s+T} = \widetilde{H}_{t,s}$. Such requirement can be easily satisfied if the Hamiltonian is periodic except for a set of slowly varying parameters $\{\mathcal{R}_t^\mu\}$; $H_{t+T}(\{\mathcal{R}_t^\mu\}) = H_t(\{\mathcal{R}_t^\mu\})$ but $H_{t+T} \equiv H_{t+T}(\{\mathcal{R}_{t+T}^\mu\})$ can be different from $H_t \equiv H_t(\{\mathcal{R}_t^\mu\})$. We define the extended Hamiltonian as $\widetilde{H}_{t,s} \equiv H_s(\{\mathcal{R}_t^\mu\})$ then it obviously shows formal periodicity $H_s(\{\mathcal{R}_t\}) = H_{s+T}(\{\mathcal{R}_t\})$.

Substituting the parameterized Hamiltonian $H_s(\{\mathcal{R}_t^\mu\})$ into Eq. (8.25), the Floquet-type operator becomes $\mathcal{H}_s(\{\mathcal{R}_t^\mu\}) = H_s(\{\mathcal{R}_t^\mu\}) - i\hbar\frac{\partial}{\partial s}$, which is now a true Floquet operator in the sense that it has the exact periodicity in the variable s. It thus follows that the eigenstates of this Floquet operator $\mathcal{H}_s(\{\mathcal{R}_t^\mu\})$ are also periodic in the variable s. We define parameterized Floquet states

$$\mathcal{H}_s(\{\mathcal{R}_t^\mu\})\,|\Phi_\alpha(s)\!:\!\{\mathcal{R}_t^\mu\}\rangle = \lambda_\alpha(\{\mathcal{R}_t^\mu\})|\Phi_\alpha(s)\!:\!\{\mathcal{R}_t^\mu\}\rangle, \tag{8.26}$$

where $\lambda_\alpha(\{\mathcal{R}_t^\mu\})$ is the αth Floquet quasi-energy. Restricting our attention to the discrete spectrum, we can further impose mutual orthonormality with respect to the inner product $\langle\!\langle\cdot\rangle\!\rangle$ defined as

$$\langle\!\langle u|v\rangle\!\rangle \equiv \int_0^T \frac{ds}{T}\langle u_s|v_s\rangle, \tag{8.27}$$

where T is the fundamental period as indicated above.

We now consider an electron-nucleus coupled system. We denote the two-time extension of the total state as $\widetilde{\Psi}_{t,s}$. The corresponding Schrödinger equation, in the nuclear coordinate representation, becomes

$$\begin{aligned} &i\hbar\left(\frac{\partial}{\partial t} + \frac{\partial}{\partial s}\right)\langle\mathbf{R}|\widetilde{\Psi}_{t,s}\rangle \\ &= \left[\sum_I \frac{1}{2M_I}\left[\frac{\hbar}{i}\frac{\partial}{\partial \mathbf{R}^I} - \frac{Q_I}{c}\mathbf{A}\right]^2 + U_{nuc}(\mathbf{R},t) + \widetilde{H}_s^{ele}(\mathcal{R}_t)\right]\langle\mathbf{R}|\widetilde{\Psi}_{t,s}\rangle, \end{aligned} \tag{8.28}$$

where $\langle\mathbf{R}|$ is the bra vector associated with the position eigenstate of nuclei $|\mathbf{R}\rangle$, M_I and Q_I are the mass and charge of the Ith nucleus and $U_{nuc}(\mathbf{R},t)$

is the nuclear potential term. **A** is the vector potential. Throughout this section, optical fields are treated within the long wavelength approximation. $\widetilde{H}_s^{ele}(\mathcal{R}_t)$ is the electronic Hamiltonian including electron-nucleus interactions. We then define the Floquet states $\{|\Phi_\alpha(s);\mathcal{R}_t\rangle\}$ as the eigenstates of the electronic Floquet operator $\mathcal{H}^{ele} \equiv \widetilde{H}^{ele} - i\hbar\frac{\partial}{\partial s}$. The parameter set $\mathcal{R}_t$ in this problem therefore includes nuclear coordinate in addition to laser parameters. The total state $\widetilde{\Psi}_{t,s}$ is then expanded in the form

$$|\widetilde{\Psi}_{t,s}\rangle = \sum_\alpha \int d\mathbf{R}|\mathbf{R}\rangle|\Phi_\alpha(s);\mathcal{R}_t\rangle\chi_\alpha(\mathbf{R},t), \tag{8.29}$$

where $\chi_\alpha(\mathbf{R},t)$ is the nuclear wavefunction associated with the electronic state $|\Phi_\alpha(s);\mathcal{R}_t\rangle$. Then the coupled Schrödinger equations for $\chi_\alpha(\mathbf{R},t)$ become

$$\begin{aligned} i\hbar\dot{\chi}_\alpha(\mathbf{R}) = &\sum_I \frac{1}{2M_I}\sum_\beta \left(\left[\frac{\hbar}{i}\frac{\partial}{\partial \mathbf{R}^I} - \frac{Q_I}{c}\mathbf{A} - i\hbar\mathbf{X}\right]^2\right)_{\alpha\beta}\chi_\beta \\ &+ U_{nuc}(\mathbf{R})\chi_\alpha + \lambda_\alpha\chi_\alpha - i\hbar\sum_\mu \dot{\zeta}^\mu \sum_\beta \mathcal{X}_{\alpha\beta}^{(\zeta^\mu)}\chi_\beta, \end{aligned} \tag{8.30}$$

where $\mathbf{X}_{\alpha\beta}^I \equiv \langle\langle\Phi_\alpha|\frac{\partial}{\partial \mathbf{R}_I}|\Phi_\beta\rangle\rangle$ are the nuclear derivative coupling terms, while $\mathcal{X}_{\alpha\beta}^{(\zeta^\mu)} \equiv \langle\langle\Phi_\alpha|\frac{\partial}{\partial \zeta^\mu}|\Phi_\beta\rangle\rangle$ are nonadiabatic coupling terms that are associated with laser parameters ζ^μ.

8.2.2.3 *Quasi-classical approximation of nuclear dynamics*

Having formulated the generalized Floquet Hamiltonian method as above, we next discuss the quasi-classical approximation on the nuclear degrees of freedom.

We first observe the close analogy of adiabatic state in the field-free dynamics and the Floquet state in the dynamics in the periodic field even in the presence of nonadiabatic transitions. One can expect a sort of mapping from field-induced dynamics to nonadiabatic dynamics of Floquet states using the field-induced PES or the quasi-energy surface (QES). The key expectation here is the validity of effective Newtonian dynamics on the nuclear degrees of freedom using the gradient force. In fact such analogy has long been discussed before the proposal of generalized Floquet method [186]. Although having some ambiguousness with respect to period-breaking nature, it has derived various important conclusions including bond-softening [148, 503] and bond-hardening [504] mechanisms.

Using generalized Floquet formulation derived above, we can now resolve these ambiguities. As we have seen in the formal theory of mixed quantum–classical representation of nonadiabatic dynamics discussed in the main text, the effective force acting on nuclei is given as

$$\mathbf{F}^{eff} \approx \langle \Phi_\alpha(t) : \{\mathcal{R}_t^\mu\}| \left(-\frac{\partial H^{\rm el}}{\partial \mathbf{R}}\right) |\Phi_\alpha(t); \{\mathcal{R}_t^\mu\}\rangle, \tag{8.31}$$

provided that the effects of nonadiabatic transitions are negligible. Further assuming that the timescale of nuclear dynamics is large compared to the fundamental period T, we can take cycle average of Eq. (8.31) to obtain

$$\mathbf{F}^{eff} \approx \langle\langle \Phi_\alpha(t) : \{\mathcal{R}_t^\mu\}| \left(-\frac{\partial H^{\rm el}}{\partial \mathbf{R}}\right) |\Phi_\alpha(t); \{\mathcal{R}_t^\mu\}\rangle\rangle = -\frac{\partial \lambda_\alpha(\{\mathcal{R}_t^\mu\})}{\partial \mathbf{R}}, \tag{8.32}$$

which is the gradient of QES.

From this derivation, we obtain three validity conditions for the gradient approximation; (a) smallness of nonadiabaticity (b) longer timescale of nuclear dynamics, together with (c) absence of purely quantum mechanical effects, such as tunneling, in the nuclear degrees of freedom, where the last one being one of the central assumptions in the MQC treatment.

Conversely, the gradient approximation breaks down around the avoided-crossing points on QESs, where strength of nonadiabatic coupling becomes large compared to the difference of adjacent quasi energies. Avoided-crossings of QES in the weak field limit, typically occur around resonant points, where the energy difference of two dipole-coupled adiabatic states equals to $n\hbar\omega$ with an integer value of n (hereafter referred to as $n\omega$ resonances). Resonant points, along with intrinsic avoided crossings on the original field-free PESs are to be treated with special care in the following analysis.

8.2.2.4 *An example of application to the field-induced bond deformation of H_2^+ and D_2^+*

We apply the method to the field-induced dynamics of H_2^+ and D_2^+, in which no nuclear derivative coupling is involved, and then in the next section we present a unified treatment of field-induced nonadiabatic dynamics and intrinsic one due to the nuclear kinetic couplings in the LiF molecule. Most of the technical details described in this section, except for the potential functions and coupling elements, are also applied to the study of LiF in the next section.

Systems and computational methods The field-induced dynamics of H_2^+ and D_2^+ has been intensively studied in the literature and hence the properties are well-known to serve as reference data. For the simplest assessment of the method, we use the two-state model proposed in Ref. [66]. The two-state models are known to sufficiently reproduce the essential effects in the field-induced bond dynamics; bond-softening and bond-hardening. We can therefore check if these two fundamental effects are correctly treated in the present method. The R-dependent electronic states are denoted as $|j; R\rangle$, with either $j = g$ or $j = u$ corresponding to the gerade and ungerade states, respectively. The molecular orientation is fixed so as the molecular axis to be parallel to the polarization vector of the applied laser in, say, x axis, and we concentrate only on the one-dimensional vibrational motion. The Schrödinger equation is

$$i\hbar\frac{\partial}{\partial t}\Psi_\alpha(R) = \sum_\beta \left[-\frac{\hbar^2}{2M}\frac{\partial^2}{\partial R^2}\delta_{\alpha\beta} + H^{ele}_{\alpha\beta}(R,t)\right]\Psi_\beta(R), \tag{8.33}$$

where M is the reduced mass of the nuclear relative motion and H^{ele} is the electronic Hamiltonian given as

$$H^{ele}(R,t) = \begin{pmatrix} U_g(R) & -\mu(R)E(t)\cos(\omega t) \\ -\mu(R)E(t)\cos(\omega t) & U_u(R) \end{pmatrix}, \tag{8.34}$$

with $E(t)\cos(\omega t)$ being the electric field, and $\mu(R)$ is the dipole matrix element between the g and u state. Here the electric field amplitude $E(t)$ is chosen to be of a single Gaussian form with the full width half maximum (FWHM) being 150 fs; $E(t) = E_0 \exp\left(-\left(\frac{t-t_c}{t_w}\right)^2\right)$ with $t_c = 162.15$ fs, $t_w = 90.08$ fs. The effective potential and the dipole matrix element are, following Ref. [66], given as

$$U_g(R) = K\left(\exp(-2D(R-R_0)) - 2\exp(-D(R-R_0))\right), \tag{8.35}$$

$$U_u(R) = K\left(\exp(-2D(R-R_0)) - 2a\exp(-D(R-R_0))\right), \tag{8.36}$$

$$\mu(R) = \mu_0 + \frac{\mu_0'}{Dy}\left(1 - \exp(-Dy(R-R_0))\right). \tag{8.37}$$

Here the parameters are set, following Ref. [66], as $D = 0.72$, $K = 0.10262$, $R_0 = 2.0$, $\mu_0 = 1.07$, $\mu_0' = 0.396$, $y = -0.055$ and $a = -1.11$, respectively. We use the atomic units throughout, except for time. The Schrödinger equation (8.33) was numerically solved using a standard grid-based split operator method [219]. Each $\chi_\alpha(\mathbf{R})$ in Eq. (8.30) is represented on a spatial grid. Although this implementation sacrifices applicability to larger systems, we can expect the most accurate results that are directly comparable to those obtained by full quantum calculations. Here the one-dimensional space is limited to the range $[0, R_{max}]$ and divided into N_S equal-spaced lattice, whose lattice points are given as $R_k = k\frac{R_{max}}{N_S}$, $k = 0, ..., N_S - 1$. The actual parameters used are $N_S = 4096$ and $R_{max} = 20$. The results were confirmed to be qualitatively independent of the system size by comparison with those of a smaller size simulation $N_S = 2048$ and $R_{max} = 16$. An imaginary potential term V_I is applied near the boundary in order to eliminate unphysical reflections by absorption;

$$V_I = \begin{cases} -iB_0 \left(\frac{R-R_b}{R_w}\right)^2 & R_a \leq R \leq R_b \\ 0 & \text{otherwise,} \end{cases} \tag{8.38}$$

where we chose $B_0 = 2.4K$ (this K is the parameter introduced in Eq. (8.35)), $R_w = 3.6$, $R_a = 14.0$ and $R_b = 18.0$.

The extended state $|\widetilde{\Psi}_{t,s}\rangle$ is propagated using the Trotter formula (see Eq. (8.39) below). We use the Fourier series expansion to represent the s-dependence of the extended state as $|\widetilde{\Psi}_{t,s}\rangle = \sum_n e^{-in\omega s}|\widetilde{\Psi}_t^F[n]\rangle$, where $|\widetilde{\Psi}_t^F[n]\rangle$ represents the nth Fourier component of the extended state vector, and n runs from $-N_w$ to N_w with a fixed large number N_w for cutoff (actually set to 50). The Fourier series is then represented in the column vector as $|\widetilde{\Psi}_t^F\rangle = \left(|\widetilde{\Psi}_t^F[-N_w]\rangle, |\widetilde{\Psi}_t^F[-N_w+1]\rangle, ..., |\widetilde{\Psi}_t^F[N_w]\rangle\right)^{\mathrm{T}}$, where the superscript T indicates the vector transposition. We then have

$$|\widetilde{\Psi}_t^F\rangle = \prod_{j=0}^{P-1} \left\{ e^{-\frac{i}{\hbar}\frac{\epsilon}{2}\mathbf{T}^{nuc}} e^{-\frac{i}{\hbar}\epsilon\mathbf{H}^{eff}(t_j)} e^{-\frac{i}{\hbar}\frac{\epsilon}{2}\mathbf{T}^{nuc}} \right\} |\widetilde{\Psi}_0^F\rangle, \tag{8.39}$$

where P is the number of the time steps in the Trotter decomposition, $\epsilon \equiv \frac{t}{P}$ is the infinitesimal time step, $t_j \equiv j\epsilon$ is the jth time point, $\mathbf{T}^{nuc}$ and $\mathbf{H}^{eff}(t_j)$ are the Fourier series representation of the nuclear kinetic term T^{nuc} and the effective electronic Hamiltonian $H^{eff} \equiv \mathcal{H}^{ele} + \mathbf{V}_I$ in a matrix form, respectively. In the actual computation, the nuclear kinetic term was calculated using the fast Fourier transformation (FFT) technique [219]. The electronic state was expanded with the Fourier-transformed diabatic basis; $e^{-in\omega s}|j; \mathbf{R}\rangle$. The total state in this representation has the form $|\widetilde{\Psi}\rangle = \sum_j \sum_k \sum_{n=-N_w}^{N_w} e^{-in\omega s}|\mathbf{R}_k\rangle|j; \mathbf{R}_k\rangle\chi_{j,k}[n]$, where $\chi_{j,k}[n]$ is the nth Fourier series component of the discretized nuclear wavefunction $\chi_{j,k} \equiv \chi_j(R_k)$.

The initial conditions of the present dynamical simulation are chosen such that the electronic state is in the ground (gerade) state and the nuclear state is at one of the vibrational eigenstates (quantum number v) of the gerade electronic state. The corresponding extended-state vector at $t = 0$ is then fixed as $\chi_{j,k}[n] = \delta_{j,g}\delta_{n,0}\phi_v(R_k)$, with the vth vibrational eigenstate wavefunction $\phi_v(R)$. As stated above, the initial choice of the extended state at $t = 0$ has arbitrariness with respect to the dependence on the variable s. The choice here corresponds to $|\widetilde{\Psi}_{t=0,s}\rangle = |\Psi_0\rangle$ (independent of s) and is found to be favorable for the present numerical implementation, since by this choice (together with the only weakly nonperiodic nature of the problem) only a small number of states in n is populated during the whole simulation. The simulation starts at $t = 0$ fs and end at $t \gtrsim 400$ fs. The duration of simulation time is taken long enough. (We recall that the pulse is Gaussian of FWHM$=$ 150 fs centered at $t_c = 162.15$ fs.)

Dissociation probability and underlying wavepacket dynamics We first show the calculated dissociation probability in Fig. 8.1. The results for H_2^+ are in good agreement with Fig. 18 of Ref. [148], in which no further analysis has been reported. As we notice in Fig. 8.1, the dissoci-

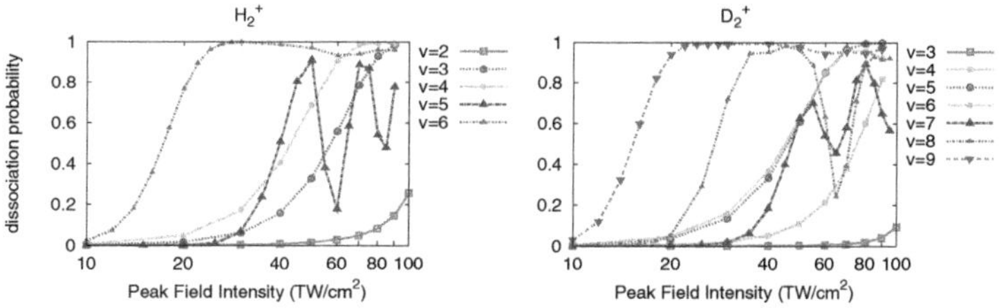

Fig. 8.1 The dissociation probability calculated from the total population lost from the simulation space $0 \leq R \leq R_{max}$ during the laser radiation. Left and right panel shows the results for H_2^+ and D_2^+, respectively. Each line shows the result obtained from the initial vibration state indicated on the right of each panel. The horizontal axis is plotted in log scale. (Reprinted with permission from K. Hanasaki *et al.*, Phys. Rev. A 88, 053426 (2013)).

ation probability versus the field intensity are classified into three overall patterns.

Pattern 1. Near-zero probability at low field followed by rapid monotonic increase at higher intensity: $v = 2, 3, 4$ in H_2^+ and $v = 3, 4, 5, 6$ in D_2^+.

Pattern 2. Non-monotonic oscillatory behavior; $v = 5$ in H_2^+ and $v = 7, 8$ in D_2^+.

Pattern 3. Similar to the pattern 1, but the increase occurs at much lower field; $v = 6$ in H_2^+ and $v = 9$ in D_2^+.

In order to analyze the dynamics behind each pattern, we show in Fig. 8.2 selected snapshots of wavepackets represented by the squared amplitude of the dominant Floquet state. The time-dependent behavior of the related QESs are also superimposed in the figure. One can observe that the

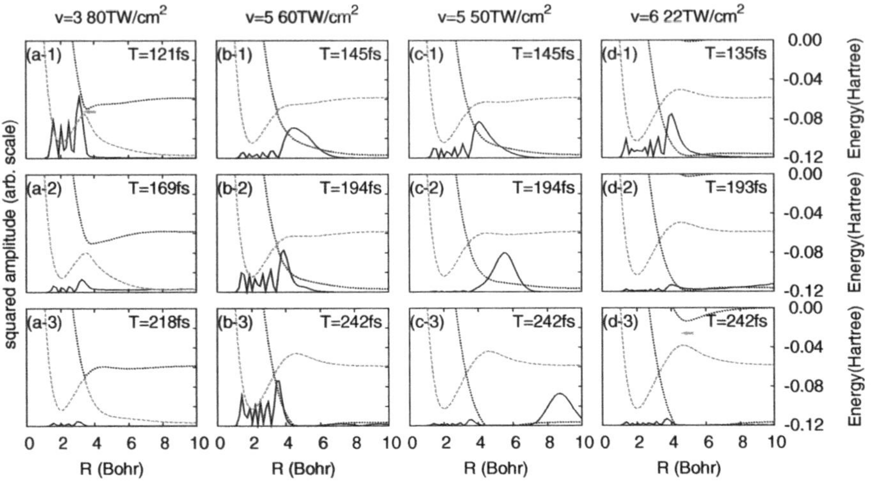

Fig. 8.2 Snapshots of the squared wavepacket amplitude and Floquet quasienergy surface. Each panel shows the snapshot at the time point which is indicated at the right top. For each panel, the black solid line represent the squared amplitude of the dominant Floquet state, and the gray dashed curve does the corresponding quasienergy surface. Another closely related QES is also plotted with black dotted line for reference. Three panels in each vertical row show snapshots of the dynamics starting from the initial vibrational state and the peak field intensity indicated at the top. We also plotted approximate position of resonant points by gray arrow. The arrow in panel (a-1) shows the 3ω resonance point while that in panel (d-3) shows 1ω resonance point. (Reprinted with permission from K. Hanasaki *et al.*, Phys. Rev. A 88, 053426 (2013)).

wavepacket motion is in accordance with the QES gradient, thereby qualitatively verifying the quasi-classical interpretation. Throughout the present simulation, the population of the $n\omega$-shifts of the dominant Floquet state was negligibly small in the entire parameter range. The population of the second-dominant Floquet state, which is distinct from the dominant, is also negligible in the parameter region in Fig. 8.2. The second-dominant state population, however, may grow to a finite value in other parameter regions (not plotted) especially around the resonant points.

Discussions below are about the mechanisms of the above phenomena, which are based on the dominant state behavior only, nonetheless characterizing the overall behavior.

Pattern 1: bond-softening

The snapshots of the dynamics of initial state $v = 3$ and the peak field intensity $I = 80$ TW cm^{-2} are shown in the three panels, (a-1) to (a-3) in Fig. 8.2, where one can see that the wavepacket moves out through deformed

QES after opening of the gap at the 3ω resonant point, which is a typical behavior in bond-softening [148]. Dynamics in other parameters categorized in this pattern also shows a similar behavior at intensities higher than an onset intensity, where a sharp increase of dissociation probability (see Fig. 8.1) occurs. We thus identify the present pattern as bond-softening.

Pattern 2: bond-hardening and inverse bond-hardening

The snapshots of the wavepacket dynamics of $v = 5$ at the peak field intensity $I = 60$ TW cm^{-2}, which is at the bottom of the dissociation probability oscillation (see Fig. 8.1), are shown in the panels (b-1) to (b-3) of Fig. 8.2. One can see that the increase of the QES local maximum near the 1ω resonant point prevents the wavepacket from dissociation, which is a typical behavior in bond-hardening [145, 148, 490, 504].

We also show, in the panels (c-1) to (c-3) in Fig. 8.2, the snapshots of the dynamics of $v = 5$ with the peak field intensity $I = 50$ TW cm^{-2}, which is at the peak of the dissociation probability oscillation. Here the development of a local maximum in QES near the 1ω resonant point seems to push out a portion of the wavepacket. Hence the underlying mechanism is the same as the normal bond-hardening seen in the dynamics at the peak field intensity $I = 60$ TW cm^{-2}, although the outcome is opposite. We thus term it "inverse bond-hardening".

Pattern 3: bond-softening (at 1ω resonant point)

The snapshots of the dynamics of $v = 6$ at the peak field intensity $I = 22$ TW cm^{-2} are shown in the panels (d-1) to (d-3) in Fig. 8.2. Here we can see the wavepacket move out through the gap formed by the 1ω resonance. The onset occurs at much lower intensity due to the stronger dipole coupling at the 1ω resonant point than that at the 3ω resonant point. Nevertheless, the essential mechanism is bond-softening, the same as the Pattern 1.

To summarize, we have thus shown that an interplay of bond-softening and bond-hardening can be clearly seen in the snapshots of time-dependent QESs and wavepacket dynamics on them, which is clearly understood in terms of the quasi-classical approximation. In fact our simulation serves as a direct real-time demonstration of possibly competing bond-softening and bond-hardening mechanisms, and moreover, has suggested a possible phenomenon, the inverse bond-hardening[2]. The additional

[2]Further discussions on physical conditions are necesssary for inverse bond-hardening, bond-softening, and bond-hardening to be observed experimentally.

calculations for D_2^+ dissociation have been found to be consistent with these interpretations.[3]

Full-quantum numerical simulation of a molecule in which an application of the generalized Floquet theory to the curve-crossing dissociation dynamics of LiF in laser fields is given in Ref. [165].

8.3 Generalization of path-branching representation for arbitrary optical and nonadiabatic transitions

We generalize the path-branching formalism of nonadiabatic dynamics, which was developed in Chapter 6, so as to treat nonadiabatic chemical reactions in external laser fields. The extension to any other types of coupling is rather straightforward.

8.3.1 *Mixed quantum-classical Hamiltonian in an optical field*

We resume with a general theory of mixed quantum–classical representation of dynamics of a system composed of fast and slow subsystems, in which the former and latter are, respectively, treated quantum and classical mechanically. In the present case, the fast and slow subsystems happen to be electronic and nuclear parts, respectively.

We start from the following full quantum mechanical molecular Hamiltonian in an optical field having the vector potential $\mathbf{A}$,

$$\mathcal{H}(\mathbf{r},\mathbf{R}|\mathbf{A}) = \frac{1}{2}\sum_k(\hat{P}_k - \frac{Z_k e}{c}A_k(\mathbf{R}))^2 + \mathcal{H}^{(el)}(\mathbf{r};\mathbf{R}|\mathbf{A}), \tag{8.40}$$

with $\mathcal{H}^{(el)}(\mathbf{r};\mathbf{R}|\mathbf{A})$ being the electronic Hamiltonian under the electromagnetic field

$$\mathcal{H}^{(el)}(\mathbf{r};\mathbf{R}|\mathbf{A}) = \frac{1}{2}\sum_j(\hat{p}_j + \frac{e}{c}A_j(\mathbf{r}))^2 + V_c(\mathbf{r};\mathbf{R}), \tag{8.41}$$

where V_c collectively denotes the Coulombic interactions of electron-electron, electron-nuclei, and nuclei-nuclei. $A_k(\mathbf{R})$ and $A_j(\mathbf{r})$ are the vector potentials for nuclei and electrons, respectively. Throughout this section, we use the mass-weighted coordinates, in which all the masses are scaled to unity. Here, $\mathbf{R} = \{R_k\}$ is the collective vector of all the nuclear positions

[3] In fact the spatial distribution of the initial wave function with respect to the position of the QES resonant points is found to have similarity among each pattern. The fact may explain the origin of each pattern, but will be discussed elsewhere.

with $\hat{\mathbf{P}} = \left\{\hat{P}_k\right\}$ being the vector of associated momenta. $Z_k e$ and c are the nuclear charge and the light velocity, respectively, with $-e$ being the electron charge. The hat as in $\hat{P}$ and so on denote the quantum operators.

In the present theoretical scheme, the Hamiltonian of Eq. (8.40), which is represented in configuration space $\{\mathbf{r}, \mathbf{R}\}$, is re-expressed in the electronic Hilbert space and nuclear configuration space, whose basis is $\{|\Phi_I(\mathbf{R})\rangle|\mathbf{R}\rangle\}$ such that [425, 492]

$$\begin{aligned}&\mathcal{H}(\mathbf{R}, elec, \mathbf{A}) \equiv \\ &\frac{1}{2}\sum_k \left\{\hat{P}_k - \frac{Z_k e}{c} A_k - i\hbar \sum_{IJ} |\Phi_I\rangle X^k_{IJ} \langle\Phi_J|\right\}^2 \\ &+ \sum_{IJ} |\Phi_I\rangle \mathcal{H}^{(el)}_{IJ} \langle\Phi_J|. \end{aligned} \tag{8.42}$$

This Hamiltonian manifests a parallelism between the electromagnetic effects and nonadiabatic coupling. We here only mention the Aharonov–Bohm effect [4] and the Longuet–Higgins [247] (or Berry [48]) phase. Also, a Lorentz-force-like term arising from the nonadiabatic coupling has been commented in Ref. [425]. To attain a mixed quantum–classical representation, we classicalize the above Hamiltonian (8.42) so as to be accessible to its (approximate) solutions more easily, by simply changing the nuclear momentum operator $\hat{P}_k$ in $\mathcal{H}(\mathbf{R}, elec, \mathbf{A})$ to its classical counterpart P_k such that

$$\begin{aligned}&\tilde{\mathcal{H}}(\mathbf{R}, \mathbf{P}, elec, \mathbf{A}) \equiv \\ &\frac{1}{2}\sum_k \left(P_k - \frac{Z_k e}{c} A_k - i\hbar \sum_{IJ} |\Phi_I\rangle X^k_{IJ} \langle\Phi_J|\right)^2 + \sum_{IJ} |\Phi_I\rangle \mathcal{H}^{(el)}_{IJ} \langle\Phi_J|. \end{aligned} \tag{8.43}$$

The tilde over $\mathcal{H}(\mathbf{R}, \mathbf{P}, elec, \mathbf{A})$ indicates the mixed quantum–classical representation.

8.3.2 *Coupled dynamics of electrons and nuclei: Generalized semiclassical Ehrenfest dynamics*

The dynamical equations of motion for electron wavepacket are derived in terms of the time dependent variational principle, giving rise to

$$\delta \int dt \langle\Phi(\mathbf{R}, t)| \left(i\hbar \frac{\partial}{\partial t} - \tilde{\mathcal{H}}(\mathbf{R}, \mathbf{P}, elec, \mathbf{A})\right) |\Phi(\mathbf{R}, t)\rangle = 0, \tag{8.44}$$

where $\tilde{\mathcal{H}}(\mathbf{R}, \mathbf{P}, elec, \mathbf{A})$ are the total Hamiltonian defined in Eq. (8.43). As usual, we expand the electronic wavepacket as

$$\Phi(\mathbf{r}, t; \mathbf{R}(t)) = \sum_I C_I(t)\Phi_I(\mathbf{r}; \mathbf{R})|_{\mathbf{R}=\mathbf{R}(t)}, \tag{8.45}$$

where $\{\Phi_I(\mathbf{r}; \mathbf{R})\}$ are basis functions such as the Slater determinants generated at each nuclear position $\mathbf{R}$, and thereby the electronic dynamics is cast to the associated coefficients $\{C_I(t)\}$. Taking account of the right variation also just as the left variation of Eq. (8.44), we obtain the coupled equations for the electron dynamics as

$$i\hbar\frac{\partial}{\partial t}C_I = \sum_J \left[\mathcal{H}_{IJ}^{(el)} - i\hbar\sum_k \left(P_k - \frac{Z_k e}{c}A_k\right) X_{IJ}^k - \frac{\hbar^2}{4}\sum_k \left(Y_{IJ}^k + Y_{JI}^{k*}\right)\right] C_J. \tag{8.46}$$

This is the generalization of the electron wavepacket theory to the dynamics in an optical field [492]. Note that the second derivative coupling elements, $Y_{IJ}^k \equiv \left\langle \Phi_I \left| \frac{\partial^2}{\partial R_k^2} \right| \Phi_J \right\rangle$, in this expression has no direct interaction with the external vector potential $\mathbf{A}$, while the first derivative coupling elements X_{IJ}^k directly couple with A_k.

We next establish the nuclear equations of motion. The equations of motion analogous to the Hamilton canonical equations of motion establishes the following set of equations. After some manipulation [423, 492, 493], we have

$$\begin{aligned}\frac{d^2}{dt^2}\mathcal{R}^k = &-\sum_{IJ}\frac{\partial}{\partial R_k}\left(|\Phi_I\rangle\mathcal{H}_{IJ}^{(el)}\langle\Phi_J|\right) - \frac{q_k e}{c}\frac{\partial}{\partial t}A_k(\mathbf{R}, t)\\ &+ i\hbar\sum_{IJ}\sum_l \dot{R}_l\left[\frac{\partial}{\partial R_k}\left(|\Phi_I\rangle X_{IJ}^l\langle\Phi_J|\right) - \frac{\partial}{\partial R_l}\left(|\Phi_I\rangle X_{IJ}^k\langle\Phi_J|\right)\right]\\ &+ \frac{e}{c}\sum_l \dot{R}_l\left(\frac{\partial}{\partial R_k}q_l A_l - \frac{\partial}{\partial R_l}q_k A_k\right),\end{aligned} \tag{8.47}$$

which is defined in the combined nuclear $(\mathbf{R}, \mathbf{P})$ and electronic Hilbert spaces. Therefore, to use this operator in practice, we should project it to a matrix form by sandwiching two electronic states, say, $\langle \Phi_I |$ and $|\Phi_J\rangle$ as

$$\begin{aligned}\mathcal{F}_{IJ}^{i_a} &\equiv \left\langle \Phi_I \left| \ddot{\mathcal{R}}^{i_a} \right| \Phi_J \right\rangle \\ &= -\sum_K \left[X_{IK}^{i_a} \mathcal{H}_{KJ}^{(el)} - \mathcal{H}_{IK}^{(el)} X_{KJ}^{i_a} \right] - \frac{\partial \mathcal{H}_{IJ}^{(el)}}{\partial R_{i_a}} \\ &\quad + i\hbar \sum_b^{nuc.} \sum_{i_b}^{x,y,z} \dot{R}_{i_b} \left[\frac{\partial X_{IJ}^{i_b}}{\partial R_{i_a}} - \frac{\partial X_{IJ}^{i_a}}{\partial R_{i_b}} \right] \\ &\quad + q_a e \left(\mathbf{E}_a + \dot{\mathbf{R}}_a \times \mathbf{B}_a \right)_{i_a} \delta_{IJ}, \end{aligned} \tag{8.48}$$

which is a natural generalization of the classical force, and is called the force matrix [423].

In case of

$$X_{\alpha\beta}^k = 0 \quad \text{and} \quad \mathbf{A} = \mathbf{0}, \tag{8.49}$$

it is most convenient to adopt the adiabatic wavefunctions $\{\psi_\alpha\}$, satisfying

$$H^{(\mathrm{el})} \psi_\alpha = E_\alpha^{ad} \psi_\alpha. \tag{8.50}$$

Then we have

$$\mathcal{F}_{\alpha\beta}^k = -\delta_{\alpha\beta} \frac{\partial E_\beta^{ad}}{\partial R_k}, \tag{8.51}$$

subject to the condition Eq. (8.49). Therefore the force matrix in the adiabatic limit reproduces the Born–Oppenheimer forces in its diagonal elements.

If a complete electronic basis was available, the first three terms in the rightmost hand of Eq. (8.48) are combined into a single piece as in the Hellmann–Feynman force

$$-\sum_K \left[X_{IK}^{i_a} \mathcal{H}_{KJ}^{(el)} - \mathcal{H}_{IK}^{(el)} X_{KJ}^{i_a} \right] - \frac{\partial \mathcal{H}_{IJ}^{(el)}}{\partial R_{i_a}} = \left(\frac{\partial \mathcal{H}^{(\mathrm{el})}}{\partial R_{i_a}} \right)_{IJ} \tag{8.52}$$

and furthermore it holds that $\left(\frac{\partial \mathcal{H}^{(\mathrm{el})}}{\partial R_{i_a}} \right)_{IJ} = \left(\frac{\partial H^{(\mathrm{el})}}{\partial R_{i_a}} \right)_{IJ}$. Thus the effect of the laser field on the force matrix through the reorganization of the electronic state should be quite limited (actually zero for a complete basis set). Only the classical Lorentz forces working on the nuclei are the dominant contribution to the force matrix.

8.4 Applications: Electron jump in laser fields

The following subsections are devoted to illustrative applications of nonadiabatic electron wavepacket dynamics to chemical reactions. By tuning pulse laser fields in a coherent or incoherent sense, one can significantly modulate the electronic transition and subsequent molecular dynamics. The correlation between the nonadiabatic term and laser fields can lead to dramatic change in chemical dynamics, which implies that nonadiabatic dynamics and intense-laser fields should be treated on an equal footing [425, 491, 492, 494]. Below we show those examples of the coupling between nonadiabatic interactions and laser fields.

8.4.1 *Control of nonadiabatic transition in NaCl*

As an example of the effect of an intense laser on chemical reactions, a possible control of nonadiabatic transition of Na + Cl by modulating its avoided crossing is studied [425]. Figure 8.3 shows the potential curves that cross each other between the lowest covalent and ionic adiabatic states. The calculation is in the level of RHF/CISD/6-31G, with 26 molecular orbitals, 91 CSFs, and 13 core orbitals. The target was set to the qualitatively correct description of only the sigma states, without careful attention to the two lowest Π states. In doing so, we chose the core orbitals up to the second highest occupied molecular orbitals.

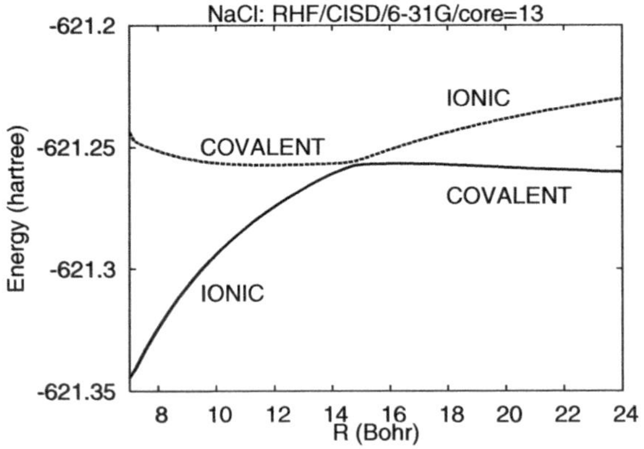

Fig. 8.3 Potential energy curves for the Na + Cl system around the avoided crossing. (Reprinted with permission from K. Takatsuka *et al.*, Adv. Chem. Phys. 144, 93 (2010)).

The initial distance of the two atoms and the collision energy is set to 22.67 (Bohr) and 1.13415 (hartree) (= 30.86 eV), respectively. Two cases of independent asymptotic electronic wavefunctions were studied, namely, the states of Na + Cl and Na^+ + Cl^-.

The vector potential in this example is

$$A(t) = A_s \exp\left(-\left(\frac{t-t_c}{t_w}\right)^2\right)\sin(\omega(t-t_c)+\delta), \tag{8.53}$$

and the associated electric field $E(t)$ is in the form

$$E(t) = \frac{A_s\omega}{c}\exp\left(-\left(\frac{t-t_c}{t_w}\right)^2\right) \times\left(\frac{2(t-t_c)}{\omega t_\omega^2}\sin(\omega(t-t_c)+\delta) - \cos(\omega(t-t_c)+\delta)\right), \tag{8.54}$$

where $\delta = 0$, $t_c = 20$ (fs), $t_w = 2.42/4.84/7.26$ (fs), $\omega = 0.0017$, and $E_s = 0.02$. The polarization vector is set parallel to the molecular axis. The field-electron interaction is treated under the velocity gauge. The peak time of the laser pulse t_c is chosen as a time at which the atoms come to the minimum energy gap point. The frequency ω is set so that $\hbar\omega$ should be equal to the minimum energy gap (at the avoided crossing).

In Fig. 8.4, we demonstrate laser control of electronic state populations in the present collision dynamics. In both cases, the laser width $t_w = 4.84$ fs was found to bring about the largest change of the electronic character. As much as about 30-40 % of modulation of the population (the product ratio) has been attained. However, the result is very sensitive to the pulse duration time, and also to the time scale of nuclear dynamics passing through the avoided crossing region. The guiding principle to control the nonadiabatic transition and the mechanism of it will be established from the view point of electron dynamics.

As an analysis of this laser control of chemical reaction in terms of electron dynamics, in Figs. 8.5 and 8.6 we present the snapshots of electron current flux and corresponding time derivative of electron density at the time 15 and 25 fs for the cases with the pulse width $t_w = 4.84$ fs starting from the asymptotic ground (Na+Cl) and first excited (Na^++Cl^-) states. Due to the non-resonance for electronic transition as well as medium strength of the laser field, complicated interference patterns are found in the spatial distribution of the flux and in the time derivative of electron density.

Through a careful systematic classification of these quantum behaviors, it will become possible to predict the correlation between the laser field type and nonadiabatic reactivity in the chemical reaction dynamics beyond the standard view based on the geometric dependence of potential energies and state couplings.

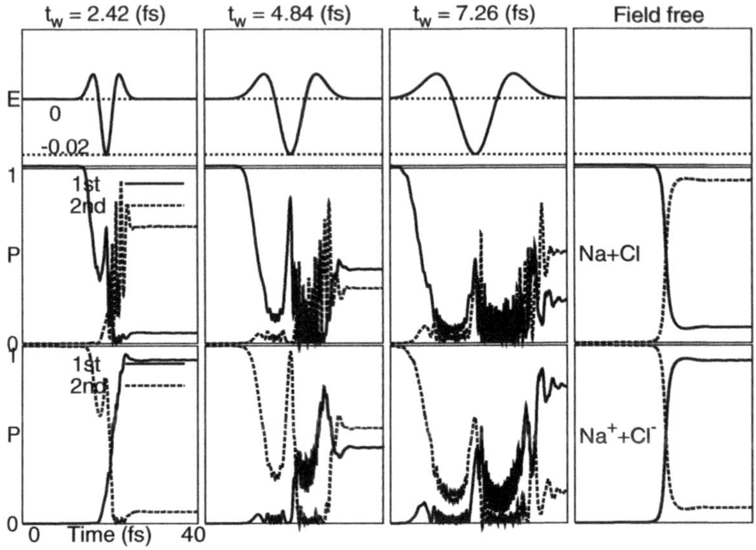

Fig. 8.4 Laser control of electronic state populations for the collisions of Na + Cl (panels of the middle row) and Na^+ + Cl^- (bottom row). The electronic populations after passing the avoided crossing are drawn. The electric fields made by the laser with different width are shown in the top row. (Reprinted with permission from K. Takatsuka *et al.*, Adv. Chem. Phys. 144, 93 (2010)).

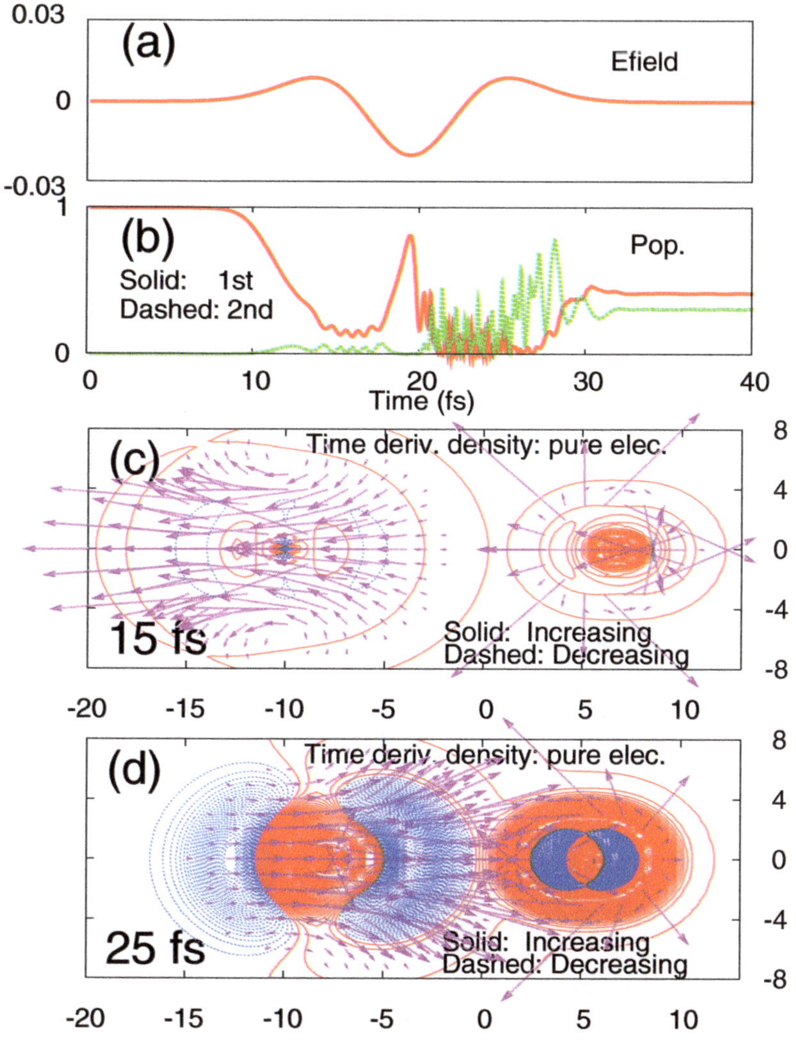

Fig. 8.5 Snapshots of electron current density and time derivative electron density from pure electronic component without that conveyed by nuclear motion, for the collisions of Na + Cl. The pulse width is 4.84 fs (corresponding to the second panel from the left, middle row in Fig. 8.4). The time dependent behavior of external electronic field and state population dynamics are given in the panels (a) and (b), respectively. The snapshots at 15 and 25 fs are shown in (c) and (d), respectively. The arrows denote electron flux density. The red sold and blue dashed contour lines in (c) and (d) show the positive and negative values of the pure electronic component of time derivative of the electron density. Here pure electronic means only the time derivative of electronic wave function was utilized in the evaluation both in flux and time derivative density.

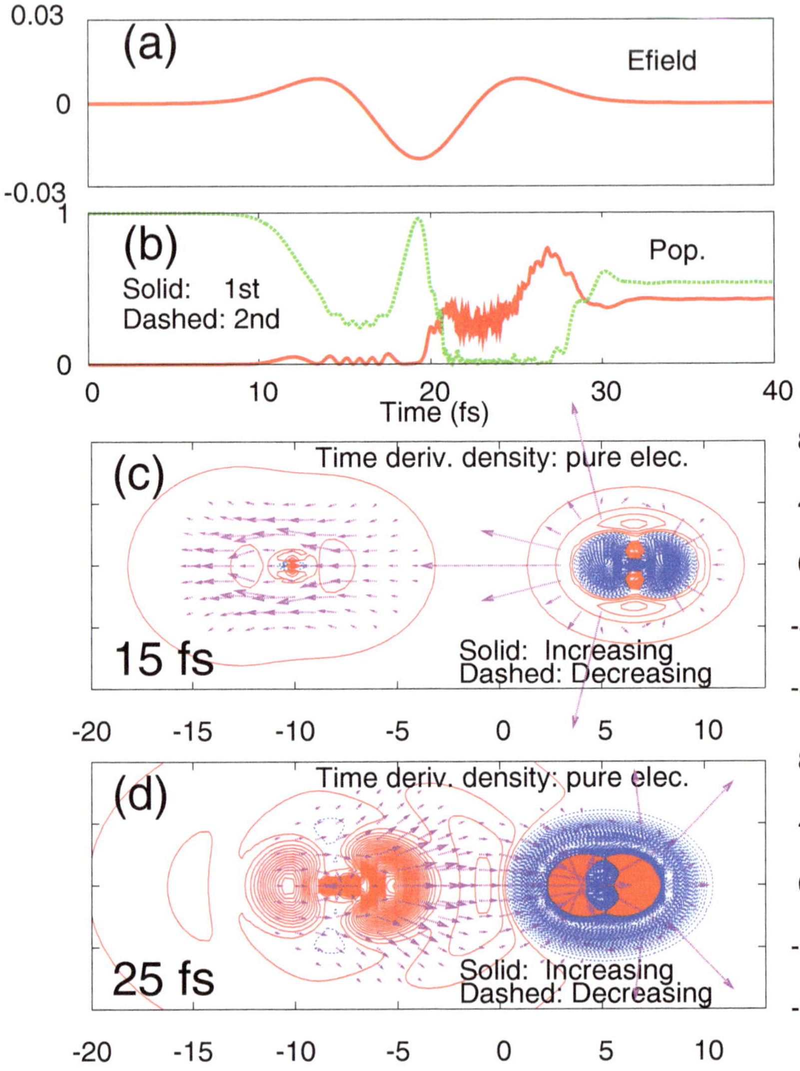

Fig. 8.6 Snapshots of electron current density and time derivative electron density from the pure electronic component without that conveyed by nuclear motion. The corresponding pulse width is 4.84 fs starting from the first excited state, $Na^{+}+Cl^{-}$ (second panel from the left, bottom row in Fig. 8.4). The time dependent behavior of external electronic field and state population dynamics are given in the panels (a) and (b), respectively. The snapshots at 15 and 25 fs are shown in (c) and (d), respectively. The arrows denote electron flux density. Here pure electronic means only the time derivative of electronic wave function was utilized in the evaluation both in flux and time derivative density.

8.4.2 *Collision of LiH + H*$^+$

A very basic and simple molecular collision, LiH + H^+ in a laser field, is investigated with the semiclassical Ehrenfest theory. The potential surfaces for $(LiH_2)^+$ system are drawn in Figs. 8.7 and 8.8.

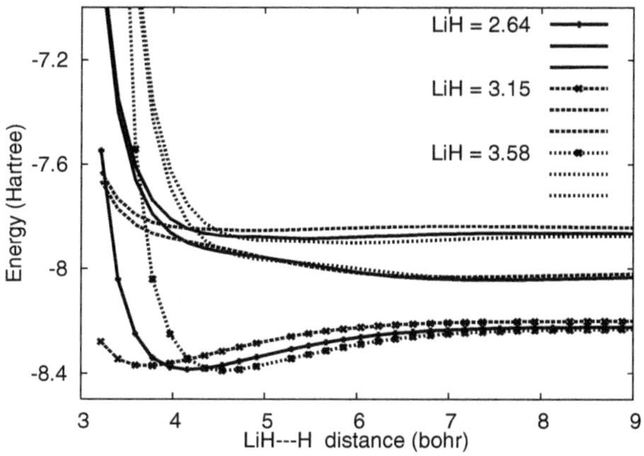

Fig. 8.7 $(LiH_2)^+$ system, adiabatic potential energy curves for the distance between the center of mass of LiH and H in a linear geometry. Three selected cases are plotted, for LiH length = 2.64, 3.15 and 3.58 bohrs. The ground state is correlated asymptotically to $(Li{-}H_{(1)})^+{—}H_{(2)}$, while the first excited state is to $(Li{-}H_{(1)}){—}H_{(2)}{}^+$. (Reprinted with permission from K. Takatsuka *et al.*, Adv. Chem. Phys. 144, 93 (2010)).

[Collinear collision of LiH + H$^+$] The collinear collision under the field free condition undergoes a simply (electronically) elastic scattering, that is

$$\mathrm{LiH} + \mathrm{H}^+ \to \mathrm{LiH} + \mathrm{H}^+. \tag{8.55}$$

This is actually the first excited state in the asymptotic region. In all the following cases, the relative kinetic energy is 0.1 hartree. The electron density map in the course of the reaction is shown in the left column of Fig. 8.9. (Note that time proceeds from bottom to top.)

The parameters of laser pulse shined on this system are $E_s = 0.02$, $\omega = 0.3$, $t_c = 7.26$ fs, $t_w = 2.42$ fs, with the polarization vector on the molecular axis. The timing is adjusted so that the laser peak becomes maximum around the time when H^+ begins to climb the gradual potential slope. The reaction has changed to

$$\mathrm{LiH} + \mathrm{H}^+ \to \begin{cases} \to \mathrm{LiH} + \mathrm{H}^+ \\ \to \mathrm{Li} + \mathrm{H}_2^+ \end{cases} \tag{8.56}$$

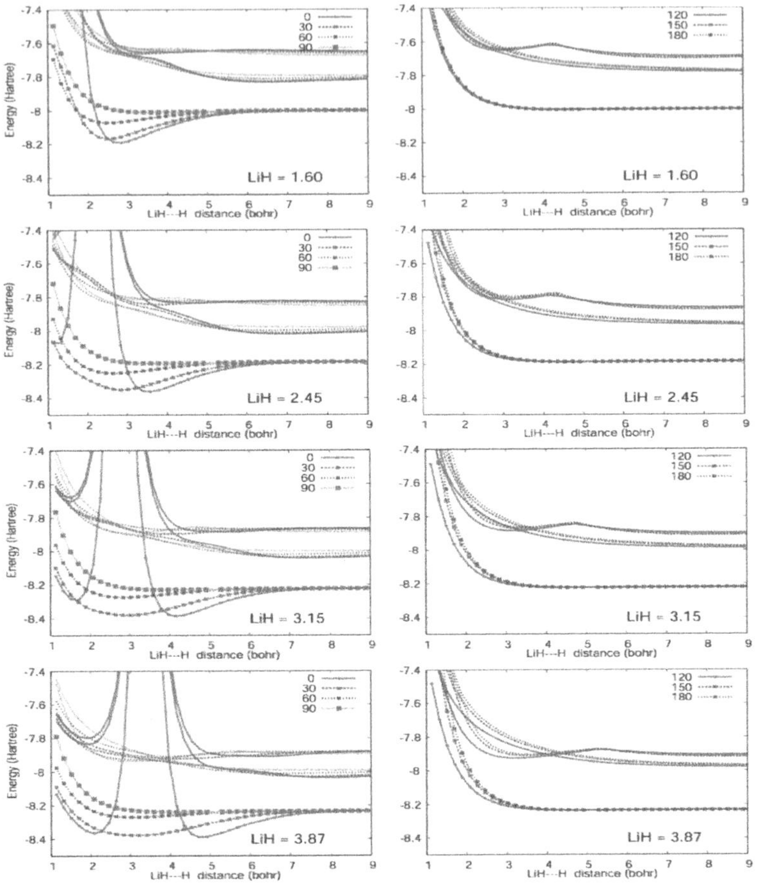

Fig. 8.8 $(LiH_2)^+$ system; adiabatic potential energy curves for the distance between the center of mass of LiH and H with various insertion angles with respect to the LiH and colliding H atom for the LiH length = 1.60, 2.45, 3.15 and 3.87 bohrs. (Reprinted with permission from K. Takatsuka *et al.*, Adv. Chem. Phys. 144, 93 (2010)).

with a large probability of the abstraction reaction. The right column of Fig. 8.9 shows the electron density of the reaction in the field. (At time $t = 0$ in this figure, a large amount of electrons has already shifted from LiH to H^+.

Some physical quantities (Mulliken charge, bond lengths, three low-lying adiabatic energies, and their difference) along the present reaction trajectory are depicted for the field-free case in Fig. 8.11 and for the reaction under laser field in Fig. 8.12. These two figures clearly illustrate how the

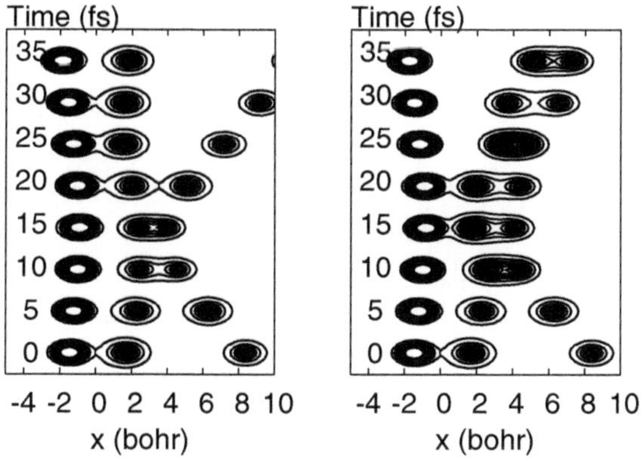

Fig. 8.9 Effect of laser on collinear collision of LiH + H^+. (Left panel) electronically elastic scattering of LiH + H^+ in the first excited state without laser. (Right panel) abstraction reaction under the laser field of E_s=0.02 a.u., ω = 0.3 a.u., t_c=7.26 fs, t_w=2.42 fs. The electric field is linearly polarized along the collision coordinate. (Reprinted with permission from K. Takatsuka *et al.*, Adv. Chem. Phys. 144, 93 (2010)).

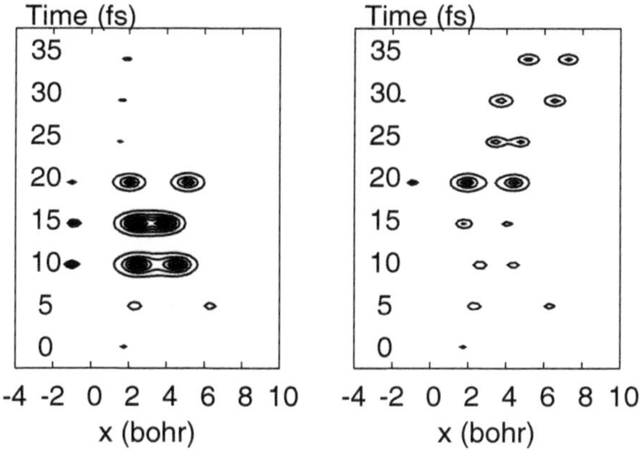

Fig. 8.10 Unpaired electron density to characterize the difference in the collinear collision of LiH+H^+ within the previous figure. (Left panel) electronically elastic scattering of LiH+H^+ in the first excited state without laser. (Right panel) abstraction reaction. (Reprinted with permission from K. Takatsuka *et al.*, Adv. Chem. Phys. 144, 93 (2010)).

laser field dramatically changed the states. Through ab initio analysis of the time-dependent electronic wavefunction, we can see what kind of difference in the electronic structure is produced by the laser. The comparison of the unpaired electron density is given in Fig. 8.10. It is very clearly shown that in the field-free case very large unpaired electron density is observed in H−H during $t = 10 - 20$ fs. This implies that the stable ground state of hydrogen molecule H−H cannot be produced. On the other hand, due to the presence of the laser, such unpairing is minimized, suggesting that two electrons can share a single molecular orbital to form a hydrogen molecule. This suggests one of the possible guiding principles to control the electronic states. The following target is naturally aimed at a method to create and/or eliminate such bonding states at desired places of a molecular system. This is principally as well as practically possible using the notion of a *dynamics filter* with respect to the laser pulse field as an external environment for molecule and the *state mixture realized in time-dependent electronic structures*.

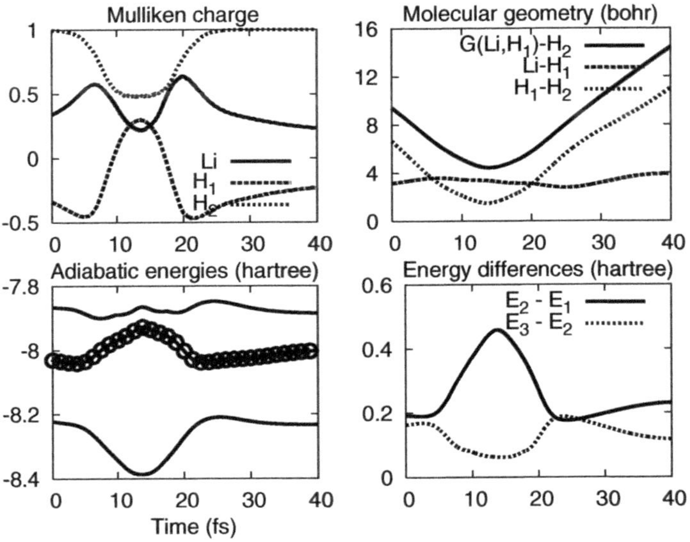

Fig. 8.11 Variation of some relevant quantities during the collinear collision of LiH + H^+ in the first excited state without external field. (Reprinted with permission from K. Takatsuka *et al.*, Adv. Chem. Phys. 144, 93 (2010)).

[Planer collisions] Let us collide H^+ with LiH with no impact parameter but different relative angles other than 180° (collinear collision). LiH is placed on the x-coordinate (Li at left and H at right). The collision angle θ is measured by an angle between the x-coordinate and the line connecting H^+ and the centroid of LiH [$(\text{LiHH})^+$ is $\theta = 0°$, while for $(\text{HLiH})^+$ $\theta = 180°$]. Three configurations, $\theta = 30°$, $60°$, and $90°$ are examined. In

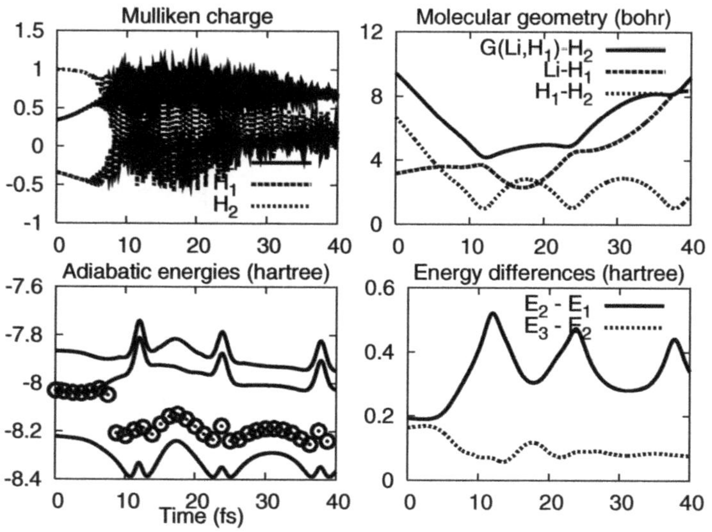

Fig. 8.12 Variation of some relevant quantities during the collinear collision of LiH + H^+ in the first excited state under the laser field of $E_S = 0.02$, $\omega = 0.3$, $t_c = 7.26$ fs, $t_w = 2.42$ fs. (Reprinted with permission from K. Takatsuka *et al.*, Adv. Chem. Phys. 144, 93 (2010)).

Fig. 8.13, we have traced the dynamics of the Mulliken charge (top row), bond lengths (the second row), bond orders (the third row), and three low-lying adiabatic energies (the bottom row). The left-most, the middle, and the right-most columns are for the collision angle of $\theta = 30°$, $60°$, and $90°$, respectively.

For these arrangements, the linearly polarized laser in the x-direction was shined as above but the Gaussian peak time t_c has been selected as 8.96, 12.10 and 13.31 fs for the angle $\theta = 30°, 60°$ and $90°$, respectively. Figure 8.14 is the counterpart of Fig. 8.13 in case of the presence of such laser fields. Three significant differences caused by the laser are (i) vibration excitation of product H–H molecule for all cases, (ii) creation of long lived $(LiH_2)^+$ compound for $\theta = 30°$ and (iii) remarkable change in the reaction channel for $\theta = 60°$. As is seen, the present presentation is not a comprehensive study of the entire feature of the reaction. For instance, the polarization vector of the laser field should play an important role. Many parameters should be scanned to make a complete list of the reaction.

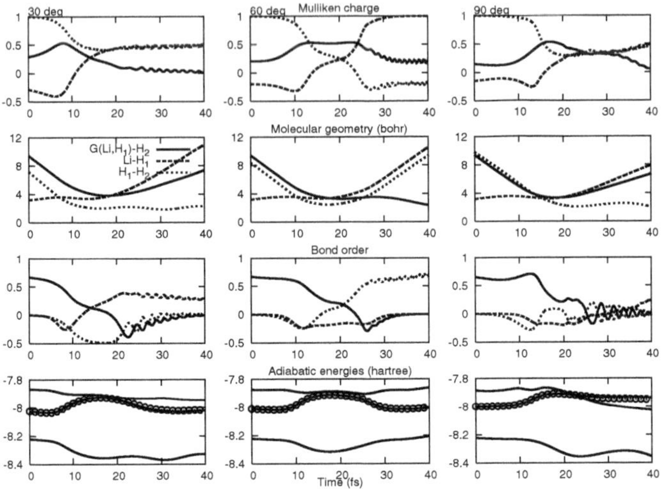

Fig. 8.13 LiH $+H^+$ inserting collision dynamics on the first excited state without external field. The circles in the panel for adiabatic energy denote mean electronic energy. (Reprinted with permission from K. Takatsuka *et al.*, Adv. Chem. Phys. 144, 93 (2010)).

8.4.3 *Diborane in laser field*

The next case study is made for diborane, $H_2BH_2BH_2$, which is well-known to have electron-deficient chemical bonding [35, 50, 69, 109]. That is, only two electrons are involved in a bond B-H-B, having an anomalous valence at the central H atom. Not only the molecular structure but the electronic structures are also quite interesting. Due to the coexistence of rich reactivity and stability strongly associated with the boron atom, this molecule appears in many processes of inorganic/organic chemical reactions [40, 231]. This molecule still has an important position in the study of the nature of chemical bonds [342, 385]. So many quantum mechanical studies, therefore, have been performed within the static Born–Oppenheimer view. We revisit this molecule from the view point of electron wavepacket dynamics coupled with nuclear motion, and attempt to probe the characteristics of this molecule by examining its response to intense laser fields. To the best of our knowledge, there is no theoretical study on the field induced electrons-nuclei coupled molecular dynamics in this system. We study this molecule in the following three aspects, (i) bond formation of diborane by collision between two monoboranes, (ii) attosecond electron wavepacket dynamics in the ground state and first excited state, (iii) induced fragmentation back to borane molecules by a linearly polarized laser.

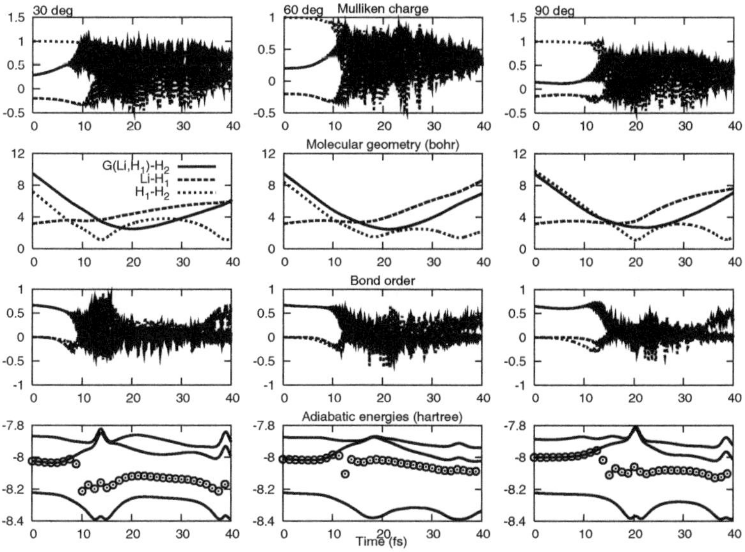

Fig. 8.14 LiH+H$^+$ insertion collision dynamics on the first excited state under the laser field. E_s, ω, and t_w are 0.02, 0.3, and 2.42 fs, respectively. The Gaussian peak times t_c are 8.96, 12.10 and 13.31 fs for the insertion angle $\theta = 30°$, $60°$ and $90°$, respectively. (Reprinted with permission from K. Takatsuka *et al.*, Adv. Chem. Phys. 144, 93 (2010)).

As the frequency of polarized pulse laser, we choose those of IR and UV nonresonant region (the laser parameters used are summarized in Table 8.1). The choice of the IR laser is due to our recognition that the vibronic (electron-phonon) coupling must be particularly crucial in the time propagation of electron wavepackets and nuclear dynamics under laser field. We will show that this is indeed the case, emphasizing the critical effect of the nuclear derivative (momentum) coupling elements.

Table 8.1 Conversion table for field parameters.

ω (a.u.)	$\hbar\omega$ (ev)	$\lambda^{a)}$ (nm)
0.057	1.55	800
0.228	6.20	200

E_s (a.u.)	I (W cm^{-2})
0.03	3.15 $\times 10^{13}$
0.10	3.50 $\times 10^{14}$

a) Optical wave length.

8.4.3.1 *Computational set up*

Laser field applied As stated above, we treat the electron-field interaction operator, V^{opt}, in the velocity form as in Eq. (8.57) below. We further

Table 8.2 Bond lengths (in units of Bohr) and bond angles (in degrees) of diborane at equilibrium geometry.

HB_1	B_1H_3	B_1B_2	H_3H_4	HB_1H	$H_3B_1H_4$	$B_1H_3B_2$
2.239	2.486	3.360	3.665	122°	95°	85°

Table 8.3 Bond orders of diborane in the first two states at equilibrium geometry of the ground electronic state.

Electronic state	HB_1	B_1H_3	B_1B_2	H_3H_4
Ground	0.858	0.399	0.284	-0.133
First excited	0.742	0.356	-0.160	-0.185

Table 8.4 Orbital energy of the canonical RHF molecular orbitals of diborane in units of Hartree.

1	2	3	4	5	6
−7.6150	−7.6145	−0.8868	−0.6386	−0.5551	−0.5397
7	HOMO	LUMO	10	11	12
−0.5188	−0.4728	0.1213	0.2199	0.2867	0.3044

apply a long wave length approximation and Coulomb gauge to the vector potential, and neglect the terms proportional to $\frac{e^2}{c^2}A^2$, which leads to the form

$$\hat{V}_{opt} = \mathbf{A}(t) \sum_j^{elec} \frac{e}{c}\hat{\mathbf{p}}_j. \tag{8.57}$$

The vector potential for a linearly polarized laser pulse employed here is of a Gaussian form

$$\mathbf{A}(t) = A_s \mathrm{exp}\left(-\left(\frac{t-t_c}{t_w}\right)^2\right) \cos(\omega(t-t_c)+\delta)\cdot\hat{u}, \tag{8.58}$$

where $\hat{u}$ denotes a unit direction vector. The symbols t_c, t_w, ω and δ are the Gaussian central time, time width, frequency, and carrier phase, respectively. This form gives the following electronic field,

$$\mathbf{E}(t) = \frac{A_s\omega}{c} e^{-[(t-t_c)/t_w]^2} \left(\sin(\omega t+\delta) + \frac{2(t-t_c)}{\omega t_\omega^2}\cos(\omega(t-t_c)+\delta)\right)\cdot\hat{u}. \tag{8.59}$$

According to the usual convention, we define the electric field strength E_s as

$$E_s \equiv \frac{\omega}{c} A_s \tag{8.60}$$

and it is the standard practice to define the field intensity as

$$I \equiv \frac{\epsilon_0\, \omega^2\, A_s^2}{2\, c} = \frac{c\, \epsilon_0\, E_s^2}{2}, \tag{8.61}$$

where ϵ_0 is the dielectric constant in the vacuum. Our practice in this article is to specify E_s first to label the field strength, and then convert it to A_s in Eq. (8.60). Then we actually use Eq. (8.58) to carry out the computations.

For a linear combination of two fields and/or for a circularly polarized laser we represent them as

$$\begin{aligned}\mathbf{A}(t) = A_{s1} \exp\left(-\left(\frac{t - t_{c1}}{t_{w1}}\right)^2\right) \cos(\omega_1\, (t - t_{c1}) + \delta_1) \cdot \hat{u}_1 \\ + A_{s2} \exp\left(-\left(\frac{t - t_{c2}}{t_{w2}}\right)^2\right) \sin(\omega_2\, (t - t_{c2}) + \delta_2) \cdot \hat{u}_2.\end{aligned} \tag{8.62}$$

In what follows, we set all the phases $\delta = 0$, $\delta_1 = 0$ and $\delta_2 = 0$ just for simplicity. Note that we can construct the circularly polarized laser pulse field by setting $A_{s1} = A_{s2}$, $t_{c1} = t_{c2}$, $t_{w1} = t_{w2}$, $w_1 = w_2$ and $\hat{u}_1 \perp \hat{u}_2$.

System studied The level of electronic states and electron wavepackets is CIS/RHF/6-31G(d) with core 1s like orbitals of boron atoms. Only the spin singlet-state manifold is considered in the electronic state calculations. The resultant number of molecular orbitals (MOs) and configuration state functions (CSFs) are 205 and 42, respectively. The profile of optimized structure is shown in Fig. 8.15. Molecular optimized geometry at the ground state is summarized in Table 8.2. The vertical excitation energy between the ground and the first excited states at this geometry is 0.3115 Hartree. The first excited state is characterized by HOMO→LUMO transition from Hartree–Fock ground configuration state and the coefficient of this configuration is 0.9808. This is qualitatively consistent with the value 0.9471 that was obtained with a rather high level calculation CISD/RHF/6-31G(d). The present computational level has been determined to compromise between computational capacity and resultant accuracy.

Some basic properties of the electronic structure are summarized below. The nodal structures of HOMO and LUMO, which are directly related to the electronic character of the first excited state, are shown in Fig. 8.16. The optical transition between the ground and first excited states is symmetry forbidden. The bond orders of HB_1, B_1H_3, B_1B_2, and H_3H_4 for ground and first excited states at the ground state optimized geometry are summarized in Table 8.3. The bond order of B_1H_3, which is far below unity, characterizes the electron deficiency of HBH chemical bonding.

The orbital energies are listed in Table 8.4. The energy gap of HOMO and LUMO is quite large, which should help diborane to be stable despite its electron deficient bonding. Besides, as observed in Fig. 8.16, the LUMO is non-bonding for the four central BH bonds but is not anti-bonding either. This constitutes a reason why the first excited state does not lead to dissociation of any bond.

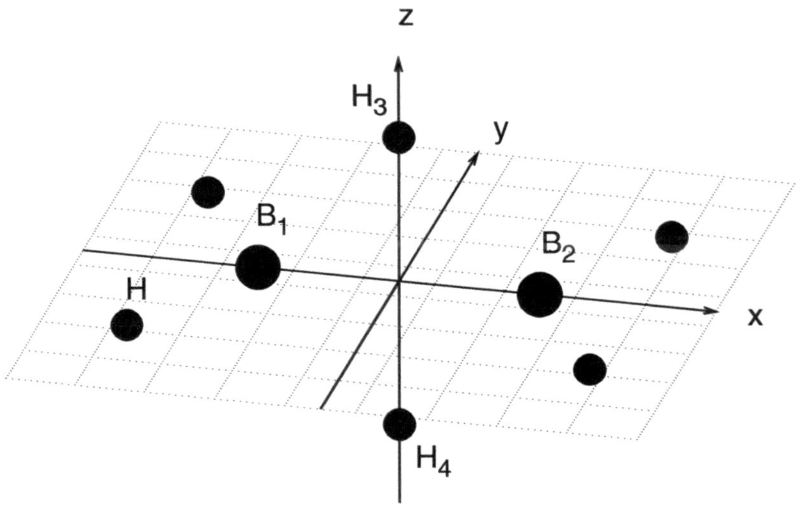

Fig. 8.15 The equilibrium structure of diborane at its ground electronic state. The hydrogen atoms other than H_3 and H_4 lie on the xy plane. (Reprinted with permission from T. Yonehara *et al.*, Chem. Phys. 366, 115 (2009)).

8.4.3.2 *Chemical bonding in diborane and its response to laser*

We study the nature of the chemical bonding of diborane from the following three aspects: (i) The attosecond electron wavepacket dynamics in two bound states, the first state (starting from the ground state) and the second state (starting from the first excited state), and their response to the circularly polarized laser. (ii) Induced fragmentation back to borane molecules by linearly polarized laser, examining how robust the electron deficient chemical bonding of diborane is.

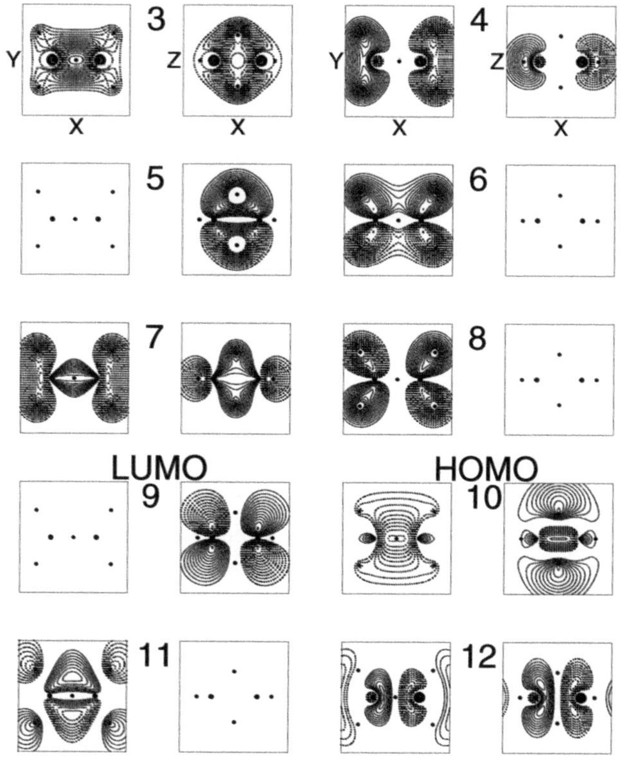

Fig. 8.16 The RHF molecular orbitals higher than the second one at the ground equilibrium structure. Each orbital is numbered as indicated in between two panels, the left of which displays its nodal pattern on the xy plane (molecular plane), and the right of which does on the xz plane (including B-B bond and vertical to the molecular plane). The HOMO and LUMO correspond to the 8th and 9th orbital, respectively. (Reprinted with permission from T. Yonehara *et al.*, Chem. Phys. 366, 115 (2009)).

8.4.3.3 *Attosecond response of the chemical bonds to circularly polarized laser fields*

We next study the response of dynamical electrons to circularly polarized pulse laser in the ground and the first-excited states of diborane.

Population dynamics in response to the laser fields First of all, we examine the population dynamics of the two low-lying states. Since we study the time-dependent response of these electronic states to laser fields, the ordering of the states with respect to the adiabatic electronic energy does not make an invariant sense. Hence we simply call the first state in tracking the dynamics of a state whose initial state is the electronic ground state before shining a laser. Likewise, our so-called second state corresponds to the first excited state. Two lasers of different frequency $\omega = 0.057$ (non-resonant IR pulse field) and $\omega = 0.228$ (non-resonant UV pulse field) are applied, varying the intensity as $E_s = 0.03$ and $E_s = 0.06$. The survival probability for each state to remain in its original electronic state is exhibited in Fig. 8.17. It is observed that the first state returns to the ground state after the IR laser is switched off irrespective of the intensity under study. A dramatic difference is noticed in the response of the second state. It is almost completely lost into the manifold of the electronic states other than the first excited state. For the UV pulse laser, the results are not as simple, but the second state is also sensitive to the laser of $E_s = 0.06$. In contrast to the IR laser of $E_s = 0.03$, the second state is found to return to the initial state with a high probability. Thus the IR laser turns out to be quite efficient to modify the electronic state, as long as the nonadiabatic coupling between the electronic and nuclear motions are explicitly taken into account (see below). Such a variety of responses of the electronic states are often observed in the realm of rather intense laser fields.

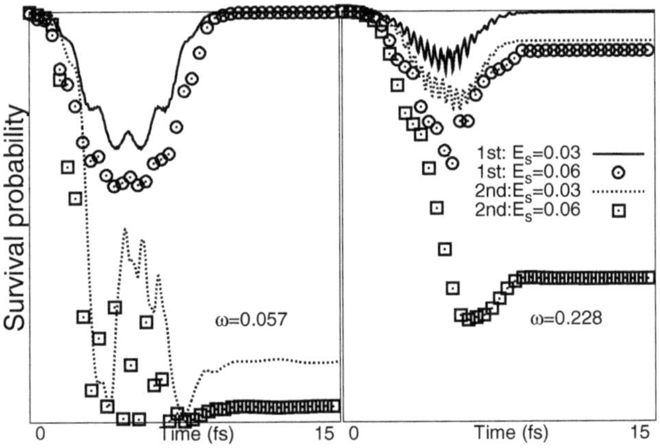

Fig. 8.17 Survival probabilities for the first and second states to remain in the gound and the first-excited electronic eigenstates, respectively, after shining the nonresonant IR (left panel) and UV (right panel) pulse lasers. Two different intensities are examined. (Reprinted with permission from T. Yonehara *et al.*, Chem. Phys. 366, 115 (2009)).

Field induced electron flux and bond order density Expecting that the electronic structure of the electron deficient chemical bonding in diborane must be flexible, we survey its laser response. Figure 8.18 presents the bond order flux and bond order density on the xz plane at the times of (a) 3.25 fs (b) 4.00 (c) 4.50 (d) 5.25 for the first and second states to respond to shining the laser of $\omega = 0.057$ a.u., E_s=0.03 a.u., $t_w = 2.42$ fs, $t_c = 4.84$ fs.

In the upper four panels in Fig. 8.18, we find two parallel flows of electrons, whose direction is approximately along some of the B-H bonds. Within 2 fs they change their directions as seen in all the snapshots. We thus have extracted an attosecond dynamics of electronic motions. In contrast to the first state, the dynamical response to field of the second state is rather complicated (the lower four panels in Fig. 8.18). It looks like a whirlpool roaming over the entire bonding region $B_1H_3H_4B_2$. This active deformation of electronic clouds as represented by these quantities indicates the flexibility of chemical bonding in the first excited state. In particular, it is a remarkable observation that the electron flux is global and collective covering the entire $B_1H_3H_4B_2$ region.

Since the flux in the second state looks like a whirlpool or rotational motion, we quantify the "angular momentum" associated with the above electron flux in the bonding area of $B_1H_3B_2H_4$. To do so, we define one particle electron angular momentum arising from the rotation flux, $\vec{L}(t)$, by

$$\vec{L}(t) \equiv \int d\vec{r}\ \vec{r} \times \vec{j}_{bond-order}(t), \tag{8.63}$$

where $\vec{r}$ and $\vec{j}_{bond-order}(t)$ are electron position vector and bond order flux, respectively. The bond order flux was calculated by using bond-order density matrix in AO base. In the volume integrals, 10 grid points were set from -2 Bohrs to 2 for x direction and 15 points for y and z directions from -3 Bohrs to 3. This space-range covers the central $B_1H_3H_4B_2$ part in the system.

Panels (a) and (b) in Fig. 8.19 represent the time variation of $\vec{L}(t)$ induced by the laser for the first and second states along with the electronic fields. The first state actually gives a clear oscillation in $\vec{L}(t)$, but interestingly, it takes mostly positive values. Especially, in the IR field case, as seen in the panel (a), $\vec{L}(t)$ is synchronized with the electric field. For instance, the maximum positions of $\vec{L}(t)$ and $E_z\,(t)$ happens to coincide almost exactly.

As for the second state, panels (b), the absolute magnitude of $\vec{L}(t)$ is about 20 times larger than that of the first system. Moreover, it takes both positive and negative values. This implies the flux changes its rotational direction rather periodically. Also, it less synchronizes with any of the electric fields. Furthermore, there is some induction time for $\vec{L}(t)$ to begin to change.

The interpretation for the physical origin of the global flow of electrons in the area of $B_1H_3H_4B_2$ and the difference between the first and second states is not easy. For the first state (initial in the ground state), the electronic rearrangement due to the laser seems to be primarily induced by mixing of (or excitation to, so to say) the LUMO (MO #9 in Fig. 8.16). Besides, as seen in the spatial distributions of MO's, it is primarily sufficient to take account of MO's #3, #5, and #7. However, none of them is optically forbidden to mix with LUMO in any direction of the polarized electric field. However, the presence of the nuclear derivative (momentum) coupling, $-i\hbar\sum_k \dot{R}_k X^k_{IJ}$ in Eq. (6.14) mixes the states of different symmetries. Then the laser fields can further mix them in a complicated manner. For the second state, two singly occupied molecular orbitals, namely HOMO and LUMO, can make the situation more complicated, absorbing more photons (recall Fig. (8.17)). We will show how crucial the presence of the nuclear derivative coupling terms are in the next subsection.

Large fluctuation of the electrons and the collective motion behind it It is expected that the electrons involved in the electron-deficient bonds should be flexibly tied with atomic nuclei. To confirm this anticipation, we measure the time dependent position of the center of mass (centroid) of the total electrons as a response to the above circularly polarized laser. The results are shown in panels (a)–(d) in Fig. 8.20. Since the origin is set to the symmetry center of the molecule, the displacement of the centroid from the center indicates the fluctuation of the induced dipole moment of the molecule.

In case of the low frequency IR laser ($\omega = 0.057$, panel (a)) the deviation of the centroid is not large at all. In the UV laser ($\omega = 0.228$, panel (b)), the electronic state is shaked to a large extent. The situation is totally different for the second state as seen in panels (c) and (d). An extremely large fluctuation of the centroid or the induced dipole moment has been realized by the laser. Even the IR laser does an efficient job. The main direction of the dynamical shift of the centroid lies in the x-axis (B-B axis) despite the application of the circularly polarized electric field. The density

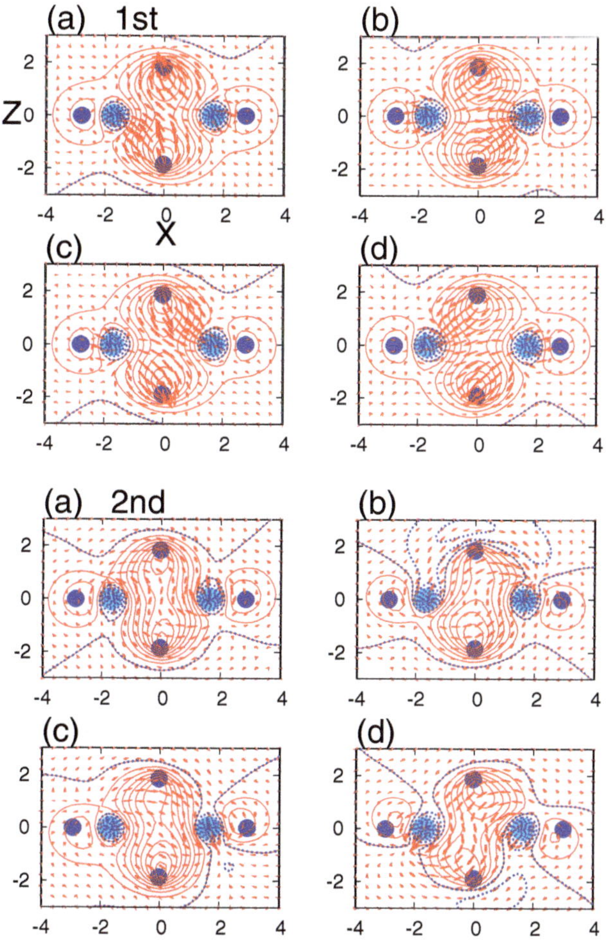

Fig. 8.18 Bond order flux (arrows) and bond order density (contour lines) plotted on xz plane induced by a pulse laser circularly polarized on this plane. The central frequency t_c and field strength E_s shined are 0.057 and 0.03, respectively. The peak time t_c and the characteristic time of pulse decay t_w are 4.84 fs and 2.42, respectively. The solid and dotted contour lines denote the positive and negative bond order densities, respectively. The increment of the contour lines is 0.01. Snapshot times are (a) 3.25 fs (b) 4.00 fs (c) 4.50 fs (d) 5.25 fs. The positions of all atoms are projected onto the the xz plane. The overlapped H atoms outside the B-B bonds are not shown. The flux vectors are multiplied by 20 for presentation. (Reprinted with permission from T. Yonehara *et al.*, Chem. Phys. 366, 115 (2009)).

matrix analysis suggests that the MO #6 and #8 are conceived to help such a large fluctuation in the x-direction.

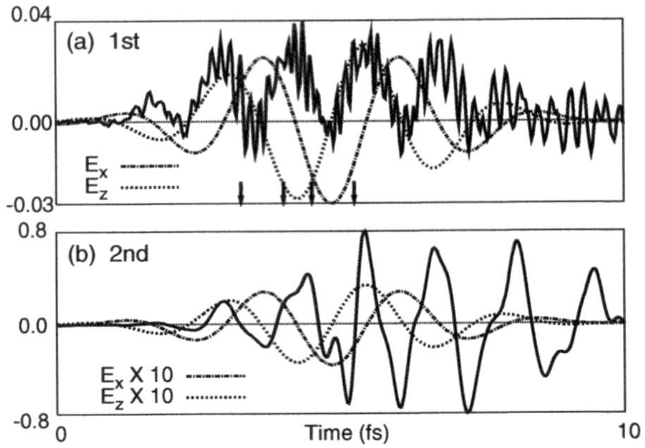

Fig. 8.19 One particle electron angular momentum (defined in the text). The y components of them are presented. The zx circularly polarized pulse laser is applied as in Fig. 8.18; $\omega = 0.057$, $E_s = 0.03$, $t_w = 2.42$ (fs), $t_c = 4.84$ (fs). The x and z components of the electric field are also superposed. The four arrows in the panel (a) denote the times for the bond order flux to be plotted in Fig. 8.18. (Reprinted with permission from T. Yonehara *et al.*, Chem. Phys. 366, 115 (2009)).

We next explore what kind of motion is actually hidden behind such large fluctuation of the electrons. In doing so, we track the dynamics of the bond order. Let $\rho^{bond}(B_1H_3)$ be the bond order of the site B_1H_3, for instance. Then, we take a difference of the bond orders of the facing sites such that

$$\Delta_{fs} = \rho^{bond}(B_1H_3) - \rho^{bond}(B_2H_4), \tag{8.64}$$

connecting the different boron atoms. Similarly, another difference

$$\Delta_{ss} = \rho^{bond}(B_1H_3) - \rho^{bond}(B_1H_4), \tag{8.65}$$

for the bonding site tied to the same boron atom (B_1 in this example). Figure 8.20, panels (e)-(f) shows these quantities with a one-to-one correspondence to panels (a)-(d), respectively. First of all, panel (a) exhibits a very clear feature that Δ_{fs} for the facing pair is almost constant, while Δ_{ss} in the same side oscillates. This implies that the bonds B_1H_3 and B_2H_4 (or B_1H_4 and B_2H_3) become stronger synchronously in phase under the circularly polarized laser field, while the bonds on the same site like B_1H_3 and B_1H_4 become strong in a manner of out-of-phase. Therefore, the centroid of the total electrons averages these motion, resulting in a very small displacement. Thus the small displacement is not inconsistent with the electron flux in the global scale as observed in Fig. 8.18. As an opposite example, let us consider the behavior in panel (h), where the facing pair Δ_{fs} oscillates and the pair in the same side Δ_{ss} remains to be rather constant. This implies that the electrons move from left side (close to B_1) to right (close to B_2) and vice versa. The centroid reflects this dynamics showing the large amplitude motion only in the x-direction. The dynamics of the bond orders in panel (c) is not as simple and is more chaotic. We thus have observed collective motions of the bonds induced by circularly polarized lasers.

8.4.3.4 *Fragmentation of diborane by intense linearly polarized pulse laser*

As the final study of the laser response of diborane, we consider its fragmentation to monoboranes. In contrast to the naive impression that the electron deficient chemical bonding must be weak, diborane is actually stable, even against laser shining. Indeed, it should be recalled that we could not prevent two monoboranes from associating to a diborane with use of an intense laser. Therefore, an attempt of laser fragmentation includes a non-trivial aspect of the diborane stability.

Dependence on the initial states and field direction Since we have observed that the circularly polarized pulse lasers do not dissociate diborane, we here use linearly polarized pulses. Again we examine the two states, the ground and first excited states as the initial electronic states. The pulse center frequency ω, strength E_s, peak time t_c, decay time t_w are 0.057 a.u., 0.1, 4.84 fs, and 2.42, respectively.

First we align the laser polarization vector to the x axis. In Fig. 8.21, we present the time propagation of three atomic distances B_1-B_2, B_1-H_3 and H_3-B_2 for the different initial states as well as different laser polarization directions. Both in panels (a) and (c), this x-polarized laser does not achieve fragmentation, although B_1-B_2 is elongated to somewhat a long distance. In contrast, the second state (the first excited state) is very stable. One major difference between the ground and first excited states is found in the bond order of B_1-B_2 (see Table 8.3). It is positive for the ground state, while negative for the first excited state. A bond having a negative bond order is less significantly weakened. Indeed, we observed much larger deformation of the bond order density in the first state than in the second state (not shown graphically). The x-polarized laser is inactive to dissociate diborane in itself, simply because no occupied state is allowed to excite to the LUMO

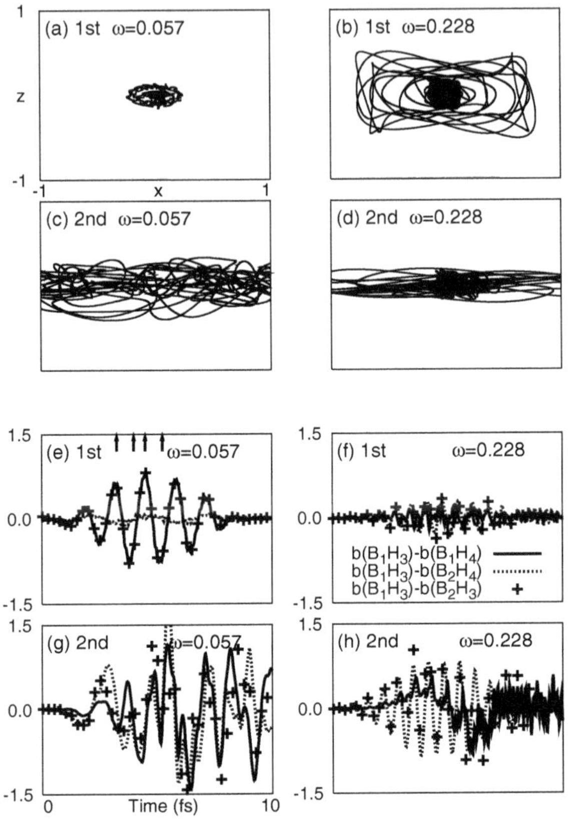

Fig. 8.20 Panel (a)-(d); dynamics of the centroid position of electrons in borane in response to the circularly polarized pulse lase, whose strength E_s, decay time t_w and peak time t_c are 0.03 a.u., 2.42 fs and 4.84, respectively. (a)-(d); relative difference of the bond orders at selected bonds in response to the same laser. The field center frequencies ω are shown in the panels. (Reprinted with permission from T. Yonehara *et al.*, Chem. Phys. 366, 115 (2009)).

in this direction as readily confirmed in Fig. 8.16. Besides, this LUMO is not dissociative. The effect of vibronic coupling will be considered later.

Then, we examine the laser polarization in the direction of $z = x$ with an anticipation that possible cleavage of two pairs of BH bonds may ignite the entire fragmentation of diborane. As clearly seen in the time-dependent variation of B_1B_2 and B_1H_3 distances in panels (b) and (d) of Fig 8.21, the same laser as above efficiently breaks the BH chemical bondings both for the first and second states. To be more precise, the B_1H_3 cleavage precedes a little earlier than the BB bond breaking. This is a clear reflection of the fact that B_1 and B_2 are bound to each other through the BHB bondings. Besides, the monotonic increase of B_1B_2 distance in time and the small change of B_2H_3 length indicate that diborane has been dissociated to two monoboranes. The bond order density in panel (b) and (d) in Fig. 8.22 clarifies how the dissociation proceed.

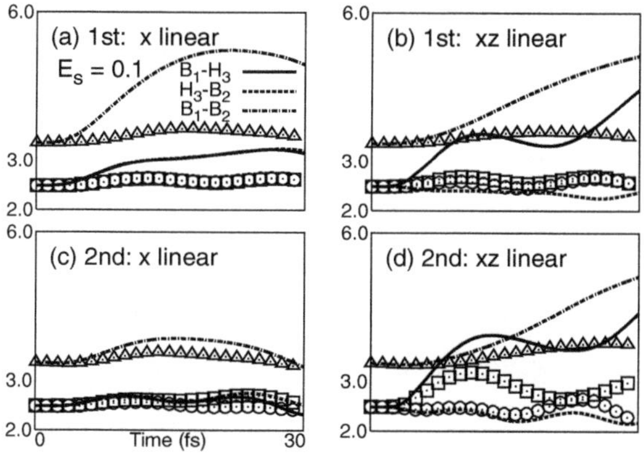

Fig. 8.21 Time propagation of the selected bond lengths, B_1H_3, H_3B_2 and B_1B_2, in response to a pulse laser linearly polarlized in the directions of x ((a),(c)) and $z = x$ ((b), (d)). These are drawn in curves. The results without the derivative coupling are also presented with squares, circles and triangles for B_1H_3, H_3B_2 and B_1B_2, respectively. (Reprinted with permission from T. Yonehara *et al.*, Chem. Phys. 366, 115 (2009)).

Strong vibronic coupling—Role of the nuclear derivative coupling in the fragmentation dynamics It turns out as above that the IR laser works well as a means to dissociate the bonds of diborane. This fact strongly suggests that the vibronic coupling should be critical in these dynamics. To clarify that this is indeed the case, we perform the semiclassical Ehrenfest calculation without the nuclear derivative coupling X^k_{IJ}. It is our experience through our previous applications of the SET to other molecular systems in the absence of laser fields that the CSF gives in general such a good diabatic representation that the role of X^k_{IJ} is often not significant. We here repeat such a comparison, using SET with and without X^k_{IJ}.

Figure 8.21 highlights that a dramatic difference is caused when X^k_{IJ} is neglected: In panels (b) and (d), we clearly observe that diborane is not dissociated at all without X^k_{IJ} for both states. The electric fields given by

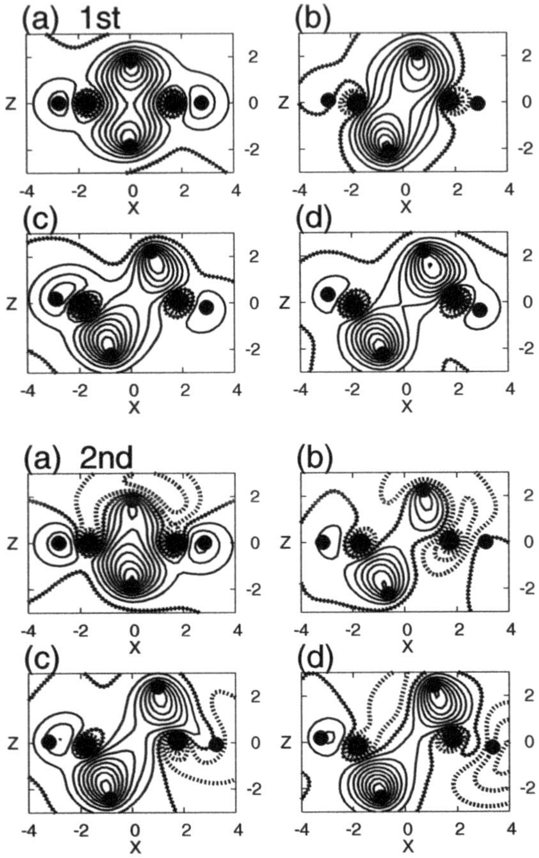

Fig. 8.22 Snapshots of bond order densities in the forced fragmentation by an intense pulse laser linearly polarized in $x = z$ direction (two parallel BH bonding direction). Upper and lower four panels correspond to the first and second states, and the panels (a), (b), (c), and (d) are shapshots at $t = 2.5$, 7.5, 10.0 and 12.5 fs, respectively. (Reprinted with permission from T. Yonehara *et al.*, Chem. Phys. 366, 115 (2009)).

the laser fields simply pull and push the nuclei and electrons in mutually opposite directions. In particular, the IR laser has no ability to make a direct transition of the electronic state to another. Therefore, without X_{IJ}^{k}, there exists only a very weak mechanism for the electrons and nuclei to mutually transfer the energy that they attain from the laser fields. Conversely, the presence of X_{IJ}^{k} can amplify the change of the electronic state and the nuclear dynamical motion.

The universality of the tendency observed here may be justified by comparing this result with those in other molecular systems. In our previous works on several isolated molecules such as methane and methyl alcohol placed in a low frequency laser field [425], we have found the same features that the absence of X^k_{IJ} can significantly underestimate the dynamics of nuclear motions. Therefore the nuclear derivative coupling strongly assists the field induced dynamics.

8.4.3.5 *Summary*

Main findings in this study are: (i) Collective motions of electron flux emerge in the central BHHB bonding area in response to the circularly polarized laser fields. The flexibility of electronic structure has been highlighted by the collective motion of flux. The other collective modes are also found in this bonding area. The origin of the dynamical stability or robustness of the electron-deficient bonding of diborane has been clarified. (ii) It is found that the laser field with a polarization along two parallel BH bonds effectively breaks them, while the laser polarization in the BB direction does not. We also revealed that the chemical bonding with positive bond order shows a sensitive response to the laser field. (iii) Nuclear derivative coupling has a critical role in the field induced fragmentation dynamics and the dynamics under IR laser field in general.

Thus the attosecond electron wavepacket dynamics coupled with nuclear motion with and without laser fields is promising as a method to analyze the nature of chemical bonds and chemical reactions in a forthcoming stage of chemistry.

8.5 Dynamics of photoionization

Electron dynamics of photoionization states in the course of nonadiabatic transition is one of the important molecular processes. It can be applied to the studies on the following dynamical processes: Tracking the history of electronic and vibrational states wandering among the excited states in laser fields with designed optical pulses. These time-dependent transient states should be able to be monitored in terms of time-resolved photoelectron spectroscopy [14, 353]. Inner shell ionization from a deep core level and the concomitant electronic state avalanches (decay) towards the hole, which may be accompanied by autoionization, is also quite interesting. In a molecular avalanche process the angular distribution of photoelectrons can be greatly modulated by nonadiabatic coupling with vibrational modes. Collision induced electron detachment is another interesting example, although it is not an optical process.

We have discussed time-resolved photoelectron spectroscopy in Sec. 3.3.4. In these studies, quantum mechanical stationary-state scattering theory is applied to calculate the photoionization transition amplitudes, in which a scattering orbital is supposed to feel the Coulombic field formed by the Hartree–Fock cation species [250]. The scattering amplitudes for the in-going boundary condition thus obtained are incorporated in the coupled Schrödinger equations for nuclear wavepackets in the Born–Huang expansion. This is quite a powerful and accurate method for single-photon ionization from electronic ground states, and much is still to be done with this approach. It is also known that the scattering amplitudes for electron scattering by electronically excited molecules are well obtained [250, 418, 419]. In these treatments there is no direct need to propagate electron wavepackets explicitly.

However, photoionization in more complicated electron wavepacket dynamics requires different ideas. The theories of exterior complex scaling [438] and infinite range exterior complex scaling [361] give quite interesting methods to impose an outgoing boundary condition on the time-dependent Schrödinger equation for electron wavepackets. In these theories, an appropriate boundary is introduced as in the R-matrix theory in which configuration space is divided into two regions; interior and exterior. The time-dependent wavefunctions inside the boundary are to be connected to the outer (exterior) counterpart to which the complex-valued coordinate rotation is applied. To the best of our knowledge, these methods have been applied only to small atoms in laser fields.

The recent progress in the theories of molecular electron dynamics not only aims at laser science but are directed to the study of chemical reaction dynamics, particularly to exploration of electronic motions that are nonadiabatically coupled with nuclear counterparts. The electronic wavefunctions in these theories are usually expanded in terms of basis functions, thereby taking an advantage of the progress in stationary-state quantum theory of molecular electronic structure. However, one of the crucial difficulties there, either in the presence or absence of laser fields, is lack of a good description of "ionizing" processes in laser-energized states, which are usually represented in terms of linear combinations of the Slater determinants of multiply excited configurations. This difficulty practically hampers the progress of the theory of electron wavepacket dynamics in high energy states created by intense lasers. To overcome we need a good and practical way to estimate how much the resultant electron-loss amounts to and moreover to identify which electronic configurations those electrons are leaving from. We discuss this aspect in this section [428].

8.5.1 *Electron flux and photoelectron signals*

We resume our discussion with a molecular placed in a laser fields as in Eqs. (8.40) and (8.41). Suppose we have $\vec{j}(\mathbf{r},t;\mathbf{R})$, as discussed in Sec. 7.1. Let $\mathbf{R}_c$ be the position of the molecular centroid. Make a sphere of a radius r_B, whose center is located at $\mathbf{R}_c$. A vector normal to the sphere is denoted as $\mathbf{r}_B$, and the polar coordinates are to be used to specify them as $\mathbf{r}_B = (r_B,\theta,\varphi)$. What we really need is the flux that is normal to the sphere, which is readily projected as

$$j_{\perp}(\mathbf{r}_B,t) = j_{\perp}(r_B,\theta,\varphi,t) = \frac{\vec{j}(\mathbf{r}_B,t)\cdot\mathbf{r}_B}{(\mathbf{r}_B\cdot\mathbf{r}_B)^{1/2}}, \tag{8.66}$$

and $j_{\perp}(r_B,\theta,\varphi,t)$ is a probability current flowing out to the angular direction (θ,φ) per unit time and per area on the sphere surface. This gives an instantaneous angle-resolved photoelectron flux in the body frame. To be a little more precise, define a tiny surface area element for integration as $dS = r_B^2 d\Omega$, with $d\Omega$ being the angle element $d\Omega = \sin\theta d\theta d\varphi$. Then, the following quantity

$$\vec{j}(\mathbf{r}_B,t)\cdot d\vec{S} = j_{\perp}(\mathbf{r}_B,t)\, r_B^2 d\Omega \tag{8.67}$$

is the number of electrons counted by a detector in a solid angle $d\Omega$ in a unit time. In addition $j_{\perp}$ is correlated with the wave number of photoelectrons as $k = j_{\perp}m/\hbar$. This assignment is valid only if the boundary at $\mathbf{r}_B$ is sufficiently far from the molecular centroid, and the wavefunction is supposed to have components bearing the information of electrons leaving out of the bounds of molecular interaction regions. Note that this boundary $\mathbf{r}_B$ is not made use of to determine the wavepackets themselves unlike in the R-matrix theory.

8.5.2 *Electron dynamics of ionizing states*

Once a complex valued electronic density matrix is in hand, it is not difficult at all to calculate the electron flux and it is not difficult either to transform the flux information to photoelectron signals (from body frame to laboratory frame), as successfully accomplished by Bandrauk et al. [499, 500]. On the other hand, it is not as simple to track the simultaneous dynamics of remaining electrons in the molecular site while photoelectrons are leaving away. Note that even if we successfully have a rescaled density $\rho(\mathbf{r},t)$ in equation (7.2) with an estimate of the flux, we can make infinitely many (untrue) electronic wavefunctions that can reproduce this same $\rho(\mathbf{r},t)$. The key to be considered is, therefore, to assign which electronic configurations the photoelectrons are leaving from and how consistently we can describe the states of the remaining electronic wavepacket. To do so, we think it to be most appropriate to use natural orbital representation of the electron flux.

8.5.2.1 *Outgoing electron flux with complex natural orbitals*

Let $\phi_\lambda(\mathbf{r},t)'s$ be the complex natural orbitals obtained as discussed in Sec. 7.2.3, with its occupation number n_λ. Let us emphasize again that the natural orbitals are in general complex valued in time-dependent situations and that the natural orbitals given by stationary-state quantum chemistry codes are generally real-valued, and therefore the resultant flux is zero. We here assume that the natural orbitals are orthonormalized. With them, the flux is given as

$$\vec{j}(\mathbf{r},t) = \frac{\hbar}{2i}\sum_\lambda n_\lambda \left[\phi_\lambda^*(\mathbf{r},t)\tilde{\nabla}\phi_\lambda(\mathbf{r},t) - \phi_\lambda(\mathbf{r},t)\tilde{\nabla}\phi_\lambda^*(\mathbf{r},t)\right]. \tag{8.68}$$

One technical matter that should be noted is that the electron flux is usually oscillatory in space-time. To avoid a possible numerical instability, we should take a short-time average of instantaneous values as

$$\vec{j}_{av}(\mathbf{r},t) \equiv \frac{1}{\Delta t}\int_{t-\Delta t/2}^{t+\Delta t/2} \vec{j}(\mathbf{r},s)\,ds, \tag{8.69}$$

which will give more stable results.

8.5.2.2 *Time-dependence of the occupation numbers*

Assume we have an electron wavepacket $\Phi(\mathbf{r}_{all},t;\mathbf{R}(t))$ and the associated natural orbitals $\{\phi_\lambda(\mathbf{r},t)\}$ as a result of electron dynamics for a short time interval $[t - \Delta t, t]$, where $\mathbf{r}_{all}$ denotes the collective representation of all electrons, that is $\mathbf{r}_{all} = \{r_j \,|j = 1, \cdots, 3N_e\}$. At the boundary position $\mathbf{r}_B$ of "molecular surface", $\vec{j}\,(\mathbf{r}_B, t)$ is evaluated with equation (8.68). (Note that for molecular systems the boundary does not have to be a spherical surface.) We further define the similar flux at $\mathbf{r}_B$ in terms of the natural orbital ϕ_λ only such that

$$\vec{j}^{(\lambda)}\,(\mathbf{r}_B, t) = \frac{\hbar}{2i}\left[\phi_\lambda^*(\mathbf{r}_B, t)\tilde{\nabla}\phi_\lambda(\mathbf{r}_B, t) - \phi_\lambda(\mathbf{r}_B, t)\tilde{\nabla}\phi_\lambda^*(\mathbf{r}_B, t)\right], \tag{8.70}$$

and the component in the normal direction as

$$j_\perp^{(\lambda)}\,(r_B, \theta, \varphi, t) = \frac{\vec{j}^{(\lambda)}\,(\mathbf{r}_B, t)\cdot\mathbf{r}_B}{(\mathbf{r}_B\cdot\mathbf{r}_B)^{1/2}}. \tag{8.71}$$

The total flux is simply reproduced such that

$$j_\perp\,(r_B, \theta, \varphi, t) = \sum_\lambda n_\lambda\,(t)\, j_\perp^{(\lambda)}\,(r_B, \theta, \varphi, t)\,. \tag{8.72}$$

The electronic population running away from the molecular site by photoionization at time t is estimated as

$$\Delta\rho\,(r_B, t) = \int_0^{2\pi} d\varphi \int_0^{\pi} \sin\theta d\theta \left[j_\perp\,(r_B, \theta, \varphi, t)\, r_B^2\right]. \tag{8.73}$$

Since this quantity is a time-density, integration of it as

$$\int_{t-\Delta t}^{t} \Delta\rho\,(r_B, s)\, ds, \tag{8.74}$$

is the population that has decreased from the sphere in this time interval.

Do the same thing for the population arising from each natural orbital as

$$\Delta\rho^{(\lambda)}\,(r_B, t) = \int_{t-\Delta t}^{t} ds\; n_\lambda\,(s) \int_0^{2\pi} d\varphi \int_0^{\pi} \sin\theta d\theta \left[j_\perp^{(\lambda)}\,(r_B, \theta, \varphi, s)\, r_B^2\right], \tag{8.75}$$

which is the loss of the population in a natural orbital density $\rho^{(\lambda)}\,(r_B, t)$. Since this quantity is equivalent to the decrease of the occupation number associated with ϕ_λ, which is denoted as $\Delta n_\lambda\,(t)$, we simply have

$$\Delta n_\lambda\,(t) = \Delta\rho^{(\lambda)}\,(r_B, t)\,. \tag{8.76}$$

Therefore due to the leak for a short while, the occupation number should be reduced to

$$n_\lambda^n\,(t) = n_\lambda\,(t) - \Delta n_\lambda\,(t)\,, \tag{8.77}$$

and accordingly the density matrix itself is also reduced to $\rho^{new}(\mathbf{r}', \mathbf{r}, t)$, that is

$$\rho^{new}(\mathbf{r}', \mathbf{r}, t) = \sum_{\lambda} \left(\sqrt{n_\lambda^n(t)} \phi_\lambda^*(\mathbf{r}', t) \right) \left(\sqrt{n_\lambda^n(t)} \phi_\lambda(\mathbf{r}, t) \right). \qquad (8.78)$$

Note that the number of leaking electrons can be fractional during a given time of photon irradiation. The fractional number is related to quantum mechanical ionization probability. Also more than one electron can be lost in the end, although the present one-electron density matrix formalism has no way to express the two electron correlation between receding electrons.

Suppose we expand the electron wavepacket $\Phi(\mathbf{r}_{all}, t; \mathbf{R}(t))$ before reducing the occupation numbers as above. Let us write it as

$$\Phi(\mathbf{r}_{all}, t; \mathbf{R}(t)) = \sum_{I} C_I^{NO}(t) \Phi^{NO}(\mathbf{r}_{all}; \mathbf{R})\big|_{\mathbf{R}=\mathbf{R}(t)}, \qquad (8.79)$$

where $\Phi_I^{NO}(\mathbf{r}_{all}; \mathbf{R})$ are

$$\Phi_I^{NO}(\mathbf{r}_{all}; \mathbf{R}) = \text{A Slater determinant consisting of } \{\phi_\lambda(\mathbf{r}, t)\}, \qquad (8.80)$$

and $C_I^{NO}(t)$ are the associated coefficients. The suffix NO on Φ_I^{NO} recalls that the determinants are made up with natural orbitals but not with the ordinary canonical molecular orbitals. We next take account of the population loss due to ionization during this time interval. Then using the above procedure, each occupation number should be changed as

$$n_\lambda(t) \rightarrow n_\lambda^n(t). \qquad (8.81)$$

Accordingly, the individual Slater determinants should be rescaled as

$$\begin{aligned} &\Phi_I^{NO,new}(\mathbf{r}_{all}; \mathbf{R}) \\ &\quad = \text{A Slater determinant consisting of } \left\{ \frac{\sqrt{n_\lambda^n(t)}}{\sqrt{n_\lambda(t)}} \phi_\lambda(\mathbf{r}, t) \right\}. \end{aligned} \qquad (8.82)$$

Then each Slater determinant should be changed as

$$\begin{aligned} &C_I^{NO}(t)\Phi_I^{NO}(\mathbf{r}_{all}; \mathbf{R}) \\ &\quad \rightarrow C_I^{NO}(t)\Phi_I^{NO,new}(\mathbf{r}_{all}; \mathbf{R}) = C_I^{NO,new}(t)\Phi_I^{NO}(\mathbf{r}_{all}; \mathbf{R}), \end{aligned} \qquad (8.83)$$

and the total wavepacket becomes

$$\Phi(\mathbf{r}_{all}, t; \mathbf{R}(t)) = \sum_{I} C_I^{NO,new}(t) \Phi_I^{NO}(\mathbf{r}_{all}; \mathbf{R})\Big|_{\mathbf{R}=\mathbf{R}(t)}. \qquad (8.84)$$

Thus the Slater determinants that contain natural orbitals of the larger leaking flux lose the more population from them. With these $\Phi(\mathbf{r}_{all}, t; \mathbf{R}(t))$

as a renewed initial condition, we proceed to the next step to take account of the dynamics starting to solve equation (6.14). In this way, the components of the total population-loss have been uniquely ascribed to the individual Slater determinants, which in turn are to be rendered to the electronic time-propagation for the next short-time step.

8.5.2.3 *Technical matters*

There are two important parameters for the present method to be numerically successful. One is the position of the boundary at which the electron flux is calculated, and the other is the number of basis functions. The theory should work better as the boundary radius (the distance between the boundary and the molecular center) is taken to be longer. In fact, exact photoelectron spectroscopic data should be obtained if a complete basis set (including plane waves or outgoing spherical waves) could be used within a very large boundary surface. On the other hand, the longer is the boundary radius, the more diffused basis functions and/or floating functions localized at many places are required, and consequently the more difficult becomes the computation. Therefore we have to optimize performance in these two factors at the level of accuracy required.

The time scale for electron dynamics studied above is very short, typically shorter than 1 femtosecond, during which the nuclei are supposed to stay still. In the above process, the electron loss has been taken into account through the change of the population of natural orbitals and subsequently the change of the coefficients of the Slater determinants. Therefore the same set of self-consistent field molecular orbitals (SCF MO) may be used for the short time period. However, for a longer time scale, it would be better to redetermine the molecular orbitals under the gradual loss of electronic population. This demands an efficient way to determine the molecular orbitals with variable and fractional occupation numbers, although this is not a new problem for quantum chemistry.

8.5.3 *Summary*

The outline of the procedure is as follows. (1) Run an electron wavepacket dynamics at a certain time, which gives rise to a complex valued electronic wavefunction. (2) Calculate a set of complex-valued natural orbitals (NO) with real-valued occupation numbers. (3) With these NOs, estimate the outgoing electron flux at a predetermined surface area of the molecule and the loss of electronic population. (4) Reduction of the electronic population from each electronic configuration is performed with a small change (decrease) of the occupation numbers, which then reorganizes the total electronic wavefunction. (5) Propagate thus reorganized electronic states (expanded in CSF) in the laser field for a next short time.

Epilogue

With the chapter "Aim of This Book: Where Are We?" we began to describe this book, which has been followed by the chapters

"Basic Framework of Theoretical Chemistry",

"Nuclear Dynamics on Adiabatic Electronic Potential Energy Surfaces",

"Breakdown of the Born–Oppenheimer Approximation: Classic Theories of Nonadiabatic Transitions and Ideas behind",

"Direct Observation of the Wavepacket Bifurcation due to Nonadiabatic Transitions",

"Nonadiabatic Electron Wavepacket Dynamics in Path-branching Representation",

"Dynamical Electron Theory for Chemical Reactions",

"Molecular Electron Dynamics in Laser Fields".

Through this path, we have tried to penetrate into a new theoretical framework for chemistry based on nonadiabatic electronic and nuclear dynamics (in laser fields). However, we here stop further exploration of the world beyond the Born–Oppenheimer paradigm. This is simply because the volume of this book is already large but not since we came to the final end of the realm. On the contrary, our journey has just begun, and we have provided with just elementary theoretical ideas, theories, practices, and so on. Nevertheless, the authors hope the readers to feel that there is a vast interesting area of molecular science that can be scrutinized only with the method of nonadiabatic electronic and nuclear dynamics, or, nonadiabatic dynamical electron theory.

Finally, we apologize that we could not introduce all the important works that have been done before. This is simply due to the volume limitation but not for any academic reasons. Also the list of references of this book is far from complete. In particular, many important works in quantum chemistry for the calculations of nonadiabatic matrix elements, which are very much involved, are missing in this volume. We hope the readers to find by themselves in the literature when they need in practical calculations.

Before concluding this volume we the authors thank our families, friends, and colleagues for having made it possible for us to keep enjoying science.

Bibliography

[1] M. Abe, Y. Ohtsuki, Y. Fujimura, Z. Lan, and W. Domcke, J. Chem. Phys. **124**, 224316 (2006).

[2] A. Accardi, I. Barth, O. Kühn, and J. Manz, J. Phys. Chem. A **114**, 11252 (2010).

[3] P. K. Agarwal, S. R. Billeter, P. T. R. Rajagopalan, S. J. Benkovic, and S. Hammes-Schiffer, Proc. Natl. Acad. Sci. U.S.A. **99**, 2794 (2002).

[4] Y. Aharonov and D. Bohm, Phys. Rev. **115**, 485 (1959).

[5] H. Akima, ACM Trans. Math. Software **17**, 341 (1991).

[6] S. Alfalah, O. Deeb, S. Zilberg, and Y. Haas, Chem. Phys. Lett. **459**, 100 (2008).

[7] A. S. Alnaser, B. Ulrich, X. M. Tong, I. V. Litvinyuk, C. M. Maharjan, P. Ranitovic, T. Osipov, R. Ali, S. Ghimire, Z. Chang, C. D. Lin, and C. L. Cocke, Phys. Rev. A. **72**, 030702(R) (2005).

[8] S. C. Althorpe and T. Seideman, J. Chem. Phys. **110**, 147 (1999).

[9] J. Alvarellos and H. Metiu, J. Chem. Phys. **88**, 4957 (1988).

[10] M. Amano and K. Takatsuka, J. Chem. Phys. **122**, 084113 (2005).

[11] K. Ando and J. T. Hynes, J. Phys. Chem. A **103**, 10398 (1999).

[12] D. Andrae, I. Barth, T. Bredtmann, H.-C. Hege, J. Manz, F. Marquardt, and B. Paulus, J. Phys. Chem. B **115**, 5476 (2011).

[13] Y. Arasaki and K. Takatsuka, Chem. Phys. **338**, 175 (2007).

[14] Y. Arasaki and K. Takatsuka, ChemPhysChem **14**, 1387 (2013).

[15] Y. Arasaki, K. Takatsuka, K. Wang, and V. McKoy, J. Chem. Phys. **112**, 8871 (2000).

[16] Y. Arasaki, K. Takatsuka, K. Wang, and V. McKoy, J. Chem. Phys. **114**, 7941 (2001).

[17] Y. Arasaki, K. Takatsuka, K. Wang, and V. McKoy, J. Chem. Phys. **119**, 7913 (2003).

[18] Y. Arasaki, K. Takatsuka, K. Wang, and V. McKoy, J. Chem. Phys. **132**, 124307 (2010).

[19] Y. Arasaki, K. Wang, V. McKoy, and K. Takatsuka, Phys. Chem. Chem. Phys. **13**, 8681 (2011).

[20] Y. Arasaki, K. Wang, V. McKoy, and K. Takatsuka, J. Phys. B **45**, 194006 (2012).
[21] A. Assion, T. Baumert, M. Bergt, T. Brixner, B. Kiefer, V. Seyfried, M. Strehle, and G. Gerber, Science **282**, 919 (1998).
[22] A. Assion, M. Geisler, J. Helbing, V. Seyfried, and T. Baumert, Phys. Rev. A **54**, R4605 (1996).
[23] G. J. Atchity and K. Ruedenberg, Theor. Chem. Acc. **97**, 47 (1997).
[24] S. H. Autler and C. H. Townes, Phys. Rev. **100**, 703 (1955).
[25] R. F. W. Bader, Chem. Rev. **91**, 893 (1991).
[26] M. Baer, Chem. Phys. Lett. **35**, 112 (1975).
[27] M. Baer, Phys. Rep. **358**, 75 (2002).
[28] M. Baer, *Beyond Born–Oppenheimer: Electronic Nonadiabatic Coupling Terms and Conical Intersections* (Wiley, Hoboken, New Jersey, 2006).
[29] N. Balakrishnan, C. Kalyanaraman, and N. Sathyamurthy, Phys. Rep. **280**, 79 (1997).
[30] A. Baltuška, T. Udem, M. Uiberacker, M. Hentschel, E. Goulielmakis, C. Gohle, R. Holzwarth, V. S. Yakovlev, A. Scrinzi, T. W. Hänsch, and F. Krausz, Nature (London) **421**, 611 (2003).
[31] P. Baltzer, L. Karlsson, B. Wannberg, D. M. P. Holland, M. A. MacDonald, M. A. Hayes, and J. H. D. Eland, Chem. Phys. **237**, 451 (1998).
[32] A. D. Bandrauk, S. Chelkowski, D. J. Diestler, J. Manz, and K.-J. Yuan, Phys. Rev. A **79**, 023403 (2009).
[33] A. D. Bandrauk, S. Chelkowski, S. Kawai, and H. Lu, Phys. Rev. Lett. **101**, 153901 (2008).
[34] A. D. Bandrauk, E.-W. S. Sedik, and C. F. Matta, J. Chem. Phys. **121**, 7764 (2004).
[35] K. E. Banyard and H. Shull, J. Chem. Phys. **44**, 384 (1966).
[36] R. Bartels, S. Backus, I. Christov, H. Kapteyn, and M. Murnane, Chem. Phys. **267**, 277 (2001).
[37] A. Bartelt, A. Lindinger, C. Lupulescu, Š. Vajda, and L. Wöste, Phys. Chem. Chem. Phys. **5**, 3610 (2003).
[38] I. Barth, H.-C. Hege, H. Ikeda, A. Kenfack, M. Koppitz, J. Manz, F. Marquardt, and G. K. Paramonov, Chem. Phys. Lett. **481**, 118 (2009).
[39] I. Barth, J. Manz, Y. Shigeta, and K. Yagi, J. Am. Chem. Soc. **128**, 7043 (2006).
[40] S. H. Bauer, Chem. Rev. **96**, 1907 (1996).
[41] T. Baumert and G. Gerber, Adv. At. Mol. Opt. Phys. **35**, 163 (1995).
[42] T. Baumert, J. L. Herek, and A. H. Zewail, J. Chem. Phys. **99**, 4430 (1993).
[43] M. H. Beck, A. Jäckle, G. A. Worth, and H.-D. Meyer, Phys. Rep. **324**, 1 (2000).
[44] M. Ben-Nun and T. J. Martínez, J. Chem. Phys. **108**, 7244 (1998).
[45] V. A. Benderskii, D. E. Makarov, and C. A. Wright, Adv. Chem. Phys. **88**, 1 (1994).
[46] V. A. Benderskii, D. E. Makarov, and C. A. Wight, Adv. Chem. Phys. **88**, 55 (1994).

[47] R. V. Bensasson, E. J. Land, and T. G. Truscott, *Excited States and Free Radicals in Biology and Medicine: Contributions from Flash Photolysis and Pulse Radiolysis* (Oxford University Press, Oxford, 1993), Chap. 5.
[48] M. V. Berry, Proc. R. Soc. London A **392**, 45 (1984).
[49] X.-B. Bian and A. D. Bandrauk, Phys. Rev. Lett. **105**, 093903 (2010).
[50] J. Bicerano, D. S. Marynick, and W. N. Lipscomb, Inorg. Chem. **17**, 3443 (1978).
[51] Ö. Birer and M. Havenith, Annu. Rev. Phys. Chem. **60**, 263 (2009).
[52] C. Z. Bisgaard, O. J. Clarkin, G. Wu, A. M. D. Lee, O. Geßner, C. C. Hayden, and A. Stolow, Science **323**, 1464 (2009).
[53] E. R. Bittner and P. J. Rossky, J. Chem. Phys. **103**, 8130 (1995), *ibid.* **107**, 8611 (1997).
[54] V. Blanchet, M. Z. Zgierski, T. Seidemann, and A. Stolow, Nature (London) **401**, 52 (1999).
[55] M. Böckmann, D. Marx, C. Peter, L. D. Site, K. Kremer, and N. L. Doltsinis, Phys. Chem. Chem. Phys. **13**, 7604 (2011).
[56] K. Bolton in *Modern Methods for Multidimensional Dynamics Computations in Chemistry*, edited by D. L. Thompson (World Scientific, Singapore,1998), p. 143.
[57] V. Bonačić-Koutecký and R. Mitrić, Chem. Rev. **105**, 11 (2005).
[58] M. Born and K. Huang, *Dynamical Theory of Crystal Lattices* (Oxford Univ. Press, Oxford, 1954).
[59] M. Born and R. Oppenheimer, Ann. Phys. (Leipzig) **389**, 457 (1927).
[60] *Proton Transfer in Hydrogen-Bonded Systems*, edited by T. Bountis (Plenum Press, New York, 1992).
[61] T. Brabec and F. Krausz, Rev. Mod. Phys. **72**, 545 (2000).
[62] A. E. Bragg, J. R. R. Verlet, A. Kammrath, O. Cheshnovsky, and D. M. Neumark, Science **306**, 669 (2004).
[63] M. Braun, C. Meier, and V. Engel, J. Chem. Phys. **105**, 530 (1996).
[64] C. Brif, R. Chakrabarti, and H. Rabitz, New J. Phys. **12**, 075008 (2010).
[65] P. Brumer and M. Shapiro, Chem. Phys. Lett. **126**, 541 (1986).
[66] F. V. Bunkin and I. I. Tugov, Phys. Rev. A. **8**, 601 (1973).
[67] I. Burghardt, L. S. Cederbaum, and J. T. Hynes, Faraday Discuss. **127**, 395 (2004).
[68] I. Burghardt and J. T. Hynes, J. Phys. Chem. A **110**, 11411 (2006).
[69] L. Burnelle and J. J. Kaufman, J. Chem. Phys. **43**, 3540 (1965).
[70] K. Burnett, V. C. Reed, J. Cooper, and P. L. Knight, Phys. Rev. A **45**, 3347 (1992).
[71] A. Carrera, I. B. Nielsen, P. Çarçabal, C. Dedonder, M. Broquier, C. Jouvet, W. Domcke, and A. L. Sobolewski, J. Chem. Phys. **130**, 024302 (2009).
[72] J. J. Carrera, X. M. Tong, and S.-I Chu, Phys. Rev. A **74**, 023404 (2006).
[73] A. Chakraborty and S. Hammes-Schiffer, J. Chem. Phys. **129**, 204101 (2008).
[74] A. Chakraborty, M. V. Pak, and S. Hammes-Schiffer, Phys. Rev. Lett. **101**, 153001 (2008).

[75] Y.-T. Chang, Y. Yamaguchi, W. H. Miller, and H. F. Schaefer III, J. Am. Chem. Soc. **109**, 7245 (1987).
[76] Z. Chang and P. Corkum, J. Opt. Soc. Am. B **27**, B9 (2010).
[77] S. Chapman, Adv. Chem. Phys. **82**, 423 (1992).
[78] E. Charron and A. Suzor-Weiner, J. Chem. Phys. **108**, 3922 (1998).
[79] S. Chelkowski and A. D. Bandrauk, Phys. Rev. A **81**, 062101 (2010).
[80] S. Chelkowski, T. Zuo, O. Atabek, and A. D. Bandrauk, Phys. Rev. A. **52**, 2977 (1995); S. Chelkowski, A. Conjusteau, T. Zuo, and A. D. Bandrauk, *ibid.* **54**, 3235 (1996).
[81] Y. Chen, P. Gross, V. Ramakrishna, H. Rabitz, and K. Mease, Chem. Phys. Lett. **252**, 447 (1996).
[82] *Femtochemistry: Ultrafast Chemical and Physical Processes in Molecular Systems*, edited by M. Chergui (World Scientific, Singapore, 1996).
[83] M. S. Child, J. Mol. Spectrosc. **53**, 280 (1974).
[84] M. S. Child, *Molecular Collision Theory* (Academic Press, New York, 1974).
[85] *Semiclassical Mechanics with Molecular Approximations* (Claredon Press, Oxford, 1991).
[86] M. S. Child, Mol. Phys. **32**, 1495 (1976).
[87] T. H. Choi and K. D. Jordan, Chem. Phys. Lett. **475**, 293 (2009); also see K. D. Jordan and M. A. Johnson, Science **329**, 42 (2010).
[88] S.-I Chu, Chem. Phys. Lett. **54**, 367 (1978).
[89] S.-I Chu, Chem. Phys. Lett. **58**, 462 (1978).
[90] S.-I Chu and D. A. Telnov, Phys. Rep. **390**, 1 (2004).
[91] T.-S. Chu, Y. Zhang, and K.-L. Han, Int. Rev. Phys. Chem. **25**, 201 (2006).
[92] D. T. Colbert and W. H. Miller, J. Chem. Phys. **96**, 1982 (1992).
[93] J. E. Combariza, N. R. Kestner, and J. Jortner, J. Chem. Phys. **100**, 2851 (1994).
[94] P. Cong, G. Roberts, J. L. Herek, A. Mohktari, and A. H. Zewail, J. Phys. Chem. **100**, 7832 (1996).
[95] D. L. Cooper, R. F. Barrow, J. Verges, C. Effantin, and J. d'Incan, Can. J. Phys. **62**, 1543 (1984).
[96] P. B. Corkum, Phys. Rev. Lett. **71**, 1994 (1993).
[97] P. B. Corkum and F. Krausz, Nat. Phys. **3**, 381 (2007).
[98] C. E. Crespo-Hernández, B. Cohen, P. M. Hare, and B. Kohler, Chem. Rev. **104**, 1977 (2004).
[99] D. S. F. Crothers, Adv. Phys. **20**, 405 (1971).
[100] B. F. E. Curchod and I. Tavernelli, J. Chem. Phys. **138**, 184112 (2013).
[101] J. M. Dahlström, A. L'Huillier, and A. Maquet, J. Phys. B **45**, 183001 (2012).
[102] K. Daigoku, S. Ishiuchi, M. Sakai, M. Fujii, and K. Hashimoto, J. Chem. Phys. **119**, 5149 (2003).
[103] C. Daniel, M.-C. Heitz, J. Manz, and C. Ribbing, J. Chem. Phys. **102**, 905 (1995).
[104] M. Dantus, R. M. Bowman, and A. H. Zewail, Nature (London) **343**, 737 (1990).

[105] M. Dantus and V. V. Lozovoy, Chem. Rev. **104**, 1813 (2004).
[106] C. E. Dateo and H. Metiu, J. Chem. Phys. **95**, 7392 (1991).
[107] J. A. Davies, J. E. LeClaire, R. E. Continetti, and C. C. Hayden, J. Chem. Phys. **111**, 1 (1999).
[108] V. Deckert, K. Iwata, and H. Hamaguchi, J. Photochem. Photobiol. A **102**, 35 (1996).
[109] D. J. DeFrees, K. Raghavachari, H. B. Schlegel, J. A. Pople, and P. v. R. Schleyer, J. Phys. Chem. **91**, 1857 (1987).
[110] A. Delon and R. Jost, J. Chem. Phys. **110**, 4300 (1999).
[111] A. Delon, R. Jost, and M. Lombardi, J. Chem. Phys. **95**, 5701 (1991).
[112] A. Delon, R. Jost, and M. Jacon, J. Chem. Phys. **114**, 331 (2001).
[113] M. J. S. Dewar, *Hyperconjugation* (Ronald Press, New York, 1962), p. 43.
[114] D. J. Diestler, Phys. Rev. A **78**, 033814 (2008).
[115] D. J. Diestler, J. Phys. Chem. A **116**, 2728 (2012).
[116] D. J. Diestler, J. Phys. Chem. A **117**, 4698 (2013).
[117] *Conical Intersections: Electronic Structure, Dynamics & Spectroscopy*, edited by W. Domcke, D. R. Yarkony, and H. Köppel, Advanced Series in Physical Chemistry, Vol. 15 (World Scientific, Singapore, 2004).
[118] N. Došlić, J. Phys. Chem. A **110**, 12400 (2006).
[119] T. J. Dudley, R. M. Olson, M. W. Schmidt, and M. S. Gordon, J. Comput. Chem. **27**, 352 (2006).
[120] T. H. Dunning, Jr., J. Chem. Phys. **55**, 716 (1971).
[121] T. H. Dunning, Jr., J. Chem. Phys. **90**, 1007 (1989).
[122] T. H. Dunning, Jr. and P. J. Hay, in *Modern Theoretical Chemistry*, edited by H. F. Schaefer III (Plenum, New York, 1977), Vol. 2, Chap. 1.
[123] O. T. Ehrler and D. M. Neumark, Acc. Chem. Res. **42**, 769 (2009).
[124] J. H. D. Eland and L. Karlsson, Chem. Phys. **237**, 139 (1998).
[125] A. T. J. B. Eppink, B. J. Whitaker, E. Gloaguen, B. Soep, A. M. Coroiu, and D. H. Parker, J. Chem. Phys. **121**, 7776 (2004).
[126] M. G. Evans and M. Polanyi, Trans. Faraday Soc. **31**, 875 (1935).
[127] H. Eyring, J. Chem. Phys. **3**, 107 (1935).
[128] M. B. Faist and R. D. Levine, J. Chem. Phys. **64**, 2953 (1976).
[129] M. D. Feit, J. A. Fleck, Jr., and A. Steiger, J. Comput. Phys. **47**, 412 (1982).
[130] H. Feshbach, Ann. of Phys. **5**, 357 (1958); Ann. of Phys. **19**, 287 (1962).
[131] A. Ferretti, G. Granucci, A. Lami, M. Persico, and G. Villani, J. Chem. Phys. **104**, 5517 (1996).
[132] I. Fischer, D. M. Villeneuve, M. J. J. Vrakking, and A. Stolow, J. Chem. Phys. **102**, 5566 (1995).
[133] S. A. Fischer, C. T. Chapman, and X. Li, J. Chem. Phys. **135**, 144102 (2011).
[134] G. Floquet, Ann. Sci. Éc. Norm. Supér. 2 **12**, 47 (1883).
[135] T. Frigato, J. VandeVondele, B. Schmidt, C. Schütte, and P. Jungwirth, J. Phys. Chem. A **112**, 6125 (2008).
[136] H. Fujisaki and K. Takatsuka, Phys. Rev. E **63**, 066221 (2001).
[137] H. Fujisaki and K. Takatsuka, J. Chem. Phys. **114**, 3497 (2001).

[138] F. X. Gadéa and T. Leininger, Theor. Chem. Acc. **116**, 566 (2006).
[139] B. C. Garrett, M. J. Redmon, D. G. Truhlar, and C. F. Melius, J. Chem. Phys. **74**, 412 (1981).
[140] B. C. Garrett and D. G. Truhlar, J. Chem. Phys. **79**, 4931 (1983).
[141] F. I. Gauthey, C. H. Keitel, P. L. Knight, and A. Maquet, Phys. Rev. A **52**, 525 (1995).
[142] R. Georges, A. Delon, and R. Jost, J. Chem. Phys. **103**, 1732 (1995).
[143] D. Geppert, A. Hofmann, and R. de Vivie-Riedle, J. Chem. Phys. **119**, 5901 (2003).
[144] D. Geppert and R. de Vivie-Riedle, Chem. Phys. Lett. **404**, 289 (2005).
[145] G. N. Gibson, M. Li, C. Guo, and J. Neira, Phys. Rev. Lett. **79**, 2022 (1997).
[146] T. J. Giese and D. M. York, J. Chem. Phys. **120**, 7939 (2004).
[147] D. Giulini, E. Joos, C. Kiefer, J. Kupsch, I.-O. Stamatescu, and H. D. Zeh, *Decoherence and the Appearance of a Classical World in Quantum Theory* (Springer, New York, 1996).
[148] A. Giusti-Suzor, F. H. Mies, L. F. DiMauro, E. Charron, and B. Yang, J. Phys. B **28**, 309 (1995).
[149] A. Gonzalez-Lafont, T. N. Truong, and D. G. Truhlar, J. Chem. Phys. **95**, 8875 (1991).
[150] M. S. Gordon, D. G. Fedorov, S. R. Pruitt, and L. V. Slipchenko, Chem. Rev. **112**, 632 (2012).
[151] T. P. M. Goumans and J. Kästner, J. Phys. Chem. A **115**, 10767 (2011).
[152] G. Grégoire, M. Mons, I. Dimicoli, F. Piuzzi, E. Charron, C. Dedonder-Lardeux, C. Jouvet, S. Martrenchard, D. Solgadi, and A. Suzor-Weiner, Eur. Phys. J. D **1**, 187 (1998).
[153] J. Grosser, O. Hoffmann, C. Rakete, and F. Rebentrost, J. Phys. Chem. A **101**, 7627 (1997).
[154] H. Haberland, C. Ludewigt, H.-G. Schindler, and D. R. Worsnop, J. Chem. Phys. **81**, 3742 (1984).
[155] M. D. Hack, A. M. Wensman, D. G. Truhlar, M. Ben-Nun, and T. J. Martínez, J. Chem. Phys. **115**, 1172 (2001).
[156] J. K. Hale and H. Koçak, *Dynamics and Bifurcations* (Springer-Verlag, New York, 1991).
[157] E. Haller, H. Köppel, and L. S. Cederbaum, J. Mol. Spectrosc. **111**, 377 (1985).
[158] N. I. Hammer, J.-W. Shin, J. M. Headrick, E. G. Diken, J. R. Roscioli, G. H. Weddle, and M. A. Johnson, Science **306**, 675 (2004).
[159] S. Hammes-Schiffer, Chem. Rev. **110**, 6937 (2010).
[160] S. Hammes-Schiffer, E. Hatcher, H. Ishikita, J. H. Skone, and A. V. Soudackov, Coord. Chem. Rev. **252**, 384 (2008).
[161] S. Hammes-Schiffer and A. V. Soudackov, J. Phys. Chem. B **112**, 14108 (2008).
[162] Y.-C. Han and L. B. Madsen, Phys. Rev. A **81**, 063430 (2010).
[163] Y.-C. Han, K.-J. Yuan, W.-H. Hu, and S.-L. Cong, J. Chem. Phys. **130**, 044308 (2009).

[164] K. Hanasaki and K. Takatsuka, Phys. Rev. A **81**, 052514 (2010).
[165] K. Hanasaki and K. Takatsuka, Phys. Rev. A **88**, 053426 (2013).
[166] E. J. Hart, Science **146**, 19 (1964).
[167] E. J. Hart and J. W. Boag, J. Am. Chem. Soc. **84**, 4090 (1962).
[168] P. Hartman, *Ordinary Differential Equations* (John Wiley & Sons, New York, 1964).
[169] K. Harumiya, I. Kawata, H. Kono, and Y. Fujimura, J. Chem. Phys. **113**, 8953 (2000).
[170] K. Harumiya, H. Kono, Y. Fujimura, I. Kawata, and A. D. Bandrauk, Phys. Rev. A **66**, 043403 (2002).
[171] S. Hayashi, J. Umemura, S. Kato, and K. Morokuma, J. Phys. Chem. **88**, 1330 (1984).
[172] H.-C. Hege, J. Manz, F. Marquardt, B. Paulus, and A. Schild, Chem. Phys. **376**, 46 (2010).
[173] T. G. Heil and A. Dalgarno, J. Phys. B **12**, L557 (1979).
[174] W. Heitler, *The Quantum Theory of Radiation*, 3rd ed. (Oxford University Press, Oxford, 1954).
[175] W. Heitler and F. London, Z. Phys. **44**, 455 (1927).
[176] T. Helgaker, P. Jørgensen, and J. Olsen, *Molecular Electronic-Structure Theory* (John Wiley & Sons, New York, 2000).
[177] E. J. Heller, J. Chem. Phys. **75**, 2923 (1981).
[178] V. H. Hellmann, *Einführung in die Quantenchemie*, (Leipzig, Wien, 1937), Chap. 8.
[179] M. Hentschel, R. Kienberger, C. Spielmann, G. A. Reider, N. Milosevic, T. Brabec, C. Corkum, U. Heinzmann, M. Drescher, and F. Krausz, Nature (London) **414**, 509 (2001).
[180] J. M. Herbert and M. Head-Gordon, Proc. Natl. Acad. Sci. U.S.A. **103**, 14282 (2006).
[181] M. S. Herman, Ann. Henri Poincaré **11**, 23 (2010).
[182] H. Higuchi and K. Takatsuka, Phys. Rev. E **66**, 035203(R) (2002).
[183] G. Hirsch, R. J. Buenker, and C. Petrongolo, Mol. Phys. **70**, 835 (1990).
[184] G. Hirsch, R. J. Buenker, and C. Petrongolo, Mol. Phys. **73**, 1085 (1991).
[185] D. M. Hirst, J. Chem. Phys. **115**, 9320 (2001).
[186] T. Ho and S.-I Chu, Chem. Phys. Lett. **141**, 315 (1987).
[187] P. von den Hoff, S. Thallmair, M. Kowalewski, R. Siemering, and R. de Vivie-Riedle, Phys. Chem. Chem. Phys. **14**, 14460 (2012).
[188] M. R. Hoffman, D. J. Fox, J. F. Gaw, Y. Osamura, Y. Yamaguchi, R. S. Grev, G. Fitzgerald, H. F. Schaefer III, P. J. Knowles, and N. C. Handy, J. Chem. Phys. **80**, 2660 (1984).
[189] A. Hofmann and R. de Vivie-Riedle, Chem. Phys. Lett. **346**, 299 (2001).
[190] T. Horio, T. Fuji, Y.-I. Suzuki, and T. Suzuki, J. Am. Chem. Soc. **131**, 10392 (2009).
[191] K. P. Huber and G. Herzberg, *Molecular Spectra and Molecular Structure. IV. Constants of Diatomic Molecules* (Van Nostrand, New York, 1979).
[192] M. H. V. Huynh and T. J. Meyer, Chem. Rev. **107**, 5004 (2007).

[193] J. Itatani, J. Levesque, D. Zeidler, H. Niikura, H. Pépin, J. C. Kieffer, P. B. Corkum, and D. M. Villeneuve, Nature (London) **432**, 867 (2004).
[194] M. Ivanov, P. B. Corkum, T. Zuo, and A. Bandrauk, Phys. Rev. Lett. **74**, 2933 (1995).
[195] S. Iwata and T. Tsurusawa, Adv. Met. Semicond. Clusters **5**, 39 (2001).
[196] G. K. Jarvis, Y. Song, C. Y. Ng, and E. R. Grant, J. Chem. Phys. **111**, 9568 (1999).
[197] A. W. Jasper, M. D. Hack, A. Chakraborty, D. G. Truhlar, and P. Piecuch, J. Chem. Phys. **115**, 7945 (2001).
[198] A. Jasper, B. K. Kendrick, C. A. Mead, and D. G. Truhlar, in *Modern Trends in Chemical Reaction Dynamics: Experiment and Theory*, edited by X. Yang and K. Liu (World Scientific, Sinapore, 2004), Part I Chap. 8.
[199] K. D. Jordan, Science **306**, 618 (2004).
[200] R. Jost, J. Nygøard, A. Pasinski, and A. Delon, J. Chem. Phys. **105**, 1287 (1996).
[201] C. Jouvet, S. Martrenchard, D. Solgadi, C. Dedonder-Lardeux, M. Mons, G. Grégoire, I. Dimicoli, F. Piuzzi, J. P. Visticot, J. M. Mestdagh, P. D'Oliveira, P. Meynadier, and M. Perdrix, J. Phys. Chem. A **101**, 2555 (1997).
[202] M. Joyeux, R. Jost, and M. Lombardi, J. Chem. Phys. **119**, 5923 (2003).
[203] R. S. Judson and H. Rabitz, Phys. Rev. Lett. **68**, 1500 (1992).
[204] T. Kato and H. Kono, J. Chem. Phys. **128**, 184102 (2008).
[205] T. Kato, H. Kono, M. Kanno, Y. Fujimura, and K. Yamanouchi, Laser Phys. **19**, 1712 (2009).
[206] A. Kelly and R. Kapral, J. Chem. Phys. **133**, 084502 (2010).
[207] A. Kenfack, I. Barth, F. Marquardt, and B. Paulus, Phys. Rev. A **82**, 062502 (2010).
[208] A. Kenfack, F. Marquardt, G. K. Paramonov, I. Barth, C. Lasser, and B. Paulus, Phys. Rev. A **81**, 052502 (2010).
[209] L. Kevan, Acc. Chem. Res. **14**, 138 (1981).
[210] M. A. Khakoo, L. Hong, B. Kim, C. Winstead, and V. McKoy, Phys. Rev. A **81**, 022720 (2010); C. Winstead, V. McKoy, and S. d'Almeida Sanchez, J. Chem. Phys. **127**, 085105 (2007); D. W. Flaherty, M. A. Kasper, J. E. Baio, D. B. Graves, H. F. Winters, C. Winstead, and V. McKoy, J. Phys. D **39**, 4393 (2006).
[211] J. Kim, S. B. Suh, and K. S. Kim, J. Chem. Phys. **111**, 10077 (1999); H. M. Lee, S. Lee, and K. S. Kim, *ibid.* **119**, 187 (2003).
[212] M. F. Kling and M. J. J. Vrakking, Annu. Rev. Phys. Chem **59**, 463 (2008).
[213] S. Klinkusch, T. Klamroth, and P. Saalfrank, Phys. Chem. Chem. Phys. **11**, 3875 (2009).
[214] P. J. Knowles and H.-J. Werner, Chem. Phys. Lett. **115**, 259 (1985).
[215] P. J. Knowles and H.-J. Werner, Chem. Phys. Lett. **145**, 514 (1988).
[216] H. Kono, Y. Sato, N. Tanaka, T. Kato, K. Nakai, S. Koseki, and Y. Fujimura, Chem. Phys. **304**, 203 (2004).

[217] H. Köppel, W. Domcke, and L.S. Cederbaum, Adv. Chem. Phys. **57**, 59 (1984); D. M. Leitner, H. Köppel, and L. S. Cederbaum, J. Chem. Phys. **104**, 434 (1996).
[218] H. Köppel, Faraday Discuss. **127**, 35 (2004).
[219] D. Kosloff and R. Kosloff, J. Comput. Phys. **52**, 35 (1983).
[220] R. Kosloff, J. Phys. Chem. **92**, 2087 (1988).
[221] P. M. Kraus, Y. Arasaki, J. B. Bertrand, S. Patchkovskii, P. B. Corkum, D. M. Villeneuve, K. Takatsuka, and H. J. Wörner, Phys. Rev. A **85**, 043409 (2012).
[222] F. Krausz and M. Ivanov, Rev. Mod. Phys. **81**, 163 (2009).
[223] V. Krishna, J. Chem. Phys. **126**, 134107 (2007).
[224] N. M. Kroll and K. M. Watson, Phys. Rev. A **13**, 1018 (1976).
[225] T. Kubař and M. Elstner, Phys. Chem. Chem. Phys. **15**, 5794 (2013).
[226] V. Kurkal, P. Fleurat-Lessard, and R. Schinke, J. Chem. Phys. **119**, 1489 (2003).
[227] L. Lain, A. Torre, R. C. Bochicchio, and R. Ponec, Chem. Phys. Lett. **346**, 283 (2001).
[228] Z. Lan, W. Domcke, V. Vallet, A. L. Sobolewski, and S. Mahapatra, J. Chem. Phys. **122**, 224315 (2005).
[229] L. D. Landau, Phys. Z. Sowjetunion **2**, 46 (1932).
[230] L. D. Landau and E. M. Lifshitz, *Quantum Mechanics: Non-Relativistic Theory* (Butterworth-Heineman, 1981).
[231] C. F. Lane, Chem. Rev. **76**, 773 (1976).
[232] P. Lankaster and K. Šalkauskas, *Curve and Surface Fitting: An Introduction* (Academic, London, 1986).
[233] B. H. Lengsfield III, P. Saxe, and D. R. Yarkony, J. Chem. Phys. **81**, 4549 (1984).
[234] B. H. Lengsfield and D. R. Yarkony, in *State-Selected and State-to-State Ion-Molecule Reaction Dynamics, Part 2: Theory*, edited by C.-Y. Ng and M. Baer, Advances in Chemical Physics, Vol. 82 (John Wiley and Sons, New York, 1992), pp. 1–71.
[235] E. Leonardi, C. Petrongolo, G. Hirsch, and R. J. Buenker, J. Chem. Phys. **105**, 9051 (1996).
[236] R. J. Levis, G. M. Menkir, and H. Rabitz, Science **292**, 709 (2001).
[237] M. Lewenstein, P. Balcou, M. Y. Ivanov, A. L'Huillier, and P. B. Corkum, Phys. Rev. A **49**, 2117 (1994).
[238] W. Li, A. A. Jaroń-Becker, C. W. Hogle, V. Sharma, X. Zhou, A. Becker, H. C. Kapteyn, and M. M. Murnane, Proc. Natl. Acad. Sci. U.S.A. **107**, 20219 (2010).
[239] A. J. Lichtenberg and M. A. Lieberman, *Regular and Chaotic Dynamics*, 2nd ed., Applied Mathematical Sciences, Vol. 38 (Springer-Verlag, New York, 1992).
[240] J. C. Light, I. P. Hamilton, and J. V. Lill, J. Chem. Phys. **82**, 1400 (1985).
[241] J. S. Lim, Y. S. Lee, and S. K. Kim, Angew. Chem. Int. Ed. **47**, 1853 (2008).

[242] *Hydrogen Transfer: Experiment and Theory*, edited by H.-H. Limbach and J. Manz, Ber. Bunsenges. Phys. Chem. **102** (1998), Issue 3, pp. 289–592 (whole issue).

[243] H. Lischka, M. Dallos, P. G. Szalay, D. R. Yarkony, and R. Shepard, J. Chem. Phys. **120**, 7322 (2004).

[244] Z. Liu and Y. Huang, Protein Sci. **23**, 539 (2014).

[245] R. M. Lobayan, R. C. Bochicchio, L. Lain, and A. Torre, J. Chem. Phys. **123**, 144116 (2005).

[246] F. London, Z. Phys. A **74**, 143 (1932).

[247] H. C. Longuet-Higgins, U. Öpik, M. H. L. Pryce, and R. A. Sack, Proc. R. Soc. London, Ser. A **244**, 1 (1958); H. C. Longuet-Higgins, Adv. Spectrosc. (N.Y.) **2**, 429 (1961); G. Herzberg and H. C. Longuet-Higgins, Discuss. Faraday Soc. **35**, 77 (1963).

[248] E. Lorin, S. Chelkowski, and A. Bandrauk, Comput. Phys. Commun. **177**, 908 (2007).

[249] E. Lorin, S. Chelkowski, and A. D. Bandrauk, New J. Phys. **10**, 025033 (2008); Comput. Phys. Commun. **177**, 908 (2007).

[250] R. R. Lucchese, K. Takatsuka, and K. McKoy, Phys. Rep. **131**, 147 (1986).

[251] A. Macías and A. Riera, J. Phys. B **11**, L489 (1978).

[252] S. Mahapatra, H. Köppel, L. S. Cederbaum, P. Stampfuß, and W. Wenzel, Chem. Phys. **259**, 211 (2000).

[253] Y. Mairesse, D. Zeidler, N. Dudovich, M. Spanner, J. Levesque, D. M. Villeneuve, and P. B. Corkum, Phys. Rev. Lett. **10**, 143903 (2008).

[254] Y. Mairesse, A. de. Bohan, L. J. Frasinski, H. Merdji, L. C. Dinu, P. Monchicourt, P. Breger, M. Kovačev, T. Auguste, B. Carrè, H. G. Muller, P. Agostini, and P. Sallerès, Phys. Rev. Lett. **93**, 163901 (2004).

[255] V. A. Mandelshtam and H. S. Taylor, J. Chem. Phys. **103**, 2903 (1995).

[256] U. Manthe and H. Köppel, J. Chem. Phys. **93**, 1658 (1990).

[257] J. Manz and L. Wöste, eds., *Femtochemistry*, (VCH, Weinheim, 1995).

[258] A. Maquet, S.-I Chu, and W. P. Reinhardt, Phys. Rev. A. **27**, 2946 (1983).

[259] N. Markova and V. Enchev, J. Mol. Struct. THEOCHEM **679**, 195 (2004).

[260] C. C. Marston and G. G. Balint-Kurti, J. Chem. Phys. **91**, 3571 (1989).

[261] T. J. Martínez, M. Ben-Nun, and R. D. Levine, J. Phys. Chem. **100**, 7884 (1996).

[262] T. J. Martínez, M. Ben-Nun, and R. D. Levine, J. Phys. Chem. A **101**, 6389 (1997).

[263] T. J. Martínez, Acc. Chem. Res. **39**, 119 (2006).

[264] F. Martin, J. Fernández, T. Havermeier, L. Foucar, Th. Weber, K. Kreidi, M. Schöffler, L. Schmidt, T. Jahnke, O. Jagutzki, A. Czasch, E. P. Benis, T. Osipov, A. L. Landers, A. Belkacem, M. H. Prior, H. Schmidt-Bocking, C. L. Cocke, and R. Dorner, Science **315**, 629 (2007).

[265] S. Martrenchard-Barra, C. Dedonder-Lardeux, C. Jouvet, D. Solgadi, M. Vervloet, G. Grégoire, and I. Dimicoli, Chem. Phys. Lett. **310**, 173 (1999).

[266] V. May and O. Kühn, *Charge and energy transfer dynamics in molecular systems*, (WILEY-VCH Verlag Berlin GmbH, Berlin, 2000).

[267] C. A. Mead, Chem. Phys. **49**, 23 (1980).

[268] C. A. Mead and D. G. Truhlar, J. Chem. Phys. **70**, 2284 (1979).
[269] R. Meana-Pañeda, D. G. Truhlar, and A. Fernández-Ramos, J. Chem. Phys. **134**, 094302 (2011).
[270] C. Meier and V. Engel, Chem. Phys. Lett. **212**, 691 (1993).
[271] C. Meier and V. Engel, J. Chem. Phys. **101**, 2673 (1994).
[272] H.-D. Meyer, F. Gatti, and G. A. Worth, eds., *Multidimensional Quantum Dynamics: MCTDH Theory and Applications*, (Wiley-VCH, Weinheim, 2009).
[273] H.-D. Meyer and W. H. Miller, J. Chem. Phys. **70**, 3214 (1979).
[274] R. E. Meyer, SIAM Review **31**, 435 (1989).
[275] S. L. Mielke, G. J. Tawa, D. G. Truhlar, and D. W. Schwenke, Int. J. Quant. Chem. Symp. **27**, 621 (1993).
[276] S. L. Mielke, G. J. Tawa, D. G. Truhlar, and D. W. Schwenke, Chem. Phys. Lett. **234**, 57 (1995).
[277] S. L. Mielke, D. G. Truhlar, and D. W. Schwenke, J. Phys. Chem. **99**, 16210 (1995).
[278] W. H. Miller, J. Chem. Phys. **53**, 1949 (1970).
[279] W. H. Miller, J. Phys. Chem. A **113**, 1405 (2009).
[280] R. Mitrić, J. Petersen, and V. Bonačić-Koutecký, Phys. Rev. A **79**, 053416 (2009).
[281] R. Mitrić, M. Hartmann, J. Pittner, and V. Bonačić-Koutecký, J. Phys. Chem. A **106**, 10477 (2002).
[282] Y. Nabekawa, T. Shimizu, T. Okino, K. Furusawa, K. Hasegawa, K. Yamanouchi, and K. Midorikawa, Phys. Rev. Lett. **96**, 083901 (2006).
[283] L. A. Nafie, J. Chem. Phys. **79**, 4950 (1983).
[284] L. A. Nafie, J. Chem. Phys. **78**, 7108 (1992).
[285] L. A. Nafie, J. Phys. Chem. A **101**, 7826 (1997).
[286] K. Nagashima and K. Takatsuka, J. Phys. Chem. A **113**, 15240 (2009).
[287] K. Nagashima and K. Takatsuka, J. Phys. Chem. A **116**, 11167 (2012).
[288] H. Nakai, M. Hoshino, K. Miyamoto, and S. Hyodo, J. Chem. Phys. **122**, 164101 (2005).
[289] H. Nakai and K. Sodeyama, J. Chem. Phys. **118**, 1119 (2003).
[290] H. Nakamura, Adv. Chem. Phys. **82**, 243 (1992).
[291] H. Nakamura, *Nonadiabatic Transition*, (World Scinetific, New Jersey, 2002).
[292] H. Nakamura and D. G. Truhlar, J. Chem. Phys. **115**, 10353 (2001).
[293] H. Nakamura and D. G. Truhlar, J. Chem. Phys. **117**, 5576 (2002).
[294] R. Nakanishi and T. Nagata, J. Phys. Chem. **130**, 224309 (2009).
[295] S. Nanbu, T. Ishida, and H. Nakamura, Chem. Sci. **1**, 663 (2010).
[296] C. Neidel, J. Klei, C.-H. Yang, A. Rouzeé, M. J. J. Vrakking, K. Klünder, M. Miranda, C. L. Arnold, T. Fordell, A. L'Huillier, M. Gisselbrecht, P. Johnsson, M. P. Dinh, E. Surand, P.-G. Reinhard, V. Despré, M. A. L. Marques, and F. Lépine, Phys. Rev. Lett. **111**, 033001 (2013).
[297] E. Neria and A. Nitzan, J. Chem. Phys. **99**, 1109 (1993).
[298] M. Nest, F. Remacle, and R. D. Levine, New J. Phys. **10**, 025019 (2008).
[299] M. Nest, Chem. Phys. Lett. **472**, 171 (2009).

[300] H. Niikura, F. Légaré, R. Hasbani, A. D. Bandrauk, M. Y. Ivanov, D. M. Villeneuve, and P. B. Corkum, Nature **417**, 917 (2002).
[301] K. Nishi, H. Sekiya, H. Kawakami, A. Mori, and Y. Nishimura, J. Chem. Phys. **109**, 1589 (1998).
[302] K. Nishi, H. Sekiya, H. Kawakami, A. Mori, and Y. Nishimura, J. Chem. Phys. **111**, 3961 (1999).
[303] K. Nobusada and K. Yabana, Phys. Rev. A **70**, 043411 (2004).
[304] P. Nuernberger, G. Vogt, T. Brixner, and G. Gerber, Phys. Chem. Chem. Phys. **9**, 2470 (2007).
[305] K. Ohmori, H. Katsuki, H. Chiba, M. Honda, Y. Hagihara, K. Fujiwara, Y. Sato, and K. Ueda, Phys. Rev. Lett. **96**, 093002 (2006).
[306] K. Ohmori, Annu. Rev. Phys. Chem. **60**, 487 (2009).
[307] M. Okuyama and K. Takatsuka, Chem. Phys. Lett. **476**, 109 (2009).
[308] M. Okuyama and K. Takatsuka, Bull. Chem. Soc. Jpn. **85**, 217 (2012).
[309] P. Gaspard, I. Burghard, eds., *Advances in Chemical Physics, Chemical Reactions and Their Control on the Femtosecond Time Scale*, XXth Solvay Conference on Chemistry, Vol. 101, (John Wiley & Sons, New York, 2009).
[310] P. Paldus, J. Chem. Phys. **61**, 5321 (1974).
[311] P. Paldus, Int. Quant. Chem. Quantum Chem. Symp. **9**, 1650 (1975).
[312] P. Paldus, Phys. Rev. A **14**, 1620 (1976).
[313] G. K. Paramonov, Chem. Phys. **338**, 329 (2007).
[314] G. K. Paramonov, Chem. Phys. Lett. **411**, 350 (2005).
[315] G. Parlant and D. R. Yarkony, Int. J. Quant. Chem. **44**, 737 (1992).
[316] M. J. Paterson, M. J. Bearpark, M. A. Robb, L. Blancafort, and G. A. Worth, Phys. Chem. Chem. Phys. **7**, 2100 (2005).
[317] P. M. Paul, E. S. Toma, P. Breger, G. Mullot, F. Augé, P. Balcou, H. G. Muller, and P. Agostini, Phys. Rev. A **292**, 1689 (2001).
[318] P. Pechukas, Phys. Rev. **181**, 174 (1969).
[319] U. Peskin and N. Moiseyev, J. Chem. Phys. **99**, 4590 (1993).
[320] J. Petersen, M. Wohlgemuth, B. Sellner, V. Bonačić-Koutecký, H. Lischka, and R. Mitrić, Phys. Chem. Chem. Phys. **14**, 4687 (2012).
[321] J. Petersen and R. Mitrić, Phys. Chem. Chem. Phys. **14**, 8299 (2012).
[322] M. L. Pons, R. Taïeb, and A. Maquet, Phys. Rev. A **54**, 3634 (1996).
[323] J. H. Posthumus, Rep. Prog. Phys. **67**, 623 (2004).
[324] E. D. Potter, J. L. Herek, S. Pedersen, Q. Liu, and A. H. Zewail, Nature **355**, 66 (1992).
[325] B. P. Prascher, D. E. Woon, K. A. Peterson, T. H. Jr. Dunning, and A. K. Wilson, Theor. Chem. Acc. **128**, 69 (2011).
[326] W. H. Press, S. A. Teukolsky, W. T. Vetterling, and B. P. Flannery, *Numerical Recipes in Fortran*, (Cambridge University Press, Cambridge, 1992).
[327] O. V. Prezhdo and P. J. Rossky, J. Phys. Chem. **100**, 17094 (1996).
[328] E. I. Proynov, J. Mol. Struct. THEOCHEM **159**, 762 (2006).
[329] M. S. Pshenichnikov, A. Baltuska, and D. A. Wiersma, Chem. Phys. Lett. **389**, 171 (2004).
[330] P. Pulay, Adv. Chem. Phys. **69**, 241 (1987).

[331] M. I. Röhr, J. Petersen, M. Wohlgemuth, V. Bonačić-Koutecký, and R. Mitrić, ChemPhysChem. **14**, 1377 (2013).
[332] K. L. Reid, T. A. Field, M. Towrie, and P. Matousek, J. Chem. Phys. **111**, 1438 (1999).
[333] D. Reignier, T. Stoecklin, P. Halvick, A. Voronin, and J. C. Rayez, Phys. Chem. Chem. Phys. **3**, 2726 (2001).
[334] F. Remacle and R. D. Levine, Proc. Natl. Acad. Sci. USA **103**, 6793 (2006).
[335] M. Richter, P. Marquetand, J. González-Vázquez, I. Sola, and L. González, J. Chem. Theory comput. **7**, 1253 (2011).
[336] D. Rolles, M. Braune, S. Cvejanovi, O. Getzner, R. Hentges, S. Korica, B. Langer, T. Lischke, G. Prüper, A. Reinköter, J. Viefhaus, B. Zimmermann, V. McKoy, and U. Becker, Nature **437**, 711 (2005).
[337] C. C. J. Roothaan, Rev. Mod. Phys. **32**, 179 (1960).
[338] F. Rosca-Pruna and M. J. J. Vrakking, Phys. Rev. Lett. **87**, 153902 (2001).
[339] J. R. Roscioli, N. I. Hammer, M. A. Johnson, K. Diri, and K. D. Jordan, J. Chem. Phys. **128**, 104314 (2008).
[340] T. S. Rose, M. J. Rosker, and A. H. Zewail, J. Chem. Phys. **88**, 6672 (1988).
[341] P. J. Rossky and J. Schnitker, J. Phys. Chem. **92**, 4277 (1988).
[342] K. Ruedenberg, Rev. Mod. Phys. **34**, 326 (1962).
[343] H. Ruppe, S. Rutz, E. Schneiber, and L. Wöste, Chem. Phys. Lett. **257**, 356 (1996).
[344] A. Saenz, Phys. Rev. A **66**, 063407 (2002).
[345] Y. I. Salamin, S. X. Hu, K. Z. Hatsagortsyan, and C. H. Keitel, Phys. Rep. **427**, 41 (2006).
[346] P. Salières, A. Maquet, S. Haessler, J. Caillat, and R. Taïeb, Rep. Prog. Phys. **75**, 062401 (2012).
[347] R. F. Salzgeber, V. Mandelshtam, C. Schlier, and H. S. Taylor, J. Chem. Phys. **109**, 937 (1998).
[348] M. Sanrey and M. Joyeux, J. Chem. Phys. **125**, 014304 (2006).
[349] G. Sansone, F. Kelkensberg, J. F. Pérez-Torres, F. Morales, M. F. Kling, W. Siu, O. Ghafur, P. Johnsson, M. Swoboda, E. Benedetti, F. Ferrari, F. Lépine, J. L. Sanz-Vicario, S. Zherebtsov, I. Znakovskaya, A. L'Huillier, M. Y. Ivanov, M. Nisoli, F. Martín, and M. J. J. Vrakking, Nature **465**, 763 (2010).
[350] F. Santoro and C. Petrongolo, J. Chem. Phys. **110**, 4419 (1999).
[351] F. Santoro, C. Petrongolo, G. Granucci, and M. Persico, Chem. Phys. **259**, 193 (2000).
[352] Y. Sato, H. Kono, S. Koseki, and Y. Fujimura, J. Am. Chem. Soc. **125**, 8019 (2003).
[353] B. Schäfer-Bung, R. Mitrić, V. Bonačić-Koutecký, A. Bartelt, C. Lupulescu, A. Lindinger, Š. Vajda, S. M. Weber, and L. Wöste, J. Phys. Chem. A **108**, 4175 (2004).
[354] S. H. Schaefer, D. Bender, and E. Tiemann, Chem. Phys. **89**, 65 (1984).
[355] S. Scheit, Y. Arasaki, and K. Takatsuka, J. Phys. Chem. A **116**, 2644 (2011).

[356] L. I. Schiff, *Quantum Mechanics*, (McGraw-Hill, New York, 1968).
[357] M. W. Schmidt, K. K. Baldridge, J. A. Boatz, S. T. Elbert, M. S. Gordon, J. H. Jensen, S. Koseki, N. Matsunaga, K. A. Nguyen, S. Su, T. L. Windus, M. Dupuis, and J. A. Montgomery, J. Comput. Chem. **14**, 1347 (1993).
[358] L. S. Schulman, *Techniques and Application of Path Integration*, (Wiley, New York, 1981).
[359] D. W. Schwenke, S. L. Mielke, G. J. Tawa, R. S. Friedman, P. Halvick, and D. G. Truhlar, Chem. Phys. Lett. **203**, 565 (1993).
[360] H. Schwoerer, R. Pausch, M. Heid, V. Engel, and W. Kiefer, J. Chem. Phys. **107**, 9749 (1997).
[361] A. Scrinzi, Phys. Rev. A **81**, 053845 (2010).
[362] M. Seel and W. Domcke, Chem. Phys. **151**, 59 (1991).
[363] M. Seel and W. Domcke, J. Chem. Phys. **95**, 7806 (1991).
[364] M. Shapiro and P. Brumer, Rep. Prog. Phys. **66**, 859 (2003).
[365] M. Shapiro and P. Brumer, *Quantum Control of Molecular Processes*, 2nd Ed. (Wiley-VCH, Wienheim, 2012).
[366] I. Shavitt, Int. Quant. Chem. Quantum Chem. Symp. **11**, 131. (1977).
[367] I. Shavitt, Int. Quant. Chem. Quantum Chem. Symp. **12**, 5. (1978).
[368] N. Shenvi, J. Chem. Phys. **130**, 124117 (2009).
[369] N. Shida, P. F. Barbara, and J. Almöf, J. Chem. Phys. **94**, 3633 (1991).
[370] J. Shirley, Phys. Rev. **138**, 979 (1965).
[371] I. A. Shkrob, W. J. Glover, R. E. Larsen, and B. J. Schwartz, J. Phys. Chem. A **111**, 5232 (2007).
[372] B. W. Shore, *Manipulating Quantum Structures Using Laser Pulses*, (Cambridge University Press, Cambridge, 2011).
[373] D. Simah, B. Hartke, and H. J. Woerner, ab initio **111**, 4523 (1999).
[374] J. Simons, J. Phys. Chem. A **11**, 6401 (2008).
[375] O. Smirnova, Y. Mairesse, S. Patchkovskii, N. Dudovich, D. Villeneuve, P. Corkum, and M. Y. Ivanov, Nature **460**, 972 (2009).
[376] F. T. Smith, Phys. Rev. **179**, 111 (1969).
[377] J. M. Smith, X. Zhang, and J. L. Knee, J. Phys. Chem. **99**, 1768 (1995).
[378] A. L. Sobolewski and W. Domcke, J. Phys. Chem. A **103**, 449 (1999).
[379] A. L. Sobolewski and W. Domcke, Chem. Phys. **259**, 181 (2000).
[380] A. L. Sobolewski and W. Domcke, J. Phys. Chem. A **105**, 927 (2001).
[381] A. L. Sobolewski, W. Domcke, C. Dedonder-Lardeux, and C. Jouvet, Phys. Chem. Chem. Phys. **4**, 1093 (2002).
[382] A. L. Sobolewski and W. Domcke, Phys. Chem. Chem. Phys. **9**, 3818 (2007).
[383] M. Spanner and P. Brumer, Phys. Rev. A **78**, 033840 (2008).
[384] M. Squillacote, J. Wang, and J. Chen, J. Am. Chem. Soc. **126**, 1940 (2004).
[385] J. F. Stanton and J. Gauss, J. Chem. Phys. **108**, 9218 (1998).
[386] H. Stapelfeldt and T. Seideman, Rev. Mod. Phys. **75**, 543 (2003).
[387] V. N. Staroverov and E. R. Davidson, Chem. Phys. Lett. **330**, 161 (2000).
[388] V. N. Staroverov and E. R. Davidson, J. Am. Chem. Soc. **122**, 186 (2000).
[389] V. N. Staroverov and E. R. Davidson, J. Am. Chem. Soc. **122**, 7377 (2000).

[390] A. Staudte, D. Pavičić, S. Chelkowski, D. Zeidler, M. Meckel, H. Niikura, M. Schöffler, S. Schössler, B. Ulrich, P. P. Rajeev, Th. Weber, T. Jahnke, D. M. Villeneuve, A. D. Bandrauk, C. L. Cocke, P. B. Corkum, and R. Dörner, Phys. Rev. Lett. **98**, 073003 (2007).
[391] J. R. Stine and R. A. Marcus, Chem. Phys. Lett. **29**, 575 (1974).
[392] G. Stock and M. Thoss, Phys. Rev. Lett. **78**, 578 (1997).
[393] G. Stock and M. Thoss, Adv. Chem. Phys. **131**, 243 (2005).
[394] A. Streitwieser, *Molecular Orbital Theory for Organic Chemists*, (John Wiley & Sons, New York, 1961).
[395] E. C. G. Stueckelberg, Helv. Phys. Acta. **5**, 369 (1932).
[396] J. T. Su and W. A. Goddard III, Proc. Natl. Acad. Sci. **106**, 1001 (2009).
[397] J. T. Su and W. A. Goddard III, J. Chem. Phys. **131**, 244501 (2009).
[398] J. E. Subotnik, J. Chem. Phys. **132**, 134112 (2010).
[399] J. E. Subotnik and N. Shenvi, J. Chem. Phys. **134**, 024105 (2011).
[400] X. Sun and W. H. Miller, J. Chem. Phys. **106**, 6346 (1997).
[401] X. Sun, H. Wang, and W. H. Miller, J. Chem. Phys. **109**, 7064 (1997).
[402] B. Sundaram and P. W. Milonni, Phys. Rev. A **41**, 6571 (1990).
[403] B. J. Sussman, M. Y. Ivanov, and A. Stolow, Phys. Rev. A **71**, 051401 (2005).
[404] B. J. Sussman, D. Townsend, M. Y. Ivanov, and A. Stolow, Science **314**, 278 (2006).
[405] B. T. Sutcliffe, J. Math. Chem. **44**, 988 (2008).
[406] T. Suzuki, L. Wang, and H. Kohguchi, J. Chem. Phys. **111**, 4859 (1999).
[407] P. Svensson, N. Bergman, and P. Ahlberg, J. Chem. Soc. Chem. Commun. **1990**, 862 (1990).
[408] H. Tachikawa, J. Chem. Phys. **125**, 144307 (2006).
[409] Y Takada and T. Cui, J. Phys. Soc. Japan **72**, 2671 (2003).
[410] K. Takahashi and K. Ikeda, J. Chem. Phys. **99**, 8680 (1993).
[411] S. Takahashi and K. Takatsuka, Phys. Rev. A **70**, 052103 (2004).
[412] S. Takahashi and K. Takatsuka, Phys. Rev. A **69**, 022110 (2004).
[413] S. Takahashi and K. Takatsuka, J. Chem. Phys. **124**, 144101 (2006).
[414] S. Takahashi and K. Takatsuka, J. Chem. Phys. **127**, 084112 (2007).
[415] S. Takahashi and K. Taktsuka, Phys. Rev. A **89**, 012108 (2014).
[416] S. Takahashi and K. Takatsuka, Phys. Rev. A **89**, 012109 (2014).
[417] K. Takatsuka, T. Fueno, and K. Yamaguchi, Theoret. Chim. Acta **48**, 175 (1978).
[418] K. Takatsuka and V. McKoy, Phys. Rev. A **24**, 2473 (1981).
[419] K. Takatsuka and V. McKoy, Phys. Rev. A **30**, 1734 (1984).
[420] K. Takatsuka and N. Hashimoto, J. Chem. Phys. **103**, 6057 (1995).
[421] K. Takatsuka, Y. Arasaki, K. Wang, and V. McKoy, Faraday Discuss. **115**, 1 (2000).
[422] K. Takatsuka, J. Chem. Phys. **124**, 064111 (2006).
[423] K. Takatsuka, J. Phys. Chem. A **111**, 10196 (2007).
[424] K. Takatsuka and A. Inoue-Ushiyama, Phys. Rev. Lett. **78**, 1404 (1997); A. Inoue-Ushiyama and K. Takatsuka, Phys. Rev. A **59**, 3256 (1999).
[425] K. Takatsuka and T. Yonehara, Adv. Chem. Phys. **144**, 93 (2010).

[426] K. Takatsuka and T. Yonehara, Phys. Chem. Chem. Phys. **13**, 4987 (2011).
[427] K. Takatsuka and S. Takahashi, Phys. Rev. A **89**, 012109 (2014).
[428] K. Takatsuka, J. Phys. B **47**, 124038 (2014).
[429] K. Takeshita and N. Shida, J. Chem. Phys. **116**, 4482 (2002).
[430] H. Tal-Ezer and R. Kosloff, J. Chem. Phys. **81**, 3967 (1984).
[431] A. Tanaka, J. Phys. A, Math. Gen. **29**, 5475 (1996); Phys. Rev. Lett. **80**, 1414 (1998).
[432] C. Tanner, C. Manca, and S. Leutwyler, J. Chem. Phys. **122**, 204326 (2005).
[433] D. J. Tannor and S. A. Rice, J. Chem. Phys. **83**, 5013 (1985).
[434] D. J. Tannor, R. Kosloff, and S. A. Rice, J. Chem. Phys. **85**, 5805 (1986).
[435] D. J. Tannor, R. Kosloff, and S. A. Rice, J. Chem. Phys. **85**, 5805 (1986).
[436] D. J. Tannor and S. A. Rice, Adv. Chem. Phys., Part 1, Vol.**70**, page 441-523 (1988).
[437] D. J. Tannor, *Introduction to Quantum Mechanics: A Time-Dependent Perspective*, (University Science Books, Sausalito, California, 2007).
[438] L. Tao, W. Vanroose, B. Reps, T. N. Rescigno, and C. W. McCurdy, Phys. Rev. A **80**, 063419 (2009).
[439] G. J. Tawa, S. L. Mielke, D. G. Truhlar, and D. W. Schwenke, J. Chem. Phys **100**, 5751 (1994).
[440] X.-M. Tong and S.-I Chu, Phys. Rev. A. **57**, 452 (1998).
[441] M. S. Topaler, M. D. Hack, T. C. Allison, Y.-P. Liu, S. L. Mielke, D. W. Schwenke, and D. G. Truhlar, J. Chem. Phys. **106**, 8699 (1997).
[442] A. Torre and L. Lain, J. Phys. Chem. A **107**, 127 (2003).
[443] N. Tsuji, S. Ishiuchi, C. Jouvet, C. Dedonder-Lardeux, M. Miyazaki, M. Sakai, and M. Fujii, ChemPhysChem **12**, 1928 (2011).
[444] J. C. Tully and R. K. Preston, J. Chem. Phys. **55**, 562 (1971).
[445] J. C. Tully, J. Chem. Phys. **93**, 1061 (1990).
[446] L. Turi, G. Hantal, P. Rossky, and D. Borgis, J. Chem.Phys. **131**, 024119 (2009).
[447] S. Ullrich, T. Schultz, M. Zgierski, and A. Stolow, Phys. Chem. Chem. Phys. **6**, 2796 (2004).
[448] H. Ushiyama, Y. Arasaki, and K. Takatsuka, Chem. Phys. Lett. **346**, 169 (2001).
[449] H. Ushiyama and K. Takatsuka, J. Chem. Phys. **115**, 5903 (2001).
[450] H. Ushiyama and K. Takatsuka, Angew. Chem. Int. Ed. **44**, 1237 (2005).
[451] H. Ushiyama and K. Takatsuka, Angew. Chem. Int. Ed. **46**, 587 (2007).
[452] R. Valero, D. G. Truhlar, and A. W. Jasper, J. Phys. Chem. A **112**, 5756 (2008).
[453] J. Vaníček, Phys. Rev. E **70**, 05520 (2004).
[454] J. Vaníček, Phys. Rev. E **73**, 04620 (2006).
[455] A. J. C. Varandas, J. Chem. Phys. **119**, 2596 (2003).
[456] A. J. C. Varandas, J. Chem. Phys. **131**, 124128 (2009).
[457] M. V. Vener and S. Iwata, Chem. Phys. Lett. **292**, 87 (1998).
[458] J. R. R. Verlet, A. E. Bragg, A. Kammrath, O. Cheshnovsky, and D. M. Neumark, Science **307**, 93 (2005).

[459] A. M. Virshup, C. Punwong, T. V. Pogorelov, B. A. Lindquist, C. Ko, and T. J. Martínez, J. Phys. Chem. B **113**, 3280 (2009).
[460] N. V. Vitanov, T. Halfmann, B. W. Shore, and K. Bergmann, Annu. Rev. Phys. Chem. **52**, 763 (2001).
[461] R. de Vivie-Riedle and A. Hofmann, in *Conical Intersections: Electronic Structure, Dynamics and Spectroscopy,* edited by W. Domcke, D. R. Yarkony, and H. Köppel, Advanced Series in Physical Chemistry, Vol. 15, page 803, (World Scientific, Singapore, 2004).
[462] R. de Vivie-Riedle, L. Kurtz, and A. Hofmann, Pure Appl. Chem. **73**, 525 (2001).
[463] G. Vogt, G. Krampert, P. Niklaus, P. Nuernberger, and G. Gerber, Phys. Rev. Lett. **94**, 068305 (2005).
[464] M. Volkhard and O. Kuhn, *Charge and energy transfer dynamics in molecular systems,* (John Wiley & Sons, 2008).
[465] H. J. Wörner, J. B. Bertrand, D. V. Kartashov, P. B. Corkum, and D. M. Villeneuve, Nature **46**, 604 (2010).
[466] H. J. Wörner, J. B. Bertrand, B. Fabre, J. Higuet, H. Ruf, A. Dubrouil, S. Patchkovskii, M. Spanner, Y. Mairesse, V. Blanchet, E. Mével, E. Constant, P. B. Corkum, and D. M. Villeneuve, Science **334**, 208 (2011).
[467] H. Wang and M. Thoss, J. Chem. Phys. **119**, 1289 (2003).
[468] F. J. Webster, J. Schnitker, M. S. Friedrichs, R. A. Friesner, and P. J. Rossky, Phys. Rev. Lett. **66**, 3172 (1991); F. J. Webster, P. J. Rossky, and R. A. Friesner, Comput. Phys. Commun. **63**, 494 (1991).
[469] H.-J. Werner and W. Meyer, J. Chem. Phys. **74**, 5802 (1981).
[470] H.-J. Werner and P. J. Knowles, J. Chem. Phys. **82**, 5053 (1985).
[471] H.-J. Werner, P. J. Knowles, G. Knizia, F. R. Manby, and M. Schütz, Wiley Interdiscip. Rev. Comput. Mol. Sci. **2**, 242 (2012).
[472] H.-J. Werner, P. J. Knowles, G. Knizia, F. R. Manby, M. Schütz, P. Celani, T. Korona, R. Lindh, A. Mitrushenkov, G. Rauhut, K. R. Shamasundar, T. B. Adler, R. D. Amos, A. Bernhardsson, A. Berning, D. L. Cooper, M. J. O. Deegan, A. J. Dobbyn, F. Eckert, E. Goll, C. Hampel, A. Hesselmann, G. Hetzer, T. Hrenar, G. Jansen, C. Köppl, Y. Liu, A. W. Lloyd, R. A. Mata, A. J. May, S. J. McNicholas, W. Meyer, M. E. Mura, A. Nicklass, D. P. O'Neill, P. Palmieri, D. Peng, K. Pflüger, R. Pitzer, M. Reiher, T. Shiozaki, H. Stoll, A. J. Stone, R. Tarroni, T. Thorsteinsson, and M. Wang, *Molpro, version 2012.1*, a package of ab initio programs, `http://www.molpro.net`,2012.
[473] J. Werschnik and E. K. U. Gross, J. Phys. B **40**, R175 (2007).
[474] A. M. Wodtke, J. C. Tully, and D. J. Auerbach, Int. Rev. Phys. Chem. **23**, 513 (2007).
[475] M. Wollenhaupt, V. Engel, and T. Baumert, Annu. Rev. Phys. Chem. **56**, 25 (2005).
[476] M. Wollenhaupt and T. Baumert, Farad. Discuss **153**, 9 (2011).
[477] K. F. Wong and P. Rossky, J. Chem. Phys. **116**, 8418 (2002); *ibid.* **116**, 8429 (2002).

[478] M. W. Wong, K. B. Wiberg, and M. J. Frisch, J. Am. Chem. Soc. **114**, 1645 (1992).
[479] G. A. Worth, H.-D. Meyer, H. Köppel, L. S. Cederbaum, and I. Burghard, Intern. Rev. Phys. Chem. **27**, 569 (2008).
[480] R. Wyatt, *Dynamics of Molecules and Chemical Reactions*, CRC Press, 1996.
[481] R. E. Wyatt, *Quantum Dynamics with Trajectories: Introduction to Quantum Dynamics*, Springer, New York, 2005.
[482] X. F. Xu, A. Kahan, S. Zilberg, and Y. Haas, J. Phys. Chem. A **113**, 9779 (2009).
[483] S. Yamabe, K. Kitaura, and K. Nishimoto, Theoret. Chim. Acta. **47**, 111 (1978).
[484] A. Yamada and S. Okazaki, J. Chem. Phys. **124**, 094110 (2006).
[485] A. Yamada and S. Okazaki, J. Chem. Phys. **128**, 044507 (2008).
[486] Y. Yamaguchi, Y. Osamura, J. D. Goddard, and H. F. Schaefer III, *A New Dimension to Quantum Chemistry: Analytic Derivative Methods in* ab initio *Molecular Electronic Structure Theory*, (Oxford university press, New York, Oxford, 1994).
[487] K. Yamanouchi, Science **295**, 1659 (2002).
[488] K. Yamanouchi, A. D. Bandrauk, and G. Gerber, eds., *Progress in Ultrafast Intense Laser Science VI*, (Springer, Berlin, 2010).
[489] S. Yang and J. D. Coe, J. Chem. Phys. **130**, 134113 (2009).
[490] G. Yao, Chem. Phys. Lett. **197**, 413 (1992).
[491] T. Yonehara, K. Hanasaki, and K. Takatsuka, Chem. Rev. **112**, 499 (2012).
[492] T. Yonehara and K. Takatsuka, J. Chem. Phys. **128**, 154104 (2008).
[493] T. Yonehara and K. Takatsuka, J. Chem. Phys. **129**, 134109 (2008).
[494] T. Yonehara and K. Takatsuka, Chem. Phys. **366**, 115 (2009).
[495] T. Yonehara and K. Takasuka, J. Chem. Phys. **132**, 244102 (2010).
[496] T. Yonehara and K. Takasuka, J. Chem. Phys. **137**, 22A520 (2012).
[497] H. Yoshida, Phys. Lett. A **150**, 262 (1990); K. Takahashi and K. Ikeda, J. Chem. Phys. **99**, 8680 (1993); R. Kosloff, J. Phys. Chem. **92**, 2087 (1988); M. Suzuki, Phys. Lett. A **165**, 387; Feit (1992); J. A. Fleck Jr. and A. Steiger, J. Comp. Phys. **47**, 412 (1982).
[498] K. J. Yuan and A. D. Bandrauk, J. Phys. B **45**, 105601 (2012).
[499] K. J. Yuan, S. Chelkowski, and A. D. Bandrauk, J. Chem. Phys. **138**, 134316 (2013).
[500] K. J. Yuan, H. Lu, and A. D. Bandrauk, ChemPhysChem **14**, 1496 (2013).
[501] N. Zamstein and D. J. Tannor, J. Chem. Phys. **137**, 22A517 (2012).
[502] N. Zamstein and D. J. Tannor, J. Chem. Phys. **137**, 22A518 (2012).
[503] A. Zavriyev, P. H. Bucksbaum, H. G. Muller, and D. W. Schumacher, Phys. Rev. A. **42**, 5500 (1990).
[504] A. Zavriyev, P. Bucksbaum, J. Squier, and F. Saline, Phys. Rev. Lett. **70**, 1077 (1993).
[505] C. Zener, Proc. R. Soc. London A **137**, 696 (1932).
[506] A. H. Zewail, *Femtochemistry: Ultrafast Dynamics of the Chemical Bond*, Vols. 1-2, (World Scientific, Singapore, 1994).

[507] J. Z. H. Zhang, *Theory and Application of Quantum Molecular Dynamics*, (World Scientific, Singapore, 1999).
[508] D. Zhong and A. H. Zewail, J. Phys. Chem. A **102**, 4031 (1998).
[509] C. Zhu, A. W. Jasper, and D. G. Truhlar, J. Chem. Phys. **120**, 5543 (2004).
[510] C. Zhu and H. Nakamura, J. Chem. Phys. **97**, 8497 (1992).
[511] C. Zhu and H. Nakamura, J. Chem. Phys. **98**, 6208 (1993).
[512] C. Zhu and H. Nakamura, J. Chem. Phys. **101**, 4855 (1994).
[513] C. Zhu and H. Nakamura, J. Chem. Phys. **102**, 7448 (1995); J. Chem. Phys. **106**, 2599 (1997).
[514] C. Zhu, H. Nakamura, N. Re, and V. Aquilanti, J. Chem. Phys. **97**, 1892 (1992).
[515] C. Zhu, S. Nangia, A. W. Jasper, and D. G. Truhlar, J. Chem. Phys. **121**, 7658 (2004).
[516] B. Zimmermann, D. Rolles, B. Langer, R. Hentges, M. Braune, S. Cvejanovic, O. Gener, F. Heiser, S. Korica, T. Lischke, A. Reinkoster, J. Viefhaus, R. Dö ner, V. McKoy, and U. Becker, Nature Phys. **4**, 649 (2008).
[517] Entire volume of *Chemical Reviews* Vol. 112 (Americal Chemical Society, 2012).

Index

www.ingramcontent.com/pod-product-compliance
Ingram Content Group UK Ltd.
Pitfield, Milton Keynes, MK11 3LW, UK
UKHW022105170926
8624IPUK00024B/136

9 789814 619646